W. MEILING and F. STARY

Nanosecond Pulse Techniques

Nanosecond Pulse Techniques

by

Dr. rer. nat. Wolfgang Meiling

and

Dr. rer. nat. Franz Stary

Deutsche Akademie der Wissenschaften zu Berlin
Central Institute for Nuclear Research
Rossendorf near Dresden
Department of Nuclear Physics

225 Figures, 16 Tables

AKADEMIE-VERLAG · BERLIN
1969

Translated by Intertext, Außenstelle Dresden

Translation by E. v. Rechenberg,
revised by H. Herzog, D. Sharp

Erschienen im Akademie-Verlag GmbH, 108 Berlin, Leipziger Str. 3–4

Lizenznummer: 202 · 100/431/69
Gesamtherstellung: VEB Druckerei „Thomas Müntzer", 582 Bad Langensalza
Bestellnummer: 5660 · ES 18 B 6, 18 B 7

On the occasion of his sixtieth birthday we wish to dedicate this book to our respected teacher, Prof. Dr. phil. habil. *Josef Schintlmeister,* the originator of multi-parameter analysis and of the $\mathrm{d}E/\mathrm{d}x$-method of particle discrimination.

PREFACE

This book intends to give a review of the present situation in the nanosecond pulse measuring technique, which deals with the generation, the measurement and the processing of electrical pulses, specifically in the nanosecond region ($1\ \text{ns} = 10^{-9}\ \text{s}$). This branch of electronics has grown out of the measuring requirements in the field of nuclear and elementary particle physics. Today improvement is still stimulated from this field, although the nanosecond pulse technique has long since gained entry in other fields. These include, above all, fast computer technology which is rapidly developing through the increasing use of integrated circuits. In computer development the large number of the same logic circuits employed have resulted in those parameters gaining in importance which play a secondary role in actual pulse measuring technique, and, for this reason, circuits which are only used in computers find no mention in this book. The same applies for problems of radar technique where, contrary to the fields mentioned, the pulse consists of micro-wave periods and consequently the principles of micro-wave technique dominate.

In a pulse circuit suitable for the nanosecond region the "path of the fast pulse" must be as short as possible to keep the influence of parasitic capacitances and inductances, and of the component limitations with respect to bandwidth to a minimum. It is characteristic that time measurements can be performed "fastest" because the starting point of the pulse can be used as time marker. Pulse-height analysis and pulse scaling, on the other hand, must necessarily be slower because the maximum of the signal under test has to pass before the measurement can be initiated. In all circuits employed for these purposes semiconductor components have completely displaced the electronic valve in recent years. Due to their differing functional behaviour and parameters these components often allow simple and more sophisticated solutions.

For this reason it is hardly surprising that, despite the close limits placed on the subject, the authors were faced with an almost uncountable number of new papers and works. Within the space available it proved impossible to give a complete list of references, and the 1,200 or more works mentioned include those published up to 1968. We have not in every case listed the chronologically first work because the dividing-line from the microsecond region is not clearly defined. Older works received consideration only in so far as they relate to important basic principles, for example, the fast-slow system according to *Bell* and *Petch*, or if especially interesting results are described or significant modifications proposed, even if these may not have gained favour in later developments. This has been done, in the last instance, to keep them from being forgotten because with the development of new components such solutions can regain their attraction at a later date. A case in point is the successful come-back of the spinthariscope as scintillation counter, the chronotron or the semiconductor detector as successor of the ionisation chamber. To make it easier for the reader to delve more deeply into a specific problem, the beginning of each section usually lists several reviews which also contain further references.

It is assumed that the reader has a general knowledge of electronic components, of circuitry for microsecond pulse technique and of the radiation detectors involved. Only those properties and characteristics are treated in detail which have importance for application in the nanosecond region, as for instance, time resolution of radiation detectors, optimum timing of a detector signal, switching behaviour of a component, properties and use of coaxial cables and transmission lines etc. Valve-equipped circuits are only touched upon in passing. In view of the steady development of components, there seemed to be little point in listing detailed data of individual components or recommending types that are in frequent use at present. Data relating to the mentioned circuits should only be considered as guide because the use of modern components often yields much better characteristics even from an older circuit version. In general, the diagram examples employed to illustrate a circuit are given in simplified presentation to make their function easier to understand.

Selection of the components is usually done according to general principle, e.g. fast switching behaviour of diodes or low capacitance of tunnel diodes, and with respect to the circuit behaviour. Calculation is mostly limited to conditions of proper setting of the working point, while, for instance, the maximum time resolution is mostly obtained by experimentation. Pulse behaviour for several ideal circuits has been calculated in detail by several workers, but it is not possible, however, to comment on them as they are both extensive and complicated. We decided though, to indicate the values of individual components in the circuit diagrams to provide at least some guide for dimensioning.

Electronic engineers working in this field will find many modifications which may serve to solve this or that practical problem, while those not quite as familiar with this subject will be enabled to rapidly reach the understanding required for the nanosecond region. Physicists, researchers and students will find this book helpful because it shows the possibilities and limitations of the nanosecond pulse technique, and offers them an opportunity to extend their knowledge of electronics. Naturally, a monograph is and cannot be a substitute for the study of the original literature mentioned, but we hope that the systematic presentation of the subject will make it easier to understand this extensive field and in addition furnish one or the other useful suggestion.

We wish to express our deep appreciation to Prof. Dr. phil. habil. *J. Schintlmeister* for his stimulating interest in our work. We also wish to thank the

"Intertext" Translating Bureau in Dresden, its translator, Mr. *E. v. Rechenberg*, and its editors, for their collaboration, and to acknowledge the assistance of Mrs. *Ch. Werner* and Mr. *H. Richter* in helping to prepare the drawings. Them and the many others who have helped with this book we herewith give our thanks and appreciation. Last, not least, we want to express our gratitude to the Akademie-Verlag for their obliging consideration given to our wishes and our desire for an early publication.

Rossendorf, 1968

Dr. rer. nat. *Wolfgang Meiling*
Dr. rer. nat. *Franz Stary*

CONTENTS

1. INTRODUCTION

In vacuum an electromagnetic wave will travel a distance of 30 cm during one nanosecond (1 ns $= 10^{-9}$ s), and in a metal conductor the distance is nearly the same. With the very much slower transport of charge carriers in electronic valves and semiconductors the distance that is travelled during the same period is at the most a few millimetres. For this reason, the transit time of an electrical signal in the conductors and the components has to be taken into consideration when constructing a circuit for work in the nanosecond region. It therefore seems to make sense to speak of a nanosecond pulse technique itself, and to separate it from the pulse technique in the microsecond region. Dimensions of most of the components and electronic apparatuses range between 1 mm and 1 m. With reference to the velocity of light in vacuum, these distances correspond to a time interval between thirty picoseconds (1 ps $= 10^{-12}$ s) and thirty nanoseconds.

With the advance of electronics into many fields of research and engineering the possibilities of taking direct time measurements in the nanosecond region also increased. They are, by nature, restricted to events occurring in small spaces, e.g. the atomic range, or to events proceeding at a very high speed, such as the propagation of an electromagnetic wave or the motion of highly accelerated elementary particles.

The nanosecond pulse technique, therefore, finds many applications in atomic, nuclear and elementary particle physics. These primarily include the measurement of very short lifetimes of excited atomic and nuclear states, and the use of time-of-flight measurements for very fast particles. The latter, for instance, serve for the determination of the neutron energy, for the discrimination of the various kinds of elementary particles and for the measurement of their lifetime values. The high operating costs of large accelerators demand that as many events as possible are processed and registered during the intensive beam pulse, where the various particle correlations and the particle trajectory are measured.

Principally two main tasks exist which require a time resolution in the nanosecond region:

(i) Determining the coincidences of events, or the measurement of time intervals between them, when they are correlated in time.

(ii) Processing and counting of like events which recur at nanosecond intervals.

To solve the first task, *Bell* and *Petch* [Be 49] introduced the so-called "fast-slow" principle, shown in Fig. 1. Here, separate branches are employed for the time measurement and for the selection of the energy of the events, and only a triple coincidence decides which event meets the required conditions with regard to energy and time, and is then admitted for registration. This made it possible, despite the limited time resolution of the circuitry employed for energy selection, to undertake time measurements in very short time ranges. Fast discriminators, which have meanwhile become available, permit a certain selection of energy before the time measuring device, but multichannel energy selection today still requires the stringent division in a fast section and a slow section.

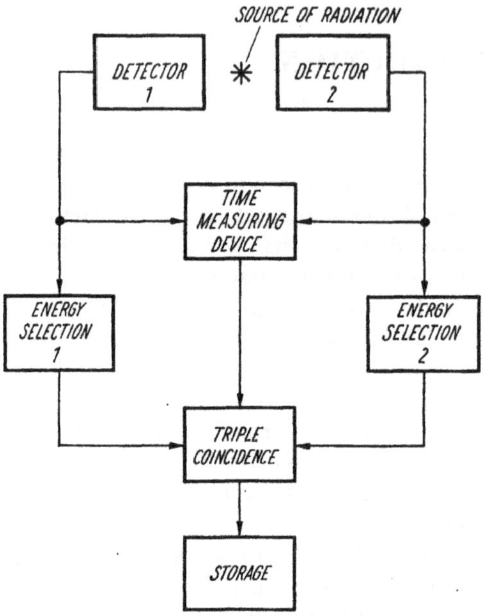

Fig. 1. Fast-slow principle. According to *Bell* and *Petch* [Be 49].

The second task demands far more from a circuit with respect to processing high pulse rates. In general, however, it is not necessary to fix the time of an event with the same precision as for the first task.

Application of nanosecond circuitry in fast computer machines also falls under the heading of (ii). The great economic importance of this new use resulted in a rapid advance in the development of specialised components which has been exceptionally beneficial for circuitry. Due to their low space and power requirements, but primarily because of higher dependability, they have practically displaced the electronic valve in nanosecond pulse technique circuitry. A review of valve-equipped pulse circuits will be found in the monographs of *Lewis* and *Wells* [Le 59], *Gruhle* [Gr 60], and *Morugin* and *Glebovich* [Mo 64].

The behaviour of semiconductor components, in some respects quite novel, frequently allows surprisingly simple solutions in circuit construction. For instance, semiconductor circuits operate on low-level signals, and distributed amplifiers are seldom needed. Trigger circuits constructed with tunnel diodes are extremely simple and are capable of handling pulse repetition rates out of reach for a valve-equipped circuit. Modern transistors have a gain-bandwidth product which by far exceeds this product of an electronic valve. The shorter signal path alone makes a semiconductor circuit better suitable for short-interval processing. In this connection, integrated circuitry can bring further advances in the future.

In general, extremely short-time events are associated with the emission of light or particles, and therefore a radiation detector is necessary to convert the received information. This applies not only for measurements in nuclear and elementary particle physics, but is also the case for high-speed photography and the use of laser. Although at present the electrical pulse is the only information carrier in an electronic computer, in future great importance will be given to light-pulse sources and photon-detectors in opto-electronic systems.

For this reason, radiation detectors with high time resolution are dealt with in greater detail in Chapter 2. Photon detectors, however, which are specifically intended for sensing modulated light [An 66; Gö 66d] have not been included. Basic circuitry, under the aspect of their application, are contained in Chapter 3; beginning with those properties of the employed components which are important for the time behaviour of a circuit. Chapter 4 describes apparatus and methods for the measurement of the time behaviour of detectors, components and circuits. The mutual cooperation of detectors and basic circuits is explained in Chapter 5 with a number of significant examples.

2. DETECTING RADIATION

2.1 PRELIMINARIES

An electrical signal must be assigned to the physical quantity to be measured by the use of electronic arrangements, unless the interesting quantity already represents the electrical signal. In the nanosecond range electromagnetic radiation or at least a particle radiation, travelling close to the speed of light, serves as information carrier. Detecting this radiation is, therefore, the beginning of the information processing chain. It is primarily the nuclear radiation which is frequently converted to visible or ultra-violet light; this is then transformed into an electrical signal by the receiving photon detector. On the other hand, an electrical pulse can be actuated directly by the radiation. In the present chapter the suggested modes of detection are dealt with in further detail and respective arrangement.

For investigations relating to the physics of atomic nuclei and of elementary particles, a considerable number of differing radiation detector devices have been designed. Of these, three groups are easily distinguishable according to their application [Bi 64], viz. dosemeters, counters and track visualisation detectors.

Dosemeters measure radiation flow through integration and are of no interest here.

Counters are used for quantum or particle detection, for the measurement of average count rate, and in many cases, for measuring the energy, the relative intensity and the kind of ionising particles. A counter also makes it possible to measure time with a very high degree of accuracy. For use in the nanosecond range scintillation counters, *Čerenkov* counters, semiconductor detectors and spark counters are most suitable.

Track visualisation detectors register the trajectory of particles, and are usually coupled with a counter assembly for background reduction. Time resolution suitable for use in the nanosecond range, however, can only be provided by scintillation chambers.

2.2 CONVERSION OF NUCLEAR RADIATION TO LIGHT

2.2.1 SCINTILLATORS

In the case of scintillation counters [Re 61; Mu 62; Bi 64] an impinging nuclear particle is first converted to light in the scintillator, which has optical contact to the photon detector via a light guide. Usually a photomultiplier serves as photon detector (see Section 2.3.1). In the scintillator, the ionising particles, i.e. electrons, protons, heavy ions and mesons, lose their energy through ionisation and excitation of the scintillator. On the other hand, non-ionising neutrons transfer their energy through recoil to the atomic nuclei or produce ionising particles through nuclear reactions. Energy of the electromagnetic γ- and X-radiation is taken over by electrons released by the photoeffect, the *Compton* process or by pair production. The generated ionising particles excite the scintillator, with the resulting energy emitted in the form of visible or ultra-violet light. When using organic scintillators this involves the molecular excitation of crystals, solutions or gases; while the crystalline structure is the prime factor in inorganic scintillators.

ORGANIC SCINTILLATORS

According to information furnished by *Birks* [Bi 64] the scintillating event involves primary and secondary processes. With a *primary process* the generation of an excited state of the so-called π-electrons is of interest in carbon compounds. The term scheme of the π-electrons shows the singlet states S_n and the triplet states T_n, with S_0 representing the ground state. De-excitation occurs with different time constants. Fast decaying components, for instance 3 ns, are primarily linked with the excitation of the second and the third singlet states (see process *1* in Fig. 2). During recombination of the ionised π-electron states, excitation to triplet states is mainly effected, from where the slow-decay components emanate, for instance with 400 ns time constant.

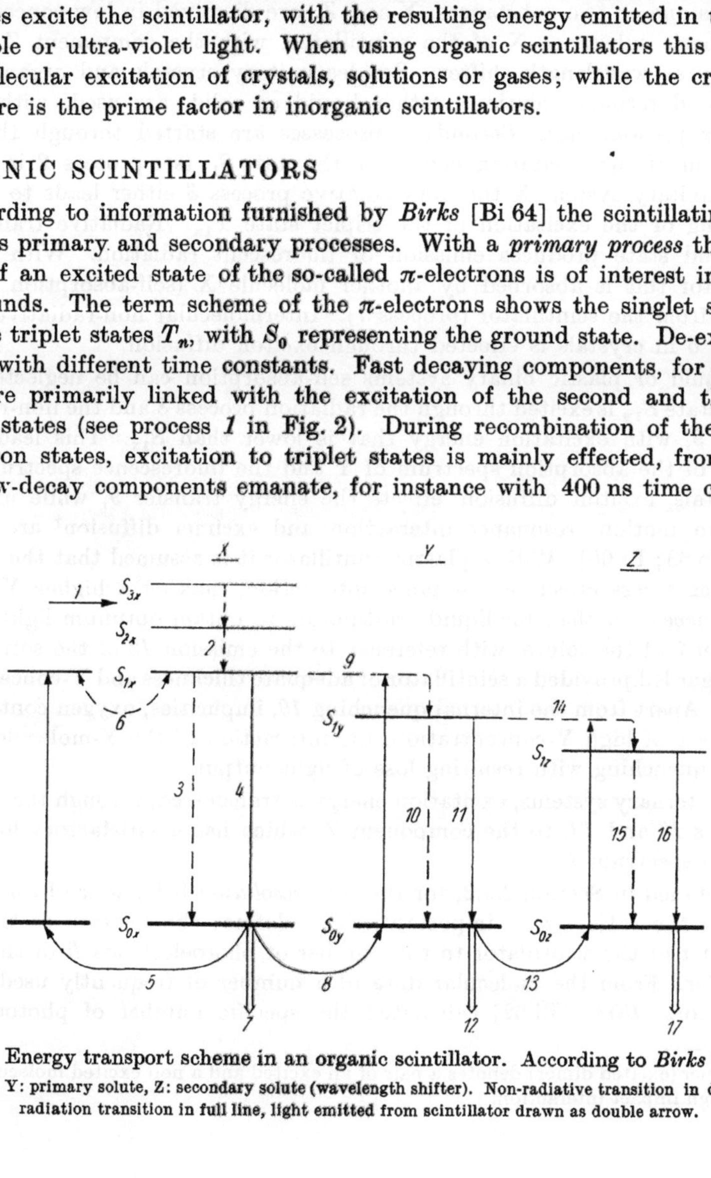

Fig. 2. Energy transport scheme in an organic scintillator. According to *Birks* [Bi 64].
X: solvent, Y: primary solute, Z: secondary solute (wavelength shifter). Non-radiative transition in dashed line, radiation transition in full line, light emitted from scintillator drawn as double arrow.

Partial excitation of high-level singlet states may cause medium speed component production, for instance 25 ns time constant, as stated by *Sjölin* [Sj 65].

During the *secondary processes*, excitation energy of the π-electrons is converted to the light emitted by the scintillator through a number of intermediate steps. To obtain high light output and adapt the emission spectrum of the scintillator to the spectral response of the employed photomultiplier, in some instances one or two substances, Y and Z, are dissolved in low concentration in the basic substance X of the scintillator, with the component Z serving expressly as wavelength shifter. Besides unitary crystals and organic glass, binary and ternary solutions, either liquid or solid, are used, with plastic solutions predominant. Secondary processes are started through the inner conversion of the excitation energy to the state S_{1x} (see process *2* in Fig. 2). In the unitary system X the non-radiative process *3* either leads to internal quenching or the excitation of the triplet state T_{1x}. Radiative transition *4* to ground state produces emission of fluorescent radiation. With a thick scintillator this is absorbed by another molecule X (self-absorption *5*) or it escapes from the scintillator (process *7*). Intermolecular non-radiative energy transfer *6* in crystals is effected through exciton diffusion.

In liquid or plastic binary systems self-absorption can be neglected. The singlet state S_{1y} is excited through the radiation process *8* and the non-radiative process *9*, with excitation energy that is lower than S_{1x}. This leads to an overlap of the absorption spectrum of Y and the fluorescence spectrum of X. In crystals, exciton diffusion effects the energy transfer *9*, while in liquids *Brown*ian motion, resonance interaction and excimer diffusion† are responsible [Vo 63; Bi 66]. With a plastic scintillator it is assumed that the polymer chain has a less effective resonance interaction, making a higher Y-concentration necessary than for liquid scintillators to obtain optimum light output. Emission *7* of the solute, with reference to the emission *12* of the solvent, can be disregarded provided a scintillator of adequate thickness and Y-concentration is used. Apart from the internal quenching *10*, impurities, oxygen content and, in the case of high Y-concentration, the interaction of the Y-molecules, cause further quenching with resulting loss of light output.

In the ternary systems, excitation energy is transferred, through the analogue processes *13* and *14*, to the component Z, which has a satisfactory long-wave emission spectrum *17*.

As detailed in Section 2.3.2, for the *time resolution* behaviour of scintillation counters the ratio τ/R is important. This defines the relation of the decay constant τ of the scintillator to the number of photoelectrons R of the photomultiplier. From the molecular data of a number of frequently used organic scintillators, *Birks* [Bi 62] calculated the specific number of photoelectrons

---

† Excimer (excited dimer) denotes a pair of an excited and a non-excited molecule formed through impact interaction.

($r = R/E$) and obtained 0·3 to 1·5 photoelectrons/keV; with E representing the energy absorbed by the scintillator. Resulting data of τ/R have been compiled in Table 1; note also Table 3, p. 21. It is seen that for a binary system, photomultipliers with a quartz window are more feasible because of their added sensitivity to the ultra-violet region of fluorescent light.

Table 1: Decay time constant τ, and its relation to the number of photoelectrons R with organic scintillators. According to *Birks* [Bi 62]

Scintillator		τ in ns	τ/R, in ps, for $E = 1$ MeV, Photomultiplier with Glass Window	Quartz Window
Unitary crystal:	anthracene	28	27·8	27·8
	p-terphenyle (TP)	4·5	7·7	5·9
	stilbene	$\leqq 3$	$\leqq 4$	$\leqq 3{\cdot}1$
Binary crystal:	anthracene (1 g/l) in naphthalene	$\approx 6{\cdot}4$	$\approx 4{\cdot}3$	$\approx 4{\cdot}2$
Binary liquid:	TP (5 g/l) in toluene	2·2	4	2·4
Binary plastic:	TP (36 g/l) in polystyrene	2·2	7·1	4·3
Ternary liquid:	TP (5 g/l) + POPOP (0·5 g/l) in toluene	$\approx 2{\cdot}2$	$\approx 2{\cdot}8$	$\approx 2{\cdot}7$
Ternary plastic:	TP (36 g/l) + POPOP (0·5 g/l) in polystyrene	3·5	5·6	5·3

For large scintillators it is more advantageous to use a ternary system with low self-absorption. Large dimensions give rise to considerable differences in length of light path and unsatisfactory time resolution [Ke 55]. For the reduction of the number of reflexions conic scintillators are more favourable [Hi 60; Lu 60; Ya 67; Gr 68]. By using only direct incidence of light on the photomultiplier [Bö 66 b], or by focusing light on the photocathode with a concave mirror [St 67b], a resolving time of less than 3 ns can also be obtained with a large scintillator, e.g. 1 m^2.

In literature it has been found that decay time and light output data are spread over a wide range [Bi 64, chapter 7, 8 and 9]. This is caused by variations in surface quenching, imperfections, impurities, self-absorption and kind of excitation. Specially with thick anthracene crystals high self-absorption causes longer decay time, compared to the molecular decay time constant $\tau_0 = 10$ ns. When using volume excitation through X-rays or γ-radiation this amounts to $\tau_v = 31$ ns. With surface excitation through ultra-violet light or α-radiation and observation of the reflected light, this comes to $\tau_{sr} = 16$ ns, while observation of the transmitted light gives $\tau_{st} = 25$ ns [Bi 64, page 215]. For comparing scintillator decay times, the constant τ_v should therefore be used.

Careful studies of the decay function through several orders of magnitude of light intensity [Ya 66; Ki 67 a] show that with the exception of anthracene, only its beginning can be approximated with an exponential function (see Fig. 3). Decay time constants obtained from the drop of initial intensity to 10% are

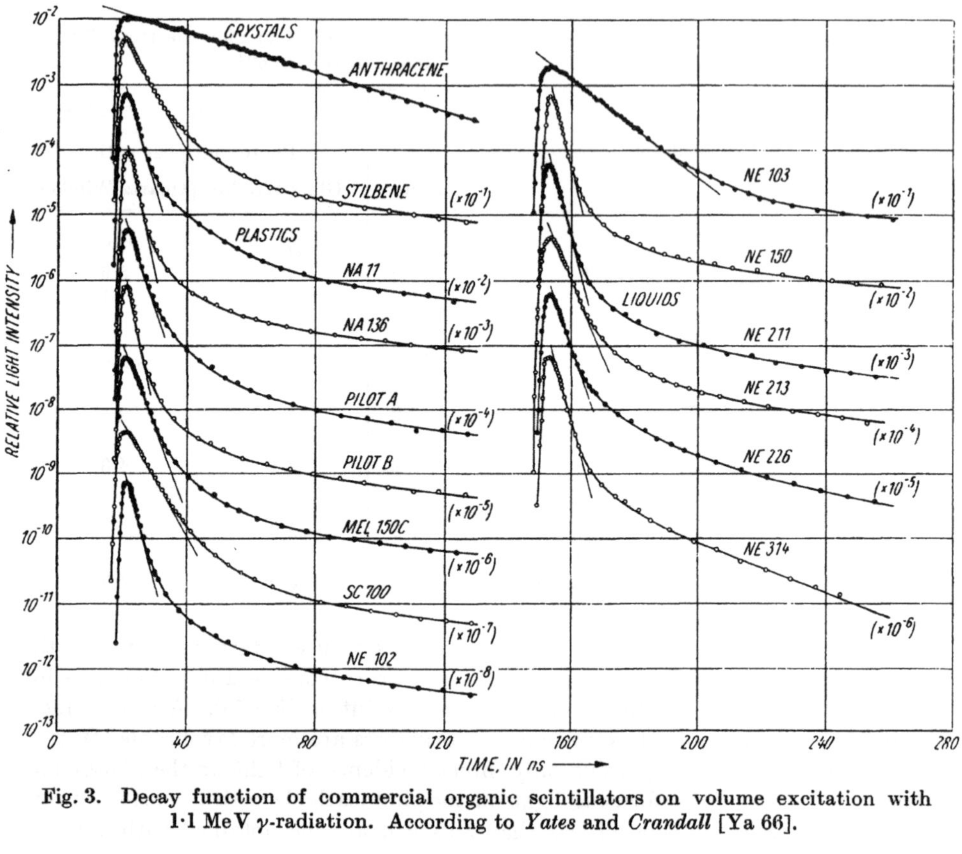

Fig. 3. Decay function of commercial organic scintillators on volume excitation with 1·1 MeV γ-radiation. According to *Yates* and *Crandall* [Ya 66].

compiled in Table 2. The lead-containing liquid scintillator NE 314 has a long-life exponential decay component. Adding lead, mercury or another heavy element causes higher γ-efficiency [Ni 63; Ts 65].

Stilbene crystals, however, show no exponential decay through 5 orders of magnitude [Bo 61]. Because of ionisation quenching of the rapid components the relative proportion of slow components increases with ionising density of the exciting radiation (see Fig. 4). This property, which is utilised for particle discrimination, is specially pronounced with the NE 213 scintillator [Pe 62; Sj 65].

Table 2: Decay time constant τ of commercial organic scintillators. According to *Yates* and *Crandall* [Ya 66] (see Fig. 3)

	Scintillator	τ, in ns
Crystal:	anthracene	29·3
	stilbene	4·9
Plastic:	Naton 11	2·6
	Naton 136	1·87
	Pilot A	2·96
	Pilot B	1·72
	Mel 150 C	3·25
	SC 700	4·86
	NE 102	2·20
	NE 103	11·34
	NE 150	2·17
Liquid:	NE 211	2·26
	NE 213	3·58
	NE 226	3·00
	NE 314	2·34

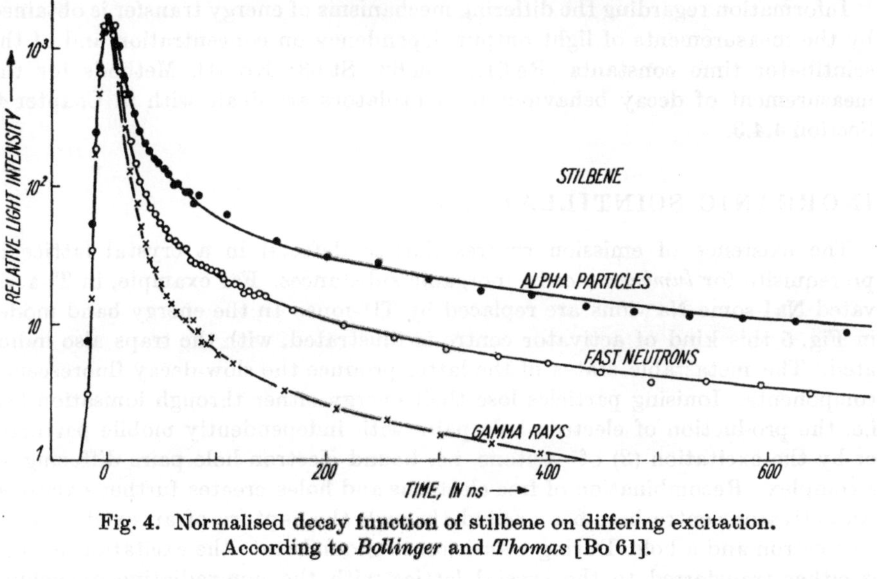

Fig. 4. Normalised decay function of stilbene on differing excitation. According to *Bollinger* and *Thomas* [Bo 61].

Scintillation rise time is also of importance for the time resolution attainable with scintillation counters, although this seldom finds consideration in theoretical approaches. With a binary system the following relation applies [Bi 64, page 228]

$$n_y = \frac{n_{0x}(w_r + w_t)}{1/\tau_x - 1/\tau_{0y}} \left(e^{-t/\tau_{0y}} - e^{-t/\tau_x}\right). \qquad (1)$$

The number of primary excited solvent molecules X is represented by n_{0x}, with n_y denoting the excited solute molecules Y, and the probability of the processes *8* and *9* in Fig. 2 is expressed by w_r and w_t, respectively. For fluorescence decay time in pure substances $\tau_{0x} > \tau_{0y}$ usually applies. With a binary system decay time τ_x of the solvent decreases with rising Y-concentration, until $\tau_x < \tau_{0y}$ obtains. Rise time is determined by the respectively lower of the two time constants τ_x, τ_{0y}; but with conventional concentration this is so low that the rise time of the photomultiplier corresponds to at least the same order of magnitude. For this reason it is not usually possible to measure the rise time of scintillations directly [Ya 66]. The convolution integral was employed by *Kramer* [Kr 67] for the indirect determination. Utilising a specialised arrangement with telescope and photodiode *McGuire* and *Palmer* [Mc 67 a] achieved a resolving time of 150 ps, which permitted the direct measurement of the rise time.

For the ternary systems, equation (1) may be used as approximation, if x is replaced by y and z is used instead of y, because with the high Y-concentration in practical use fluorescence of the basic binary system increases rapidly and decays with τ_{0y}. Rise time of a ternary system is determined by the higher of the two time constants τ_x or τ_y.

Information regarding the differing mechanisms of energy transfer is obtained by the measurements of light output dependency on concentration and of the scintillator time constants [Re 61 a; Bu 62; St 63; Ko 64]. Methods for the measurement of decay behaviour in scintillators are dealt with in Chapter 4, Section 4.4.3.

INORGANIC SCINTILLATORS

The existence of emission centres (lattice defects) in a crystal lattice is prerequisite for *luminescence* in inorganic substances. For example, in Tl-activated NaI some Na^+-ions are replaced by Tl^+-ions. In the energy band model in Fig. 5 this kind of activator centre is illustrated, with the traps also indicated. The metastable states of the latter produce the slow-decay fluorescence components. Ionising particles lose their energy either through ionisation (*1*), i.e. the production of electron-hole pairs with independently mobile partners, or by the excitation (*2*) of excitons, i.e. bound electron-hole pairs diffusing as a complex. Recombination of free electrons and holes creates further excitons. An activator centre becomes excited through the capture of an exciton or of an electron and a hole. During transition to ground state the excitation energy is either transferred to the crystal lattice with the non-radiative quenching process (*3*) or it is emitted as fluorescent radiation (*4*) of the activator centre.

Properties of inorganic scintillators are different from those of organic scintillators, as their density is high and they have elements of high atomic number Z, with consequent higher γ-efficiency. Their decay time constant, however, is higher by the factor of 100. For the test work in the nanosecond

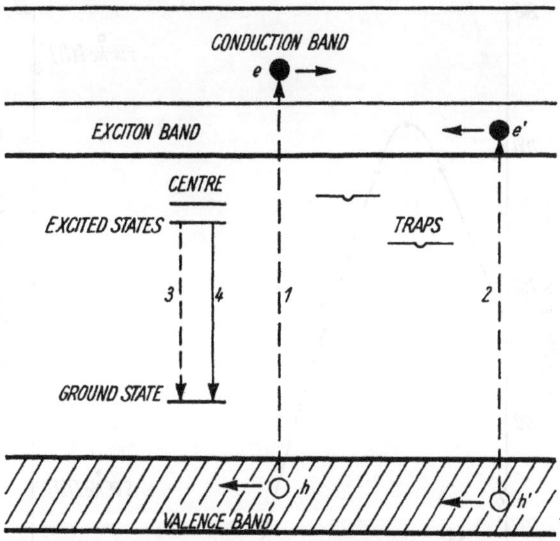

Fig. 5. Energy band model of an activated inorganic scintillator with independently mobile electron-hole pair ($e-h$) and bound electron-hole pair ($e'-h'$; exciton).

range NaI(Tl) is used (with $\tau = 230$ ns), which has twice the light output of anthracene. Fluorescent radiation of the Tl^+-activator is not absorbed by the basic lattice, in consequence low self-absorption obtains, analogue to the binary organic system. Due to rapidly increasing concentration quenching, optimum Tl^+-concentration amounts to 0·13 mole percent. Cooled unactivated NaI crystals at a temperature of liquid nitrogen (≈ 80 °K) have a higher light output and lower time constant with consequent better time resolution than activated NaI (see Fig. 6 and Table 6, p. 44).

Besides some slow decaying components [Bi 64, chapter 11; Po 66], with Tl^+-concentrations of $\gtrsim 0{\cdot}1\%$, decay function also shows a fast-decay component with $\tau \approx 10$ ns [Eb 54]. Formation of Tl^+-dimer centres† is seen as the cause [Sc 64; Sc 66]. Although the relative part of the dimer only amounts to 1%, because of its high cross-section for exciton trapping, it contributes 25% of the fluorescence yield. Due to the interaction with the Tl^+-ion only a fraction of it is emitted as fast-decay component. Crystals having a higher relative dimer concentration could possibly bridge the gap existing between high time-resolution organic scintillators and the inorganic scintillator with satisfactory energy resolution and γ-efficiency. However, *Hurley* et al. [Hu 67 a] did not find any evidence of a fast component in NaI (Tl).

† Bound Tl^+-ion pair on immediately adjacent Na^+-lattice sites.

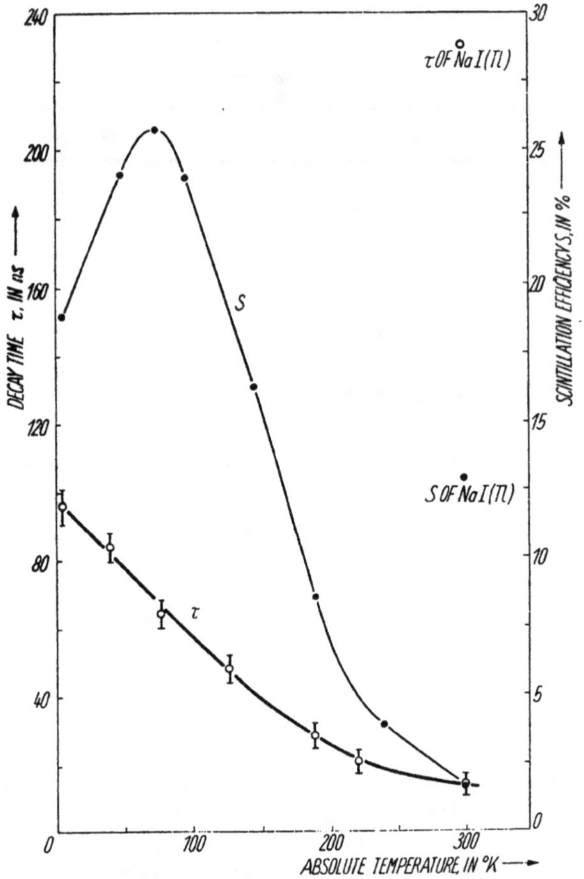

Fig. 6. Temperature dependence of γ-efficiency S and of decay time constant τ of unactivated NaI. Data of NaI(Tl) with optimum Tl-concentration at room temperature have been entered for comparison. According to *Sciver* and *Bogart* [Sc 58].

GAS SCINTILLATORS

Primarily inert gas is used for gas scintillators [Mu 62; Ba 63; Bi 64, chapter 14], where the rigid impacts of ionising particles ionise and excite the gas atoms. Recombination and de-excitation is effected through systems of optical, metastable and resonance states involving excited ions and dimers. This produces a very complex emission spectrum of lines, bands and continua located primarily in the ultra-violet region, which requires photomultipliers with quartz windows or wavelength shifters. Apart from the internal quenching in the atom, collision quenching on impact of gas atoms occurs which makes intensity and decay time constant greatly dependent on gas pressure. Rise of the latter causes a material reduction of the life-time of metastable states. Observed decay time τ represents the mean value of all frequency components weighted with respective relative intensity and relative detection efficiency. Decay time is not only influenced by pressure and temperature, but also depends on the purity of the gas, the properties of the wavelength shifter and of the photomultiplier. At atmospheric pressure, values of $\tau < 10$ ns were obtained. Of any of the inert gases xenon gives the highest light output. Liquid and solid xenon with a light output approaching that of NaI(Tl), is specially favourable for γ-detection because of its high density and atomic number ($Z = 54$). An argon — 10% nitrogen mixture gives higher efficiency than pure oxygen, while its decay time constant is 1·6 ns compared with 4·7 ns [He 67].

Liquid helium can be utilised as neutron polariser and scintillator, and gives a resolving time of 6 ns in coincidence with a plastic scintillator [Si 62]. Difficulties, however, arise from transfer of light at low temperatures and the low sensitivity of the photomultipliers in the ultra-violet region.

Gas scintillators with light amplified through an electric field compare with the good energy resolution of a proportional counter and show high time resolution because the primary component is retained [Bi 64, chapter 14].

2.2.2 *ČERENKOV* DETECTORS

ČERENKOV COUNTERS

For the detection of high-energy particles travelling close to the velocity of light *Čerenkov* counters are extensively employed. A number of reviewing papers describe properties and use [Je 53; Ma 54; Hu 60; Li 61; Mo 62; Ka 66c; Ne 66]. In principle the *Čerenkov* counter is similar to a scintillation counter. It consists of a light-transparent radiator with a high index of refraction, $n = c/v_L$, of an optical system for light transmission and of one or more photomultipliers. When a charged particle passes the radiator at a velocity of $v = \beta c$, faster than that of light v_L in the medium, an electromagnetic shock wave results (see Fig. 7). The opening angle Θ of the light cone travelling along the particle trajectory is defined by

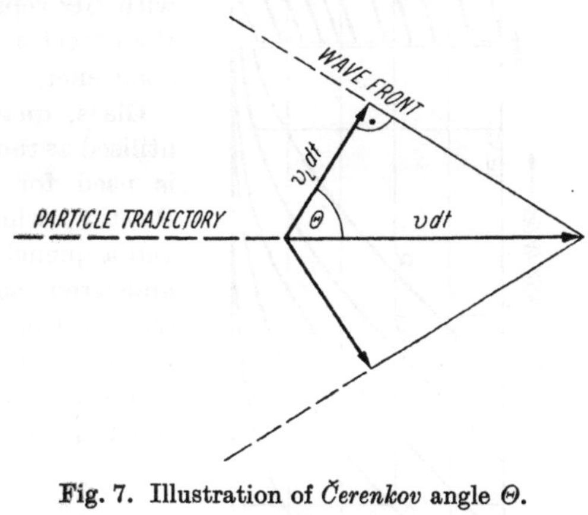

Fig. 7. Illustration of *Čerenkov* angle Θ.

$$\cos \Theta = \frac{v_L}{v} = \frac{1}{\beta n} \tag{2}$$

which enables the particle velocity v to be determined by measuring this *Čerenkov* angle.

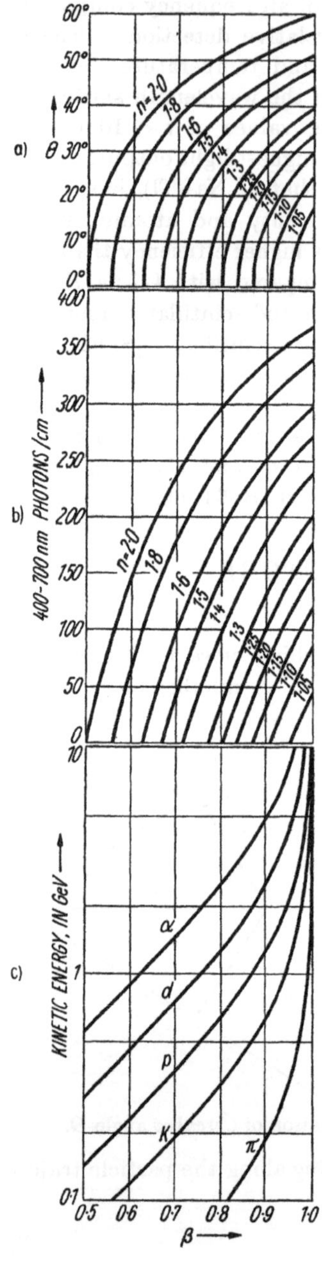

Fig. 8. Dependence of *Čerenkov* angle (*a*) and light output (*b*) with different refractive indices, and of the energy of different particles (*c*) on relative particle velocity. According to *Moyer* [Mo 62].

Light emission occurs primarily in the visible and near ultra-violet region, with light emission intensity described by

$$\frac{d^2N}{dx\, d\nu} = \frac{2\pi Z^2}{137\, c} \sin^2 \Theta\,, \tag{3}$$

where $d^2N/dx\, d\nu$ is the number of photons per centimetre and frequency interval unit, c the velocity of light in vacuum and Z the number of electron charges of the particle. Differently charged particles consequently produce different pulse height values. The diagrams *a* and *b* in Fig. 8 show data for some values of n and for the visible spectrum between 400 and 700 nm, that have been obtained from the equations (2) and (3). The relation of kinetic energy E and the relative velocity β for π- and K-mesons, protons (p), deuterons (d) and alpha particles is illustrated in diagram *c* of Fig. 8. An approximation ($\beta \approx 1$) of the relative energy resolution is given by

$$\frac{\Delta E}{E} \approx E^2 \tan \Theta\, \Delta\Theta \tag{4}$$

with $\Delta\Theta$ representing the angular resolution of the detector. Thus, high light output excludes good energy resolution and vice versa.

Glass, quartz, lucite, water etc. are usually utilised as radiator material. Lead-loaded glass is used for total absorption spectrometers. Plastics, which contain a wavelength shifter and a quenching agent for the suppression of undesired scintillations, are also employed [Ni 63]. Gas radiators are very advantageous. For this purpose, non-scintillating gases with a low atomic number are used which have the required refractive index at a moderate pressure, e.g. CO_2, ethylene and others. Through the low index of refraction, $n \approx 1$, detection of only high speed particles ($\beta \rightarrow 1$) becomes feasible. The refractive index n and consequently the energy range sensed by the detector is easily changed by varying the gas pressure. Very good energy resolution is obtained through the small *Čerenkov* angle ($\Theta \approx$ 5 to 10°), as equation (4) shows. A velocity resolution of $\Delta\beta \lesssim 10^{-3}$ is obtained [Yu 61]. According to the light transfer system employed, there are two types of *Čerenkov* counters: the non-focussing and the focussing counter.

Analogue to scintillation counter design, with the *non-focussing counter* (threshold detector) the radiator is mounted in a container with reflecting walls and has direct optical contact with the photomultiplier. Above the particle velocity threshold $\beta_{min} = 1/n$, given by the refractive index n of the radiator, the *Čerenkov* light will always be accepted by the photomultiplier. Sometimes an upper velocity limit is introduced by the total reflexion occurring above Θ_{max} [Hu 60].

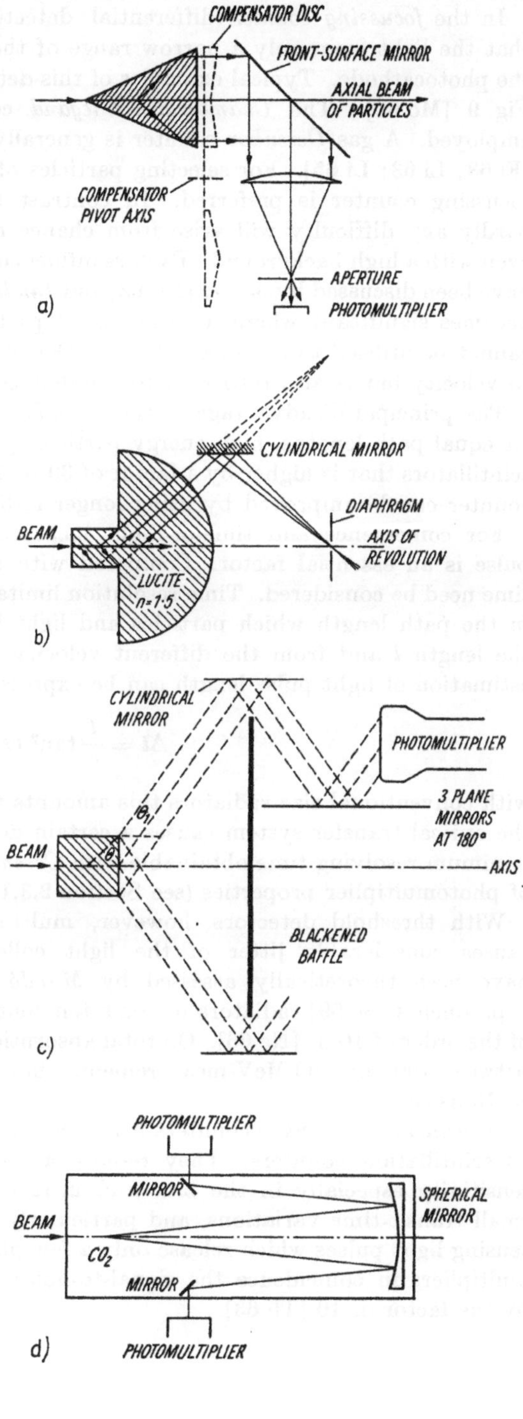

Fig. 9. Focussing counter according to *Getting-Jelley* (*a*), *Marshall* (*b*), *Chamberlain-Wiegand* (*c*) and *Kinsey* (*d*). According to *Moyer* [Mo 62].

In the *focussing counter* (differential detector) mirrors and lenses ensure that the light from only a narrow range of the *Čerenkov* angle is focussed on the photocathode. Typical examples of this detector construction are shown in Fig. 9 [Mo 62]. The *Chamberlain-Wiegand* counter (*c*) is most frequently employed. A gas *Čerenkov* counter is generally designed, as shown in Fig. 9*d* [Fi 63; Li 63; Li 65]. For selecting particles of a narrow velocity interval the focussing counter is preferred. In contrast to scintillation counters, then, hardly any difficulty will arise from chance coincidence and pile-up effects, even with a high background. Factors influencing attainable velocity resolution have been discussed by *Moyer* [Mo 62] and *Lindenbaum* [Li 61]. Good resolution becomes significant where for short-lived particles the time-of-flight method cannot be utilised to measure velocity. Detector sensitivity responds not only to velocity but is also restricted to the defined direction of the particles.

The principal disadvantage of the *Čerenkov* counter is its low light output. At equal path lengths, high energy particles produce a light output in plastic scintillators that is higher by a factor of 30 to 50. This condition in a *Čerenkov* counter can be improved by using longer radiators.

For coincidence and time-of-flight measurements the length of the light pulse is an essential factor. Compared with scintillators, no inherent decay time need be considered. Time resolution limitation derives from the difference in the path length which particles and light have to travel in a radiator of the length l and from the different velocity of the particles and light. An estimation of light pulse length can be expressed by

$$\Delta t = \frac{l}{v} \tan^2 \Theta; \tag{5}$$

with conventional size radiators this amounts to $\Delta t = 10$ to 100 ps. Although the optical transfer system causes a certain deterioration in performance, the minimum resolving time obtainable with *Čerenkov* counters is mainly a matter of photomultiplier properties (see Section 2.3.1).

With threshold detectors, however, multi-reflexion at the radiator walls causes considerable jitter of the light collecting time. These influences have been theoretically assessed by *Mandó* [Ma 54a]. In agreement with experiments [Se 59] radiators of some ten centimetres cause time fluctuations of the order of 10 ns [Ne 66]. On total absorption of γ-radiation with an energy between 100 and 500 MeV measurements obtained a resolving time of $2\,\tau_0 = 8$ to 10 ns [Al 65].

In principle, *Čerenkov* counters make the same demands on photomultipliers as scintillation counters. They require a large cathode face, high cathode sensitivity especially in the blue and ultra-violet spectrum, high gain, only small transit-time variations, and particularly a good signal-to-noise ratio for sensing light pulses which release only a few photoelectrons. Using two photomultipliers in coincidence the signal-to-noise ratio can be improved at least by the factor of 10 [Th 63].

ČERENKOV CHAMBERS

If *Čerenkov* light is focussed with a lens located coaxially to the particle trajectory a ring is produced in the focal plane. According to

$$d = 2 f \tan \Theta \tag{6}$$

and given focal length f, the diameter d of the ring depends on the *Čerenkov* angle Θ and thus on particle velocity. The photomultiplier can be replaced by an image intensifier [Hu 60; Ro 60; Bi 62a; Ir 66] having the properties and of a design as described in Section 2.3.3. At the output of the image intensifier a ring image of the *Čerenkov* cone of sufficient intensity is produced to permit photographic recording. Position of the ring gives additional information with respect to the direction of the particle trajectory. If velocity along the path in the radiator changes, this causes a widening of the ring. A change of direction caused by curving of the trajectory in a magnetic field blurs the outline of the ring at opposing sections. Parallel shifting of the trajectory, however, has no effect on the ring image.

As is the case with scintillation chambers, *Čerenkov* chambers act as track visualisation detectors with high time resolution. Image registration can be triggered through a counter telescope which gates the image intensifier. Limits of performance of *Čerenkov* chambers have been assessed by *Roberts* [Ro 60], and properties and application have been reviewed by *Medvedev* [Me 66].

2.2.3 SCINTILLATION CHAMBERS

These are very similar in design to the *Čerenkov* chambers [Za 55; Za 56]. Along the trajectory in a scintillator, high energy particles produce a weak light trace, which is photographed, and subsequently evaluated, after suitable amplification with an image intensifier (see Section 2.3.3). For stereo reproduction two intensifiers are used. Essentially, scintillation material must meet the same requirements as for a scintillation counter, viz. high light output, short decay time, high transparency with reference to its own light and a spectral distribution of the emitted light that corresponds to the spectral sensitivity of the input photocathode of the image intensifier. According to scintillator shape and type of light transfer system, there are two kinds of scintillation chambers.

a) The *homogeneous chamber* as illustrated with the example in Fig. 10 [Gi 62; Gi 63], where a massive block is used as scintillator, with an arrangement of lenses employed to transfer the light to the four-stage image intensifier tube. The electronic setup sketched into the block diagram serves to trigger the image intensifier on coincidence of the two scintillation counters *Sc* (see Section 2.3.3). To make the trajectory of the respective particle visible, trigger delay must be less than the decay time of the scintillator or of the phosphorescent screen of the image intensifier, which acts as image store. For detecting rare

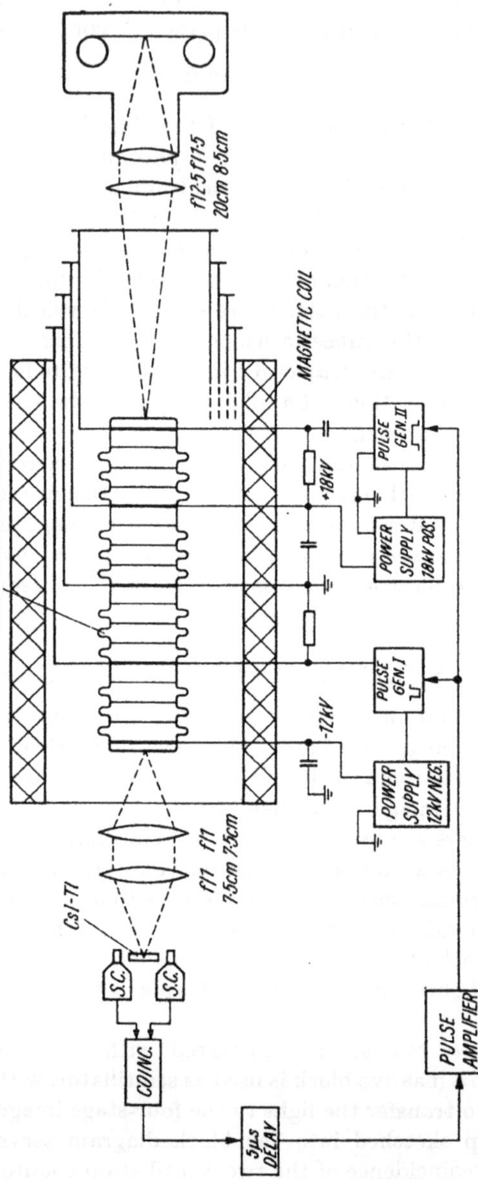

Fig. 10. Construction of a homogeneous scintillation chamber. According to *Gildemeister* and *Giese* [Gi 63].

events against a high background the intensifier tube is gated through the photocathode, with rise time and trigger delay together amounting to 60 ns.

A good space resolution, homogeneity and isotropic sensitivity are among the advantages of the homogeneous chamber. Further, brightness of particle trace depends on the ionising density of the particles. Although NaI crystals of 10 cm × 15 cm × 15 cm [Bu 62a] have already been used, as well as combinations of smaller crystals [Jo 62], depth of the chamber is restricted by the focussing depth of the optical system.

b) *Fibre chambers* of any desired focussing depth, although with much less favourable spatial resolution, are obtainable if the massive block scintillator is replaced by plastic scintillator fibres of 0·5 to 1 mm diameter [Re 61; Bu 65]. For stereoscopic reproduction they are arranged in an alternate pattern, at right angles to each other, and coupled either directly or through glass fibres to the photocathodes of the two image intensifiers. For satisfactory spatial resolution, the image intensifier must be provided with a face plate of thin glass or glass fibre. Chamber size is restricted by the light absorption of the plastic fibres and the difficulty of fibre bundle production, which contains 2×10^5 threads for a chamber of 13 cm × 13 cm × 25 cm [Ca 62; Bu 62a]. However, the short decay time results in a resolving time of a few nanoseconds [La 62]. Further references are quoted by *Medvedev* [Me 66] who gives some examples of use.

Despite the construction of specialised image intensifiers [Bu 62a], scintillation chambers are waning in importance. They are displaced by spark chambers which are also triggered with little time delay by the respective event (see Section 5.3). A spark chamber has the advantage that it is simpler and cheaper to make and that larger dimension can be used. However, it has no isotropic sensitivity and gives no information with respect to ionising density.

2.3 CONVERSION OF LIGHT TO ELECTRICAL SIGNALS (PHOTON DETECTORS)

2.3.1 PHOTOMULTIPLIERS

At present, the most frequently employed photon detector is the photomultiplier because it is the only device having sufficient sensitivity and time resolution for detecting nuclear radiation with the aid of scintillation counters or *Čerenkov* counters. With photomultipliers and the image intensifier described in Section 2.3.3 the external photoelectric effect is utilised to convert light to an electrical signal. Amplification is usually handled through secondary emission, with the primary electrons emanating from the photocathode focussed on the first dynode by an electron-optical system. Every impinging primary electron releases a few secondary electrons. A series of subsequently located dynode stages are then used to amplify the electrical signal. A terminating collecting electrode (the anode) serves to deliver the output signal of the photomultiplier [Ro 53; Ha 57; Sh 58; Ma 62; Bi 64; So 65; Ne 66, chapter 6].

This section deals with the "short-time photomultiplier", that is a multiplier suitable for the measurement of nanosecond light pulses. In general, this type of multiplier must meet the following demands:

(i) Little time-jitter, i.e. the shape of the leading edge and the location of the centroid of the output pulse remain sensibly constant where accurate time interval measurements in the nanosecond range are significant.

(ii) Delivery of narrow output pulses for recording events with intervals around 10 ns.

(iii) High amplification and low noise, if only a few photoelectrons are released due to the small amount of light from a pulse.

(iv) Good linearity, if the charge of the output pulse has to be proportional to the amount of light from the incident light pulses.

(v) An output pulse that is not accompanied by a prepulse or an after-pulse, which will interfere with satisfactory resolving time and lead to double counts.

PHOTOCATHODE

The number of photoelectrons N_e released from the photocathode is proportional to the number of impinging photons N_{ph}, as expressed with

$$N_e = \eta \, N_{ph} \; . \tag{7}$$

This proportionality of light-intensity changes has been verified experimentally, up to frequencies of 26 GHz [Jo 65]. The quantum yield factor η is greatly influenced by the wavelength of light. If a high quantum yield is desirable, i.e. as high a number of photoelectrons as possible, the spectral sensitivity of the photocathode has to be adapted to the spectrum of incident light. This applies, for example, for the measurement of energy and time with scintillation counters.

Properties of the various types of photocathodes have been reviewed [Ha 57; Sp 63]. It was found that maximum sensitivity of most photocathodes obtains in the blue region, reaching to near ultra-violet. Quantum yield is $\eta = 10$ to 15%. Expressed as photoelectron current per unit of light intensity, this corresponds to a cathode sensitivity of 40 to 60 μA/lm. Improved Cs–Sb cathodes, type S 11, permitted an increase in quantum yield to 30% [St 66]. Near the long-wave region quantum yield decreases sharply, and even with the specialised infra-red-sensitive cathode, type S 1, amounts to less than 1%.

For the time resolution of scintillation counters discussed in Section 2.3.2, a major point, after absorption of energy E in the scintillator, is the number of photoelectrons R which are emitted at the photocathode and actually reach the first dynode. Thus, R depends on the radiation energy, the light output of the scintillator, the quantum yield of the photocathode and the transfer coefficient to the first dynode.

Respective data regarding the specific number of photoelectrons, as expressed by

$$r = R/E \tag{8}$$

are compiled in Table 3. It will be noticed that considerable spread is present due to differing parameters, e.g. cathode sensitivity and measuring error within wide limits.

Table 3: Specific number of photoelectrons r obtained by measurement

Scintillator	Photo-multiplier	Cathode Sensitivity	$r = R/E$ in keV^{-1}	Method of Measurement†	Authors
NaI(Tl)	RCA 5819	60 μA/lm	5	a	*Borkowski* and *Clark* [Bo 53]
NaI(Tl)			3 to 5		*Sharp* and *Thomson* [Sh 58]
NaI(Tl)	RCA 6342 A		2·5	b, c	*Currie* et al. [Cu 61]
NaI(Tl)	DuMont 6363	91 μA/lm	10	b	*Bollinger* and *Thomas* [Bo 62]
NaI(Tl)			1·63	b	*Braunsfurth* and *Körner* [Br 65]
NaI(Tl)	DuMont 6363	average	6·4		*Baker*, in [Ly 66]
NaI(Tl)	RCA 6810 A	average	4·8		*Baker*, in [Ly 66]
NaI(Tl)	RCA 8575		7·85 to 9·3	b, c	*Lynch* [Ly 66]
Anthracene			2		*Sharp* and *Thomson* [Sh 58]
Anthracene			1·4		*Currie* et al. [Cu 61]
Anthracene	RCA 6810 A		1·0		*Ruby* et al. [Ru 63]
Stilbene			0·8		*Sharp* and *Thomson* [Sh 58]
NE 102			0·5		*Currie* et al. [Cu 61]
Naton 136	XP 1020		0·4	b	*Bertolini* et al. [Be 66]

† a) Comparison of the height of single electron response with the spectrum of a NaI(Tl) scintillator, b) from energy resolution statistics, c) from time resolution statistics.

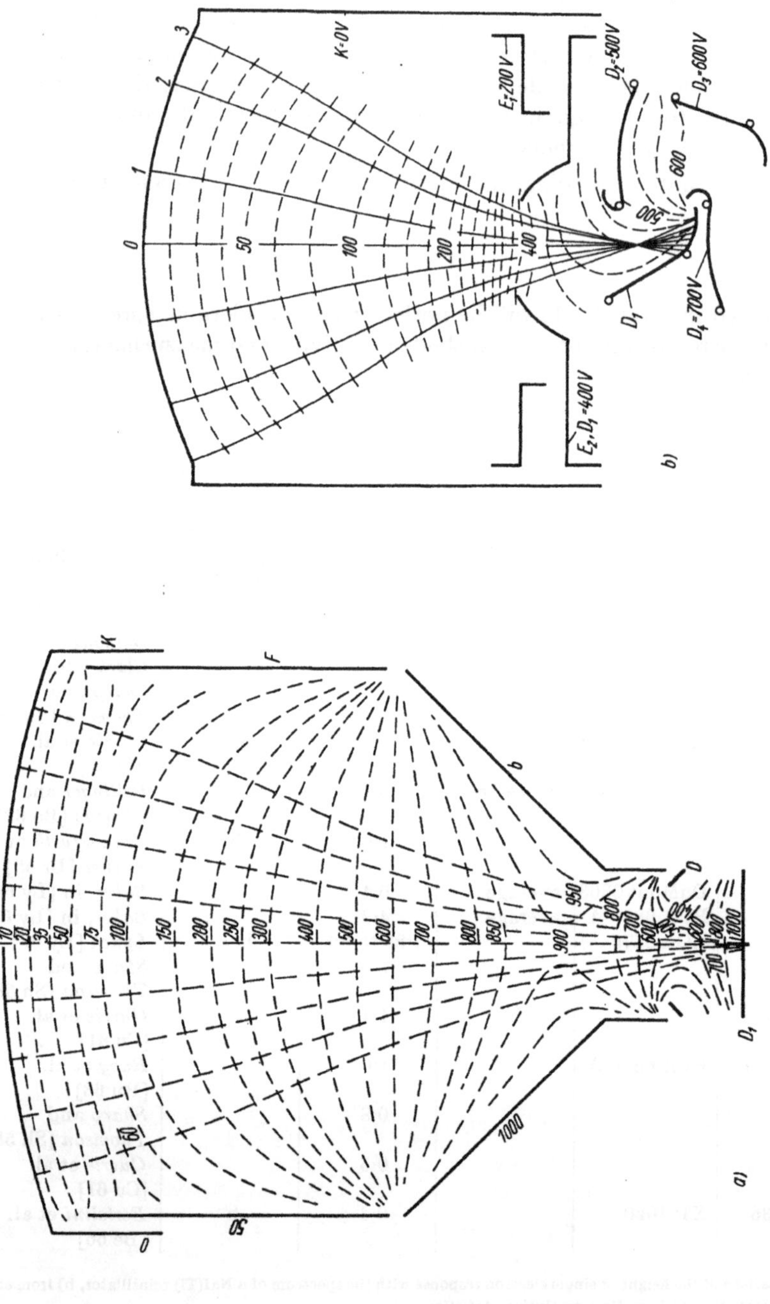

Fig. 11. Transfer Systems in Short-Time Photomultipliers.

a) 56 AVP [VALVO-Manual]. *b)* K 14 FS 50 [Gö 62]. *c)* XP 1020 [VALVO-Manual].

K cathode, E_1, *F*, F_1, F_2 focussing electrodes, *b*, E_2 accelerating electrodes, *D* deflecting electrode, D_1 input of dynode system.

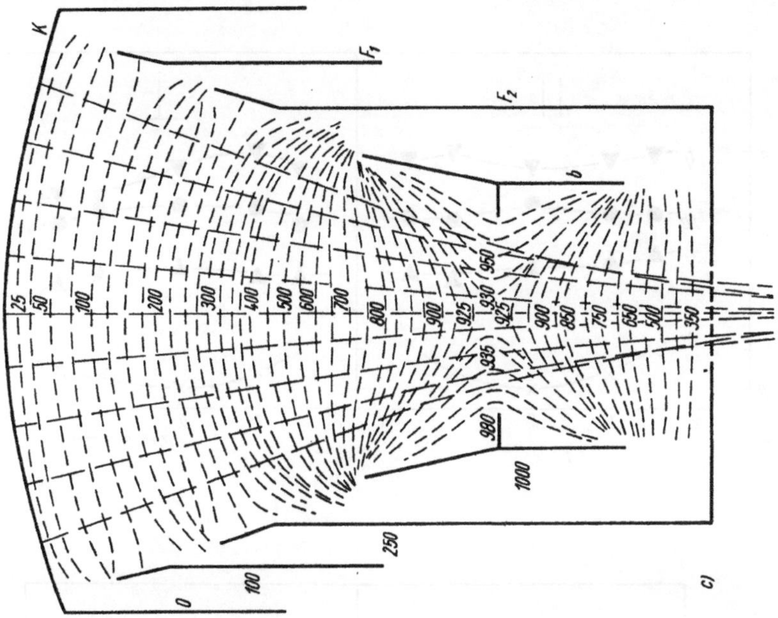

Eig. 11 c

ELECTRON-OPTICAL TRANSFER SYSTEM

To a considerable extent, photoelectron transfer to the first dynode depends on conditions of potential in the cathode region and on the wavelength λ of the incident light. For example, at a wavelength of $\lambda = 400$ nm in the 56 AVP system 65 to 75% of the photoelectrons reach the first dynode, while 95% arrive at the first stage in the XP 1020 system; at a wavelength of $\lambda = 600$ nm this amounts to respectively 80 to 90% or approaches 100% [Pi 64]. Apart from the desirability that all photoelectrons arrive at the first dynode stage, the short-time photomultiplier must meet the added condition of an equal transit time to the first dynode for any photoelectron, irrespective of its location of emission, its initial velocity and direction. In practice, this is closely approached by the use of the characteristic-curved photocathode and with the assistance of auxiliary focussing and accelerating electrodes. Examples of transfer arrangements showing field lines and electron trajectories are illustrated in Fig. 11 *a*, *b* and *c*. Types 56 AVP and K 14 FS 50 use a triode system, while type XP 1020 operates with a tetrode system. Fig. 12 *a* and *b* show the relation of transit time to location of emission, with reference to the cathode centre $x = 0$. From the rotation-symmetric field of the cathode region the photoelectrons are transferred to the cylindrical field of the dynode system causing a difference in transit time of emitted photoelectrons along a line parallel or vertical to the symmetry plane of the dynode system. Rise time, half width

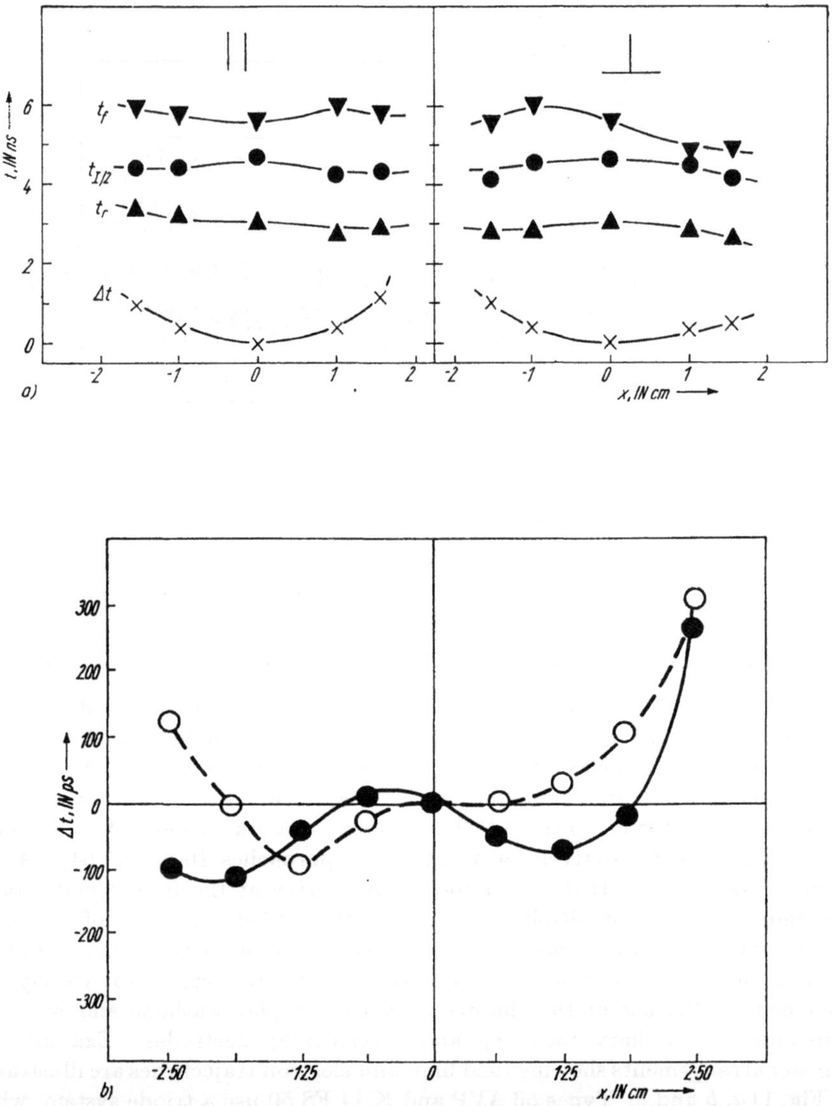

Fig. 12. Dependence of transit-time difference cathode-dynode 1 on emission location, parallel and vertical to the symmetry plane of the dynode system.

a) 56 AVP [Ce 60]. *b)* C 70045 A [Bi 64a].

and decay time of the output pulse, with reference to the location of the light spot on the photocathode, are also included in Fig. 12*a*.

So-called prepulses in the transfer system of a photomultiplier may be caused when the passage of light through the semitransparent photocathode gives rise to photoelectrons at the focussing and accelerating electrodes, and these reach the first dynode in a shorter transit time than the regular electrons emitted by the cathode. Depending on conditions of potential, prepulses appear before the actual pulse with a lead of up to 10 ns and can disturb time resolution to a considerable extent. Prepulse occurrence could be observed with type 56 AVP [Bo 64; Ya 66], ФЭУ-36 [Bo 64], XP 1020 [Be 65] and ФЭУ-30 [Av 66; La 67], while type K 14 FS 50 had no prepulses (see Fig. 11*b* and 13*c*). Elimination of the photoelectric sources will avoid the prepulses. Emission of these undesirable electrons has no influence on the time resolution of the scintillation counter, if the distribution of potential in the transfer system provides for these electrons to be subsequently transferred to the dynode system as regular photoelectrons. Nevertheless, such spurious emission in the transfer system represents an added source of noise [Sh 58; De 64].

DYNODE SYSTEM

In conventional photomultipliers, the dynode system consists of 6 to 14 reflexion dynodes, and additional electrodes which serve to focus the electron beam, besides reducing space charge effects and transit time fluctuations. Total amplification can attain levels to 10^9. Properties of the dynode material have been discussed, among others, by *Hartmann* and *Bernhard* [Ha 57]. In the focussed dynode systems, the secondary electrons are preferably aligned to the centre of the following dynode. Unfocussed arrangements, such as the Venetian blind dynode system and the sector dynodes, cannot be used in short-time photomultipliers [Sh 58].

Some focussed dynode systems are illustrated in Fig. 13. The circular system shown in Fig. 13*a* was used in type 931 A and type 1 P 21 photomultipliers, which were among the first having high time resolution [Be 52]. The linear arrangement shown in Fig. 13*b* is generally preferred and used, for instance, in the 56 AVP and XP 1020 photomultipliers. *Pietri* demonstrated its focussing function [Pi 62; Pi 67]. The meander system illustrated in Fig. 13*c* allows a compact configuration of the dynode system and creates high field strength in front of the odd-number stages through dynode overlapping [Gö 62]. Fig. 13*d* presents part of the photomultiplier C 70045 A with its photocathode located on the side. This multiplier has a rise time that is shorter than 1 ns (see Table 4, p. 28).

Single electron response (SER) is caused by a single photoelectron and serves to assess the properties of the dynode system and of the collecting electrode (anode). Typical rise time and half-width of the SER is around 1 to 3 ns, with a symmetric bell-shaped pulse preferred. Rise time of the SER represents

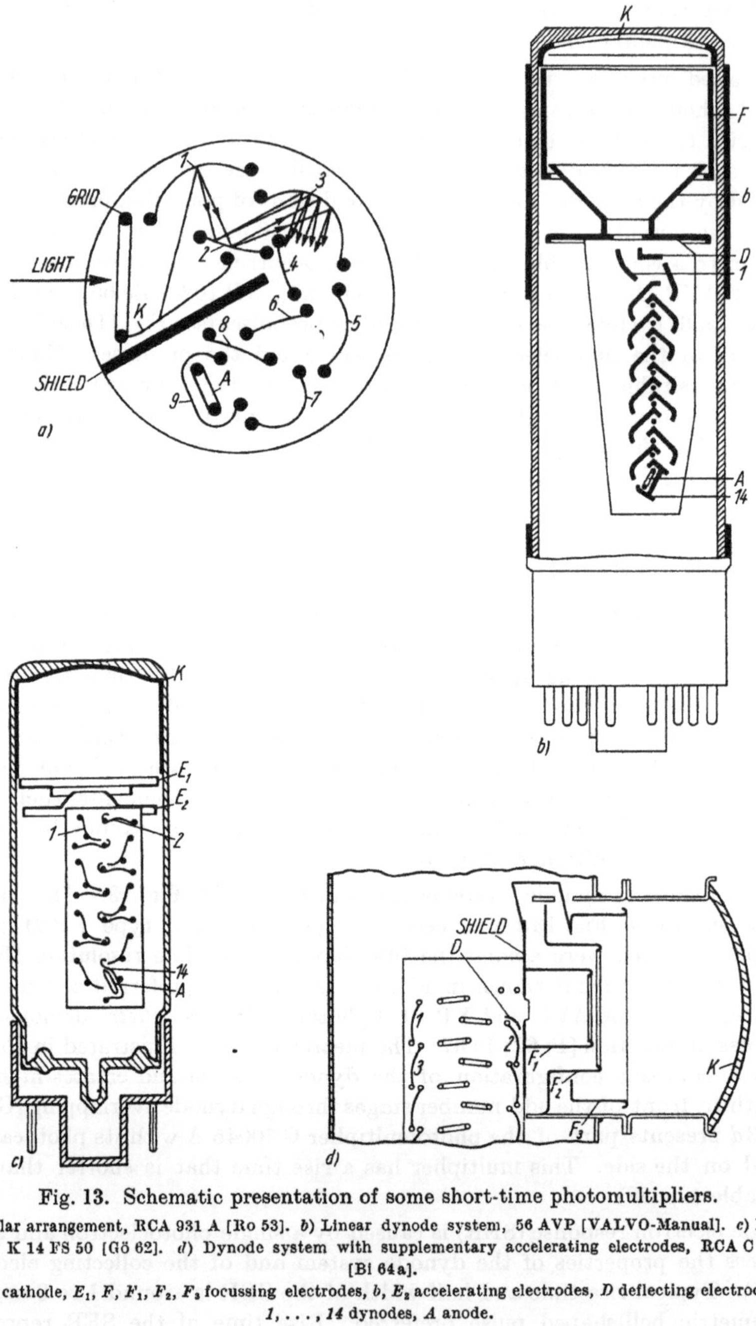

Fig. 13. Schematic presentation of some short-time photomultipliers.

a) Circular arrangement, RCA 931 A [Ro 53]. *b)* Linear dynode system, 56 AVP [VALVO-Manual]. *c)* Meander system, K 14 FS 50 [Gö 62]. *d)* Dynode system with supplementary accelerating electrodes, RCA C 70045 A [Bl 64a].

K cathode, E_1, *F*, F_1, F_2, F_3 focussing electrodes, *b*, E_2 accelerating electrodes, *D* deflecting electrode, *1, . . . , 14* dynodes, *A* anode.

the lower limit of this parameter for the multiplier output pulse. On short-time illumination of the photocathode, a nearly symmetric current pulse, composed of many superposed SER pulses, should be expected at the multiplier output. Rise time and width of the delivered output pulse are determined by transit time fluctuation in the transfer system and by the shape of the single electron response. On high current intensity, space charge effects may exert an influence on the last dynode stages which extends decay time of the current pulse.

COLLECTING ELECTRODE (ANODE)

Usually the collecting electrode or anode is designed in the shape of a grid and placed in front of the last dynode to keep capacitance low. For short rise time it is feasible to use a coaxial connector as anode output. With a non-integrated current pulse at the photomultiplier output this has proved of advantage. It has been reported that the use of a 75 Ω coaxial anode output on type ФЭУ-36 multipliers reduced the rise time from the former 2·1 to 2·3 ns to a level of 1·1 to 1·5 ns, with pulse width respectively reduced from 6 to 4 ns [Be 65a]. Type ФЭУ-72 is an improved version of ФЭУ-36 and equipped with a coaxial anode output [Vo 67].

ADJUSTMENT OF OPTIMUM OPERATIONAL BEHAVIOUR

Parameters for some of the short-time photomultipliers have been summarised in Table 4. Data of pulse width, rise time and decay time represent general values because of their dependence on multiplier electrode potentials, and because a length of the light pulse used for measurement could not in every case be obtained that underpassed the transit time fluctuation in the multiplier.

Many manufacturers recommend two different potential distributions for short-time photomultipliers. Either the same voltage at each stage of the dynode system is used for maximum amplification, or progressive voltage increments at the last 3 to 6 dynode stages are resorted to for high peak currents and maximum range of proportionality. Usually, the voltage between the cathode and dynode 1 is 2 to 4 times the average potential difference of the dynode stages, and amounts to 300 to 600 V. Potentials of the transfer system and of the last dynode stage exert a considerable influence on the parameters of the multipliers. These voltages are therefore individually balanced in each unit. Careful potential adjustment of the transfer system improves the multiplier energy resolution [Ga 58; Gr 64] and tends to reduce noise [Pa 62; Pa 62a]. For example, with type 56 AVP and type 58 AVP multipliers, individual optimising of the electrode voltages in the 22 respectively 14 examined units obtained amplification and linearity data which equalled or bettered advertised specifications. With the standard voltage divider recommended by the manufacturer this could not be obtained in every case [Be 63;

Table 4: Short-Time

Type	6810 A[a]	56 AVP[a]
Manufacturer	RCA	VALVO
Cathode diameter, in mm	42	42
Input system	triode	triode
Number of stages	14, linear	14, linear
Coaxial anode output	—	—
Operating voltage, in kV	max. 2·4	2·5; max. 2·6
Gain	$>10^7$	$\geqq 10^8$
Maximum current pulse, in A		0·5 to 1
Rise time t_r, in ns, from 10 to 90%[b]	3 to 7[d] ; 2·7[c]	2; 2·2[d]; 2·1[e]
Fall time t_f, in ns, from 90 to 10%[b]	5 to 22[d] ; 8·5[c]	2·2[d]; 3·5[e]
Pulse width $t_{1/2}$, in ns, at 50%[b]	4 to 13[d] ; 11·5[c]	4; 3·2[d]; 4·25[e]
Pulse base, in ns[b]	15[c]	7·5[e]; 9[f]
Transit time fluctuation, in ns[c]	20 mm: 3	border: $\leqq 0{\cdot}5$
Total transit time, in ns		36
Single electron response (SER)	$t_r \approx t_f = 2{\cdot}6$ ns, $t_{1/2} = 4{\cdot}0$ ns [Ce 60]	$t_r \approx t_f < 2$ ns, $t_{1/2} \approx 2{\cdot}7$ ns [Ce 60] *Gauss*ian [Hy 64]
Remarks		

[a] Data primarily from manufacturers specifications.
[b] On illumination with a light pulse which has a short length with reference to time fluctuation in the photomultiplier.
[c] Transit-time difference in cathode-dynode 1 region, measured with light spot at specified distance from cathode centre.
[d] According to *Cernigoi* et al. [Ce 60], with *Čerenkov* radiation and various amplification factors.

Be 64]. Through a similar adjustment, favourable conditions can be established for minimum rise time and width of the output pulse, and for maximum time resolution [Ec 63; Bo 64; Bo 64a; Le 64; Vi 64]. In most cases advertised specifications can be obtained with respect to the last itemised parameters.

A light pulse generator is usually employed for optimising the operational conditions. Multiplier pulse shape and time resolution, with reference to electrode voltage, are analysed with the aid of an oscilloscope and a time analyser, respectively (see Section 4.4). As long as the shape of the output pulse is not disturbed by space charge effects, optimising can be effected with maximum height of the integrated pulse.

When compared to average current level of the multiplier anode the current through the dynode voltage divider chain should be high at least in the last dynode stages. Otherwise dynode potential shifting with consequent alteration of the shape and height of the pulse will occur. This is of particular importance for operation with pulsed accelerators [Ce 60a]. *Barna* describes a voltage

Photomultipliers

58 AVP[a]	K 14 FS 50	ФЭУ-36	XP 1020[a]	C 70045 A[i]	XP 1141[a]
VALVO	Carl Zeiss Jena	Soviet Union	VALVO	RCA	VALVO
110	42	34	42	>30	42·5
tetrode	triode		tetrode		
14, linear	14, meander	13, linear	12, linear	14	7, linear
—	—	—	100 Ω	50 Ω	—
2·4; max. 3·0	max. 2·8	max. 2·9	2·5; max. 3·0	4·8	6·7
max. 10^9	10^8	10^6 to 10^8	max. 10^9		10^6
0·5 to 1		1	0·5 to 1		3 to 5
2; 2 to 2·8[l]	2·5; 2·1[g]	1·5 to 2·5	<1·8; 1·6[h]	0·45; 0·7[k]	1·9
	3·1[g]				
4	4·2[g]		4	0·84; 2[k]	3·2
	9·0[g]		7[f]		
50 mm: 1			18 mm: <0·2	25 mm: <0·3	
45	40		28	32	16
			$\propto t^{3\cdot7}\, e^{-1\cdot7\,t}$		
			[Gr 66]		
		Lateral cathode and anode connection	With 50 Ω output as type XP 1021	Lateral cathode. *Gaussian* transit-time distribution in transfer region	Low-ohmic photocathode

e *Eckardt* et al. [Ec 62] with H_2-light pulse generator.

f *Pietri* [Pi 62].

g *Görlich* et al. [Gö 62].

h *Pietri* [Pi 64].

i Data according to *Birk* [Bi 64a], with t_r and $t_{1/2}$ measured with *Čerenkov* radiation.

k *Miehe* et al. [Mi 66], with Hg-relay.

l *Hasselbach* and *Mack* [Ha 67b].

divider having transistors with an output impedance corresponding to a 200 mA dynode voltage divider chain [Ba 63a]. It is further necessary to provide for by-pass capacitors in at least the last 3 or 4 dynode stages. For this purpose capacitors of more than 0·1 μF are connected to the voltage divider resistors. A good by-pass to earth of the dynode pulses is obtained by connecting low-inductivity ceramic capacitors directly to the socket pins (see [Gi 66], for example). Inductance and capacitance of the dynode leads represent resonant circuits which cause pulse-top ringing of 100 to 200 MHz [Ec 62; Ra 63; Be 65a]. To obtain a defined pulse shape, by-pass capacitors and damping resistors (100 Ω) are interposed in the dynode leads [Vy 61; Av 66] and a coaxial anode output resorted to [Hy 64; Be 65a]. Resonant circuit damping becomes important when the multiplier is illuminated with long light pulses.

If the cathode of the photomultiplier is connected to earth potential the multiplier pulses have to be taken off with an isolating capacitor because of the positive high-voltage at the anode. Another disadvantage is the possible occurrence of spurious pulses in the circuitry from the power supply. On the other hand, with the anode on earth potential and negative high-voltage applied to the cathode, the subsequent circuit can be connected through d.c. coupling. However, this may give rise to spurious pulses caused by wall charging at the cathode end [Ko 66a].

With multipliers having a coaxial anode output, the inner conductor is connected to the anode and the shield to the last dynode. When the output signal is passed to the subsequent circuitry through a coaxial cable, its outer conductor is usually connected to the earth potential. In that case, the divider resistor between the last dynode and the anode would be short-circuited by the input impedance R_i of this subsequent circuit, necessitating d.c. voltage decoupling through an isolating capacitor or a transformer. Some of these arrangements are shown in Fig. 14 [Ha 65].

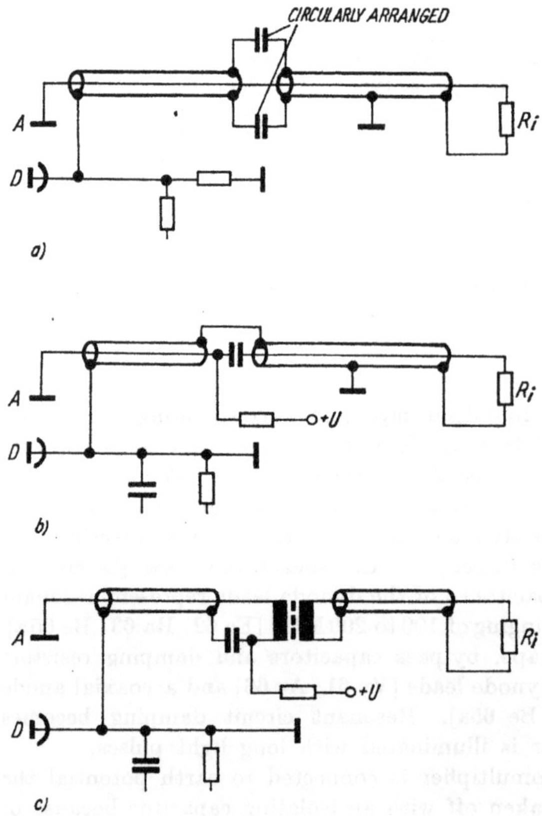

Fig. 14. Output circuits of photomultipliers with coaxial anode output.

A anode, *D* last dynode, *U* separately applied voltage between *D* and *A*.

CONTROLLED PHOTOMULTIPLIERS

In many instances it is desirable to close and open the photomultiplier for a short period of time. With the described construction a control pulse is used to effect a change of the potential of the focussing electrodes [Si 63; Ke 64] or of one of the dynodes, for example, with type ФЭУ-36, the 2nd or the 4th dynode [Vy 63; Vy 65] or the 8th and 9th dynode [Kl 67].

Photocathodes with a control grid inserted directly behind them also have been designed [Wa 61; Vy 66]. For multipliers of this type one volt is sufficient to obtain full control action. Difficulties arise, however, from high internal impedance of the photocathode which reaches megohm level. Together with the grid-cathode capacitance, a significant time constant of 0·1 to 1 μs has to be considered for control.

This difficulty can be overcome by using *two control grids* (Fig. 15), as in the photomultiplier SKP-1 from VEB Carl Zeiss Jena [Gö 64; Gö 66]. Grid (*1*) accelerates the electrons and screens grid (*2*) from the cathode. As a scintillation counter, SKP-1 has already been operated with a control frequency of 150 MHz [Mu 65; Po 65]. It has further uses in laser application.

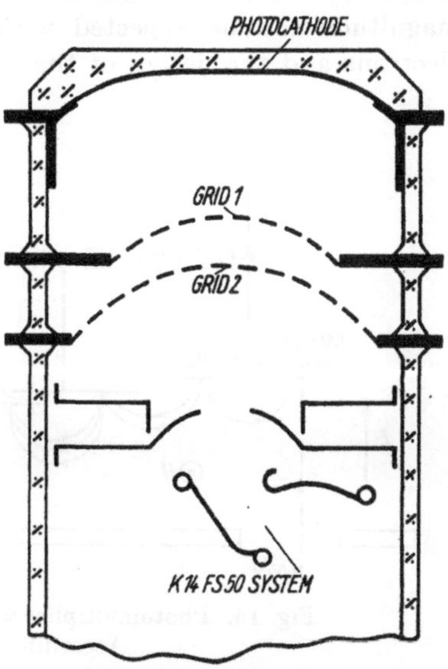

Fig. 15. Control system of the photomultiplier SKP-1. According to *Görlich* et al. [Gö 64].

PHOTOMULTIPLIERS WITH TRANSMISSION DYNODES

Development work on photomultipliers concentrates on further reduction of transit-time fluctuations. Work in this direction aims to improve the transfer system, increases the gain of the first dynode, reduces the total number of dynodes and simultaneously increases the operating voltage. Considerably better time resolution can be expected when the reflexion dynodes are replaced by plane parallel transmission secondary emission dynodes, where the secondary electrons pass through the dynode to the following dynode. An example is the 7-stage type WX 30006 photomultiplier which has a rise time of less than 0·5 ns at an operating voltage of 32 kV [Sa 64]. The alkali-halide transmission dynodes, however, still have a relatively short lifetime.

CROSSED FIELD MULTIPLIERS

The use of homogeneous electric and magnetic fields, located vertically to each other, has already been proposed for focussing by *Kubeckii* [Ku 30] in 1930 and has been utilised several times since [Sm 51; Go 61; Be 63a]. The principal design of a photomultiplier with *static crossed fields* is shown in Fig. 16. Potential of the various dynodes is adjusted to obtain homogeneous conditions for the electric field $\boldsymbol{E}$ between the dynodes and the earthed electrode. Initially, the electrons are rapidly accelerated, but slowed down by the curved trajectory in the magnetic field $\boldsymbol{H}$. Effective accelerating voltage is $|\boldsymbol{E}|\,\Delta y$ and transit-time per stage is proportional to Δy. When compared with conventional construction, a reduction of transit-time fluctuation by one order of magnitude can be expected with this design [Po 66a]. Transfer of photoelectrons and the design of the collecting electrode is still difficult.

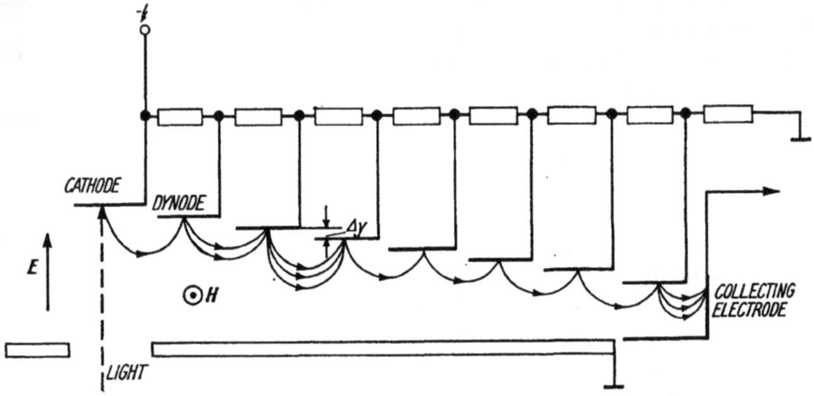

Fig. 16. Photomultiplier with crossed electric and magnetic fields. According to *Belov* et al. [Be 63a].

Interesting properties are observed with a photomultiplier háving *dynamic crossed fields* where the electric field is produced, for example, by a micro-wave generator [Ga 62]. In this case, phase focussing of the secondary electrons is effected causing a series of brief samples to appear at the output with a spacing equal to the micro-wave period. The envelope of the samples reproduces the shape of the light signals, provided their length is greater than the micro-wave period. In the event of periodic input signals with a frequency that is higher than the micro-wave frequency, the envelope shows the difference frequency values. Gain factors up to 10^7 can be obtained with this system [Ga 62].

HYBRID PHOTOMULTIPLIERS

Amplification in these photomultipliers is not effected through secondary electron multiplication, but the electrons are accelerated to about 10 kV and injected into a silicon-diode [Ka 66; Wo 66]. Because the production of a charge-carrier pair in silicon requires only around 3·5 eV, a gain of the order of 1,000 is obtained. When using a silicon-transistor instead of the diode as collecting electrode [Wo 66] the total gain includes transistor gain β and amounts to about 10^6. Transit-time fluctuations remain small because of the high field strength in the transfer system. "Fast" photomultipliers can therefore be expected, especially when using diodes, giving a linear delivery to 20 A [Ka 66].

2.3.2 TIME RESOLUTION IN SCINTILLATION COUNTERS

For work in the nanosecond range the photomultiplier has gained a prominent place over all other radiation detectors. Apart from the parameters of the photomultiplier itself, attainable time resolution is further affected by the properties of the light pulse and by the time analyser unit attached to the multiplier. In the following discussion scintillators are treated as light pulse sources, but the subject matter can also be applied analogously for other sources.

The theoretical assessment of the time resolution of scintillation counters deals principally with two problems: viz. which part of the multiplier output pulse gives the best time resolution; and what level of statistical accuracy is obtainable in time measurement compatible with scintillation counter properties and the attached time analyser.

Mathematical treatment is very complicated because of the multiplicity of statistical processes in the scintillator and the multiplier that have to be taken into consideration. In this section results and conclusions from theoretical analysis, without detailing calculations, are discussed and a comparison is made with experimental data.

PULSE SHAPE OF A SCINTILLATION COUNTER WITH NO TRANSIT-TIME FLUCTUATIONS

Assuming a scintillation counter with an ideal photomultiplier having no transit-time fluctuation, the course of the current pulse at the multiplier output is the same as the decay function of the scintillator. Consequently an approximation of the anode current pulse can be expressed for organic crystals by

$$I(t) = \frac{R}{\tau} e^{-t/\tau}, \tag{9}$$

for inorganic crystals and organic solutions by

$$I(t) = \frac{R}{\tau - \tau_1} (e^{-t/\tau} - e^{-t/\tau_1}) \tag{10}$$

(see Section 2.2.1); with τ_1 the rise time constant and τ the decay time constant of the scintillator. Total number of photoelectrons emitted per scintillation is defined by

$$\int_0^\infty I(t)\, dt = R\,. \tag{11}$$

For the sake of simplicity, the multiplier gain is entered as equal to unity, as in the equations (9) and (10).

The load resistance of the multiplier anode consists of the parallel-circuited ohmic resistor R_a and the stray capacitance C_s (Fig. 17). Voltage $U(t)$ across the load resistance is given by the differential equation

$$I(t) = I_R(t) + I_C(t) = \frac{U(t)}{R_a} + C_s \frac{dU(t)}{dt} . \tag{12}$$

With time constant $\tau_a = R_a C_s$ and equation (10) this yields

$$U(t) = R \frac{R_a}{\tau_a - \tau} \left[\frac{\left(1 - \frac{\tau_1}{\tau_a}\right) e^{-t/\tau} - \left(1 - \frac{\tau_1}{\tau}\right) e^{-t/\tau_a} - \tau_1 \left(\frac{1}{\tau} - \frac{1}{\tau_a}\right) e^{-t/\tau_1}}{\left(1 - \frac{\tau_1}{\tau_a}\right)\left(1 - \frac{\tau_1}{\tau}\right)} \right]. \tag{13}$$

With $\tau_1 = 0$, equation (9) results in

$$U(t) = R \frac{R_a}{\tau_a - \tau} (e^{-t/\tau} - e^{-t/\tau_a}) . \tag{14}$$

Fig. 17 gives a schematic representation of the results from equations (13) and (14). Fluctuations of electron transit time in the photomultiplier result in a finite rise time t_r of the multiplier pulse. Consideration of these fluctuations qualitatively leads to the same results as the finite rise time τ_1 of the scintillation pulse.

When R_a lies between 50 and 100 Ω, in most cases $\tau_a \ll \tau$ is applicable for the anode time constant, and the multiplier current pulse is not integrated — this is designated as "current mode". In this case, the rise time of the current and voltage pulses is primarily determined by the higher of the two values τ_1 and t_r. With the anode time constant $\tau_a \gg \tau$ the current pulse is integrated, which is defined as "voltage mode". Rise time of the voltage pulse is then of the same magnitude as the decay time constant τ of the scintillator, provided the rise time t_r of the multiplier is less than τ.

RESPONSE OF TIME ANALYSER (MACHINE TIME)

In ideal presentation, response of the time analyser represents machine time (see Section 3.4.2). Machine time defines the moment of arrival of an accepted needle pulse which would yield the same measuring result as the actual pulse of the scintillation counter. Here the following modes apply:

a) *Utilising Part of the Total Pulse*

(i) Current pulse integration at the multiplier output ($R_a C_s \gg \tau$).

Centroid Response. The voltage pulse cuts off a valve that conducts in rest state and has a linear I_a-U_g characteristic. The anode current of the valve is integrated with the capacitance C_a. The point where the linear rise crosses the abscissa represents the machine time t_c. In this case, it is equal to the centroid t_c of the current pulse up to the cut-off of the limiter

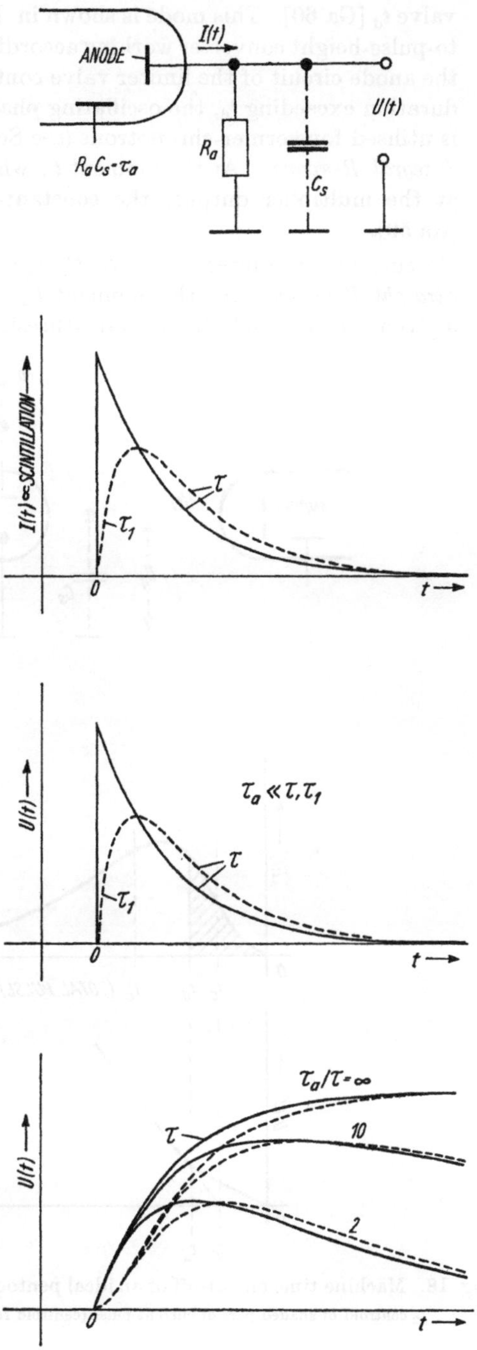

Fig. 17. Qualitative behaviour of anode current and voltage in a scintillation counter with ideal photomultiplier. Rise time constants of scintillator, full line $\tau_1 = 0$, broken line $\tau_1 > 0$.

valve t_d [Ga 60]. This mode is shown in Fig. 18 and corresponds to a time-to-pulse-height converter working according to the start-stop principle. When the anode circuit of the limiter valve contains a resonant circuit with a cycle duration exceeding t_d, the oscillating phase also depends on t_c [Co 57]. This is utilised for vernier-chronotrons (see Section 3.4.5).

Integral Response. At the moment t_i, where a given charge C has collected at the multiplier output, the constant-threshold trigger circuit responds [Ga 59].

(ii) No current pulse integration ($R_a\, C_s \ll \tau$).

Straight Response. At the moment t_s, where the multiplier pulse attains a given current level, the constant-threshold trigger circuit responds [Hy 64].

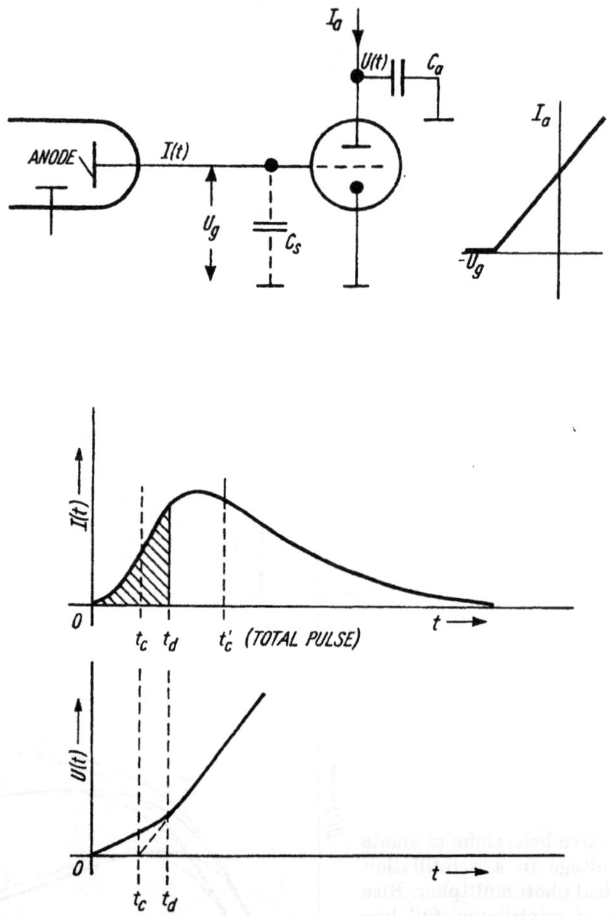

Fig. 18. Machine time on cut-off of an ideal pentode. According to *Gatti* and *Svelto* [Ga 60].
t_c centroid of shaded part of current pulse (centroid response), t_d moment when valve cuts off.

b) Utilising Total Current Pulse

Centroid response mode is feasible in this case only when using a resonant circuit. Machine time then corresponds to the centroid t_c of the total pulse. Integral and straight response are modified in such a manner that triggering is always effected when a constant fraction of the maximum current or of the voltage height is reached.

Against the non-linear methods *a*), the linear mode *b*) has the advantage that machine time is not affected by the pulse height fluctuations of the multiplier. The crossover point of the suitably differentiated multiplier pulse is also independent of pulse height. Two points on the pulse curve determine this machine time. Fig. 19 shows some examples. Zero-crossing of a pulse shaped by two short-circuited cables (double transit time = Δt) occurs at the moment when condition

$$A\,f(t) - 2\,A\,f(t - \Delta t) = 0$$

or (15)

$$f(t) = 2\,f(t - \Delta t)$$

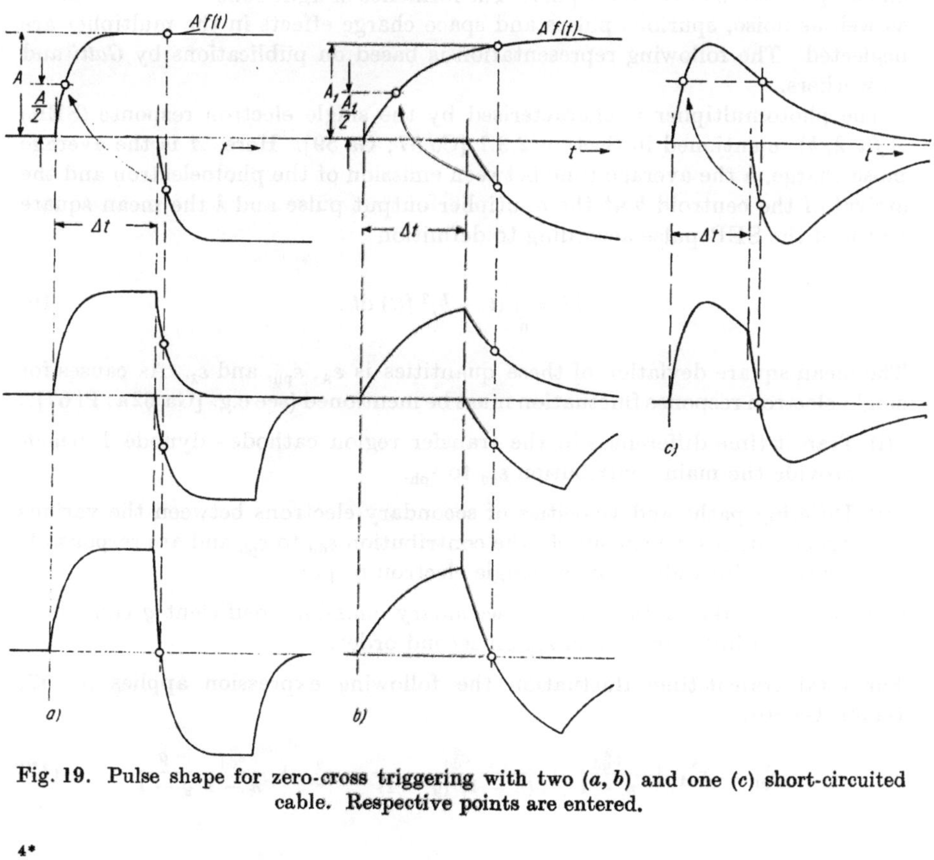

Fig. 19. Pulse shape for zero-cross triggering with two (*a*, *b*) and one (*c*) short-circuited cable. Respective points are entered.

applies, as shown in Fig. 19 *a* and *b*. If Δt is long compared to the rise time of the pulse $A\,f(t)$, triggering is effected at 50% of final height A with this mode (see Fig. 19 *a*). With a lower Δt the significant points are located at A_1 and $A_1/2$ (see Fig. 19*b*). When shaping a decaying pulse by a single short-circuited cable two significant points occur at the same fraction of maximum height (see Fig. 19*c*).

CONTRIBUTIONS TO TRANSIT-TIME FLUCTUATION

In most cases it is the variance ε^2 or the mean square deviation ε of the machine time t_c, t_l and t_s with reference to the parameters of the scintillator and of the photomultiplier that is determined when calculating time resolution. If the coincidence curve of two identical scintillation counters is plotted, full width at half-maximum amounts to $2\,\tau_0 = \sqrt{2}\times 2{\cdot}36\,\varepsilon$ if *Gaussian* distribution obtains. Statistical characteristics of the scintillation counter output pulses result from the behaviour of the scintillation pulse and from the fluctuation of transit-time and amplification of the multiplier. The influence of light collection and transfer as well as noise, spurious pulses and space charge effects in the multiplier are neglected. The following representation is based on publications by *Gatti* and co-workers.

The photomultiplier is characterised by the single electron response (SER) $A\,f(t, \lambda, h)$, mentioned in Section 2.3.1 [Co 57; Ga 59]. Here, A is the average pulse charge, h the average time between emission of the photoelectron and the arrival of the centroid b at the multiplier output pulse and λ the mean square width of the SER pulse according to definition

$$\lambda^2 = \int_0^\infty (t - b)^2 f(t)\,\mathrm{d}t\,. \tag{16}$$

The mean square deviation of these quantities is ε_A, ε_{ph}, and ε_λ. As causes for single electron response fluctuation must be mentioned (see e.g. [Ga 62a; Pi 67]):

(i) Transit-time differences in the transfer region cathode—dynode 1 which provide the main contribution ε_{cd} to ε_{ph}.

(ii) Differing paths and velocities of secondary electrons between the various dynode stages, which supply the contribution ε_{dd} to ε_{ph} and are responsible for the finite width λ of the single electron response.

(iii) Statistical fluctuations of the secondary emission coefficient g contribute to ε_A and influence ε_λ and ε_{ph} on second order.

For total transit-time fluctuation the following expression applies [Co 57; Ga 59; Ga 66]:

$$\varepsilon_{ph}^2 = \varepsilon_{cd}^2 + \frac{\varepsilon_{dd}^2}{n-1} + \frac{\varepsilon_{dd}^2}{(n-1)\,(g-1)} = \varepsilon_{cd}^2 + \frac{\varepsilon_{dd}^2}{n-1}\,\frac{g}{g-1} \tag{17}$$

with n being the number of dynode stages. Applying *Poisson* statistics the mean square deviation of amplification (see, e.g. [Co 57]) can be written

$$\varepsilon_A^2 = \frac{1}{g-1}. \tag{18}$$

To keep fluctuations of the mean gain A of a single photoelectron small, it is therefore necessary to have a high secondary emission coefficient g at the first dynode. *Gatti* and his co-workers use $g = 4$ in most of their calculations. Width of the single electron pulse is [Co 57]

$$\lambda^2 = (n-1)\,\varepsilon_{dd}^2. \tag{19}$$

Experimental results show that the single electron response usually has a constant shape approaching symmetry and a width of 1 to 2 ns (see Table 4; [Ce 60; Ga 62a; Hy 64]). A behaviour proportional to $t\,e^{-xt}$ has also been reported [Ha 56]; while multiplier XP 1020 SER-shape was approximated by $t^{3.7}\,e^{-1.7\,t}$ [Gr 66]. In the calculations, the shape of the single electron response is expressed approximately by a *Gauss*ian distribution or a more or less symmetric function (see Table 5). Mean square deviation ε_λ of single electron pulse width is neglected in the calculations.

STEPS OF APPROXIMATION FOR CALCULATING TIME RESOLUTION

Before detailing the results from time resolution calculations a few remarks regarding transit-time jitter seem in order. On the average, photoelectron emission from the cathode shows approximately the same time behaviour as the light pulse of the scintillator. Using $I(t) = (R/\tau)\,e^{-t/\tau}$ as a rough approximation, the mean time interval of the emitted photoelectron amounts to τ/R, with $t \approx 0$. When the time interval τ/R is large against the transit-time fluctuations ε_{ph} and the width λ of the single-electron pulse of the photomultiplier, the variance, for instance, of the machine time t_l, is obtained by adding the contribution ε_{ph}^2 to the time fluctuation of the emission of the C-th photoelectron.

On *condition* $\tau/R \lessgtr \lambda$ the various single-electron pulses $f(t)$ overlap. Mean output pulse shape $\overline{i(t)}$ is obtained by convolution and expressed with

$$\overline{i(t)} = \int_0^t I(t-t')\,f(t')\,dt' \equiv I(t) * f(t). \tag{20}$$

Single-electron pulse overlapping causes a charge at the multiplier output of differently weighted $C' > C$ photoelectrons, which corresponds to the total charge of single-electron pulses produced by C photoelectrons.

When *condition* $\tau/R \lessgtr \varepsilon_{ph}$ applies, primarily because of the transit-time fluctuations of photoelectrons in the cathode—dynode 1 region, the single-electron pulses may overtake each other. Time of arrival of the C-th single-electron pulse at the output will consequently fluctuate less than the time of arrival of the single-electron pulse produced by the cathode emitted C-th photoelectron. This represents a certain compensating effect for transit-time fluctuations [Ga 64].

Taking single-electron pulses having a width of $\lambda = 0$, the convolution integral gives the average shape of the output pulse [Hy 64]

$$I^*(t) = I(t) * p(t) , \tag{21}$$

where $p(t)$ denotes probability-density function of the transit-time of single-electron pulses, and $\varepsilon_{\mathrm{ph}}$ the mean square deviation of $p(t)$. It is clear that accurate knowledge of $p(t)$ is only of importance when there is evident probability of single-electron overtaking. $I^*(t)$ can be considered as *equivalent illumination function* of the cathode of a photomultiplier showing no more transit-time jitter.

In a similar manner as with equation (20), the influence of finite width of a single-electron pulse (SER) can be obtained by convolution [Ga 64, Hy 64]

$$\overline{i(t)} = I^*(t) * f(t) . \tag{22}$$

Equivalent illumination function $I^*(t)$ can be measured directly when using the single-electron method (see Section 4.4.2; [Bi 64a; Be 66; Be 66a]). Function $p(t)$ can be determined with the aid of equation (21) provided the light pulse shape $I(t)$, especially the time constant of the leading edge, is known with sufficient accuracy. *Gauss*ian distribution is proposed for $p(t)$ in several papers [Ga 64; Hy 64]. With the multiplier C 70045 A, de-convolution of the somewhat asymmetric function $I^*(t)$ also indicates the presence of a *Gauss*ian distribution with reference to $p(t)$ [Bi 64a].

The advantage of the considerations according to equations (21) and (22) is that calculation can be based on the directly measurable equivalent illumination function and that only the shape of the single-electron response which can be obtained needs consideration.

In accordance with respective assumptions, the various calculations of the machine time fluctuation ε can be grouped essentially as follows (see Table 5):

Group I: Single-electron pulses neither overtake nor overlap

$$\tau/R > \varepsilon_{\mathrm{ph}}, \tau/R > \lambda ,$$

i.e. according to *Post* and *Schiff* [Po 50], variance of the time of arrival of the C-th photoelectron in an ideal multiplier amounts to

$$\varepsilon_{\mathrm{I}}^2 = \frac{\tau^2}{R^2} C \left(1 + \frac{2(C+1)}{R} + \cdots\right) . \tag{23}$$

Further equations are given in Table 5.

Group II: Single-electron pulses do not overtake, but overlap

$$\tau/R > \varepsilon_{\mathrm{ph}}, \tau/R \lesseqgtr \lambda .$$

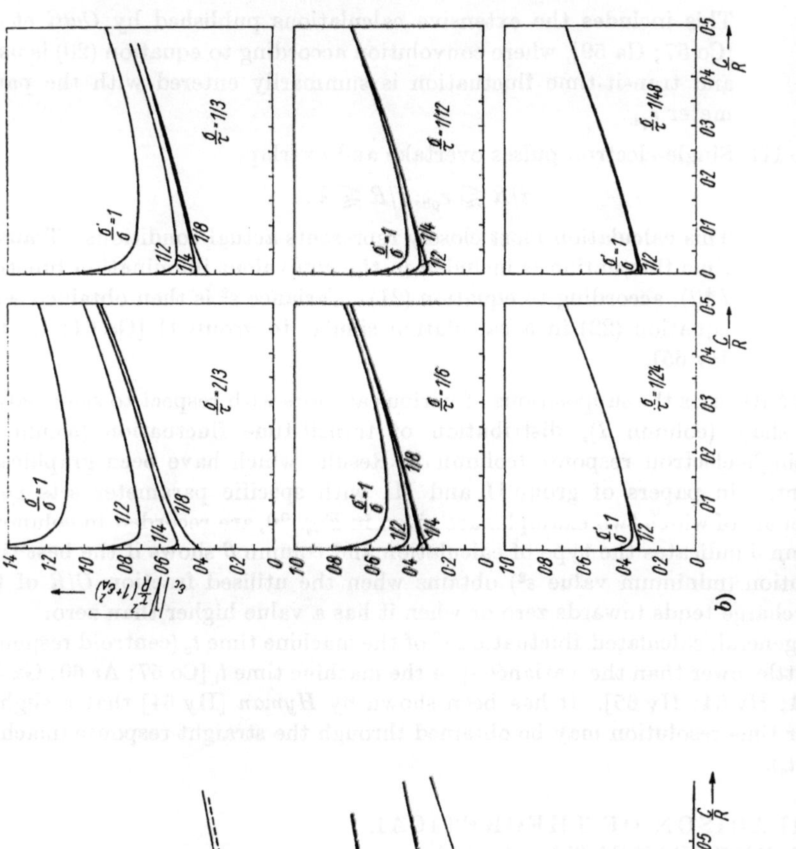

Fig. 20. Calculated fluctuation of machine time t_C as function of C/R for various parameters of the scintillation counter (see Table 5).

a) According to *Gatti* and *Svelto* [Ga 64], broken line $p_2(t)$, full line $p_1(t)$. b) According to *Hyman* [Hy 65], $\gamma = 5{\cdot}0$.

This includes the extensive calculations published by *Gatti* et al. [Co 57; Ga 59], where convolution according to equation (20) is used and transit-time fluctuation is summarily entered with the parameter ε_{ph}.

Group III: Single-electron pulses overtake and overlap

$$\tau/R \lesseqqgtr \varepsilon_{ph}, \tau/R \lesseqqgtr \lambda .$$

This calculation most closely represents actual conditions. Transit-time fluctuation is included in the equivalent illumination function $I^*(t)$, according to equation (21). Variance ε^2 is then obtained with equation (22) in a calculation similar to group II [Ga 64; Hy 64; Hy 65].

Table 5 itemises the suppositions of various authors with respect to scintillation pulse shape (column *1*), distribution of transit-time fluctuation (column *2*) and single-electron response (column *3*). Results which have been graphically presented in papers of group II and III with specific parameter sets τ/ε_{ph}, λ/ε_{ph} et al., of which two examples are given in Fig. 20, are recorded in column *4*. Column *5* indicates the type of calculation and column *6* shows if the best time resolution (minimum value ε^2) obtains when the utilised fraction C/R of the total charge tends towards zero or when it has a value higher than zero.

In general, calculated fluctuation ε_c^2 of the machine time t_c (centroid response) is a little lower than the variance ε_l^2 of the machine time t_l [Co 57; Ar 60; Ga 59; Ga 64; Hy 64; Hy 65]. It has been shown by *Hyman* [Hy 64] that a slightly better time resolution may be obtained through the straight response (machine time t_s).

COMPARISON OF THEORETICAL AND EXPERIMENTAL DATA

When undertaking a comparison of the theoretical value of the time resolution with data obtained in experiments, too close quantitative agreement should not be expected. In most cases, parameters required for calculation cannot be measured directly and will, at times, be found only by comparison with experimentally obtained resolving time data [Ba 63b]. Further, the shape of the scintillation pulse used for calculations seems too simplified, and machine time is not as sharply defined with the time analysers employed as in theory. Another point is the additional contributions contained in measured time resolution values which do not enter into theory, for instance, collection time of light in the scintillator [Co 67b; Ya 67]. *Cocchi* and *Rota* [Co 67b] used the Monte-Carlo method to calculate the collection time of a cylindrical scintillator. This revealed that with increased length of the scintillator the collection time cannot be ignored in timing experiments in the range of a few hundred picoseconds. In many cases, measurements obtained better time resolution data than could be expected from theoretical calculation.

Table 5: Some Calculations

0	Scintillation Pulse	Photomultiplier: Transit-Time Fluctuation	Photomultiplier: Single-Electron Response	
Group I $\frac{\tau}{R} > \varepsilon_{ph}$, $\frac{\tau}{R} > \lambda$, C integer	$I(t) = \frac{R}{\tau} e^{-t/\tau}$	$\varepsilon_{ph} = 0$ $\frac{\tau}{R} \gg \varepsilon_{ph} \neq 0$	$\lambda = 0$ $\varepsilon_A^2 \neq 0$	ε ε
	$I(t) = \frac{R}{\tau} \cdot 1(t)$	$\varepsilon_{ph} = 0$	$\lambda = 0$	ε ε
	$I(t) = \frac{R}{\tau} \cdot 1(t)$	$\frac{\tau}{R} \gg \varepsilon_{ph} \neq 0$	$\varepsilon_A^2 \neq 0$ $\lambda = 0$	ε ε
Group II $\frac{\tau}{R} > \varepsilon_{ph}$, $\frac{\tau}{R} \lessgtr \lambda$	$I(t) = \frac{R}{\tau} \cdot 1(t)$	$\varepsilon_{ph} = 0$	$\varepsilon_A = 0$ $\lambda = 0$ and $\lambda \neq 0$	ε
	$I(t) = \frac{R}{\tau} e^{-t/\tau}$	$\varepsilon_{ph}^2 = \varepsilon_{cd}^2 + \frac{\varepsilon_{dd}^2}{g-1}$	$f(t) = \frac{\sqrt{5}}{\lambda}\left(e^{-\frac{\sqrt{5}}{2\lambda}t} - e^{-\frac{\sqrt{5}}{\lambda}t}\right)$ $\varepsilon_A^2 = \frac{1}{g-1}$	ε
	$I(t) = \frac{R}{\tau} e^{-t/\tau}$	$\varepsilon_{ph}^2 = \varepsilon_{cd}^2 + \frac{\varepsilon_{dd}^2}{g-1}\left(1 + \frac{2}{g}\right)$	$f(t) = \frac{3}{8!\,\lambda}\left(\frac{3\,t}{\lambda}\right)^8 \cdot e^{-3\,t/\lambda}$ $\varepsilon_A^2 = \frac{1}{g-1}$	ε
Group III $\frac{\tau}{R} \lessgtr \varepsilon_{ph}$, $\frac{\tau}{R} \lessgtr \lambda$	$I(t) = \frac{R}{\tau} e^{-t/\tau}$	$p(t) \propto e^{-t/\tau_2}$ $I^*(t) = \frac{R}{\tau - \tau_2}(e^{-t/\tau} - e^{-t/\tau_2})$	$\lambda = 0$	[illegible]
	$I(t) = \frac{R}{\tau} \cdot 1(t)$	$p(t) = \frac{1}{\varepsilon_{ph}} e^{-t/\varepsilon_{ph}}$ $I^*(t) = \frac{R}{\tau}(1 - e^{-t/\varepsilon_{ph}})$	$f(t) = \frac{3}{8!\,\lambda}\left(\frac{3\,t}{\lambda}\right)^8 \cdot e^{-3\,t/\lambda}$	
	$I(t) = \frac{R}{\tau} e^{-t/\tau}$	$p_1(t) = \frac{1}{\varepsilon_{ph}} e^{-t/\varepsilon_{ph}}$ $p_2(t) = \frac{2}{\varepsilon_{ph}^2}\, t\, e^{-\sqrt{2}t/\varepsilon_{ph}}$	$f(t) = \frac{3}{8!\,\lambda}\left(\frac{3\,t}{\lambda}\right)^8 \cdot e^{-3\,t/\lambda}$	
	$I(t) = \frac{R}{\tau} e^{-t/\tau}$ $I(t) = \frac{\gamma}{\gamma-1}\frac{1}{\tau}(e^{-t/\tau} - e^{-\gamma t/\tau})$	$p(t)$ = clipped *Gaussian* distribution (σ')	$f(t)$ = clipped *Gaussian* distribution (σ)	

[a] Variance s^2 of machine time t_c, t_i and t_s. [b] Does not apply for low C/R. [c] $C/R \to 0$ when no transit-time jitter. [d] Example in I

Ausschlagtafel 5660 Meiling/Stary

of Scintillation Counters

Results [a]	Calculation Mode	Optimum Time Resolution	Authors
$\frac{+1)}{R}+\cdots\Big]$ $\frac{+1)}{R}+\cdots\Big]\left[1+\varepsilon_A^2\left(1-\frac{1}{C}\right)^2\right]+\varepsilon_{ph}^2$	analytical	$C\to 1$	*Post* and *Schiff* [Po 50] *Gatti* and *Svelto* [Ga 62a]
$\frac{}{C}+\frac{1}{2C^2}\Big)$	analytical	$C\to 1$	*Post* and *Schiff* [Po 50] *Colombo* et al. [Co 57] *Gatti* and *Svelto* [Ga 62a]
$\Big)^2\Big]\frac{\tau^2}{R^2}C+\varepsilon_{ph}^2$ $\frac{3}{C}+\frac{1}{2C^2}\Big)+\frac{\varepsilon_{ph}^2}{C}\Big]\left[1+\varepsilon_A^2\left(1-\frac{1}{C}\right)^2\right]$	analytical	$C\to 1$	*Gatti* and *Svelto* [Ga 62a; Ga 64]
	Monte Carlo for $C\leqq 4$	$C\to 0$	*Arecchi* et al. [Ar 60]
$\frac{\lambda}{\varepsilon_{ph}};\frac{C}{R}\Big)$	analytical [b]	$\frac{C}{R}\neq 0$	*Colombo* et al. [Co 57]
$;\frac{\lambda}{\varepsilon_{ph}};\frac{C}{R}\Big)^e$	analytical [b]	$\frac{C}{R}\neq 0$	*Gatti* and *Svelto* [Ga 59]
P for arrival of C-th electron, full width mum of P-curve, for constant R	analytical	$\frac{C}{R}\to 0$	*El-Wahab* and *Kane* [El 62]
$;\frac{\lambda}{\varepsilon_{ph}};C\Big)^e$	Monte Carlo for $C\leqq 4$	$\frac{C}{R}\to 0$	*Gatti* and *Svelto* [Ga 64]
$_{c,i}\left(\frac{\tau}{\varepsilon_{ph}};\frac{\lambda}{\varepsilon_{ph}};\frac{C}{R}\right)^e$ ference for p_1 and p_2 [d]	analytical [b]	$\frac{C}{R}\to 0$	*Gatti* and *Svelto* [Ga 64]
$\left(\frac{C}{R}\right)$ for $R=100$ [c]	Monte Carlo and analytical	$\frac{C}{R}\neq 0$ [c]	*Hyman* et al. [Hy 64]
$^{1/2}F_{c,i}\left(\frac{\sigma'}{\sigma};\frac{\sigma}{\tau};\gamma;\frac{C}{R}\right)^{d\,e}$	analytical	$\frac{C}{R}\neq 0$ and $\frac{C}{R}\to 0$	*Hyman* [Hy 65]

ner of writing indicates that for calculating ε_c function F_c, ε_i function F_i etc. applies.

From the presentation of $\varepsilon^2 = f(\tau/R)$ of Table 5 it will be seen that low time fluctuation is obtainable only with fast-decay scintillators and a high photoelectron yield. Some characteristic values of the mean photoelectron interval τ/R and of the number of emitted photoelectrons are given in Tables 1, p. 7, and 3, p. 21. With an energy of 100 keV organic scintillators show a mean photoelectron interval of 0·05 ns. Comparison with the approximations $\lambda \approx 1$ ns and $\varepsilon_{ph} \approx 0{\cdot}5$ ns shows that the close approach to the conditions of the results of group III is actually derived from these preconditions. The same also applies for "slow" scintillators; for example, with 100 keV and about 5 photoelectrons/keV the NaI(Tl) scintillator shows a mean photoelectron interval of $\tau/R \approx 0{\cdot}5$ ns. This leads to the conclusion that the derived equation (23) of *Post* and *Schiff* applies for "slow" scintillators only at an energy below 100 keV. To obtain optimum time resolution the use of short-time photomultipliers is recommended when NaI(Tl) scintillators are utilised [Ga 64].

In many tests during experiments with organic scintillators or a light pulse generator optimum time resolution could be observed where $C/R \neq 0$ [Ba 63b; Sc 63; Bo 64a; Pr 64; Br 65; Be 66; Mc 67], although this is not in aggreement with all calculations (see Table 5). This is connected with the behaviour of the equivalent illumination function. With a slow rise of the function $I^*(t)$, caused either by the finite rise of the scintillator pulse $I(t)$ or by the respective transit-time distribution $p(t)$, or by both, optimum time resolution occurs with $C/R \neq 0$ [Ga 64; Hy 64; Hy 65; Be 66]. Using an illumination function $I^*(t)$ obtained by measurement the minimum value of the time fluctuations was always found on condition $C/R \neq 0$ [Be 66; Be 66a; Be 67a]. With increasing width λ of the single-electron pulse, however, this optimum value of C/R decreases, while λ has little influence on the obtained optimum time resolution [Be 66]. With decreasing time constant τ the optimum C/R value shifts to higher levels [Ba 63b; Br 65]. *Bertolaccini* et al. [Be 67a] obtained good agreement between experimental time resolution and theoretical calculation. This holds when calculation is based on the measured equivalent illumination function, on single-electron response measurements, on gain fluctuation and on the selected part of the spectrum. Investigations, however, did not include very small total numbers R of photoelectrons.

To illustrate the time resolution at present obtainable with two scintillation counters, some of the experimental data have been given in Table 6, p. 44. Note also Fig. 21, p. 46. Utilising light pulse generators, still lower values of full width at half-maximum $2\,\tau_0$ of the coincidence curves are obtained [Ag 62; Bo 64a; Mi 66]. Significant machine time of the various experiments can be assigned only approximately to one of the modes t_c, t_i or t_s, and frequently ranges between t_i and t_s.

Table 6 also contains data of the slope of the coincidence curve presented as *instrumental half-time* $T_{1/2\,\mathrm{instr}}$. It is mainly this value which determines the lower measuring limit of the lifetime of the excited states of nuclei (see Section 5.1). It will also be evident from Table 6 that for fast-decay scintillators

Table 6: Time Resolution Obtained with Two Scintillation

Authors	Test Conditions		
	Machine Time	Photo-multiplier	Scintillator
Braunsfurth and *Körner* [Br 65]	$\approx t_c$	56 AVP	NaI(Tl) NaI(Tl) NaI(Tl) + NaI, 80 °K
Lynch [Ly 66]	$\approx t_s$ to t_i	C 31000 +XP 1020	NaI(Tl), 25 mm × 25 mm + Naton 136 NaI(Tl), 0·5 mm × 25 mm + Naton 136 NaI(Tl), 6 mm × 22 mm + Naton 136
Bartl and *Weinzierl* [Ba 63b]	t_i	56 AVP	NE 102 stilbene
Schwarzschild [Sc 63]		56 AVP	Naton 136 13 mm × 13 mm
Flauger and *Schneider* [Fl 64]		56 AVP	Naton 136
Present et al. [Pr 64]†	$\approx t_i$	XP 1020	Naton 136
Bertolini et al. [Be 66]	$\approx t_s$	XP 1020	Naton 136
Miehe et al. [Mi 66]	$\approx t_i$	C 70045 A	Naton 136, 20 mm × 10 mm

† See Fig. 21, p. 46.

optimum time resolution is obtained on $C/R \approx 0{\cdot}1$, which approximates theory. This means that maximum time resolution is not assured at the moment of zero-crossing of a pulse as shown in Fig. 19, p. 37. It is merely good time resolution that is obtained over a wide range of scintillation counter pulse height.

With NaI(Tl) scintillators asymmetric coincidence curves are generally observed if one scintillation counter is set on high energy and the other on a low energy of the order of 30 to 100 keV. In this case the lower-energy scintillation counter determines the slope of the coincidence curve that is less steep. This well-known effect in experiments can be verified and explained from statistics of time resolution [Br 65; Ly 66; We 66; Ha 67].

The influence of noise, spurious pulses and multiplier pulse pile-up tend to form a slowly decaying component (tail) at the slope of the coincidence curves, which cause considerable annoyance, for example, for lifetime measurements. Special blocking means can suppress most of the pile-up effects (see Section 3.5.3

Counters ($2\tau_0$ = full width at half-maximum of coincidence curve)

Energy	Optimum C/R	Coincidence Curve: $2\tau_0$, in ns	Coincidence Curve: $T_{1/2\,\text{instr}}$, in ps
1·17 + 1·33 MeV	0·005	0·80	90
511 keV	0·004	1·30	165
511 keV	0·08	1·08	145 and 110
{ 511 + 511 keV		0·52	120
{ 110 + 511 keV		1·2	340
{ 130 + 511 keV		0·85	170
{ 31·6 keV + 600 keV (e^-) ^{137}Cs		1·6	630
900 keV } ^{60}Co	0·2	0·48	
900 keV } ^{60}Co	0·09	0·53	
900 keV } ^{60}Co	0·1 to 0·2	0·32	37
300 keV } ^{60}Co		0·43	48
75 keV } ^{60}Co		0·75	100
900 keV ^{60}Co		0·24	28
900 keV ^{60}Co	≈0·06	0·21	29
940 keV ^{60}Co		0·16 to 0·20	
930 keV ^{60}Co	≈0·06	0·185	23
290 keV ^{22}Na	≈0·09	0·290	60

[Sc 63; Br 65; Mo 67]). Noise influence and pile-up probability at the multiplier output are reduced by the use of low time-constants $R_a C_s$, amounting to some nanoseconds. Apart of utilising low noise and may be cooled photomultipliers, the use of trigger stages (machine time t_i or t_s) is recommended, because the low pulses which do not reach the threshold of the trigger stage will be kept away from the time analyser.

From theoretical analysis it can be concluded that $\varepsilon^2 \propto 1/R$ applies for the time fluctuations, and consequently $2\tau_0 \propto 1/\sqrt{E}$ for the full width at half-maximum of the coincidence curves. This corresponds to $2\tau_0 \to 0$ for $E \to \infty$. Repeated studies of the energy dependency of $2\tau_0$ have been undertaken and the obtained results verified in the main [Br 65; Be 66a]. A series of tests disclosed, however, that resolving time does not decrease proportional to $1/\sqrt{E}$. Causes for this behaviour could not be clearly defined [Cu 61; Sc 63; Pr 64; Ga 64].

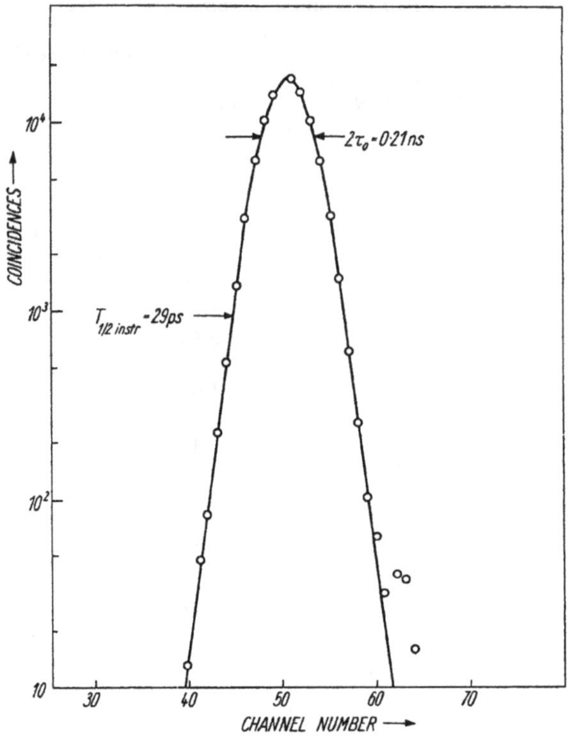

Fig. 21. Coincidence curve with two multipliers XP 1020 and ^{60}Co radiation. According to *Present* et al. [Pr 64].

In final evaluation, it can be stated that theoretical calculations give a good qualitative presentation of experimental facts. Quantitative agreement is still unsatisfactory with respect to the magnitude of optimum time resolution and the portion C/R of the total charge utilised for the purpose. Deviations are of the order of 50 to 100%. For more accurate calculation, scintillator and photomultiplier parameters have to be better known and require further extensive research and investigation, for example, with the aid of light pulse generators.

RECENT CALCULATIONS OF TIME RESOLUTION

Apart from the work of the groups *Gatti* and *Svelto*, and of *Hyman* (cp. Table 5), further calculations have been published based on other methods and assumptions. *Euling* [Eu 64] and *Donati* et al. [Do 67] viewed the statistical events in photomultipliers as time-dependent branching processes. *El-Wahab* and *Kane* [El 62] calculated the differential probability of the arrival of the C-th photoelectron under the assumptions of exponential decaying light emission of the scintillator and of an exponential function of the transit time spread in the photomultiplier focussing system (cp. Table 5). More sophisticated calculations were obtained by adding two further exponential functions for the finite rise time of scintillator light pulses and for the transit time spread [El 67; El 68]. Results received from the four consecutive exponential decays show optimum time resolution at $C/R \neq 0$, depending on selected parameters. They can be well fitted to experimental values for the dependence of time resolution on C/R without having to consider the various machine times. *Sigfridsson* [Si 67] used a similar statistical method.

With the Monte-Carlo method the time resolution of scintillation counters can be more accurately determined in the case of a low average number $\bar{R}$ of photoelectrons ($\bar{R} < 100$) than through analytical calculation [Ar 60; Ga 64; Hy 64; Co 67c]. The reason is that the Monte-Carlo method is better adapted to the actual processes than the analytical calculation with continuous variables. Another convenience is that complicated physical relations and the influence of omissions and simplifications afford a better presentation. With $\bar{R} > 100$ photoelectrons good agreement is usually obtained between the results from analytical calculation and from the Monte-Carlo method [Hy 64], while with a lower $\bar{R}$ considerable differences exist [Ar 60; Ga 64; Co 67c]. The time resolution with $\bar{R} = 10$, 20, 50 and 100 was calculated by *Cocchi* and *Rota* [Co 67c] using the measured equivalent illumination function $I^*(t)$ and the approximately symmetric SER function $f(t)$ according to *Gatti* and *Svelto* ([Ga 59; Ga 64], see Table 5). Only such events were accepted where the total charge at the multiplier anode is in the selected interval. This corresponds to a pulse height selection during the measurement and is easily handled with the Monte-Carlo method. Results given for the straight response of the time analyser with $\bar{R} = 10$ and $\bar{R} = 20$ show a slight improvement of time resolution with a more defined pulse height selection, but with increasing counting loss.

With low $\bar{R}$ values the Monte-Carlo calculations show marked individual differences of the waveforms of anode current pulses [Hy 64]. With this in view, *Rota* et al. [Ro 67] proposed double sampling to improve the time resolution with a low $\bar{R}$. In that case, an event is accepted for registration only when the time difference of the arrival at two threshold levels is in the selected time interval.

2.3.3 IMAGE INTENSIFIERS

As with the photomultiplier, the external photoelectric effect is utilised in the photoelectronic image intensifier to convert light into liberated electrons to which energy is fed [Mc 61; Ec 62a; Me 66]. The main point is to retransform the optical image produced on the photocathode through an electron-optical system into an amplified optical presentation that permits visual observation or photographic recording. Apart from the amplification of a weak-intensity image, the invisible infra-red, ultra-violet and X-ray images are frequently transferred to the visible spectral region.

Image intensifiers can be opened and closed by means of electrical pulses. Along with many other applications, intensifiers are consequently suitable to serve as rapid shutter device for high-speed photography and as nuclear particle track detector, such as *Čerenkov* chamber (see Section 2.2.2) and scintillation chamber (see Section 2.2.3).

The photocathode must have sufficient surface area and yield a high quantum output in the desired spectral region. The registration of single electrons must be possible if nuclear radiation is to be proved. This requires an amplification of about 10^7. If the image intensifier is triggered by the significant event itself, intermediate image storing up to 1 μs is necessary to compensate for trigger time delay. A geometric resolution of at least 10 to 20 lines/mm is required, which approximates the resolution of the human eye.

SINGLE-STAGE IMAGE INTENSIFIERS

Single-stage tubes, as shown in Fig. 22*a*, are primarily used as infra-red image converters. The photoelectrons emitted by the photocathode are accelerated to a level of 5 to 25 kV by a homogeneous electric field and excite the luminescent screen. A homogeneous magnetic field serves for focussing, and electron-optical imaging is scaled 1:1. Utilising electrostatic or short magnetic lenses the image can be reduced or enlarged by the factor of 2 to 7 [Ec 62a]. Electrostatic focussing has the advantage that the operating voltage need not be stabilised and that the display arrives without rotational displacement.

Light is radiated by the luminescent screen in all directions. A reflecting layer of thin aluminium on the inside increases the light output, nevertheless, part of the light is lost due to the optical imaging aperture. On a photographic plate, therefore, light intensity is higher by one order of magnitude only as compared to the photocathode. Attempts to attain higher values through partial light feedback have become unnecessary due to the development of the multi-stage image converter.

With total operating voltage applied in form of a pulse an opening time of 100 ns can be obtained, but the photocathode resistance mentioned in Section 2.3.1 still represents a source of interference. Image converters with built-in control grid require only 100 to 300 V for control and have an opening time down to 10 ns. Using a special design with two grids and low electrode capacitance a reduction of the blocking voltage to about 5 V in some instances has been attained [Mc 61]. If still shorter opening time is desirable, electric or magnetic

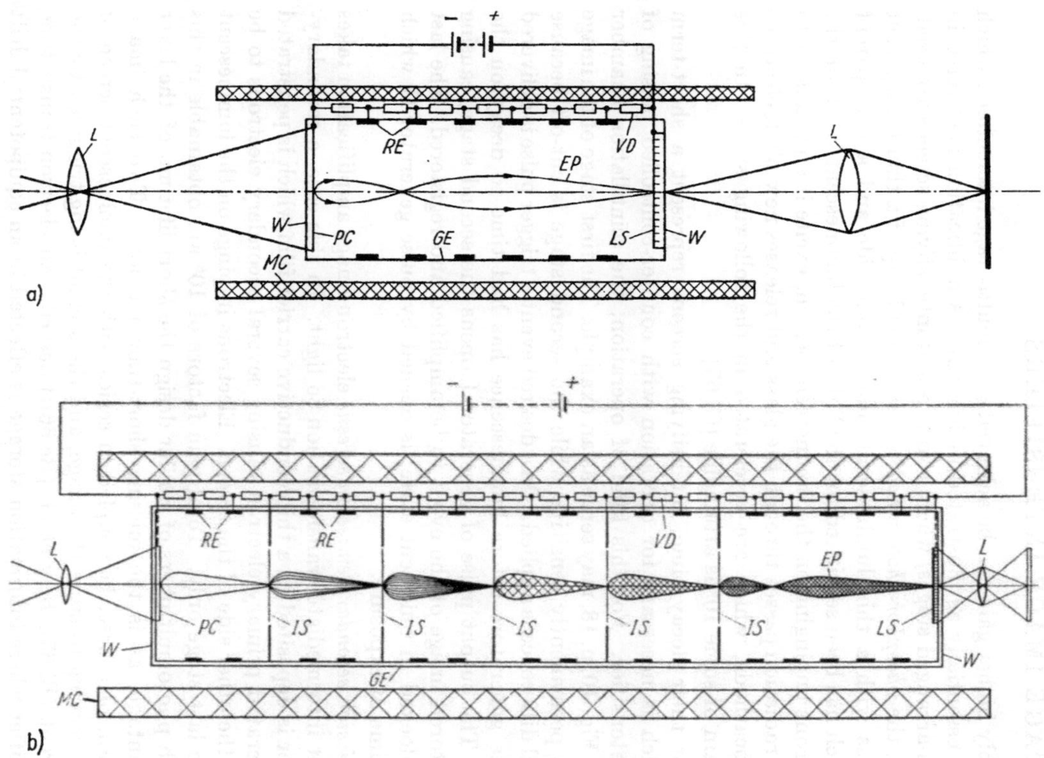

Fig. 22. Construction (schematic) of a single-stage (*a*) and of a multi-stage (*b*) image intensifier. According to *McGee* [Mc 61].

GE glass envelope, *W* window, *PC* photocathode, *L* optical lens, *VD* voltage divider, *LS* luminescent screen, *MC* magnetic focussing coil, *EP* electron path, *IS* intensifying screens either phosphor-photocathode screen or transmission secondary-emission screen, *RE* ring electrode for homogenising the electrostatic field of acceleration.

deflection is used for sweeping the electron beam across a diaphragm. Image converters of this type are employed for high-speed photography (see Fig. 23, p. 53).

MULTI-STAGE IMAGE INTENSIFIERS

Considerably higher gain can be achieved with multi-stage intensifiers, which are designed essentially as illustrated in Fig. 22*b*. Amplification is effected in several series-arranged stages, where two types of intensifying screens are used. One of these, the *phosphor-photocathode screen*, is made up of a thin transparent panel of glass with a thin luminescent film on one side and a transparent cathode, which has been sensitised for the light of the luminescent layer, on the other. Electrons impinging on the luminescent screen, excite it and cause it to radiate. This radiation passes through the glass and releases new photoelectrons at the photocathode, which are accelerated in the following stage. In this manner, a gain of some 10^6 is attainable [Gi 63].

Because of their decay time, intensifying screens represent a short-term memory which is necessary for operation with counter-controlled pulsing of the image intensifier. For this kind of operation, the scintillation chamber illustrated in Fig. 10, p. 18, may serve as an example. The first stage of the image intensifier is permanently sensitive while the second stage is cut-off because zero potential difference is applied. On a desired event a trigger pulse is delivered to both pulse generators before luminescence has had time to decay on the first screen. The output pulse of generator I opens the second stage, causing the briefly stored image of the event to be amplified and registered. The last stage, also blocked in quiescent state, is opened by pulse generator II which determines time of exposure.

On *transmission secondary-emission screens* electron image amplification takes place without intermediate transformation to light. In this case, a secondary-emission layer is deposited on a thin conductive carrier film which is penetrated by the accelerated primary electron causing several secondary electrons to be liberated on the other side of the screen. Electrons impinge on the luminescent screen at the last stage only. Total gain factors of 10^5 are obtainable in this way. As with photomultipliers of similar design the short lifetime of the layer is a disadvantage, as is the missing short-time memory. Two such image converters can, of course, be coupled in series with the luminescent screen of the first unit serving to store the image, and the second being triggered by the registered event [Bi 63]. *Berg* et al. [Be 66d] describe an electron transit-time image intensifier where information storage is effected in an equipotential drift section through a period of 200 ns.

With the so-called *channelled secondary-emission image intensifier* the interesting experiment was undertaken to assign an individual secondary-emission multiplier to every image element. The unsatisfactory geometric resolution of at the best 0·25 mm is compensated by the relatively low unstabilised operating voltage and the fact that no magnetic field is needed [Mc 61; Fl 62].

HIGH-SPEED PHOTOGRAPHY

To enable maximum time resolution photography [Co 63; Hu 66a; Co 67a] rotating mirrors, *Kerr* cells or image converters are used. Three modes of operation apply, viz. *single-frame cameras* for a single picture; *multiple-frame cameras* for making a series of pictures of the event, and *streak cameras* where an image-strip is produced and moved rapidly at right angle across the stationary film. This produces a two-dimensional picture displaying the light distribution in time across the one dimensional slit image. A modification of this principle is the *electron-optical chronography* [Za 55]. In this case image converters are employed and a circle or ellipse represents the time base. This makes synchronisation of the start of the sweep with the mostly unique event unnecessary.

A comparison of attained resolving time with principal limits is made in Table 7 (s. next page). In order to observe events through time ranges of 10^{-9} to 10^{-14} s electron-optical chronography seems to offer the only possibility, as the table shows. This system converts light signals into electronic signals, which is followed by the time analysis of the latter. The lower limit of the resolving time results only from the frequency of the light that acts as information carrier, modulated by space and time, during image transmission. An analysis undertaken by *Zavoisky* and *Fanchenko* [Fa 61; Za 65] disclosed that the inherent time resolution of the imaging optic, of the photocathode† and of the electron-optical system should permit time measurements down to some 10^{-14} s. However, a precondition is the presence of a high field strength near the photocathode to allow rapid acceleration of the photoelectrons to a velocity greater than their escape velocity to prevent the influence of initial velocity dispersion.

An example of the construction of an electron-optical chronograph is shown in Fig. 23, p. 53 [Za 65]†§. Recording is not made on a photographic film, but with the aid of a television pick-up tube (*TV*) with its output signal either viewed on a monitor screen (*MS*) or recorded on a tape (*TR*). In the time-analysing image converter (*IC*) a well-defined beam of electrons passes through the slotted diaphragm(*D*). Using a shutter voltage of 1·5 kV between the plates (P_1) and (P_2) the beam current can be cut off completely. To prevent the image from blurring during the rise time of the shutter voltage the same deflection voltage, but with reverse polarity, is applied to the compensating plates (P_3) and (P_4), which in some cases are adjustable to eliminate deviations [Hu 62]. Multi-frame pictures with intervals of 10 to 100 ns can be obtained by using stepwise increments of the deflection voltage at the plate pairs (P_5, P_6) and (P_7, P_8), disposed vertically to each other. For still shorter time only continuous deflection can be established by applying phase-displaced sinusoidal voltages to both plate

† However, compare [Du 66a].
†† For a discussion of valve performance see [Da 67; De 67a].

Table 7: Resolving Time in High-Speed Photography

Type of Camera	Limit Given by Principle		Presently Obtained Values	
	minimum resolving time, in s	cause of limitation†	resolving time, in s	mode of operation
Rotating mirrors	10^{-9} to $2 \cdot 10^{-10}$	tensile strength of material	10^{-7}††	multiple frame
			$5 \cdot 10^{-9}$††	streak
Kerr cell	10^{-10} to 10^{-12}	relaxation time of molecular orientation	10^{-8}††	single frame
Image converter	10^{-14}	frequency of light, physical time resolution of conversion process	10^{-8} to 10^{-9}††	single frame
			10^{-7} to 10^{-8}†	multiple frame
			10^{-12} to 10^{-13}†	streak

† According to ***Zavoisky*** and ***Fanchenko*** [Za 65].
†† According to ***Huston*** [Hu 62].

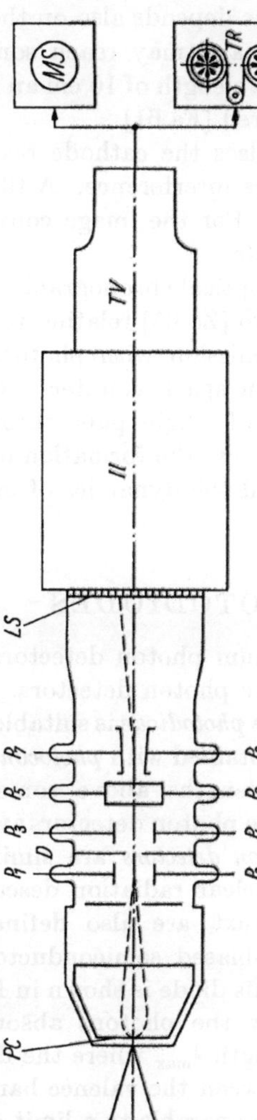

Fig. 23. Electron-optical chronograph with TV scanning. According to *Zavoisky* and *Fanchenko* [Za 65].

O light source,
L optical lens system,
AF variable attenuation filter,
IC time-analysing image converter,
PC photocathode,
LS luminescent screen,
$P_1, \ldots, P_6$ deflection plates,
D shutter slotted diaphragm,
II image intensifier,
TV television pickup tube,
MS monitor screen,
TR tape recorder.

pairs. In consequence, resolving time depends also on the length and the space resolution of the produced ellipses, and may reach some picoseconds. With a resonance cavity deflector for a wavelength of 10 cm an instrumental resolving time of $1{\cdot}5 \times 10^{-13}$ s could be measured [Fa 61].

With very short intense light pulses the cathode resistance and the space charge of the image converter causes interference. A filter (*AF*) is sometimes necessary to reduce light intensity. For the image converter (*II*) any of the described multi-stage tubes is suitable.

Regarding application of electron-optical chronography a number of examples are given by *Zavoisky* and *Fanchenko* [Za 65] relating to plasma spectroscopy, for the analysis of dark electron emission with photocathodes and of spark discharges as found, for instance, in spark counters. Other interesting uses are also mentioned, e.g. the passage of a light pulse through dispersing media; the onset of diffraction and interference, the formation of optical non-linearity under the effect of a laser beam; and the dynamics of molecular reorientation processes.

2.3.4 PHOTODIODES

In contrast to the described vacuum photon detectors, the internal photoelectric effect is utilised in solid-state photon detectors. For work with nanosecond light pulses at present only the ***photodiode*** is suitable, because the required gain-bandwidth product cannot be obtained with ***photoconductors***. Nevertheless, in the infra-red region with wavelengths above some micrometres photoconductors represent the only suitable photon detectors available today [An 66].

Photodiodes or *p-n junction photon detectors* are similar in construction to the semiconductor detectors for nuclear radiation described in Section 2.4.1. The terms used in the following text are also defined there. Essentially the photodiode represents a reserve-biased semiconductor diode. The energy band system of the p-n junction of this diode is shown in Fig. 28. Freely mobile electron-hole pairs are produced by the photons absorbed in the material. The upper limit of the light wavelength λ_{max} where the detector still responds, is determined by the energy gap between the valence band and the conduction band (see Fig. 27). Silicon diodes are operable to a limit of about $\lambda_{max} = 1\ \mu m$, while with germanium diodes the value reaches $\lambda_{max} = 1{\cdot}7\ \mu m$. At the lower limit of the detectable light wavelength absorption takes place primarily in the material located over the depletion layer, where the charge carriers recombine before they are collected.

Due to the resulting concentration gradient electrons and holes diffuse slowly into the depletion layer where no free charge carriers are present. The high field strength in the depletion layer (see Fig. 29*a*) accelerates the charge carriers produced in or diffused into the layer, causing them to drift rapidly through this region and become active externally as a pulse of the reverse current. To obtain the fast response of the photodiode, diffusion effects should be prevented from exerting an influence by placing the light absorption in a region where high field strength prevails. For complete absorption a minimum thickness of the sensitive layer is necessary and in most cases a large sensitive area is desirable. Drift time is small if the barrier is thin and a high reverse voltage is supplied to give maximum drift velocity to the charge carriers. On the other hand, diode capacitance should be as low as possible and this requires a thick barrier and a small diode area. These conflicting demands have produced different types of diodes, construction and properties of which have been compiled by *Anderson* and *McMurtry* [An 66]. Fundamental properties of photodiodes and phototransistors are also discussed by *Soboleva* et al. [So 65]. Compared to most vacuum photon detectors, no or only little current amplification is obtainable with the photodiode. This restricts their usefulness considerably, despite their at times fast action.

PHOTODIODES WITHOUT CURRENT AMPLIFICATION

p-n Diodes

To obtain complete light absorption in the depletion layer and to attain short drift time with moderate capacitance, the junction is produced near the surface, with the thickness of the depletion layer adapted to the maximum frequency to be handled. In silicon the optimum value is $w = 10\,\mu$m with a cut-off frequency of 3 GHz. Unfortunately, however, the size of the depletion layer depends on the applied reverse voltage (see Fig. 28, p. 61).

Surface-Barrier Diodes

When a semiconducting layer is replaced by a thin metal film a barrier is formed very close to the surface. If an antireflexion film is added quantum yields of up to 70% can be obtained with silicon diodes.

p-i-n Diodes

Size of the depletion layer becomes independent of reverse voltage if a compensated i-layer is interposed between the p- and the n-layer (see Fig. 27, p. 60) on which the depletion layer spreads. Constant field strength prevails in the depletion layer (see Fig. 29*b*, p. 63). The diode design illustrated in Fig. 24 has a very low series resistance and a low series inductance. With short light pulses an output rise time underpassing 1 ns can be measured [Ri 62]. Rise time is composed of two components consistent with the differing mobility of the electrons (μ_e) and of the holes (μ_h). An output pulse shape calculated on the non-factual assumption [Lu 64] of uniform excitation of the total sensitive volume through a light pulse described by a step-function is shown in Fig. 25. *Lucovsky* et al. [Lu 64] took absorption into consideration, which effects the exponentially decreasing excitation, although only for sinusoidal modulated light. In this case, the diode responds more rapidly if light impinges on the p-layer, which causes the electrons to pass through practically the entire i-layer, with the electrons, however, showing twice the mobility as the holes.

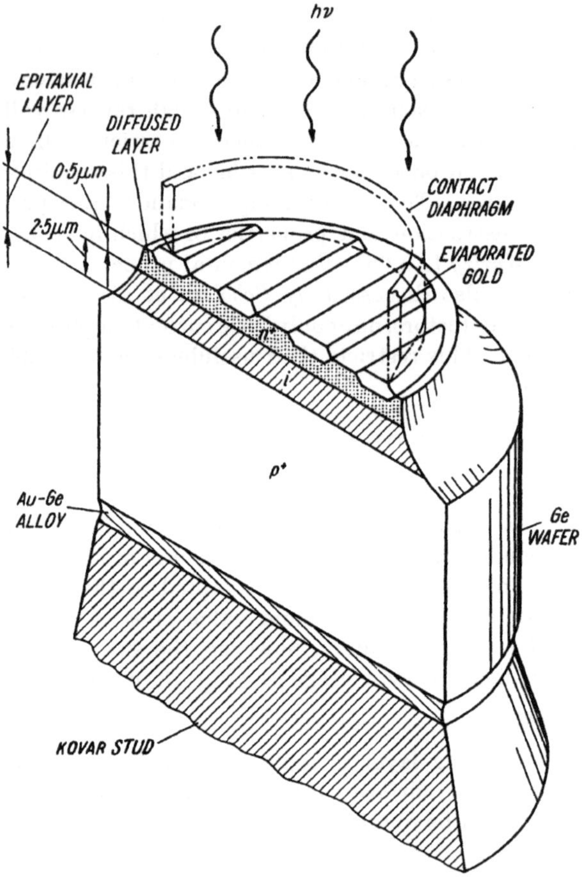

Fig. 24. Construction of a p-i-n photodiode. According to *Riesz* [Ri 62].

Hetero-Junction Diodes

Penetration depth of long-wave infra-red radiation often is greater than the optimum thickness of the depletion layer consistent with frequency behaviour. For this reason it is more advantageous to arrange for the light incident along the plane of the depletion layer, rather than the vertical. However, a more elegant method is offered by the hetero-junction diode. This consists, for example, of a gallium-arsenide p-layer with a band gap of 1·39 eV and a germanium n-layer having a band gap of 0·66 eV [Re 63]. Photons with energy located between the band gap of the two layers penetrate the GaAs practically without absorption, but are heavily absorbed by the Ge material within a layer of 1 μm. Consequently, a depletion layer of little thickness is needed, which

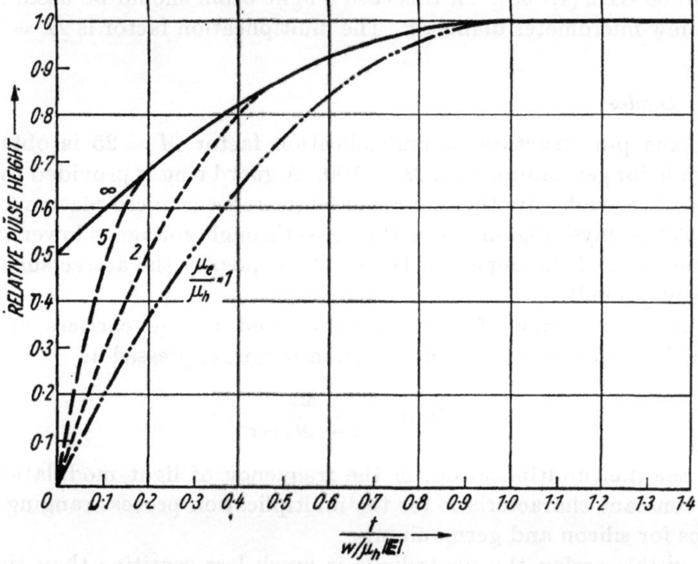

Fig. 25. Rise of the output pulse of a p-i-n photodiode on homogeneous volume excitation with a step-type light pulse, referred to different carrier mobility values. According to *Lucovsky* et al. [Lu 64].

w thickness of i-layer, μ_e mobility of electrons, μ_h mobility of holes, E field strength in i-layer.

must not be produced directly below the surface of the diode. Hetero-junction diodes work very rapidly and are insensitive against surface effects. With a p⁺-layer (see Section 2.4.1) of $GaAs_xP_{1-x}$ ($x = 0{\cdot}8$) and an n⁺-layer of GaAs, for $\lambda = 0{\cdot}78$ to $0{\cdot}87\,\mu m$, a total efficiency of 50% and a rise time of $t_r = 3{\cdot}2$ ns was measured with circuit impedance furnishing a considerable contribution [Ra 64].

PHOTODIODES WITH CURRENT AMPLIFICATION

At a very high reverse voltage electric breakthrough occurs. Multiplication of the charge carriers is observed just under breakthrough voltage level which may be explained by the action of impact ionisation [Mi 55]. When charge carriers are accelerated, for instance, in germanium to 1·5 eV, new electron-pairs are produced by them. Photodiodes showing such avalanche multiplication have a gain-bandwidth product reaching 100 GHz.

Point Contact Diodes

If arsenic-doped gold foil is brought in contact, for example, with an i-layer deposited on p⁺-germanium, this produces an active volume of 4 to 5 μm diameter. A diode of this type has very low capacitance which does not influence response time. Drift time is less than 40 ps and permits cut-off frequencies from 25 to 50 GHz [Ki 62]. In this case a light beam should be used, however, of only a few micrometer diameter. The multiplication factor is $M = 10$.

Avalanche Diodes

With plane p-n structures a multiplication factor $M = 25$ is obtained for silicon, while for germanium it is $M = 100$. A guard ring is provided to prevent surface breakthrough. Another source of interference is micro-plasma formation in the depletion layer region where the breakthrough voltage is lower than that of the other area of the depletion layer. Consequently the active surface must be sufficiently small.

Finite collecting time of secondary-produced charge carriers causes the frequency dependence of the multiplication factor, expressed as

$$M(\omega) = \frac{M_0}{1 + M_0 \, \omega \, \tau} \tag{24}$$

with M_0 the d.c. multiplication, ω the frequency of light modulation, and τ the time constant characteristic for the multiplication process ranging between 1·5 to 3 ps for silicon and germanium.

In the visible region the photodiode is much less sensitive than the photomultiplier, but in the infra-red region between 1 and 1·7 μm the Ge-avalanche diode has proved superior as photon detector.

Photodiodes are smaller, less expensive and more reliable than photomultipliers although their low gain, weak output signal and the small size of their active region are serious drawbacks. For this reason, they are specially suited for detecting well-defined light beams of sufficient intensity, for instance, in optical communication systems using a laser beam. *Denton* et al. [De 66] describe an optical pulse-code modulation system where the Ge-avalanche diode serves for the detection of laser pulses of 0·6 ns width arriving at a repetition frequency of 224 MHz.

BEAM-OF-LIGHT TRANSISTOR

According to a proposal made by *Rediker* et al. [Re 63] a luminescence diode (Fig. 26 — see Section 4.2.3) and a hetero-junction diode can be combined to form an opto-electronic component which acts like a transistor [Fa 63]. The GaAs diode acts as emitter and is operated in forward direction. The infra-red light it emits is absorbed in the hetero-junction diode which acts as collector diode. The advantage of such a beam-of-light transistor is that its material is characterised by short-lifetime carriers, that the base is passed with the velocity of light and that carriers are produced in the very thin collector layer. Light collection is much better at the collector end if a lasing diode is used as emitter [Ru 65].

Of similar construction is the p-n-i-n diode developed by *Meyerhofer* et al. [Me 67c] which represents a light-activated light-coupled switch. Its current-voltage characteristic exhibits a current-controlled negative resistance region with an off-to-on ratio as high as 10^6 and a turn-on time down to 10 ns.

The use of injection laser diodes for high-speed logic gates in computers displays interesting aspects discussed, for instance, by *Nishizawa* [Ni 67a].

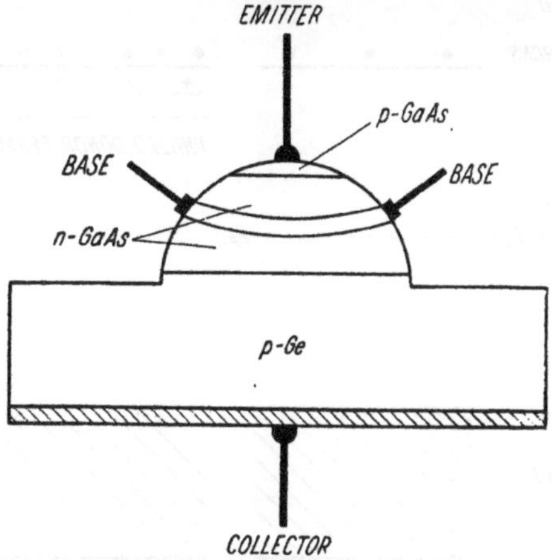

Fig. 26. Construction of a beam-of-light transistor. According to *Rediker* et al. [Re 63].

2.4 CONVERSION OF NUCLEAR RADIATION TO ELECTRICAL SIGNALS

2.4.1 SEMICONDUCTOR DETECTORS

In the semiconductor detector, impinging radiation liberates charge carriers through ionisation ("solid-state ionisation chamber"). Due to their energy resolution, which is far superior to a scintillation counter, these detectors have gained a prominent place in experimental nuclear physics [Ma 63; Ta 63; De 64a; Ew 64; Ju 65; De 66b; Ne 66; Go 66a; Ak 67].

SEMICONDUCTOR LAYERS

Semiconductor detectors and the photodiodes discussed in Section 2.3.4 are single crystals of tetravalent silicon or germanium in layers of various kinds of conduction. Group III—V compounds, such as GaAs [Ri 67b], are used for photodiodes also. Fig. 27*a* shows the band model of an ideal-state crystal having intrinsic conduction due to thermal excitation of the electron-hole pairs, where electrons are lifted out of the fully occupied valence band into the empty conduction band where they can move freely. In the valence band electrons leave positively charged holes which can also move freely. Between the two bands is an energy gap, the width of which determines the minimum energy necessary for the production of an electron-hole pair. Probability of electron population of respective energy states in the valence band and the conduction band is assessed with the aid of *Fermi*'s theory of electron gas. The so-called *Fermi* energy has a population probability of 50%. With intrinsic conduction

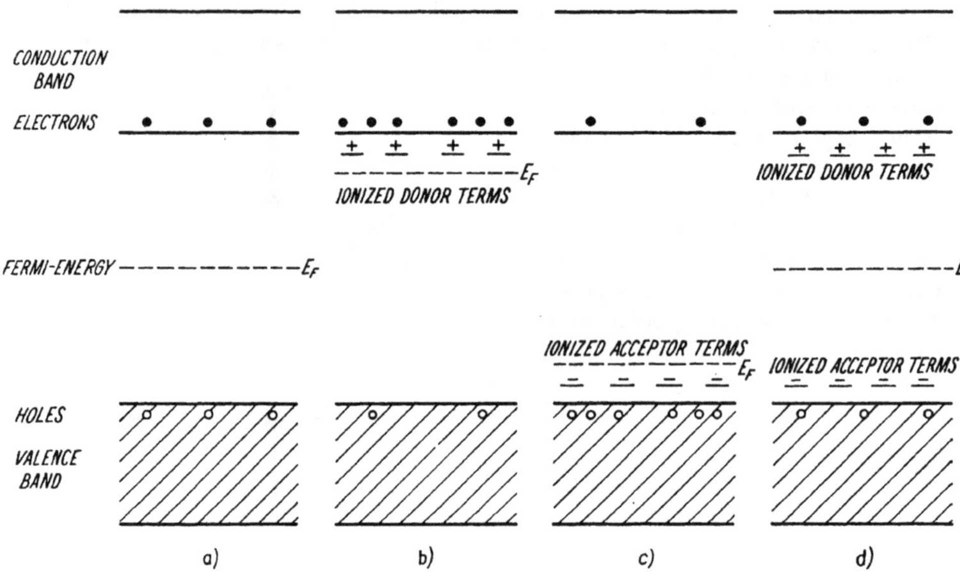

Fig. 27. Band model of semiconductor layers and location of *Fermi* energy E_F.
a) Intrinsic conduction in ideal crystal. *b*) Conduction of excess electrons (n). *c*) Conduction of excess holes (p). *d*) Intrinsic conduction through compensation (i-layer).

this is located in the middle of the energy gap. In case of pentavalent atoms, e.g., phosphorus as donors, because of the thermal ionisation of the latter, the doped material has an excess of free electrons (n-conduction) which appear as majority carriers (see Fig. 27*b*). Trivalent atoms, e.g., gallium, act as acceptors. In this case thermal ionisation produces an excess of moving holes, representing majority carriers, and causes p-type conduction (see Fig. 27*c*). *Fermi* energy is shifted due to the incorporated donors and acceptors, which result in a very small number of minority carriers, i.e. holes in case of n-conduction and electrons in case of p-conduction. Concentration of the impurity atoms, for example, lies between 10^{12} to 10^{16} per cubic centimetre.

If lithium atoms are introduced in p-conducting material as donors positively charged lithium ions are formed. These occupy interstitial sites which consequently makes them easily movable under the influence of an electric field (drift). A progressively advancing compensation of negatively charged acceptors through the crystal is effected which is self-controlled by the space charge. These compensated i-layers have intrinsic conduction and a high specific resistance, amounting to ϱ (Si) $= 2{\cdot}3 \times 10^5$ Ωcm for silicon (see Fig. 27*d*). Because of the narrow energy gap germanium has a low specific resistance at room temperature, amounting to ϱ (Ge) ≈ 50 Ωcm; while at 77 °K this is ϱ (Ge) $= 5 \times 10^4$ Ωcm. Under low temperature conditions back diffusion of lithium is prevented.

ASYMMETRIC P-N JUNCTION

When on a low-doped p-layer a highly doped n^+-layer† is deposited (see Fig. 28) the majority carriers, present in high concentration, diffuse to the adjacent layer where they recombine. This causes the formation of a depletion layer, primarily in the p-region, having a thickness w_0, where free carrier

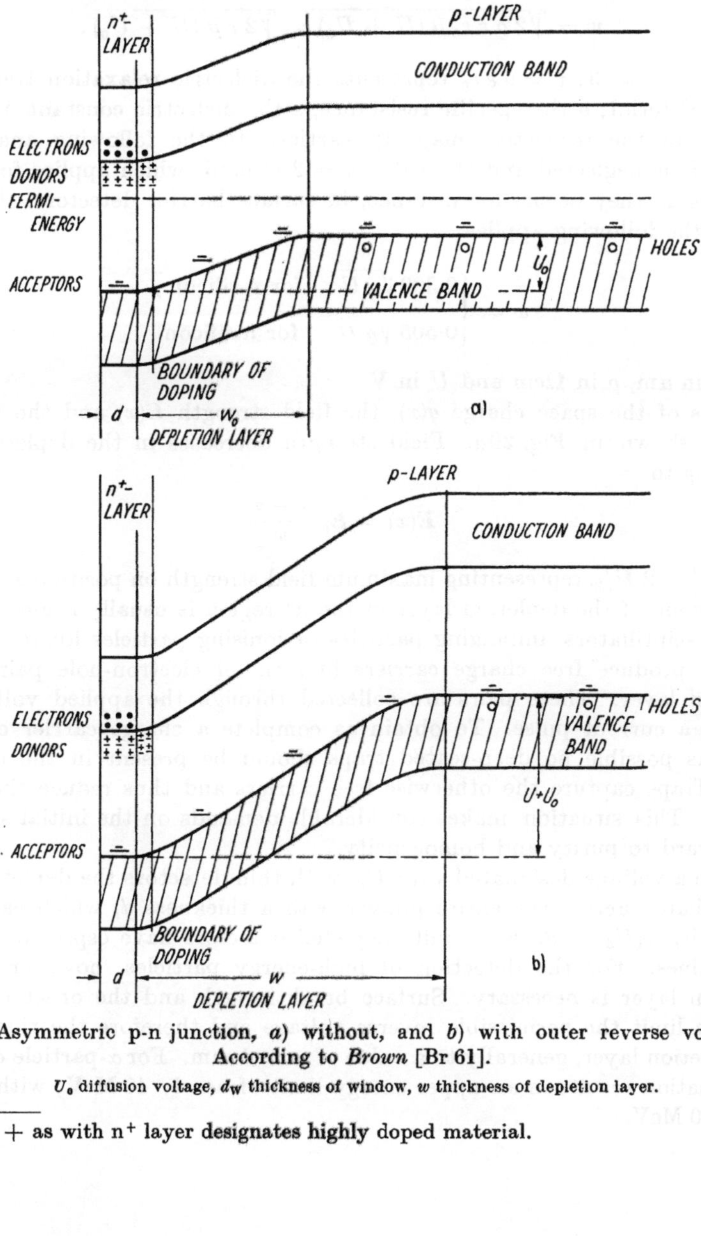

Fig. 28. Asymmetric p-n junction, *a*) without, and *b*) with outer reverse voltage U. According to *Brown* [Br 61].

U_0 diffusion voltage, d_W thickness of window, w thickness of depletion layer.

---

† Suffix + as with n^+ layer designates highly doped material.

concentration is reduced by several orders of magnitude. Remaining ionised acceptors and donors cause a difference in potential which is sufficient to raise the *Fermi* energy of both layers to the same level. Addition of an external voltage U of the same polarity to this diffusion voltage U_0 (0·5 to 0·7 V in silicon) causes more charge carriers to be transferred, with consequent increase of the thickness of the depletion layer, thus

$$w = \sqrt[n]{2\,\varrho\,\varepsilon\,\varepsilon_0\,\mu\,(U + U_0)} = \sqrt[n]{2\,\tau\,\mu\,(U + U_0)}\,, \tag{25}$$

where $n = 2$ to 3; $\tau = \varrho\,\varepsilon\,\varepsilon_0$ represents the dielectric relaxation time of the p- or n-material; ϱ the specific resistance, ε the dielectric constant and μ the mobility of the respective majority carrier. In the following analysis U_0 against U is neglected and the value $n = 2$ is used, which applies for abrupt junctions as they occur, for instance, in surface barrier detectors. In consequence the following applies:

$$w = \begin{cases} 0{\cdot}309\,\sqrt{\varrho\,U} & \text{for p-silicon}\,, \\ 0{\cdot}505\,\sqrt{\varrho\,U} & \text{for n-silicon} \end{cases} \tag{25a}$$

with w in μm, ϱ in Ωcm and U in V.

Curves of the space charge $q(x)$, the field strength $\boldsymbol{E}(x)$ and the potential $\varphi(x)$ are shown in Fig. 29*a*. Field strength decreases in the depletion layer according to

$$\boldsymbol{E}(x) = \boldsymbol{E}_0\,\frac{w - x}{w} \tag{26}$$

with $|\boldsymbol{E}_0| = 2\,U/w$ representing maximum field strength on position $x = 0$. The narrow zone of the depletion layer in the n^+-region is usually neglected.

As in scintillators, impinging particles or ionising particles liberated in the material produce free charge carriers in form of electron-hole pairs in the depletion layer. These pairs are collected through the applied voltage and produce a current pulse. To obtain as complete a charge carrier collecting action as possible no deep-seated traps should be present in the depletion layer. Traps capture the otherwise free carriers and thus reduce their mean lifetime. This situation makes considerable demands on the initial substance with regard to purity and homogeneity.

Above a voltage designated with U_d, with thin detectors the depletion layer is distributed across the entire p-layer with a thickness d, which can be expressed by $w(U_d) = d$. Such fully depleted detectors have especially low rise time values. For the detection of high-energy particles, however, a thick depletion layer is necessary. Surface breakthrough and the onset of carrier injection limit the permissible reverse voltage and therefore the thickness of the depletion layer, generally to a value of $w \approx 1$ mm. For α-particle detection this situation results in an upper energy limit of nearly 40 MeV; with protons this is 10 MeV.

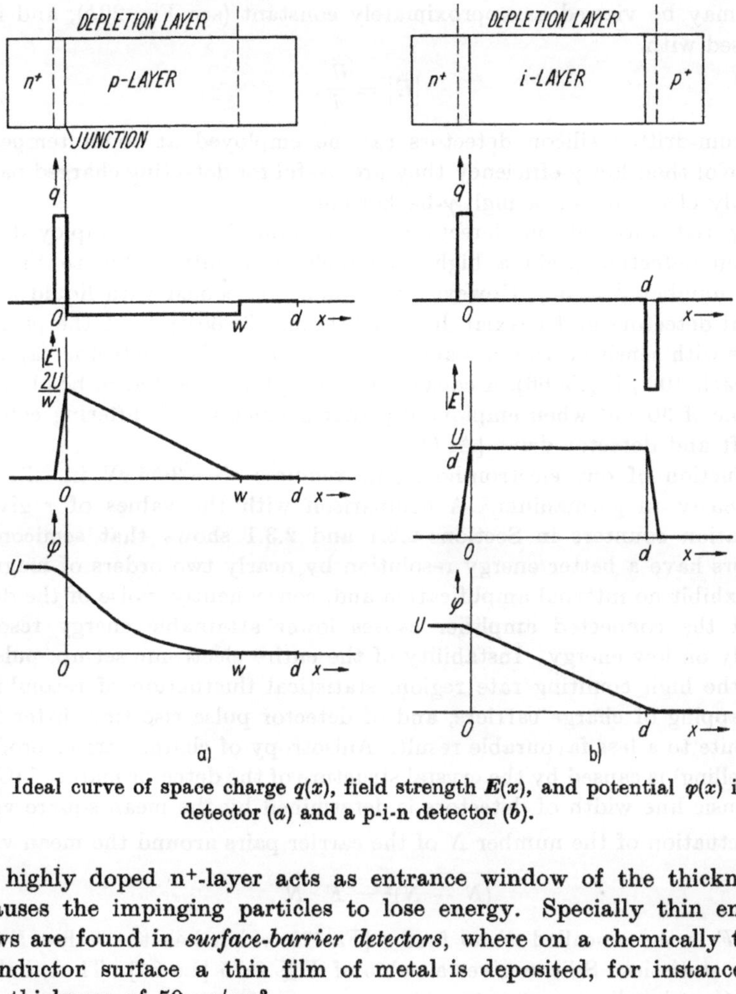

Fig. 29. Ideal curve of space charge $q(x)$, field strength $E(x)$, and potential $\varphi(x)$ in a p-n detector (*a*) and a p-i-n detector (*b*).

The highly doped n^+-layer acts as entrance window of the thickness d_w and causes the impinging particles to lose energy. Specially thin entrance windows are found in *surface-barrier detectors*, where on a chemically treated semiconductor surface a thin film of metal is deposited, for instance, gold with a thickness of 50 µg/cm².

DRIFTED P-I-N DETECTORS

The detection of long range particles, for example, particles with high energy, or even γ-radiation, requires a depletion layer that is specially thick. This can be attained if ϱ is very high, as equation (25) shows. For the purpose, a thick compensated i-layer is located between the highly doped n^+- and p^+-layer, and this p-i-n detector is driven with a voltage $U \gg U_d$. This gives a field strength of

$$|E(x)| = \frac{U - U_d}{d} + \frac{2\,U_d}{d^2}(d - x) \tag{27}$$

which may be viewed as approximately constant (see Fig. 29*b*), and is then expressed with

$$|\boldsymbol{E}| = \frac{U}{d} \,. \tag{27a}$$

Lithium-drifted silicon detectors can be employed at room temperature. Because of their low γ-efficiency they are useful for detecting charged particles, primarily electrons, on a high γ-background.

Compared with silicon detectors, germanium detectors employed for γ-radiation detection yield a higher photoelectron output due to the higher atomic number $Z = 32$. However, they must be cooled with liquid air. Cylindrical detectors and coaxial drift of lithium [Ta 65] permit the production of units with sensitive volumes amounting to 50 cm^3, but detector capacitance may reach 100 pF [Ta 66]. A capacitance of 5 pF is expected to be obtained at a volume of 30 cm^3 when employing planar geometry and differing contacting for drift and detector drive [Ar 66].

Production of one electron-hole pair requires $\varepsilon' = 3{\cdot}55$ eV in silicon and $\varepsilon' = 2{\cdot}85$ eV in germanium.† A comparison with the values of r given for scintillation counters in Sections 2.2.1 and 2.3.1 shows that semiconductor detectors have a better energy resolution by nearly two orders of magnitude. They exhibit no internal amplification and, consequently, noise of the detector and of the connected amplifier causes lower attainable energy resolution, specially on low energy. Instability of the entire electronic set-up, pulse pile-up in the high counting rate region, statistical fluctuation of recombination and trapping of charge carriers, and of detector pulse rise time jitter further contribute to a less favourable result. Anisotropy of charge carrier production (channelling) is caused by the crystal structure of the detector material [Gi 66a].

Intrinsic line width of detectors is determined by the mean square value of the fluctuation of the number N of the carrier pairs around the mean value $\overline{N}$

$$\overline{(N - \overline{N})^2} = F \cdot \overline{N} \,, \tag{28}$$

where F is the so-called *Fano* factor [Fa 47] which has the value $F = 1$ in *Poisson* statistics. Silicon gives a value of $F \lesssim 0{\cdot}15$ [La 64]. Thus, halfwidth of the *Gauss*ian lines comes to

$$\Delta E_{\mathrm{i}} = 2{\cdot}355 \sqrt{E\,\varepsilon'\,F} \,. \tag{29}$$

A comparison of the measured halfwidth ΔE and the intrinsic line width ΔE_{i} is given in Fig. 30 [He 66]. Data for ΔE_{i} were obtained by comparing the width of calibration lines ΔE_{E} of a pulse generator, thus

$$(\Delta E)_{\mathrm{i}}^2 = (\Delta E)^2 - (\Delta E)_{\mathrm{E}}^2 \,. \tag{30}$$

Observed energy dependence of ΔE_{i} gives a value of $F \lesssim 0{\cdot}15$ also for germanium [Ma 66].

† For temperature dependency measurements see [Pe 68].

It will be seen that with low energies the internal resolution can be neglected with reference to noise influence. With cooled field-effect transistors in the charge-sensitive pre-amplifier stage a halfwidth of $\Delta E = 0{\cdot}75$ keV was obtained at an energy of $E = 100$ keV [He 66].

For higher photoelectron output in γ-radiation detection attempts have been made to replace germanium with a detector material having a still higher atomic number. At present the semiconductor properties of suitable material are still far from satisfactory [Ma 66a].

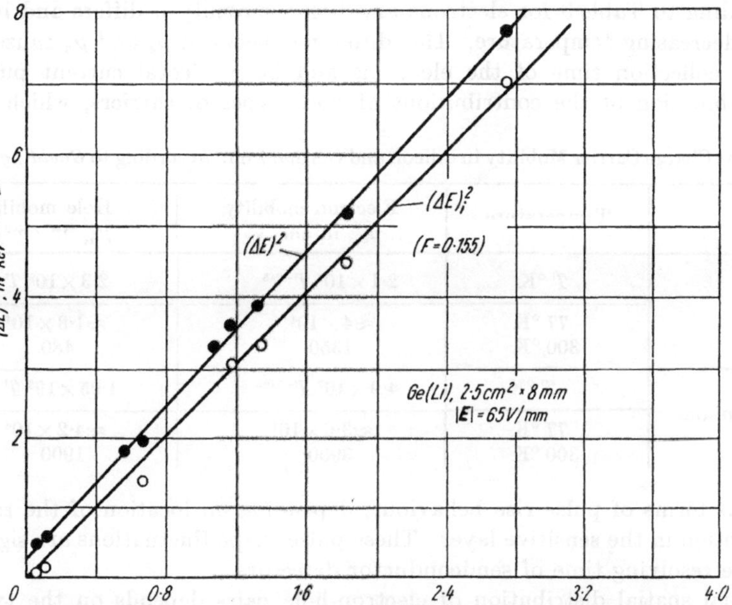

Fig. 30. Energy dependence of measured half-width ΔE and of internal line width ΔE_i of a Ge(Li) detector. According to *Heath* et al. [He 66].

DETECTOR PULSE RISE TIME

In time measurements with scintillation counters, resolving time is determined primarily by the relation τ/R of the scintillator decay time to the number of photoelectrons emitted in the multiplier (Section 2.3.2). With 100 keV, values of 25 ps to 1 ns are obtained for τ/R. The location of radiation absorption in the scintillator is relatively immaterial. If with semiconductor detectors a similar quantity is to be defined, for instance, with the collecting time t_{coll} of the charge carriers, the 100 times higher number R of the produced carriers would give a value of $t_{coll}/R \approx 1$ ps, which can be neglected compared to other influences.

In general, for the *collection time* applies

$$t_{\text{coll}} = \int_{x_1}^{x_2} \frac{\mathrm{d}x}{v(\boldsymbol{E})} \tag{31}$$

with x_1 defining the location of radiation absorption and x_2 the limit of the electric field the charge carriers pass through. For a moderate field strength drift velocity is proportional to field strength $\boldsymbol{E}$:

$$\boldsymbol{v}(\boldsymbol{E}) = \mu\, \boldsymbol{E}(x)\,. \tag{32}$$

According to Table 8 for electrons and holes, mobility μ differs and increases with decreasing temperature. The difference between μ_e and μ_h causes a differing collection time of the electrons and holes. Total current pulses are a combination of the contributions of both types of carriers, which lead to

Table 8: Charge Carrier Mobility in Silicon and Germanium. According to *Goulding* [Go 66a]

	Temperature	Electron mobility μ_{e_0}, in cm²/Vs	Hole mobility μ_{h_0}, in cm²/Vs
Silicon	T °K	$2{\cdot}1\times 10^{9}\ T^{-2\cdot5}$	$2{\cdot}3\times 10^{9}\ T^{-2\cdot7}$
	77 °K 300 °K	$\approx 4\times 10^{4}$ 1350	$\approx 1{\cdot}8\times 10^{4}$ 480
Germanium	T °K	$4{\cdot}9\times 10^{7}\ T^{-1\cdot66}$	$1{\cdot}05\times 19^{9}\ T^{-2\cdot33}$
	77 °K 300 °K	$\approx 3{\cdot}6\times 10^{4}$ 3900	$\approx 4{\cdot}2\times 10^{4}$ 1900

various forms of pulse rise behaviour, depending on location of the radiation absorption in the sensitive layer. These pulse shape fluctuations are significant for the resolving time of semiconductor detectors.

Initial spatial distribution of electron-hole pairs depends on the type and the energy of absorbed radiation. *Stopping time* at a range of 0·5 mm amounts to about 3 ps for electrons and 26 ps for protons and α-particles [To 61].

During the primary ionisation process electrons are lifted from the low states in the valence band to the high states in the conduction band. In the secondary process the excess energy serves to produce further charge carrier pairs and to excite the phonon states of the crystal lattice. This requires a time amounting to 1 ps [To 61], which can be ignored with reference to collection time, the same as the stopping time.

In detecting heavy particles, such as protons, α-particles and fission fragments, only little energy reaches the carriers during the primary impact and their range is consequently short. This leads to formation of a cylinder of dense plasma with a carrier concentration of 10^{17} cm^{-3} in the case of 10 MeV protons and 10^{19} to 10^{20} cm^{-3} if fission fragments are involved; this is higher than impurity concentration by several orders of magnitude. Electric field strength is consequently considerably reduced and the high carrier density can leak only gradually into the surrounding electric field through carrier diffusion and "erosion" of the boundary. This results in delaying the carrier collection by the *plasma time* t_p, which, for instance, yielded a value of 10 ns with ^{252}Cf fission fragments by measurement [Mi 60]. For very low field strength *Meyer* and *Langmann* [Me 65] found the relation

$$t_p = \frac{\alpha}{E_0^2 + E^2}. \tag{33}$$

The factor $\alpha = 1{\cdot}9 \times 10^{-2}\,\mathrm{V^2 s/cm^2}$ was found experimentally and for $|E_0| \approx 300\,\mathrm{V/cm}$ was selected. In the case of low field strength plasma time dominates any of the other influences (see Fig. 38, p. 76), especially when particle incidence is at the rear of the p-n detector where field strength tends towards zero, as shown in Fig. 29*a*. For fission fragments plasma time is about 4 times longer than for α-particles [Me 66a]. An improved version of plasma time calculation was carried out by *Tove* and *Seibt* [To 67b].

COLLECTION TIME IN P-I-N DETECTORS

For the calculation of the pulse rise the p- and n-layer contributions can be neglected and the detector may be considered as a capacitor with a constant field strength [see equation (27a)]. Either the energy balance [Go 66a] or the *Ramo* theorem [Ca 63; Ba 66; Pi 66] is used

$$\frac{\mathrm{d}(\Delta Q)}{Q} = \frac{\mathrm{d}x}{d} \tag{34}$$

with d the thickness of the detector material and $\mathrm{d}(\Delta Q)$ the change of the influence charge if the total charge $Q = (E/\varepsilon')\, e$ produced in the detector is displaced by $\mathrm{d}x = v\,\mathrm{d}t$. Assuming that the entire radiation absorption occurs at position $x = x_0$ equations (31), (32) and (34) result in

$$\left.\begin{aligned}
\left(\frac{\Delta Q}{Q}\right)_e &= \frac{\mu_e |E|\, t}{d} && \text{for} \quad 0 < t < \frac{x_0}{\mu_e |E|}, \\
\left(\frac{\Delta Q}{Q}\right)_e &= \frac{x_0}{d} && \text{for} \quad t \geqq \frac{x_0}{\mu_e |E|}, \\
\left(\frac{\Delta Q}{Q}\right)_h &= \frac{\mu_h |E|\, t}{d} && \text{for} \quad 0 < t < \frac{d - x_0}{\mu_h |E|}, \\
\left(\frac{\Delta Q}{Q}\right)_h &= \frac{d - x_0}{d} && \text{for} \quad t \geqq \frac{d - x_0}{\mu_h |E|}.
\end{aligned}\right\} \tag{35}$$

For the case $\mu_e = 3\,\mu_h$ (Si at 300 °K) Fig. 31*a* shows the various contributions and the total pulse as

$$\frac{\Delta Q}{Q} = \left(\frac{\Delta Q}{Q}\right)_e + \left(\frac{\Delta Q}{Q}\right)_h. \tag{36}$$

Maximum collection time results when carrier production takes place on the border of the i-layer ($x_0 = 0$ or d). In Fig. 31*b* the total pulse is shown on condition of equal carrier mobility $\mu_e = \mu_h$, which is closely approximated with germanium at 77 °K, according to Table 8. Condition of point absorption is well met for γ-energies with E_γ located between 100 keV and 2 MeV because the range of the produced *Compton*- or photoelectrons can be neglected. On the other hand, γ-absorption is so low that any parameters x_0/w occur with the same probability. Thus, absorption relation differs entirely from those found in the photodiode (see Section 2.3.4). With an energy of $E_\gamma < 100$ keV γ-absorption, and at $E_\gamma > 2$ MeV the electron range must be taken into account.

Pulse rise data for γ-radiation incident parallel to the direction of $\boldsymbol{E}$ have been calculated by *Balland* et al. [Ba 66; Pi 66; Ba 67d; Ba 67e], with electrons having a finite range R. For R the length of the track projection, referred to $|\boldsymbol{E}|$, has to be taken in the case of *Compton* electrons. The curve for $R = d/2$ with parameter x_0/d is shown on Fig. 31*c*, which corresponds to $E_\gamma \approx 1{\cdot}2$ MeV with $d = 2$ mm. In the case of $(x_0 + R) < d$ electrons are stopped entirely in the sensitive layer, while $(x_0 + R) > d$ results in a charge loss because the produced carriers recombine rapidly in the adjacent n^+- and p^+-layers.

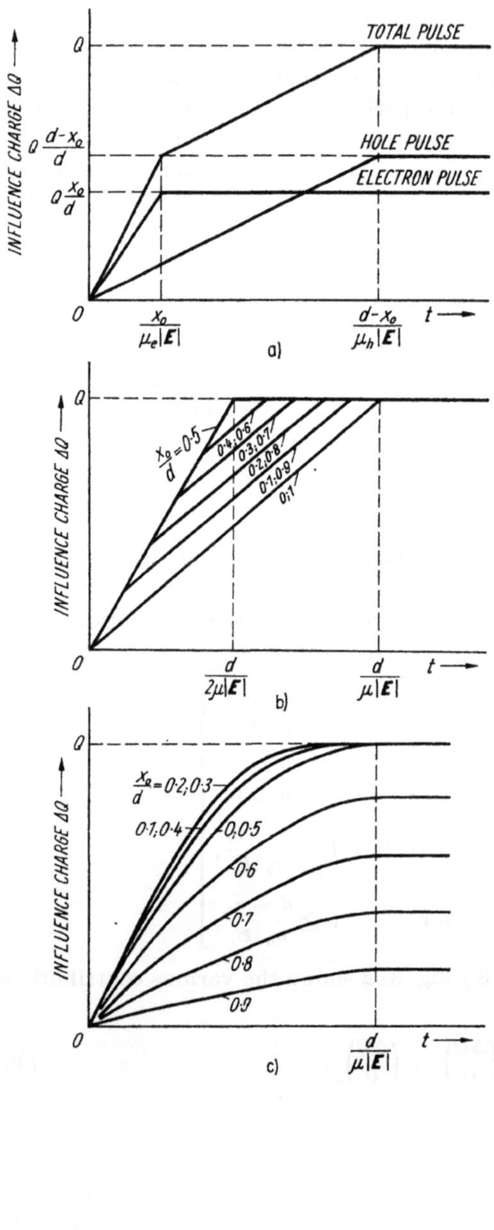

Fig. 31. Pulse rise with a p-i-n detector. According to *Balland* et al. [Ba 66; Pi 66].

a) Electron range $R = 0$, carrier mobility $\mu_e = 3\,\mu_h$, position of radiation absorption x_0. *b*) $R = 0$, $\mu_e = \mu_h$, parameter x_0/d. *c*) $R = d/2$, $\mu_e = \mu_h$, parameter x_0/d. With $x_0/d > 0{\cdot}5$ charge loss is caused by incomplete stopping in sensitive layer.

Poenaru [Po 67] calculated the current and charge pulse shapes in a coaxial Ge(Li) detector.

Particles having different ranges also result in varying pulse rise values which can be utilised for particle discrimination. Respective calculations have been undertaken by *Ammerlan* et al. [Am 63].

In all these calculations the constancy of carrier mobility was assumed. Above a velocity of 10^6 to 10^7 cm/s, however, this does not apply [Ry 53]. At 77 °K carrier mobility in germanium is so high that saturation velocity $v = 1{\cdot}5\times10^7$ cm/s is reached already above $|\boldsymbol{E}| = 1$ kV/cm, which then must be inserted in place of $\mu\,|\boldsymbol{E}|$ [Go 66a].

COLLECTION TIME IN P-N DETECTORS

Assuming that at $x = x_0$ the total charge Q is produced in the p-layer, the field strength behaviour shown in Fig. 29*a*, p. 63 [equation (26)] causes increasing velocity of electrons during transit according to equation (32), while hole velocity consistently decreases. On condition of constant carrier mobility — using equation (34) with w in place of d — the *Ramo* theorem† yields a charge pulse [To 64] of

$$\left.\begin{aligned}\left(\frac{\Delta Q}{Q}\right)_e &= \frac{w-x_0}{w}\left(e^{t/\tau_e}-1\right) && \text{for} \quad 0 < t < \tau_e \ln\frac{w}{w-x_0}\,,\\ \left(\frac{\Delta Q}{Q}\right)_e &= \frac{x_0}{w} && \text{for} \quad t \geqq \tau_e \ln\frac{w}{w-x_0}\,,\\ \left(\frac{\Delta Q}{Q}\right)_h &= \frac{w-x_0}{w}\left(1-e^{-t/\tau_h}\right).\end{aligned}\right\} \tag{37}$$

Here, the collection time constants are $\tau_{e,\,h} = w^2/2\,U\,\mu_{e,\,h}$. The hole collection time constant $\tau_h = \varrho\,\varepsilon\,\varepsilon_0$ is termed the dielectric relaxation time of the p-material. The respective contributions and the total pulse, according to equation (36), are given in Fig. 32.

When the detector consists of n-material, electrons and holes exchange function and the dielectric relaxation time of the n-material signifies the electron collection time constant τ_e. In every case, the rise time t_r depends on the specific resistance ϱ of the material:

$$t_r \approx 2{\cdot}2\,\tau = 2{\cdot}2\,\varrho\,\varepsilon\,\varepsilon_0 = 2{\cdot}2\,\varrho\,(\Omega\text{cm})\ \text{ps}\,. \tag{38}$$

In general, particle range R should not be ignored with reference to thickness of depletion layer w. Besides w, U, and type of conduction of the detector, the pulse rise depends also on direction, length and the rate of ionisation loss $\mathrm{d}E/\mathrm{d}x$ of the particle track. For a particle incidence parallel to the direction

† This may be derived also from the energy balance of a plate capacitor, however, with due consideration of the space charge, otherwise somewhat differing terms will result [To 61; Go 66a].

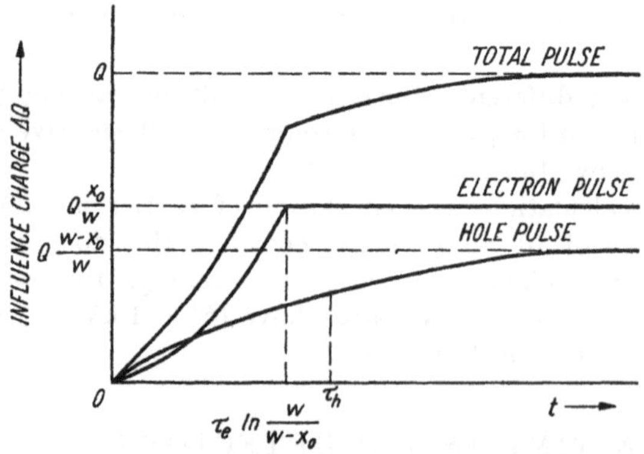

Fig. 32. Pulse rise with a p-n detector where absorption occurs in the p-layer.
Particle range $R = 0$, carrier mobility constant, position of radiation absorption x_0.

of the field and with dE/dx at constant level, the influences were taken into account by *Tove* and *Falk* [To 61; To 64], and the resulting data plotted in the nomogram shown in Fig. 33.

If particle range exceeds depletion layer thickness, diffusion of the charge carriers produced outside of the depletion layer results in an extended component of the pulse rise [Ta 67]. According to equation (25), adequate thickness of the depletion layer requires a high voltage, especially with a low ϱ. But near the junction the field strength rises to such an extent that charge carrier velocity exceeds thermal equilibrium. Thus, mobility of such "hot carriers" depends on field strength. Data about the character of this dependency found in literature are contradictory [Me 65; Fa 65]. For silicon *Meyer* and *Langmann* [Me 65] employ the relation

$$\mu = \frac{\mu_0}{\eta\,|\boldsymbol{E}| + e^{-\varkappa |\boldsymbol{E}|}} \tag{39}$$

with μ_0 denoting mobility at zero field strength and the parameter values $\eta_e = 1{\cdot}7 \times 10^{-4}$ cm/V, $\eta_h = 0{\cdot}9 \times 10^{-4}$ cm/V and $\varkappa = 0{\cdot}85 \times 10^{-4}$ cm/V obtained by experiment. Collection time calculated according to this expression agrees fairly well with experimental data (Fig. 38, p. 76). Mobility reduction with increasing field strength results in a considerable increase of the rise time in detectors having low specific resistance [Fa 65].

Collection time is shortest when a voltage of $U \gg U_d$ is used [Me 65], as with p-i-n detectors,

$$t_{\min} = \frac{0{\cdot}8\,\eta\,d}{\mu_0}\,. \tag{40}$$

Collection time is independent of ϱ and with n-silicon amounts to 10 ps per micrometre thickness of the layer.

Conclusions regarding range or direction of incident particles may also be drawn from the pulse rise in p-n detectors. Calculations given by *Alberigi-Quaranta* et al. [Al 65a] take the inconstancy of the carrier mobility into consideration. Further, position of particle absorption in the detector can also be determined from the rise time with satisfying accuracy [La 63; Lu 65].

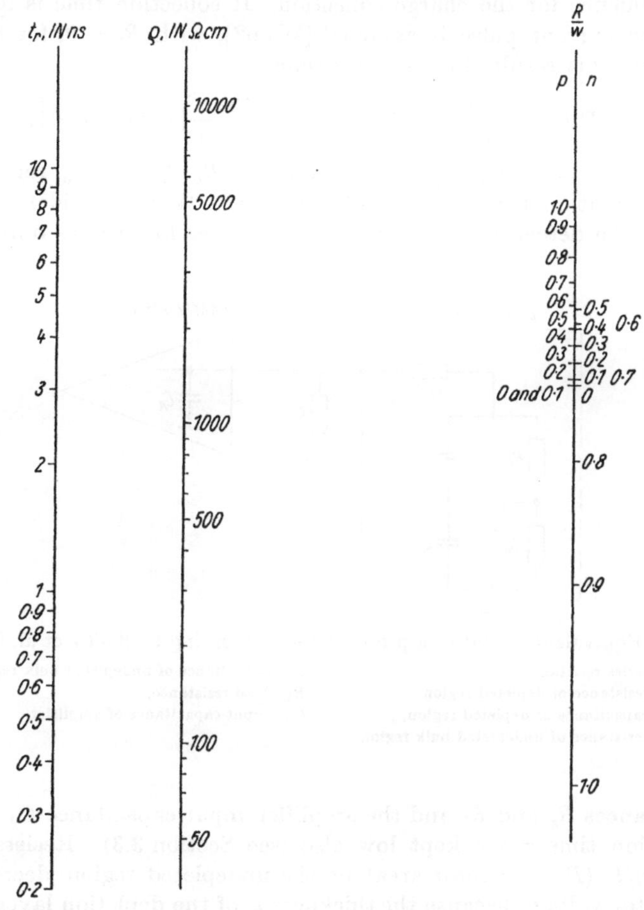

Fig. 33. Nomogram of rise time t_r with a p-n detector on finite particle range R. According to *Tove* and *Falk* [To 64].

EQUIVALENT CIRCUIT OF THE P-N DETECTOR

For the calculation of the shape of the voltage pulse from $Q(t)$ the equivalent circuit of the detector shown in Fig. 34 has to be considered [Wa 62]. Resistance R_B of the depletion layer and the load resistance R_L usually are very high and may be neglected. *Balland* et al. [Ba 66], however, take R_L into consideration. The series resistance R_s can be kept low, especially with surface barrier detectors ($R_s \approx 0$). Its influence has been given attention by *Walter* et al. [Wa 62]. Although some factors are disregarded, the obtained expression for the pulse shape still is relatively complicated [Wa 62; Si 63a; Ba 66]. For calculation it is usually assumed that, referred to thickness of the depletion layer, particle range is small, so that one time constant remains only as of primary influence for the charge collection. If collection time is ignored, i.e. if a δ-shaped current pulse is assumed [Wa 62], with $R_s = 0$, for the voltage characteristic this results in the expression

$$U(t) = \frac{Q}{C_A + C_B}\left[1 - \left(1 - \frac{C_b}{C_b + C_s}\right) e^{-t/(\tau + R_b C_s)}\right]. \tag{41}$$

In this case $C_s = C_A C_B/(C_A + C_B)$ and $\tau = R_b C_b = \varrho \varepsilon \varepsilon_0$ represents the dielectric relaxation time of the bulk material. When considering the time constant of the detector circuit, rise time assumes low values only when the

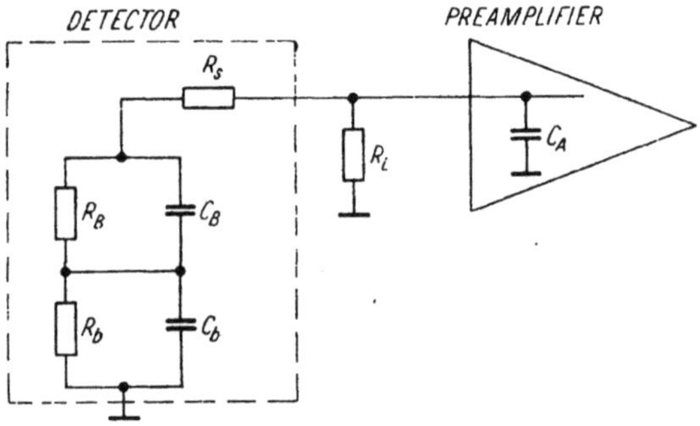

Fig. 34. Equivalent circuit of a p-n detector. According to *Walter* et al. [Wa 62].

R_s series resistance,
R_B resistance of depleted region,
C_B capacitance of depleted region,
R_b resistance of undepleted bulk region,
C_b capacitance of undepleted bulk region,
R_L load resistance,
C_A input capacitance of amplifier.

series resistances R_s and R_b and the amplifier input capacitance C_A as well as the relaxation time τ are kept low also (see Section 3.3). Resistance $R_b = \varrho (d - w)/F$ (F = detector area) of the undepleted region decreases with rising detector voltage because the thickness w of the depletion layer increases. For this reason driving the detector in fully depleted or overdepleted state is preferable.

Poenaru [Po 67a] used the *Laplace* transformation to calculate the voltage and the current waveforms for various load impedances. This involved attaching different preamplifiers to the detector output, viz. voltage sensitive — charge sensitive — current sensitive, as well as connecting two preamplifiers at the same time, with current and charge sensitive character, respectively.

EXPERIMENTAL RESULTS

p-i-n Detectors. As equation (37) indicates, short collection time is obtained only with thin detectors at high field strength [Ku 66]. Some experimental results on silicon detectors are illustrated in Fig. 35 [Wa 64], which agree well with calculated data. The considerable increase of carrier mobility at low temperatures results in a decrease of rise time. In coincidence measurements *Sakai* [Sa 63] observed the reduction of instrumental half-life (see definition in Section 2.3.2) from 30 ns at room temperature to 7·5 ns at −193 °C.

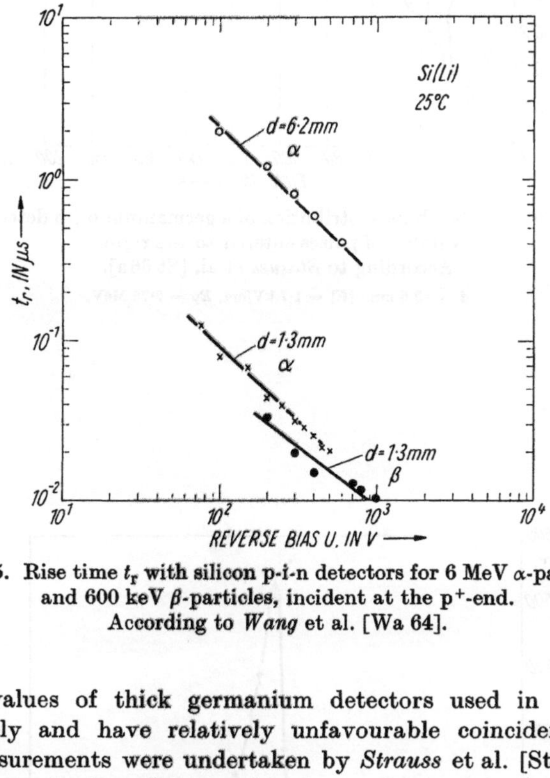

Fig. 35. Rise time t_r with silicon p-i-n detectors for 6 MeV α-particles and 600 keV β-particles, incident at the p^+-end. According to *Wang* et al. [Wa 64].

Rise time values of thick germanium detectors used in γ-spectroscopy fluctuate widely and have relatively unfavourable coincidence resolution. Extensive measurements were undertaken by *Strauss* et al. [St 66a]. For the purpose, the output pulses of the semiconductor detector is clipped by a short-circuited cable causing the height of the shaped pulse to depend on the steepness of the rise. Rise time fluctuations result in a pulse height spectrum at a specific clipping time as shown in Fig. 36*a*. Changing the clipping time scans the entire pulse shape and an image is obtained which corresponds to Fig. 31, p. 68. Abundance distribution of the 50% points of the pulse rise, important for zero-cross timing (see Section 2.3.2), is plotted in Fig. 36*b*. Because of the wide fluctuations of the rise time the first part of the leading edge should preferably be used for timing (see Section 3.4.2). This means, the threshold

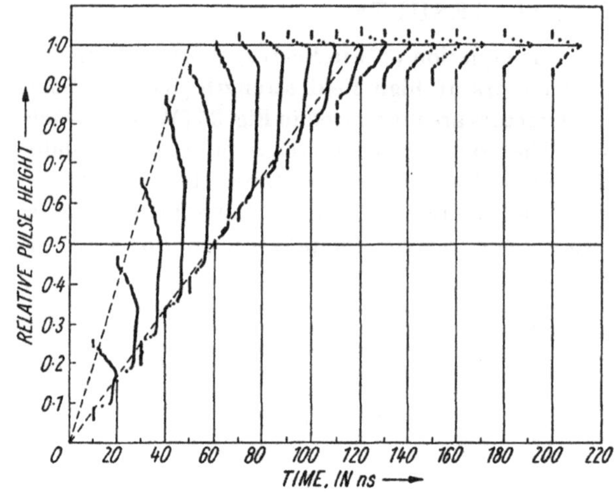

Fig. 36a. Pulse shape distribution of a germanium p-i-n detector. Number of pulses entered to the right. According to *Strauss* et al. [St 66a].
$d = 10{\cdot}6$ mm, $|\boldsymbol{E}| = 1{\cdot}7$ kV/cm, $E_\gamma = 2{\cdot}75$ MeV.

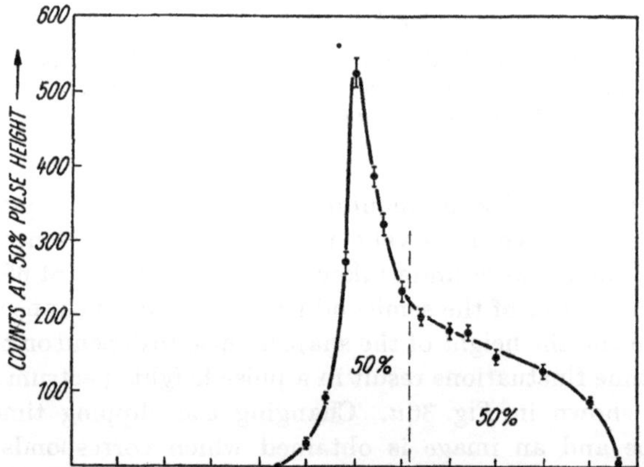

Fig. 36b. Abundance distribution of intersecting points with zero-cross timing (at 50% level). According to *Strauss* et al. [St 66a].

of the discriminator has to be adjusted as low as the noise level permits. The frequently employed zero-cross timing is unsuitable in this case (see Fig. 127, p. 202). A very favourable improvement of coincidence resolution is obtained if the level of the utilised limiter is reduced from 50% to 10% and 4% of the pulse height, of which Fig. 37 gives an impressive illustration [Gr 66a].

With time constants of the amplifier adjusted to optimum value with reference to minimum noise influence, the rise time fluctuation can contribute to the fluctuation of pulse height in such a measure that it assumes the same magnitude as the statistical contribution [St 66a; Do 66]. In coincidence work there is always a certain difficulty in picking off a fast signal at the detector without influencing the energy resolution. In Section 3.4.2 various methods of this timing pick-off are discussed.

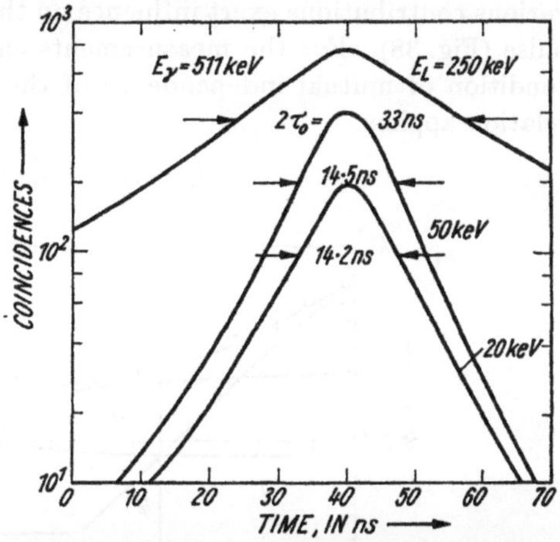

Fig. 37. Coincidence curves obtained with two coaxial germanium p-i-n detectors with differing limiting height E_L. $E_\gamma = 511$ keV (^{22}Na). According to *Graham* et al. [Gr 66a].

Coaxial detectors are especially suitable for coincidence measurements because of their high efficiency, although rise time is subject to considerable fluctuations due to the inhomogeneous field near the closed end [Gr 66a]. Coaxial detectors that are open on both ends give better results [Ma 66b]. Resolution curves obtained with a scintillation counter and a Ge(Li)-detector are shown in Fig. 127. At an energy level of $E_\gamma = 500$ keV and a limitation at 13% a resolving time of $2\tau_0 = 9$ ns is obtained. Even commercially available detectors yield $2\tau_0 = 5{\cdot}9$ ns†. Planar detectors give the lowest resolving time values. *Ewan* et al. [Ew 66], for instance, got $2\tau_0 = 3{\cdot}9$ ns with a plastic scintillator and a germanium p-i-n detector of 5 mm thickness at $E_\gamma = 511$ keV, and *Balland* et al. [Ba 66; Pi 66] obtained $2\tau_0 = 2{\cdot}8$ ns with a 2 mm detector. With ^{60}Co-radiation and selected maximum energy $2\tau_0 = 1{\cdot}1$ ns could be measured [Ba 66; Pi 66].

p-n Detectors. These detectors show markedly lower rise time values and consequently yield a better time resolution. A compromise must be made, however, between the thickness of the depletion layer required for total particle stop and the rise time. Because of the short collection time the equivalent circuit of the detectors is of special significance and must receive consideration. Comparisons of experimental and calculated values have confirmed the validity

† ORTEC information, e.g. in Rev. Sci. Instr. **37** (1966), No. 10.

of the assumed equivalent circuit [Si 63a; Fa 65a; Ba 66]. With surface barrier detectors made of silicon having a resistivity of $\varrho = 60\ \Omega$cm a rise time of $t_r \approx 2$ ns† could be measured [Si 63a]. The extensive measurements of *Meyer* and *Langmann* [Me 65; Me 66a] show specifically the manner in which the various contributions exert influence on the resulting rise time t_r of the detector pulse (Fig. 38). For the measurements short-range α-particles were used. On condition of mutual independence of the various contributions, the following relation applies:

$$t_r^2 = t_{coll}^2 + t_p^2 + t_{RC}^2 + t_A^2 \tag{42}$$

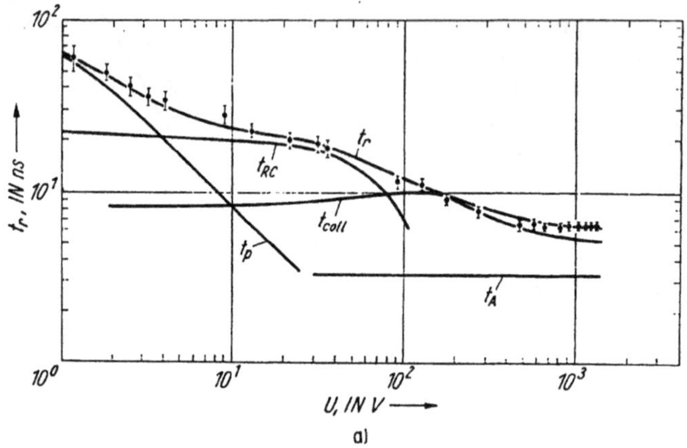

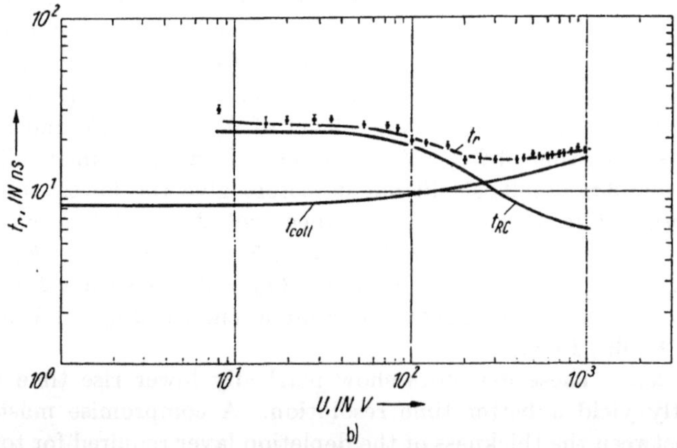

Fig. 38. Comparison of experimental and theoretical values of the rise time in p-n detectors of n-silicon for α-particles of short range. According to *Meyer* and *Langmann* [Me 65].
a) $\varrho = 7$ kΩcm, $d = 0{\cdot}48$ mm. *b)* $\varrho = 8{\cdot}3$ kΩcm, $d = 1{\cdot}5$ mm.

† And about 10 ns with ϱ in the order of 10 kΩ cm [To 67a].

with t_{coll} the collection time where the inconstancy of carrier mobility has received consideration [according to equation (39)]; t_p the plasma time [according to equation (33)] and t_A the rise time of the measuring apparatus. Neglecting C_b also, calculation of t_{RC} from equation (41) results in

$$t_{RC} = 2{\cdot}2\, R_b \frac{C_A\, C_B}{C_A + C_B}\,. \tag{43}$$

The good agreement that has been attained is an indication that all essential contributions received consideration. From Fig. 38 and equation (43) it is evident that the resistance of the bulk in the undepleted region exerts a decisive influence on the rise time with not fully depleted detectors. With very low field strength, however, plasma time predominates. It will be seen from Fig. 38*b* that with high voltage the decrease of carrier mobility can effect an increase of collection time and thus total rise time.

In coincidence measurements with a plastic scintillation counter and a surface barrier detector a resolving time of $2\,\tau_0 \approx 0{\cdot}5$ ns and an instrumental half-life of 70 ps could be obtained [Ne 65; Fo 66]. With two surface barrier detectors and α-radiation of 10 MeV *Wahl* [Wa 64a] received resolving time values down to $2\,\tau_0 \approx 0{\cdot}2$ ns. Coincidence curves measured with fission fragments [Me 66a] are given in Fig. 126, p. 201. Timing with a low threshold yielded a minimum resolving time of $2\,\tau_0 = 90$ ps. In this resolving time region some otherwise secondary influences gain in importance. To the width of the coincidence curve the angular-dependent time-of-flight of fission fragments contributes 50 ps and the fluctuation of the pulse height across the selected window contributes 17 ps.

DETECTORS WITH CURRENT AMPLIFICATION

In a similar manner as with photodiodes the avalanche mechanism is also utilised in semiconductor detectors to obtain internal amplification and a higher output pulse [Mi 62; Hu 65; Hu 66; Ma 66a]. Rise time of the detector pulse is of the order of one nanosecond. Gains up to 10^3 have already been obtained with short-range particles, however, the factor is strongly influenced by the range of the particles.

2.4.2 SPARK COUNTERS

When ionising radiation passes through a gas electrons are liberated. If the electric field, which extracts the electrons, transfers sufficient energy to the primary electrons to cause impact ionisation the collected charge exceeds primary ionisation. With increasing field strength this gas amplification rises rapidly and above a gas-filling-dependent minimum value leads to a discharge. This either occurs along the anode wire of the cylindrical counter device (GM counter, BF_3 counting tube) or as spark breakdown between anode and cathode in a spark counter. Pulse height at the detector output is independent of primary ionisation and therefore of radiation energy. In GM counting tubes the primary electrons first have to reach the region of maximum field strength in the immediate vicinity of the anode wire, which leads, however, to statistically fluctuating time delay between primary ionisation and discharge, ranging between 0·1 and 0·5 μs [Pi 49]. With a homogeneous field between plane-parallel electrodes better time resolution can be expected because every electron produced in the primary process is located immediately in range of high gas amplification [Ma 48].

Properties of parallel plate spark counters were first described by *Keuffel* [Ke 48; Ke 49], *Madansky* and *Pidd* [Ma 48; Pi 49; Ma 50]. One of these first spark counters with parallel electrodes is shown in Fig. 39.

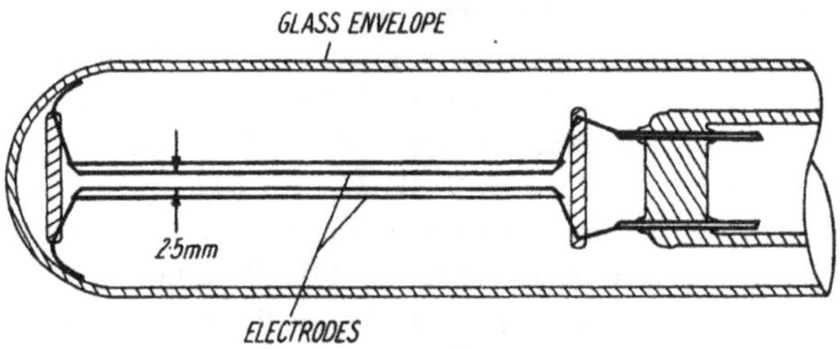

Fig. 39. Cross-section of a parallel-plate spark counter. According to *Keuffel* [Ke 49].

Spark breakdown can be caused by a single primary electron. Through collision ionisation an electron avalanche builds up until the internal field causes further ionisation advance to deteriorate after the critical quantity of about 10^8 electrons has been attained. The recombination of ions in the avalanche causes light quanta to be emitted, which release further avalanche events, preferably in the amplified field at the head of the primary avalanche, in direction of the anode. In this manner a discharge streamer is formed at a velocity which exceeds that of corresponding electron mobility. When the streamer reaches the anode practically the total difference of potential is applied between the primary avalanche and the cathode because of the good conductance of the streamer. This causes the formation of another discharge streamer at the cathode which leads to complete breakdown. With parallel plate avalanche counters discussed in the next Section field strength remains at such a low level that the avalanche cannot attain the critical level needed for a breakdown.

The higher the working voltage of the counting tube, with reference to static breakdown voltage, the shorter will be the built-up time of a spark discharge. Overvoltage level cannot be arbitrarily increased because spontaneous discharge probability then becomes too high. With smaller electrode areas the use of a higher overvoltage level is possible. As Table 9 illustrates, narrower electrode spacing and a higher gas pressure also contribute to the reduction

Table 9: Full Width at Half-Maximum $2\tau_0$ of Coincidence Curves Obtained with Two Parallel-Plate Spark Counters

Electrodes		Gas Filling	Breakdown Voltage	Overvoltage	$2\tau_0$	Authors
Spacing in mm	Size in cm²		in kV	in kV	in ns	
2·5	35	0·5 at Ar +6 Torr Xylen		0·25 to 0·9	5 to 17†	*Keuffel* [Ke 49]
			4	0·3 0·9	18 6	*Madansky* and *Pidd* [Ma 50]
2	50	Ar + alcohol, 60 Torr	2·2	0·5	5	*Robinson* [Ro 53a]
0·2	0·75	Ar + ether, 3 to 13 at			(0·4)	*Babykin* et al. [Ba 56]
0·2		Ar + ether, 13 at	3·6††		≈0·1	*Zavoiskii* and *Smolkin* [Za 56a]
0·28		3·7 at Ar, + 600 Torr O_2		0·8 to 2·8	0·4 to 1·35	*Artemenkov* and *Babykin* [Ar 58]
0·15		3 at Ar + 400 Torr O_2	1·4	2	0·17	*Artemenkov* and *Babykin* [Ar 60a]
0·1		20 at He + 0·5 at O_2	4††		≈0·01	*Skachkov* [Sk 61]

† Time interval where 50% of total coincidences occur.
†† Working voltage.

of time delay and consequently resolving time. The table contains the half-width $2\,\tau_0$ of the coincidence curves obtained with two equivalent parallel-plate spark counters. Oxygen addition evidently influences the time resolution favourably [Sk 61]. The full width at half-maximum of about 0·01 ns obtained by *Skachkov* seems to be the best resolving time attained with two radiation detectors so far (Fig. 40 [Sk 61]). Although with parallel-plate spark counters

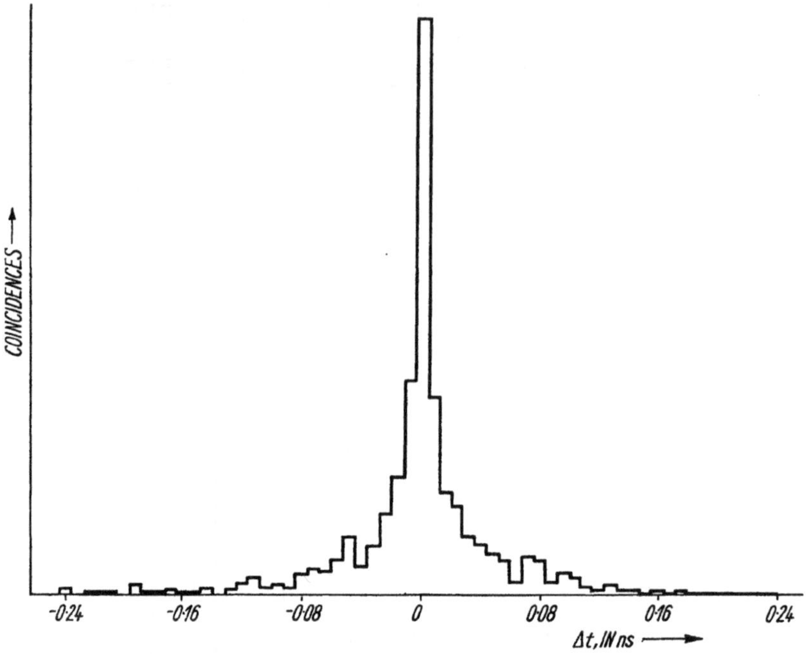

Fig. 40. Coincidence curve obtained with two parallel-plate spark counters. According to *Skachkov* [Sk 61].

a phantastic time resolution can be obtained, for short-time measuring work they have not gained much headway, especially because of their long dead time which amounts to about 1 to 10 ms. A spark counter can give a particular service because it permits the localisation of ionising particle tracks with a high degree of accuracy and makes it visible, as *Keuffel* has pointed out [Ke 49]. Resulting development produced a variety of types of spark chambers which are utilised for the track measurement of high-energy particles and have thus gained an important place in elementary particle physics (see Section 5.3).

2.4.3 PARALLEL-PLATE AVALANCHE COUNTERS

This avalanche counter was used by *Christiansen* [Ch 52] for nanosecond measurements and corresponds in design to the parallel-plate spark counter, but is driven with a voltage level below spark breakdown. Construction is shown schematically in Fig. 41 [Bl 63]. Similar systems are described in a number of papers [He 61; Dr 64; Kr 65; Kr 65a; Kr 67a]. The electrode system consists of an anode block and a cathode of plastic foil on which a film of gold has been evaporated. On this, the radioactive source is deposited. Electrode spacing is 2 mm, but spacings of 0·5 mm [Kr 65a], 1 mm [Kr 65; Kr 65a] and 1·5 mm [Bl 61] have been used also. With a filling of acetone vapour at a pressure of 50 to 100 Torr electron response

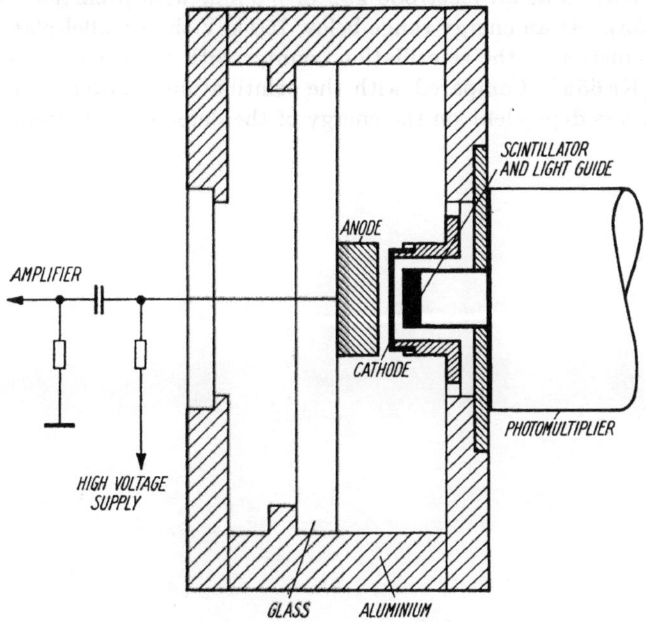

Fig. 41. Schematic drawing of a parallel-plate avalanche counter. According to *Bloess* and *Münnich* [Bl 63].

probability comes to nearly 100%. For γ-quanta, however, this is only 0·1% [Bl 61; Kr 65a]. The various designs operate at a voltage ranging from 1·5 to 2 kV, which is usually only some ten volts below the static breakdown level. In this manner maximum gas amplification is obtained without the occurrence of a spark breakdown which would extend the dead time of the counter by more than 10^4. Mean height of the released avalanche depends on the position of formation and reaches maximum value when it starts near the cathode. Therefore, a continuously decreasing pulse height spectrum is obtained at the output which has an average pulse height of 10 mV [Kr 65a]. There is, however, no proportionality to the energy of the incident electrons. This absence of energy resolution represents a serious drawback for many kinds of test work.

Time fluctuations of the output pulse are caused by the differing paths of the avalanche and by a time jitter during the initial stages of avalanche formation. Both contributions are nearly proportional to the distance between plates

[Kr 65a]. Because maximum transit time of an avalanche amounts to about 5 ns a time resolution can be expected in the nanosecond range. For reducing apparatus influence, in consequence of the different pulse height, only a small range is usually taken from the output pulse spectrum. In coincidence with a scintillation counter the following half-width values were obtained: $2\,\tau_0 \approx 5$ ns at an electron energy of 2·9 keV [Bl 61], $2\,\tau_0 = 1{\cdot}85$ ns at about 8 keV [Bl 63] and $2\,\tau_0 = 0{\cdot}37$ ns at an electrode gap of 0·5 mm with β-particles of 300 keV max. [Kr 65a]. At an energy range below 150 keV the parallel-plate avalanche counter is superior to the scintillation counter with respect to attainable time resolution [Kr 65a]. Compared with the scintillation counter, time resolution is markedly less dependent on the energy of the detected radiation.

3. CIRCUITRY

3.1 COMPONENTS

3.1.1 RESISTANCE-CAPACITANCE-INDUCTANCE NETWORKS

RESISTORS

In the high-frequency range the parallel capacitance and the series inductance of a resistor represent a source of disturbance. The equivalent circuit depicted in Fig. 42 [Me 62; Me 65a] applies for a wavelength that is high against the length l of the resistor unit. From this the effective resistance R' and the reactance X' may be approximated with

$$\left.\begin{aligned} R' &\approx R\left(1 + \omega^2 C (2L - R^2 C)\right), \\ X' &\approx \omega (L - R^2 C), \end{aligned}\right\} \quad (44)$$

with R the d.c. value of the resistor and $\omega = 2\pi f$ the angular frequency.

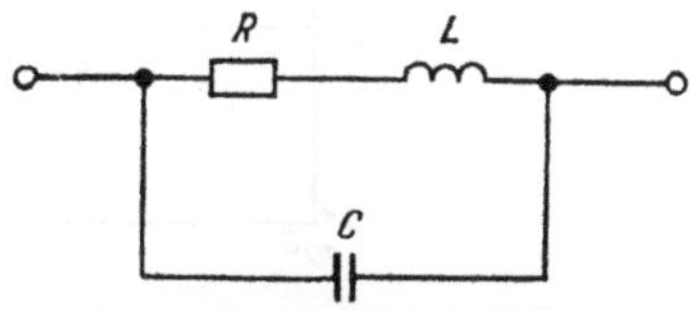

Fig. 42. Equivalent circuit of a resistor for high frequencies.

Wire-wound resistors are characterised by very high inductance and capacitance values and are therefore not suitable for use with high frequencies. In addition the skin effect (see Section 3.1.2) makes itself noticeable with thin wire. Bifilar-wound resistors have a low inductance, and for this reason, for example, shunts of bifilar constantan strip having a resistance of some hundredth ohms and an inductance in the nanohenry range, are used for the analysis of high-current discharge phenomena in the kA-region [Kü 59; An 61].

For conventional application film resistors with a high specific resistance and a thin film of carbon or metal are employed, where interference from the skin-effect is rare. For balancing the resistance value a spiral is, in most cases, ground-in which, however, increases the inductance; but this can be improved by using a meander configuration. The so-called vhf film-type resistors have a smooth cylindrical surface and very low inductance. Parallel capacitance, which ranges from 0·3 to 0·5 pF with a 0·5 W resistor in axial connection, remains as a major source of disturbance. With high frequencies, this capacitance results in the decrease of the effective resistance R', especially when R has a high value (Fig. 43). In a low-ohmic circuit this consequently has little influence.

Particularly for the reflexion-free termination of a transmission line with a characteristic impedance Z (see Section 3.1.2) a resistor is required having a reactance $X' = 0$. This is valid if $L = R^2 C$. Such termination resistors are inserted in a coaxial shielding system, where the maximum permissible frequency then depends on the dimensions of the system. If shielding has the characteristic impedance of $Z_s = Z/\sqrt{3}$ the reactive component for wavelengths above $\lambda = 15\,l$ can be neglected. This corresponds to a maximum frequency of about 2 GHz [Ab 61; Me 62]. A shielding system with $Z_s = Z/\sqrt{5}$ and an

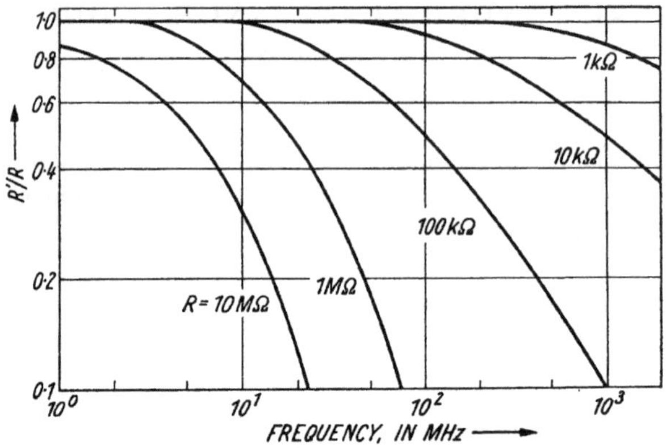

Fig. 43. Frequency dependence of the effective resistance R' due to parallel capacitance with 0·5 W resistors, having a d.c. resistance R. According to *Meinke* and *Gundlach* [Me 62].

additional inductivity in series, which is represented by the enlarged diameter D of part of the shielding, according to equation (52), is suitable for a range down to $\lambda = 6\,l$ (see Fig. 44*a*). An exponentially tapered coaxial line end with inbuilt terminating resistance, dimensioned as shown in Fig. 44*b*, yields a particularly satisfactory frequency behaviour in a range down to $\lambda = l$, i.e. frequencies to about 10 GHz.

CAPACITORS

The conducting films of a capacitor together with the connecting leads represent an inductance which must be taken into account in high frequency work. Losses additionally occurring in the dielectric are taken care of in form of a series resistance in the equivalent circuit (Fig. 45), which also contains the resistance of the capacitor films and of the terminals. On high frequencies, the contribution of the insulation resistance parallel to the capacitance can be neglected against the influence of the above contributions [Me 62]. Reactance is described with

$$j\,X = -j\,\frac{1}{\omega\,C'}\,, \tag{45}$$

where $C' = C/(1 - \omega^2\,L\,C)$ is the effective capacitance of the capacitor, which is higher against the rated value of C. Relation between capacitance, inductance and the frequency, with capacitance deviation amounting to 1%, is given in

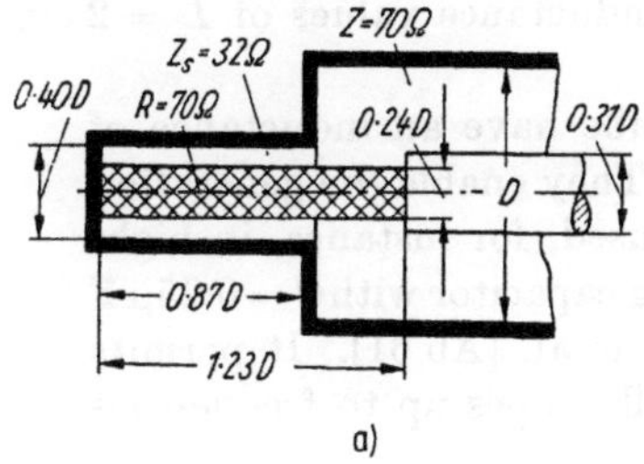

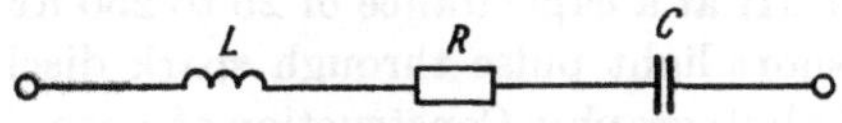

Fig. 45. Equivalent circuit of a capacitor for high frequencies.

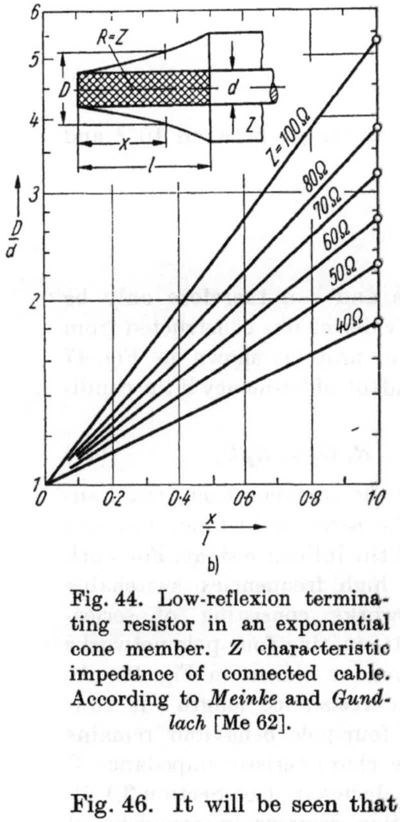

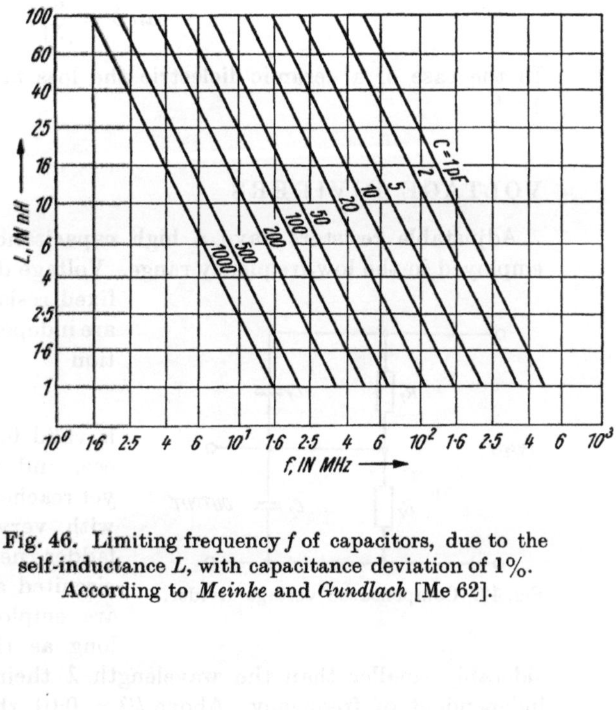

Fig. 44. Low-reflexion terminating resistor in an exponential cone member. Z characteristic impedance of connected cable. According to *Meinke* and *Gundlach* [Me 62].

Fig. 46. Limiting frequency f of capacitors, due to the self-inductance L, with capacitance deviation of 1%. According to *Meinke* and *Gundlach* [Me 62].

Fig. 46. It will be seen that for the frequency components relating to nanosecond pulses the self-inductance must be situated in the nanohenry region. Wound paper capacitors have a relatively high inductance, which can be lowered to a certain extent, however, by incorporating several leads at the side. Wound paper capacitors and block capacitors (mica insulated) have values of $L = 10$ to 40 nH. Ceramic capacitors are constructed of only one element and can be kept very compact because of the high permittivity of the material. Arranging the connections at one end brings an improvement.

Tube-type and disk-type ceramic capacitors have inductance values of $L = 2$ to 10 nH.

Wound capacitors in shape of a short coaxial tube have an inductance of only 1 nH at a capacitance of 25 to 250 nF [Pr 67]. They enable the generation of a short light pulse through spark discharge, as used, for instance, in high-speed photography. Construction of a coaxial coupling capacitor with $C = 0{\cdot}25\,\mu F$ and $Z = 100\,\Omega$ has been described by *Abercrombie* et al. [Ab 61]. It permits d.c. voltage isolation to 300 V with high-speed oscilloscopes up to frequencies of 3 GHz.

When good contacting and low-resistance films are provided, dielectric loss is of primary influence for the loss factor

$$\tan \delta = \frac{R}{|X|}. \tag{46}$$

In the case of a ceramic dielectric the loss factor ranges between 10^{-3} and 10^{-4}.

VOLTAGE DIVIDERS

Adjustable resistors have a high capacitance and can therefore only be employed in the low-frequency range. Voltage divider chains constructed from fixed resistor units as shown in Fig. 47 are independent of frequency if the condition

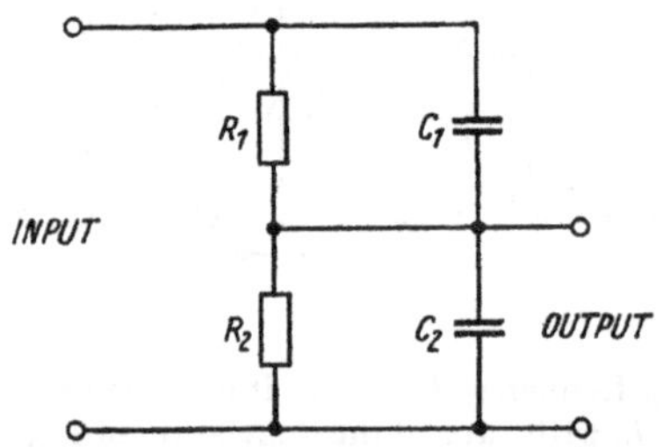

Fig. 47. Compensated voltage divider.

$$R_1\,C_1 = R_2\,C_2 \tag{47}$$

is valid for the compensation capacitances, and the series inductance has not yet reached the influence stage. For work with very high frequencies switchable ladder networks consisting of series-circuited attenuating four-pole networks are employed as shown in Fig. 48. As long as the resistance length l is considerably smaller than the wavelength λ their four-pole behaviour remains independent of frequency. Above $l/\lambda = 0{\cdot}01$ the characteristic impedance Z becomes complex and the attenuation factor α decreases (see Section 3.1.2). If the resistors are arranged coaxially attenuation remains independent of frequency up to $l/\lambda = 0{\cdot}4$. For the purpose, the conductive surroundings of the branch-arm resistors in the T-network (Fig. 48*b*) must have the characteristic impedance $Z_W \approx 0{\cdot}5\,W$, while in the Π-network (Fig. 48*a*) this must amount to $Z_W \approx W$. Optimum characteristic impedance Z_R of the conductive surrounding of the series-arm resistor is dependent on attenuation and may be taken from Fig. 48*c*. Very high attenuation factors can be obtained with asymmetric Π-networks [St 62]. With an attenuation factor of $n = U_i/U_0 = 100$ no rise time deterioration ($t_r = 1{\cdot}5$ ns) of a step pulse could be observed, although the divider resistors were not coaxially arranged.

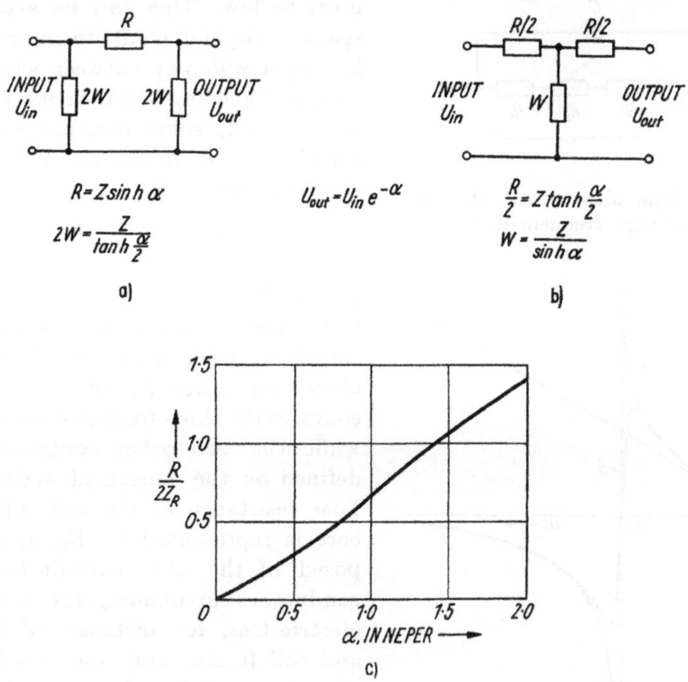

Fig. 48. Symmetric four-pole attenuators. According to *Meinke* and *Gundlach* [Me 62].
Z characteristic impedance of the four-pole, α four-pole attenuation.
a) Π-element. *b)* T-element. *c)* Characteristic impedance Z_R of the conducting surroundings of the longitudinal resistor of a coaxially mounted Π- or T-element with reference to d.c. attenuation α.

COILS

The equivalent circuit of a coil which approaches high frequency validity fairly closely is given in Fig. 49*a*. Neglecting the loss resistance R the inductance together with the self-capacitance represents a resonant circuit having the resonant frequency

$$\omega_r = \frac{1}{\sqrt{L\,C}}. \tag{48}$$

The reactance curve is given in Fig. 49*b*, which shows that the coil has inductive behaviour below resonance frequency.

$$j\,X = j\,\omega\,L' \tag{49}$$

applies for the reactance, with the effective inductance $L' = L/(1 - \omega^2 L C)$ increased against L on low frequency. Above resonance frequency the coil behaviour becomes capacitive. To permit operation as far below resonance frequency as possible, self-capacitance must be low. This can be attained by spacing the adjacent turns far apart, leaving a wide gap between winding and core, and keeping components, surrounding the coil, some distance away, e.g. shielding. The figure of merit Q of a coil is defined with

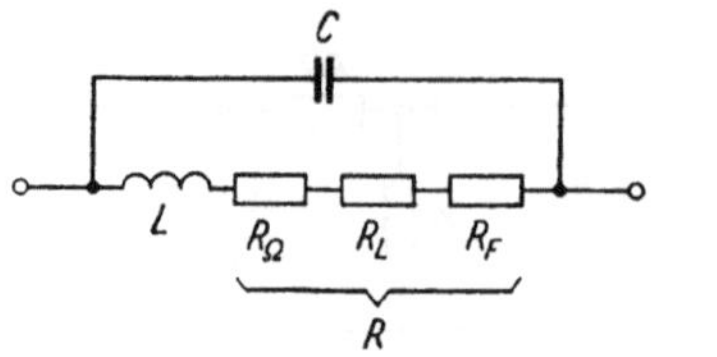

Fig. 49*a*. Equivalent circuit of a coil for high frequencies.

$$Q = \frac{1}{\tan \delta} = \frac{\omega L}{R}, \tag{50}$$

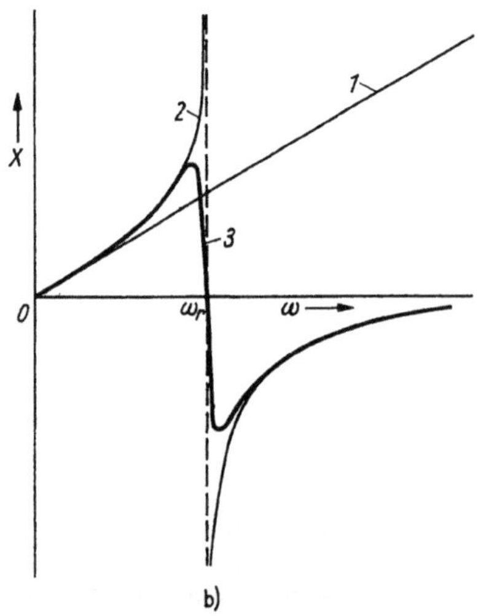

Fig. 49*b*. Frequency dependence of the reactance of a no-loss coil without (*1*) and with (*2*) consideration of the self-capacitance, and of a lossy coil (*3*). According to *Meinke* and *Gundlach* [Me 62].

where R is the loss resistance of the coil, which is made up of a number of contributions as shown in Fig. 49*a*. The ohmic resistance R_Ω of the conductors contains the skin effect as main influence, while the two other contributions are defined by the consumed active power. Loss resistance of the coil without the core is represented by R_L, and is composed of the eddy current loss in the conductors surrounding the coil, the dielectric loss, for instance, of insulation and coil frame, and, on very high frequencies, radiation loss must be added. Loss in the ferrite core of the coil is defined by R_F and consists of hysteresis loss, eddy current loss and persistence loss. These increase with frequency. Although the figure of merit of a coil can be improved by using a high-permeability core to obtain a higher inductance L with the same dimensions, the loss factor also increases, especially on high frequencies. Very high frequencies require very low inductance, obtainable by using short and thick conductors for the coil where frequently the ferrite core is omitted to offset the disadvantage. Thus, very high figures of merit can be simply obtained, for instance, $Q = 2{,}000$ at 100 MHz and $Q = 5{,}000$ at 1 GHz [Me 65a].

For a plain toroid the inductance L can be expressed by

$$L = 2\pi D\left(\ln\frac{8D}{d} - 2\right) \tag{51}$$

with L in nH, d and D in cm, and where D represents the diameter of the ring and d the diameter of the wire.

Straight conductors yield still lower inductance, although they are very sensitive to the influence of the surroundings. Adjacent conductors reduce inductance, and wide spacing is therefore an important consideration for coil shielding. On the other hand, a reduced inductance is desirable when incorporating a resistance in a coaxial shielding structure, for instance, with a four-pole attenuation network. In some cases a very low inductance of nanohenry range is obtainable by utilising a short-circuited section of coaxial cable (see Section 3.2.2). This must have a length l which is shorter than the shortest occurring wavelength. An approximation of this inductance can be written

$$L = 2\,l \ln \frac{D}{d} \tag{52}$$

with L in nH and l in cm, and where D represents the outer diameter and d the inner diameter of the coaxial cable.

TRANSFORMERS

Pulse transformers can advantageously replace a circuit used for the following functions:

a) transformation of current, voltage or impedance;

b) reversal of pulse polarity;

c) isolation and pulse coupling of circuits with different potential or where a high d.c. current passes through one of them.

Fundamental properties are described, for example, by *Millman* and *Taub* [Mi 56]. With an ideal transformer the ratio n for the current i or the voltage u is

$$n = \frac{u_s}{u_p} = \frac{i_p}{i_s} = \sqrt{\frac{L_s}{L_p}} = \frac{N_s}{N_p}\,. \tag{53}$$

The subscripts p and s define values of the primary and secondary ends of the transformer, respectively. L denotes inductance and N the number of turns. Thus, a load resistance R_s connected to the secondary subjects a generator to a load in the same manner as a resistance R_s/n^2. With a practical transformer the influences of the coupling factor $K = M/\sqrt{L_p L_s} \neq 1$ (M = mutual conductance) and of the stray capacitance, especially with high frequencies, have to be taken into account, although the factor K deviates at most 1% from unity in a good pulse transformer. The equivalent circuit of Fig. 50*a* consists of the ideal transformer having a ratio n, the effective primary inductance $L = K^2 L_p$ in parallel to it, and the leakage inductance $L_\sigma = (1 - K^2) L_p$ located in series. With short-circuited secondary the effective impedance of the transformer is determined by L_σ and with the secondary on open circuit by L. The series resistance R_a includes the output resistance of the generator and the line resistance of the primary winding. The shunt-resistance $R_b = R_c + R_s/n^2$ represents the loss resistance R_c describing the loss in the magnet core, and the resistance R_s of the secondary end transferred to the primary. Apart from the load resistance, the line resistance of the secondary winding is also contained in R_s. The effective total capacitance $C = C' + n^2 C_s$ includes

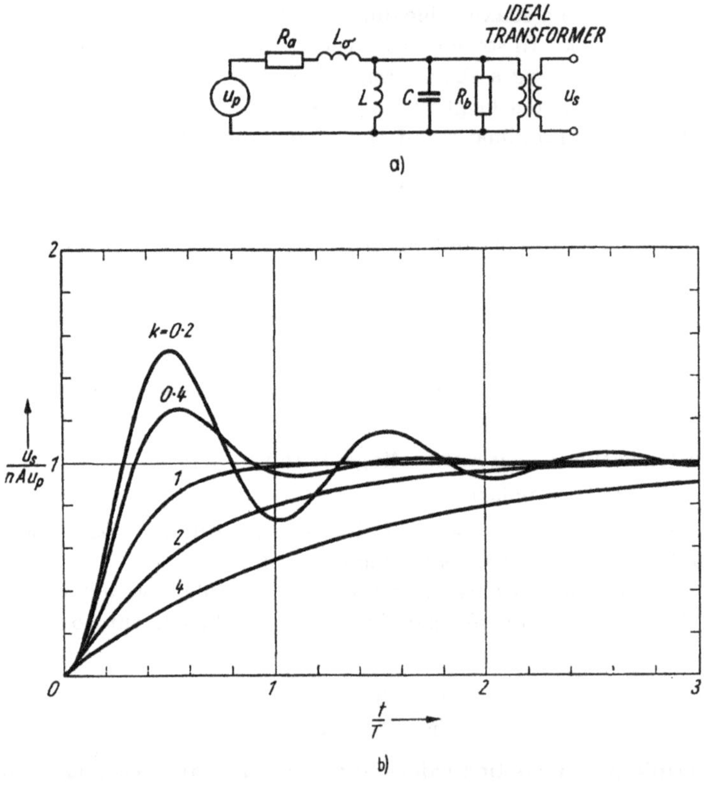

Fig. 50. Pulse Transformer. According to *Millman* and *Taub* [Mi 56].
a) Equivalent circuit. *b*) Transient response.

the self-capacitance C' of the transformer and the stray capacitance C_s of the secondary end transferred to the primary. In the main, the capacitance between windings C_0 determines the self-capacitance C'. Irrespective of the kind of circuitry $C' \approx n^2 C_0$ applies [Mi 56].

Due to mathematical complexity, for pulse signals calculation is usually handled separately for the rise-time response, the response during the flat top interval and the response after pulse termination. In Fig. 50*b* the response with a step-type input pulse of a height u_p is shown, with the influence of L neglected.

For the attenuation factor A, the damping constant k and the period T the following relations apply:

$$\left.\begin{aligned} A &= \frac{R_b}{R_a + R_b}, \\ k &= \left(\frac{R_a}{L_\sigma} + \frac{1}{R_b\, C}\right)\frac{T}{4\pi}, \\ T &= 2\pi\sqrt{L_\sigma\, C\, A}\,. \end{aligned}\right\} \tag{54}$$

Critical damping of the resonant circuit formed by the leakage inductance and the shunt capacitance is obtained with $k = 1$. In this case the rise time is expressed by

$$t_r = 0{\cdot}53\, T = 3{\cdot}35\sqrt{L_\sigma\, C\, A}\,. \tag{55}$$

For a steep pulse rise L_σ and C must be kept low. Because capacitances are transferred with n^2, $t_r \propto n$ applies, therefore, large step ratios should be avoided. For the construction of a low stray-capacitance winding the same aspects apply as for coils. In general, a small-dimensioned pulse transformer is preferable. Core material having a high permeability with high frequencies must be used to attain a coupling factor $K \approx 1$ with consequent low L_σ.

For the flat top behaviour of the pulse L_σ and C are of little importance. With the time constant of the differentiating network resulting from R_a, R_b and L, the flat top droops. This is expressed by

$$\tau_d = \frac{L\,(R_a + R_b)}{R_a\, R_b}\,. \tag{56}$$

A high inductance primary winding is therefore necessary for the transmission of long pulses.

Kohn [Ko 62] uses a ratio of $n = 3$ as a favourable compromise between rise time, pulse shape and amplification for the coupling of the various stages of a transistor amplifier. With a transformer having a diameter of 2·4 mm and a width of 2 mm the winding capacitance is 0·5 pF and the leakage inductance amounts to 100 nH. Primary inductance with 1 μH is sufficient to transmit 5 ns pulses.

Stegmeier [St 65a] showed that winding capacitance has little influence in the case of impedances that do not exceed some 100 Ω. Leakage inductance is another factor influencing the rise time and can be kept low by stranding $(n + 1)$ wires. Then, one represents the secondary winding, while the n others form the primary in series connection.

PULSE INVERTER

Pulse polarity reversion is simple with a coaxial cable terminated on both ends with its characteristic impedance [Le 59; Mo 64; Ba 66a], where it is only necessary to earth the outer sheath at the input and the inner conductor at the output. The desired opposed polarity pulse is then picked-off at the outer sheath. This results in a noticeable droop of the flat top section in long pulses. The reason is the impedance ωL of the line sheath which is parallel to the characteristic impedance Z_0 of the line, resulting in a load resistance at the output that is short-circuited for low frequencies through the line sheath. Increasing the outer conductor inductance L will bring some improvement with low frequencies, which is effected by winding the transmission line on a ferrite core. To ensure a low rise time of the pulses at the load resistance as low a capacitance as possible must be provided for between the windings and against the earthed shielding by a wide spaced construction, especially near the output of the transmission line. Equation (56) applies for the differentiating constant, with $R_a = R_b = Z_0$. *Balland* et al. [Ba 66a] showed the advantage of distributing the winding on two ferrite cores.

High-frequency behaviour can be improved by using a helical line structure and inserting it into a cylindrical shielding member. Together with the inductance of the helix the stray capacitance results in a high characteristic impedance Z. In place of ωL the impedance Z then is parallel to the characteristic impedance Z_0 of the line which prevents droop. However, this causes mismatching and leads to a stepwise decrease of the pulse height with pulses of a length above the double transit time in the helical line. By inserting magnetic material, inductance and, consequently, transit time and characteristic impedance of the helix can be increased, making the reflexion factor smaller and giving longer steps.

With sufficiently short pulses a transmission line inverter will prove an advantage because its rise time depends primarily on the quality of the line (see Section 3.1.2). This can be assembled easily; it is linear and permits high-power transfer. Relatively large dimensions and the limited impedance of the available transmission lines represent a certain drawback. Transit time in the line leads to a signal delay and represents a frequent disturbance. Sometimes, however, this is utilised in the circuit function (see Section 3.7.2 — Fig. 175, p. 285). *Abercrombie* et al. [Ab 61] describe line inverters having an upper frequency up to 1·5 GHz, with a droop of 10% with 10 μs pulses.

TRANSMISSION LINE TRANSFORMERS

According to the principle of cable inverters the so-called stacked-line transformer can be constructed [Le 59], where two lines are circuited in parallel at the input (impedance $Z_0/2$) and are arranged in series at the output (impedance $2\,Z_0$). This results in the transformation of the impedance by the factor 4. Line transformers of this type are constructed also with strip lines and twisted pair lines [Ru 59; Go 62a; Ta 63a]. Bifilar wires of 0·2 to 0·05 mm diameter and defined characteristic impedance allow the construction of miniature nanosecond transformers in sizes of some millimetres [Bo 67]. They are suitable for demonstrating how the transmission line form is derived from the conventional type pulse transformer (Fig. 51 *a* and *c*). These arrangements offer the advantage

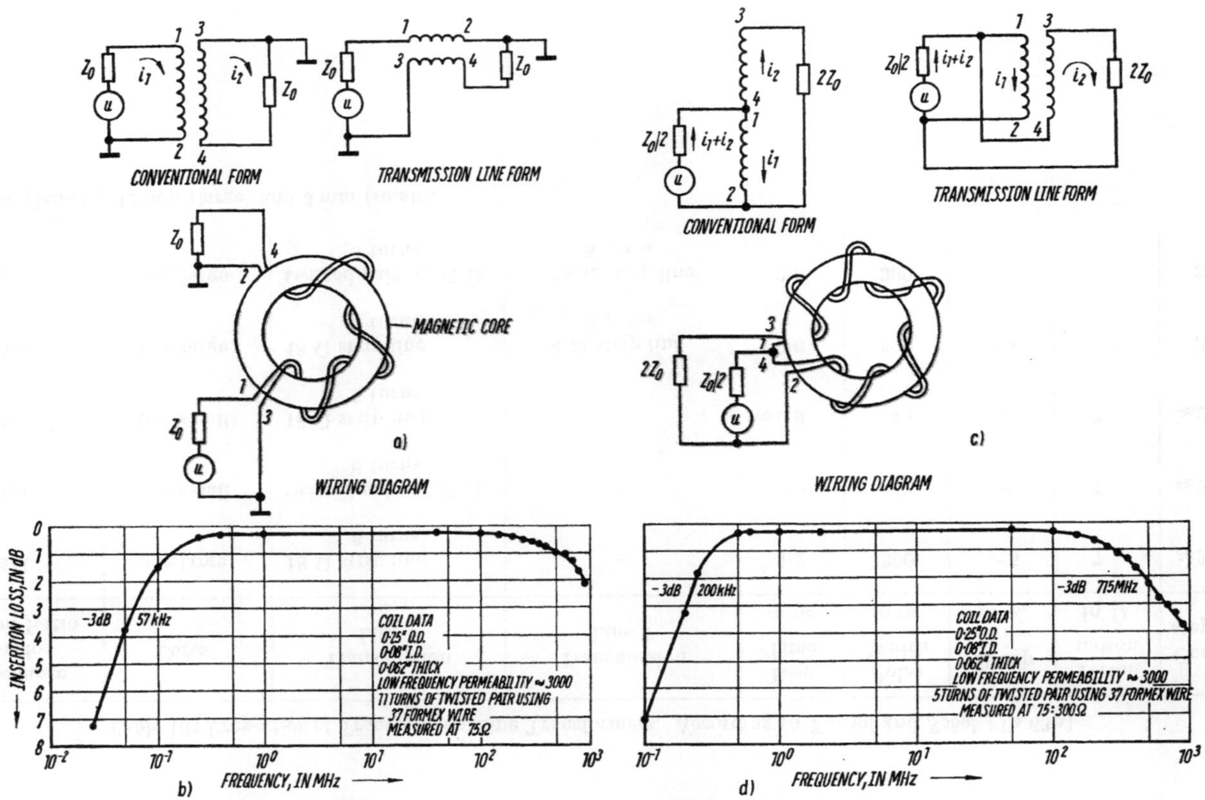

Fig. 51. Bifilar-wound pulse transformers, showing the transition to the transmission line transformer. According to *Ruthroff* [Ru 59].

a) Reversing transformer. *b*) Insertion loss versus frequency of reversing transformer. *c*) 4:1 impedance transformer. *d*) Insertion loss versus frequency of 4:1 impedance transformer.

Table 10: Properties of Transmission Line Transformers. According to *Tansal* and *Sobol* [Ta 63a]

Impedance Transformation Ratio	Cores†	Transmission Line 1	Transmission Line 2	Rise Time in ns	Pulse Width in ns	Droop in %	Termination in Ω	Current step up
4:1	one large	18 Ω strip line 6 turns	–	1·2	300	<5	7	≈2
4:1	one small	twisted pair ≈ 35 Ω 6 turns	–	1·0	30	<10	7	≈2
4:1	one small	18 Ω strip line 5 turns	–	≈0·9	40	≈0	7	≈2
16:1	two large	18 Ω strip line 6 turns	4 Ω strip line 5 turns	2·0	300	<8	1·3	3·6
16:1	two large	twisted pair ≈ 35 Ω 6 turns	14 Ω strip line 5 turns	2·0	300	<10	1·2	3·6

† Outer diameter 12 mm (large) and 3 mm (small).

that interwinding capacitance becomes part of the characteristic impedance of the transmission line represented by the primary and the secondary windings. Resonance with leakage inductance cannot occur here. Good coupling is obtained due to the narrow spacing of the windings. Construction data and frequency behaviour of these transformers are illustrated in Fig. 51*b* and *d*. Utilising the suggested principle, quite a number of other transformer problems can be handled [Ru 59]. Theory and application is reviewed in detail by *Matick* [Ma 68a]. As an example, design and properties of balanced-to-unbalanced (balun) transformers for a pulse rise time of 1 ns are described and a 9:1 impedance transformer with 4 ns rise time is detailed.

In integrated circuit combinations the pulse transformer is very useful to adapt the high impedance (≈50 Ω) of the external transmission line to the low impedance of the circuit in the order of 1 Ω. For this purpose the use of strip lines is most suitable (see Section 3.1.2), because it enables the establishment of very low characteristic impedance values. Single transformers permit an impedance transformation of 4:1. For a higher ratio transformer cascade arrangements are employed. The data given by *Tansal* and *Sobol* [Ta 63a] in Table 10 were obtained by using strip line of copper-clad teflon and ferrite cores of different diameters.

3.1.2 TRANSMISSION LINES AND PLUG CONNECTORS

SKIN EFFECT

Nanosecond pulses contain very high-frequency components and require consideration of the line resistance increase due to the skin effect [Me 62]. With increasing distance x from the conductor surface current density j decreases according to

$$j(x) = j_0 \, e^{-x/s} \tag{57}$$

where $s = \sqrt{\varrho / \pi f \mu \mu_0}$ represents the depth of penetration, ϱ the specific resistance, f the frequency and $\mu \mu_0$ the permeability of the conductor material. With ferromagnetic conductors the skin effect is particularly pronounced. Consequently, the effective cross-section of the conductor decreases with high frequency, while its resistance increases with $\sqrt{f}$. On the other hand, the influence from the difference in conductor material is not as pronounced as is the case with low frequencies. Resistance of a round conductor in dependence of frequency is given in Fig. 52. To obtain a low line resistance it is therefore feasible to use a thick conductor with a smooth surface, which may have to be coated with a good conducting layer, such as silver. Dirt or oxidation of the surface leads to practically no resistance increase, because the penetration depth of an oxide film is large due to its very high resistivity. This permits practically the entire current to penetrate to the metal surface.

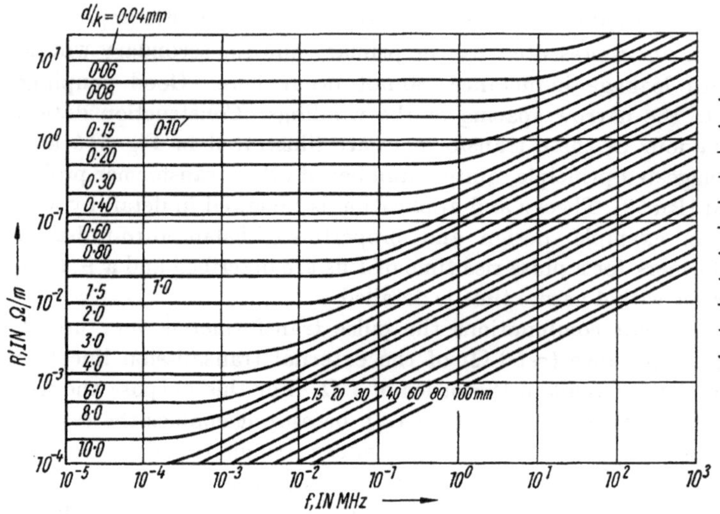

MATERIAL	k
SILVER	1·00
COPPER	1·03
GOLD	1·17
ALUMINIUM	1·37
ZINC	1·97
BRASS	2·20
PLATINUM	2·62
MANGANIN	5·20
CONSTANTAN	5·57
CARBON	≈50

Fig. 52. Resistance R of round conductors of different materials, having a diameter d referred to frequency f. According to *Meinke* and *Gundlach* [Me 62].

TRANSMISSION LINES

Transmission of nanosecond pulses over a long distance is possible only with a matched transmission line, because with a conventional wire connection self-induction and capacitance to ground cause intolerable pulse distortion already with conductor lengths of a few centimetres. Rectilinear uniform lines are usually employed, where cross-section and permittivity are constant over the whole length of the line. Properties of these lines are detailed in the high-frequency engineering textbooks [Vi 60; Me 62, chapter C]. Their behaviour when transmitting nanosecond pulses has been presented by *Lewis* and *Wells* [Le 59] and thoroughly dealt with by *Morugin* and *Glebovich* [Mo 64]. Pulse transmission is effected through travelling waves in principal or transverse electromagnetic mode, where the electric and magnetic field strength is located at right angles to each other and directed vertical to the direction of the propagation of the waves. The higher modes, significant for the wave propagation in a wave guide, have a lower cut-off frequency f_c. For the mode of the next higher order (H_{11}) with coaxial cables the relation [Le 59]

$$f_c \approx \frac{2c}{\pi(d + D)} \tag{58}$$

applies, where d is the inner diameter, D is the outer diameter of the two conductors. Cut-off frequency f_c amounts to around 10 GHz. Thus, the frequency components of the pulses to be transmitted in general underpass this cut-off frequency.

Practically zero-field condition obtains in the conductor interior because of the skin effect, which causes the magnetic field lines in the rectilinear uniform line to become equal to the equipotential lines of the electric field on the principal mode. Interior inductance of the conductor may consequently be neglected.

Series inductance, shunt capacitance and loss resistance are distributed across the whole length of the line. The equivalent circuit of a unit length of line (Fig. 53) consequently contains the capacitance per unit length C', the inductance per unit length L' and, in series with it, the resistance per unit length R', which represents the sum of the resistances of all conductors and is caused primarily by the skin effect. Because of its large surface the contribution of the outer conductor in a coaxial cable is frequently small. Radiation loss also contributes, for instance, with a strip line when shielding is inadequate. If magnetic loss occurs in the dielectric this is added also to R'. Dielectric loss results in the shunt conductance per unit length $G' = \omega\, C' \tan\delta$, which becomes effective only on high frequency.

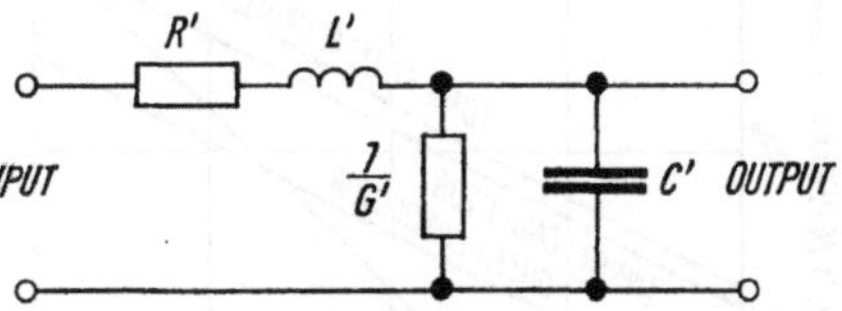

Fig. 53. Equivalent circuit of a unit length of cable.

For wave propagation the differential equation

$$\frac{d^2u}{dx^2} = \gamma^2\, u \tag{59}$$

holds containing the complex propagation constant

$$\gamma = \alpha + j\,\beta\,. \tag{60}$$

The expression for attenuation per unit length α and the phase shift per unit length β is written

$$\frac{\alpha}{\beta} = \sqrt{\pm\frac{R'\,G' - \omega^2\,L'\,C'}{2} + \frac{1}{2}\sqrt{(R'^2 + \omega^2\,L'^2)\,(G'^2 + \omega^2\,C'^2)}}\,. \tag{61}$$

In a rough approximation, the low-attenuation line with $R' \ll \omega\,L'$ and $G' \ll \omega\,C'$ can be expressed with

$$\left.\begin{aligned} \alpha &\approx \frac{R'}{2}\sqrt{\frac{C'}{L'}} + \frac{G'}{2}\sqrt{\frac{L'}{C'}} = a\sqrt{f} + b\,f\,, \\ \beta &\approx \omega\sqrt{L'\,C'}\,. \end{aligned}\right\} \tag{61a}$$

Thus the attenuation per unit length, specifically with high frequencies, can be split approximately in a longitudinal attenuation determined only by the line resistances and the transverse attenuation influenced by the dielectric. Relative contribution and frequency behaviour of these components are given in Fig. 54. Below some hundreds MHz, longitudinal attenuation rising proportional with $\sqrt{f}$ dominates. Above 1 GHz the transverse attenuation increasing proportionally to f represents the deciding influence. For commercially available cables, *Morugin* and *Glebovich* [Mo 64] give an attenuation of $\alpha = 1{\cdot}9$ to

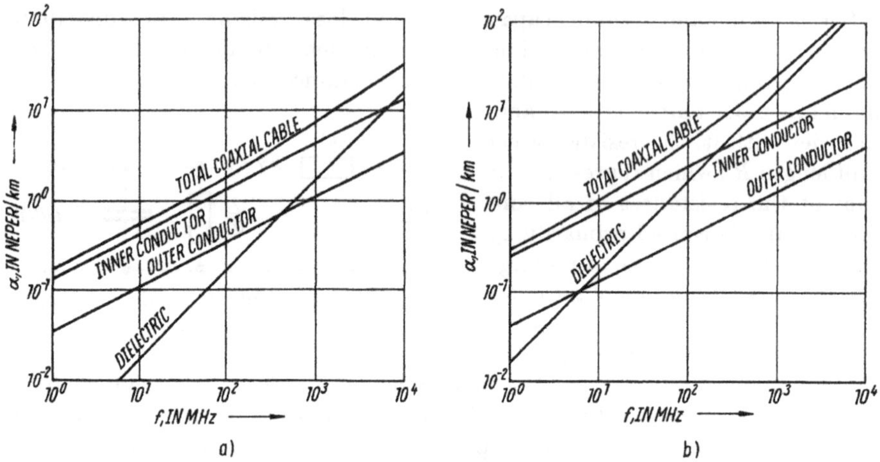

Fig. 54. Attenuation constant α contributions in coaxial cables.
According to *Meinke* and *Gundlach* [Me 62].

a) Trolitul-air dielectric, effective $\varepsilon = 1{\cdot}2$, $\tan\delta = 10^{-4}$, $d = 5$ mm, $D = 18$ mm. *b*) Completely filled with dielectric material, $\varepsilon = 2{\cdot}4$, $\tan\delta = 7\times10^{-4}$, $d = 2{\cdot}7$ mm, $D = 16$ mm; 1 Neper = 8·686 decibel (dB).

2·7 dB/m in the 10 GHz range. Using a dielectric design needing only very little material will obtain a very low attenuation, with the inner conductor held by disks, beads or a spiral made of trolitul, polyethylene, foam material or ceramic (Fig. 54*a*). In the 1 GHz region attenuation of such a cable amounts, for instance, to $\alpha = 0{\cdot}025$ to $0{\cdot}1$ dB/m [Br 65a]. Cables completely filled with dielectric material (Fig. 54*b*) have a higher electric strength. With an inner conductor made up of helical strips and insulating intermediate layers the influence of the skin effect is less [Me 62].

Considerable reduction of attenuation is obtained when a normal coaxial cable is cooled to 77 °K [Gl 66] because the specific resistance of the conductor is notably lower and dielectric loss is reduced even by two orders of magnitude against that on room temperature. Reduced attenuation, especially with a short cable, leads to a rise time which is lower by one order of magnitude because of the high influence exerted by the dielectric loss, as shown with equation (67). The use of superconducting cables has the advantage that even with a long cable attenuation is very low. Such cables have a dielectric loss of nearly zero and a conductivity that is higher by many orders of magnitude than that at room temperature. Special care, however, must be exercised with plug connectors and terminating resistors because reflected pulses are not damped but frequently travel back and forth in the line and cause considerable distortion. *Cummings* and *Wilson* [Cu 66] tested flexible subminiature cables where, for instance, one had an inner conductor of Nb—25% Zr alloy and the pulse rise time amounted to $t_r = 0{\cdot}65$ ns, irrespective of the length of the cable (up to 274 m). A strip line arrangement is described by *Glebovich* [Gl 67] which operates at 4·2 °K. It consists of a quartz cylinder on which Nb-Zr wire is wound together with a lead strip. On a delay of 50 ns this gives an upper cut-off frequency of 35 GHz.

Relations for phase velocity v_p and group velocity v_g are

$$v_p = \frac{\omega}{\beta}, \quad v_g = \frac{d\omega}{d\beta}. \tag{62}$$

Both velocities are equal to each other only when attenuation is negligible. Time delay per unit length of a cable is expressed with

$$T' \approx \sqrt{L'\,C'}. \tag{63}$$

At every point of the cable a relation of voltage to current exists which is termed characteristic impedance and written

$$Z = \sqrt{\frac{R' + j\,\omega\,L'}{G' + j\,\omega\,C'}}. \tag{64}$$

On negligible attenuation this is real and independent of frequency, becoming

$$Z \approx \sqrt{\frac{L'}{C'}}. \tag{64a}$$

Of the many possible transmission line arrangements [Le 59; Vi 60; Me 62] in nanosecond pulse technique the coaxial cable is preferred and the stripline is also used to some extent. As variable delay line, the open rectangular line is also employed (helical line according to *Bell* et al. [Be 52]). Characteristic impedance of these arrangements is noted in Fig. 55 (s. next page).

In copper-conductor coaxial cables minimum attenuation is found with a diameter ratio $D/d = 3{\cdot}6$. With air as dielectric this relation gives a characteristic impedance of $Z = 77\ \Omega$, while the use of an $\varepsilon = 2{\cdot}3$ dielectric results in $Z = 50\ \Omega$. A higher characteristic impedance is preferable with nanosecond valve circuitry. In transistor circuits the frequency dependence of the input impedance R_i of the transistors does not disturb cable matching, provided the characteristic impedance is $Z \gg R_i$ in the common-base circuit having an impedance of $R_i \approx 10\ \Omega$ and $Z \ll R_i$ in the common-emitter circuit with R_i in the order of 500 Ω. Thus a cable having a characteristic impedance of 50 Ω represents an acceptable compromise [Su 62].

Strip lines have the advantage that they can be constructed for a wide range of characteristic impedances. In some cases both conductors may be accessible permitting the introduction of continuously variable tapping points and shorting bridges. Their constructions make them particularly suitable for

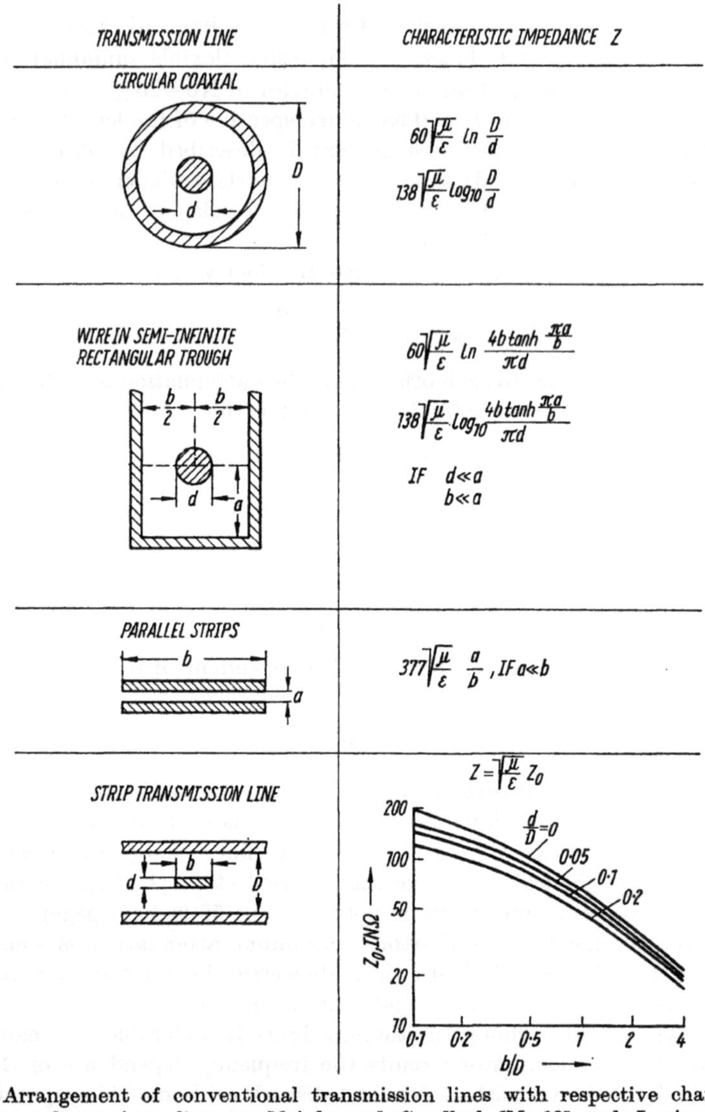

Fig. 55. Arrangement of conventional transmission lines with respective characteristic impedance values. According to *Meinke* and *Gundlach* [Me 62] and *Lewis* and *Wells* [Le 59].

printed circuits. Incomplete shielding of plain micro-strips, however, causes frequency dependent radiation loss and interference signal pick-up. For this reason shielded strip transmission lines are employed. Their design and characteristic impedance are shown in Fig. 55. Width of the outer strip should be about $b + 2\,D$ with irradiated polyolefin ($\varepsilon = 2{\cdot}3$ and $\tan\delta = 10^{-4}$), for instance, used as dielectric [Da 66]. Properties of these transmission lines have been reviewed by *Morugin* and *Glebovich* [Mo 64] and *Swift* [Sw 67]. Especially low longitudinal attenuation is obtainable with low characteristic impedance strip lines and wide spacing between strip components. With frequencies above 1 GHz transverse attenuation due to dielectric loss predominates, similar as with coaxial cables.

PULSE DISTORTION IN TRANSMISSION LINES

When line loss is taken into account, a frequency dependence of attenuation and phase velocity which distorts the pulse travelling through the line is derived from equation (61) and (62). Because mathematical calculation of this problem is difficult, it is treated in high-frequency approximation, i.e. $f \gtrapprox 100$ kHz. As seen in Fig. 54, for frequencies below some hundreds MHz it is sufficient when the longitudinal attenuation due to the skin effect receives consideration only [Wi 57; Fi 60; Go 62b]. If the input signal $U_i(t)$ represents a voltage step of unit height, it assumes the shape illustrated in Fig. 56 after travelling through a line with the length l; in this case referred to a U.S.S.R. product type PK-50-11-13 with $Z = 50\ \Omega$ and α (10 GHz) $= 2{\cdot}4$ dB/m. After an initially steep rise, the pulse height increases only very slowly and reaches its final height after a very long time. It seems therefore not feasible to use conventional rise time definition (rise of 10 to 90%) to characterise a transmission line. The period of rise to 50% of the final pulse height as definition for the characteristic time τ of a line seems to serve the purpose better.

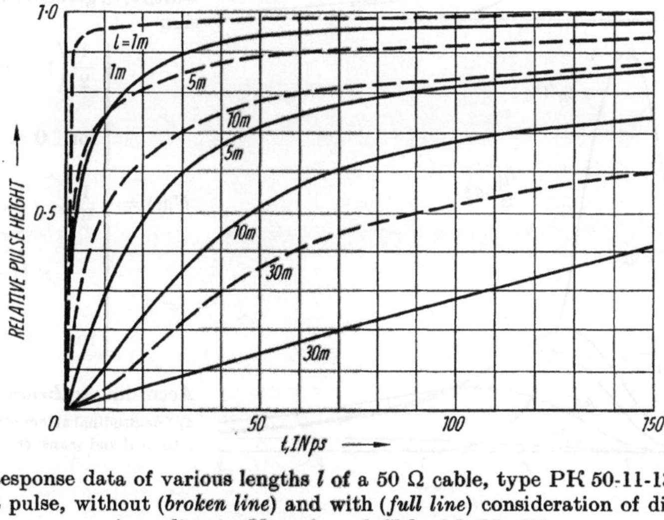

Fig. 56. Response data of various lengths l of a 50 Ω cable, type PK 50-11-13, on a step-type input pulse, without (*broken line*) and with (*full line*) consideration of dielectric loss. According to *Morugin* and *Glebovich* [Mo 64].

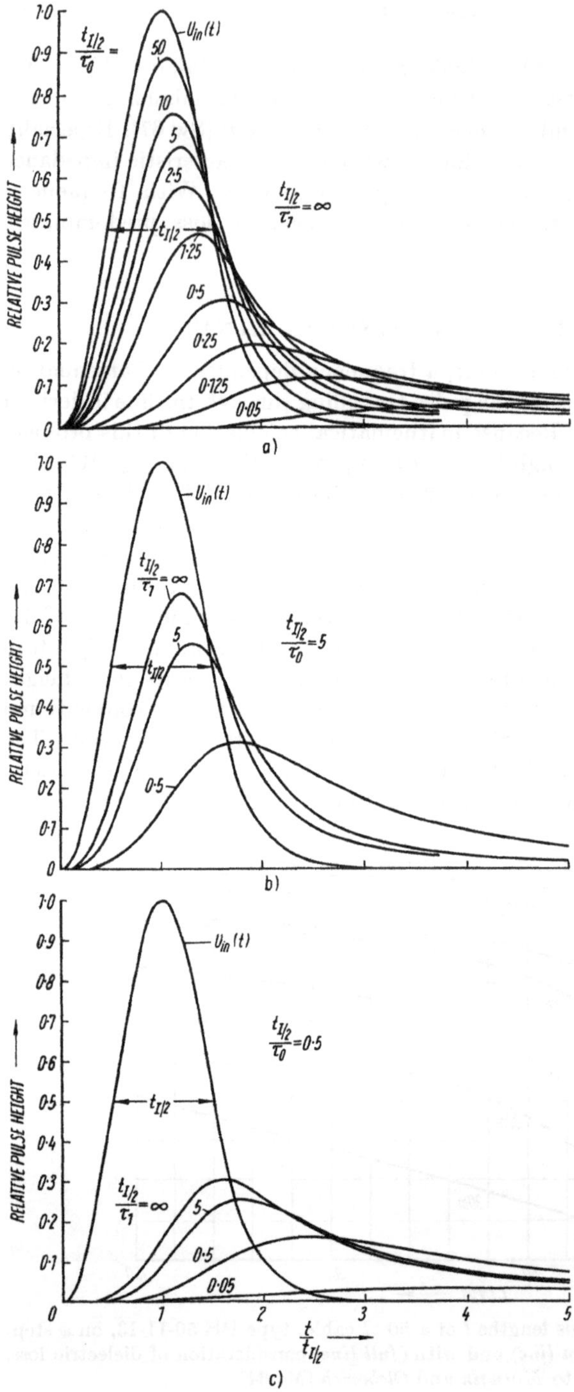

Fig. 57. Response curves of a cable having the characteristic time τ_0 (longitudinal attenuation only) and τ_1 (transverse attenuation only) on a photomultiplier pulse with the half-width $t_{I/2}$, given by the expression

$$U_i(t) = \begin{cases} \frac{1}{2}\left(1 - \cos\frac{\pi}{t_{I/2}}\,t\right) \\ \text{for} \quad 0 < t \leqq \frac{3}{2}\,t_{I/2} \\ \frac{1}{2}\exp\left\{-\frac{\pi}{t_{I/2}} \times \right. \\ \qquad \left. \times \left(t - \frac{3}{2}\,t_{I/2}\right)\right\} \\ \text{for} \quad t > \frac{3}{2}\,t_{I/2} \end{cases}$$

According to *Brianti* [Br 65a].

a) Longitudinal attenuation. b, c) Longitudinal and transverse attenuation.

In Fig. 56 the influence of transverse attenuation caused by dielectric loss is also seen. This further flattens the pulse rise and should be taken into account for rise time values of the input pulse below 1 ns, especially with a short transmission line [Gl 64]. With strip transmission lines an analogue behaviour is found [Mo 64].

For the calculation of transmitting properties of transmission lines for the two limiting cases, *Brianti* [Br 65a] derived the analytical expression $v(t)$ for the response to a δ-shaped input pulse separately, viz. for longitudinal attenuation [$b\,f = 0$ in equation (61a)] and for transverse attenuation ($a\sqrt{f} = 0$). Response function $v(t)$ is then obtained with the aid of the convolution integral from both components. For an input pulse of the analytically or numerically given term $U_\mathrm{i}(t)$ from further convolution

$$U_0(t) = \int_0^t U_\mathrm{i}\,(t - t')\; v(t')\;\mathrm{d}t' \tag{65}$$

is obtained for the shape of the output pulse $U_0(t)$. Distortion of photomultiplier pulses, which is of special significance for practical application, of full width at half-maximum $t_{1/2}$ is shown in Fig. 57 for various values of longitudinal attenuation (*a*) and with reference to different values of transverse attenuation (*b* and *c*). For the two extremes the frequency-independent characteristic time can be expressed with

$$\left.\begin{aligned} \tau_0 &= \frac{a^2\, l^2}{\pi} \quad \text{for} \quad b\,f = 0\,, \\ \tau_1 &= \frac{b\,l}{\pi} \quad \text{for} \quad a\sqrt{f} = 0\,. \end{aligned}\right\} \tag{66}$$

In the medium frequency range the longitudinal attenuation $a\sqrt{f}$ and the length l of the cable determine the magnitude of τ_0 (Fig. 58). With 1 GHz the broken line indicates that the attenuation α measured at this frequency level may deviate from the value $a\sqrt{f}$ derived by extrapolation from the low frequency region. The characteristic time τ_1 can be calculated from the magnitude of the deviation $\alpha - a\sqrt{f} = b\,f$. In case the relation

$$\frac{\tau_1}{\tau_0} = \frac{b}{a^2\, l} \tag{67}$$

tends toward zero the longitudinal attenuation needs only be considered. If the relation assumes a high value, transverse attenuation must also be taken into account. This applies specifically to short transmission cables of $l < 10$ to 30 m, where dielectric loss exerts a relatively greater influence on pulse shape (see Fig. 56).

Although deterioration of the pulse rise as described is a disadvantage, loss in cables can also serve to improve pulse behaviour, for example, in *ferrite lines*, where the formation of shock waves is utilised [Mo 64]. Rectangular ferrite rings are threaded on the inner conductor. Magnetisation of the ferrite

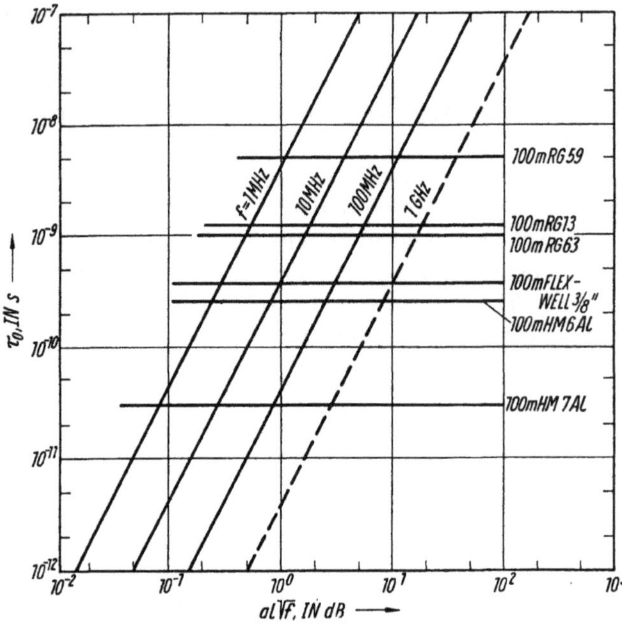

Fig. 58. Characteristic time τ_0 of cables (rise time of a step pulse to 50% of total height) in dependence of longitudinal attenuation at various frequencies. According to *Brianti* [Br 65a].

changes with the alteration of the current passing through the inner conductor. With very rapid remagnetisation, in the order of $>10^8$ to 10^9 Oe/s, loss is produced in the ferrite of a magnitude which decreases with approaching saturation. Thus, the initial part of the pulse serves to magnetise the ferrite and is lost. The pulse gets shorter with increasing length of the cable. Pulse steepening results because the loss during the initial period of the pulse is higher and with increasing saturation of the ferrite the input pulse nearly retains its initial height. Obtainable final steepness is given by the relaxation time of the ferrite.

In contrast to other methods of pulse formation the rise time of the shock waves decreases with increasing pulse height. In principle a slope of 10^{12} to 10^{14} V/s can be attained, although this necessitates a current on the order of 10 to 1,000 A for remagnetisation. In combination with thyratrons ferrite cables allow the generation of high-voltage pulses with a height of some kilovolt and a rise time underpassing 1 ns (see Section 4.1.3). Maximum pulse repetition, however, is restricted to about 100 kHz because of the temperature rise in the ferrite due to energy loss. Electric strength of the cable represents the practical limitation of the steepness of the leading edge.

With a slow remagnetisation amounting to $<10^7$ to 10^8 Oe/s the inconstancy of the permeability $\mu(\boldsymbol{H})$ of the ferrite can be utilised for steepening the leading edge. Depending on the obtained magnetic field strength a different transit time $T = \sqrt{\mu(\boldsymbol{H})}\sqrt{L_0 C}$ of the respective part of the pulse rise results; with L_0 the inductance of the cable without ferrite. This produces a steeper or flatter leading edge, depending on which section of the hysteresis curve the pulse is passing through. A biasing magnetising current can be employed for setting the suitable working point on the hysteresis curve. In a similar manner the hysteresis curve $\varepsilon(\boldsymbol{E})$ of ferroelectrics or the voltage dependent capacitance $C(U)$ of semiconductor diodes may be utilised for pulse steepening. Employing lumped delay lines with semiconductor diodes has the advantage that the pulse repetition rate can amount to 10 MHz [Ow 67]. *Mullick* [Mu 67] gives the theory of shock wave formation in transmission lines with non-linear distributed capacitance and conductance.

By appropriate choice of the functions $L(I)$ and $C(U)$ there is a class of non-linear lines whose characteristic impedance remains constant on a phase velocity variable by the bias current. For small signals this represents a variable delay line [Fa 67].

VARIABLE DELAY LINES

In many cases it is desirable to vary the delay time of pulses caused by their transit time in the cable (see Chapter 4 and 5). This is handled with arrangements where cables of different length are interposed in the signal path, sometimes with the aid of coaxial switches. The various stages are constructed in such a manner that delay cable connections are always reflexion-free.

Continuous variation of the delay, to about 100 ns, is made possible with *helical lines* [Be 52; Sp 62; Co 66b], which are constructed as shown in Fig. 59*a*. The inner conductor is placed in a rectangular trough milled into a rotary metal cylinder of about 10 cm diameter (for calculation of characteristic impedance see Fig. 55). For the pulse pick-off a slide contact is employed which is fastened to a moving carriage. Because of the large winding diameter the behaviour of the line is similar to a homogeneous line structure. Air as dielectric suffices for calculation consideration because the influence of the few insulator stubs used for holding the inner conductor is negligible. Radiation loss of the open rectangular line is the main attenuation factor. Shielding the entire helical line will afford some reduction of the loss. The attenuation of such a shielded line having a characteristic impedance of $Z = 76\ \Omega$ and a maximum delay of $T = 70$ ns is plotted with reference to frequency in Fig. 59*b* [Sp 62]. An advantage is the wide range of variability of the delay. Infeed of pulses must be effected at the tapping point through a cable with a characteristic impedance $Z/2$, otherwise reflexion will occur. *Trombone-type* decimetre extension lines with air as dielectric in construction as shown in Fig. 59*c* can be utilised much more conveniently. Diameter ratio is selected to warrant the same characteristic impedance on any adjustment of the line ($Z = 70\ \Omega$ and $60\ \Omega$ respectively). However, an upper frequency limit of 3 GHz is given due to reflexion from the

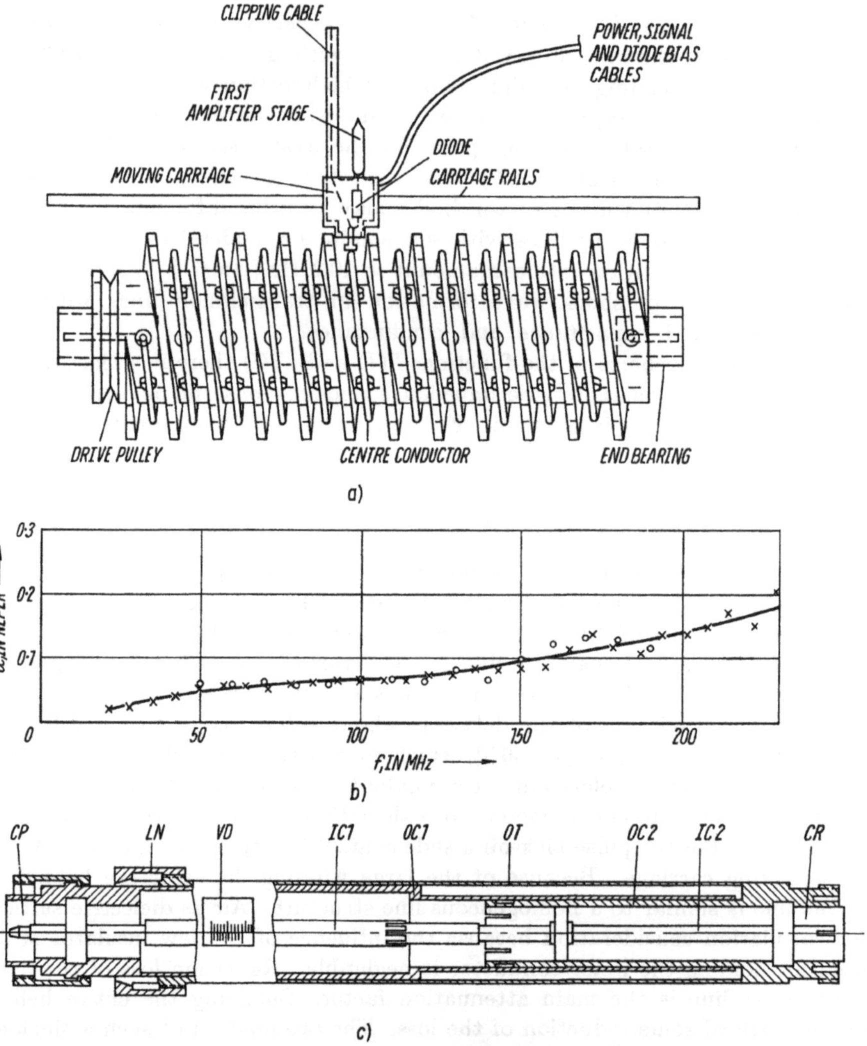

Fig. 59. Variable delay lines.

a) Helical line. According to *Bell* et al. [Be 52]. *b*) Attenuation in a helical line with $Z = 76\ \Omega$, $T = 0$ to 70 ns. According to *Sprinz* and *Brunner* [Sp 62]. *c*) Decimetre extension line (trombone-type) with $Z = 70\,(60)\ \Omega$, $T \approx 1{\cdot}2$ to $1{\cdot}6$ ns. VEB Rafena Radeberg, G.D.R.

IC 1 inner conductor, *OC* 1 stepped outer conductor } tightly joined,
IC 2 stepped inner conductor, *OC* 2 outer conductor, } tightly joined,
CP connector plug,
CR connector receptacle,
LN locking nut,
VD vernier dial,
OT outer tube.

point where one cross-section changes to another and from the supporting disks. Delay time can be adjusted with accuracy to less than 1 ps. Lines of this type have the drawback of a rather narrow range of varying the delay time, i.e. $\Delta T \approx 0{\cdot}4$ ns.

Ferrite transmission lines are very conveniently varied through influencing the loss with the magnitude of the bias magnetising current which changes the time location of the leading edge of the pulse [Me 65b]. For instance, with a cable of 20 m length delay time can be changed by 100 ns with a current alteration of 1 A. The temperature coefficient of the ferrite, however, can lead to delay time instability, especially with a high current level. Pulse reflexion contributes also to a change of the magnetic bias and, therefore, the delay time [Mo 64].

PLUG CONNECTORS

For connecting two transmission lines having the same characteristic impedance, plug connectors are employed which are designed to avoid the formation of reflexions. For the purpose, the influence of the skin effect is taken into account to ensure short and uniform current paths at the contacting points. Cross-section transitions and insulating supports are constructed to keep the reflexion factor r (see equation (102) in Section 3.2.2) below the level of the limiting curve relating to the figure of merit of the respective plug connection, as shown in Fig. 60. The low-inductance connection of the outer conductor of the cable to the plug socket member is of great significance for the shielding quality of a connector. This is characterised by the value of the coupling resistance $Z_c = u_c/i$, where u_c respresents the voltage measured on the outside between the ends of the outer conductor of the plug connection and i is the current flowing in the inner conductor. With a good quality connector the coupling resistance amounts to $Z_c \approx 10^{-4}\ \Omega$. A detailed review regarding the properties of transmission line components has been undertaken by *Meinke* and *Gundlach* [Me 62, chapter E].

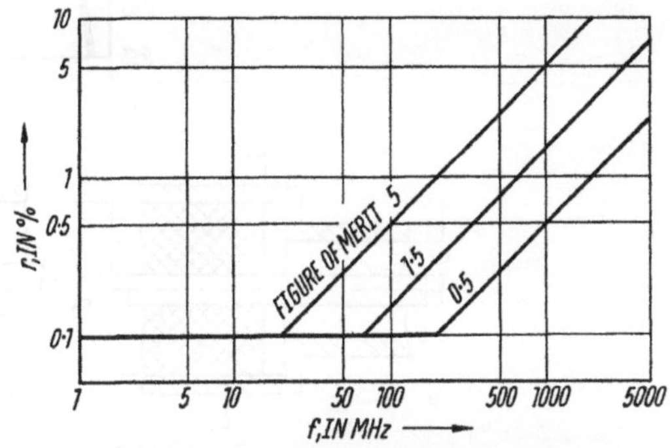

Fig. 60. Limit curves of the reflexion factor for plug connectors of various figures of merit. According to *Meinke* and *Gundlach* [Me 62].

Insulating disks used as support for the inner conductor act as additional transverse capacitance. With thin disks the reflexion factor can be approximated with

$$r \approx \frac{\pi h}{\lambda_0} (\varepsilon - 1) \tag{68}$$

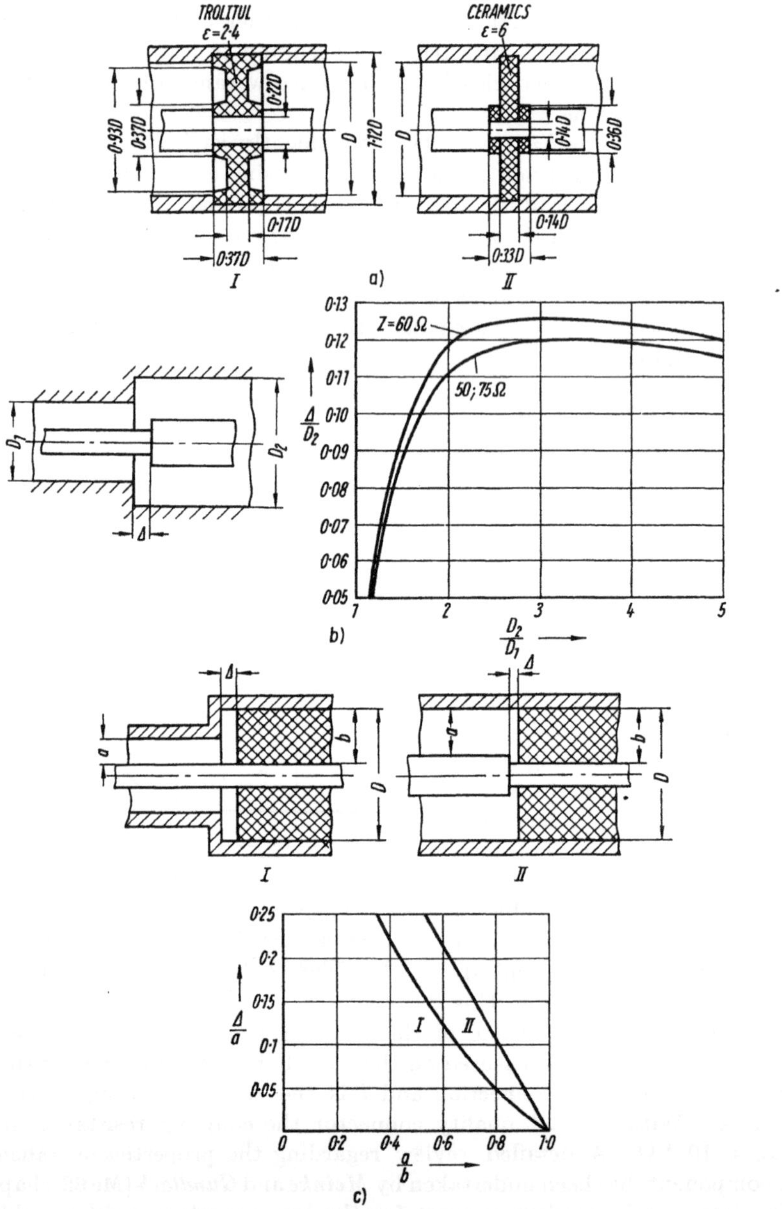

Fig. 61. Reflexion-free inhomogeneities in coaxial lines.
Compensation of the transverse capacitance through a longitudinal inductance.
According to *Meinke* and *Gundlach* [Me 62].

a) Insulation support for $Z = 70\ \Omega$. *b*) Change of cross-section. *c*) Abrupt transition of dielectric constant.

with h the thickness of the disk, λ_0 the wavelength in air and ε the relative dielectric constant of the disk material. Ceramic pin supports with a low dielectric constant yield specially low reflexion factor values. For instance, supports for strip lines using air as dielectric can be located in a zero-field area and will then not cause any disturbance. A longitudinal inductance may be utilised to compensate the transverse capacitance by either reducing the diameter of the inner conductor or increasing the outer conductor diameter at the location of the support, or by altering both diameters as shown in Fig. 61*a*. The reflexion factor of a compensated support increases to the square when approaching the limit wavelength $\lambda_k = 2\,\pi\,h\,\sqrt{\varepsilon}$. In consideration of mechanical strength an excessive reduction of the inner conductor diameter is avoided. Instead the effective dielectric constant is reduced by using a suitable disk profile or the recess on the inner conductor is lengthened (see Fig. 61*a*).

With a constant characteristic impedance, $d_1/D_1 = d_2/D_2$ applies for the cross-section transition. It must be remembered that the electrically effective step is displaced with regard to the geometric step. In the equivalent circuit this also acts as an additional transverse capacitance when $\lambda_0 > 4\,D$ holds. This capacitance becomes frequency-dependent in the case of still higher frequencies. Transverse capacitance is reduced and widely frequency-independent compensated by displacing the transition point of the outer conductor against the point of the inner conductor by the distance Δ, as shown in Fig. 61*b*. A conical transition of conductor cross-sections increases the electric strength.

In the case of a transition, for example, of a plug connector with air as dielectric to a transmission line having a dielectric material, either the inner conductor diameter must be reduced or the diameter of the outer conductor must be increased to obtain the same characteristic impedance. A reflexion-free transition is established when displacing the cross-sectional alteration by the distance Δ with respect to the dielectric bounding surface as illustrated in Fig. 61*c*. With $\lambda_0 > 5\,D$ a practically frequency-independent Δ is obtained.

Branchings, used for example at the pulse generator output, can be kept reflexion-free if they act like a termination resistor Z at any branch end (see Section 3.2.2). This is realised by coaxially mounted matching resistors R which have the value $R = Z/3$ in the case of two outputs (see Fig. 62*a*) and a value $R = Z/2$ in the case of three outputs (see Fig. 62*b*).

At times the characteristic impedance Z of the plug connector coincides only approximately with the impedance Z_0 of the cable. Provided $2\,T < t_r$ applies, the reflexion factor in this case is [Le 59]

$$r = \frac{2\,T}{t_r}\,\frac{Z - Z_0}{Z + Z_0}\,. \tag{69}$$

Impedance matching must be the more accurate the lower is the pulse rise time t_r. A certain mismatching can be tolerated if the transit time T in the plug connector is very low. Experiments with a number of American plug connectors confirm this relation [Sc 64a]. With pulses having a rise time of $t_r = 0{\cdot}4$ ns, reflexion factors were obtained by measurement amounting to $r = 2$ to 20%. A type BNC plug connector and the cable, type RG 55 A/U, gave the value of $r = 1\%$. *Weinschel* [We 67] gives a comprehensive survey regarding properties and test methods of standard precision coaxial connectors in current use in USA.

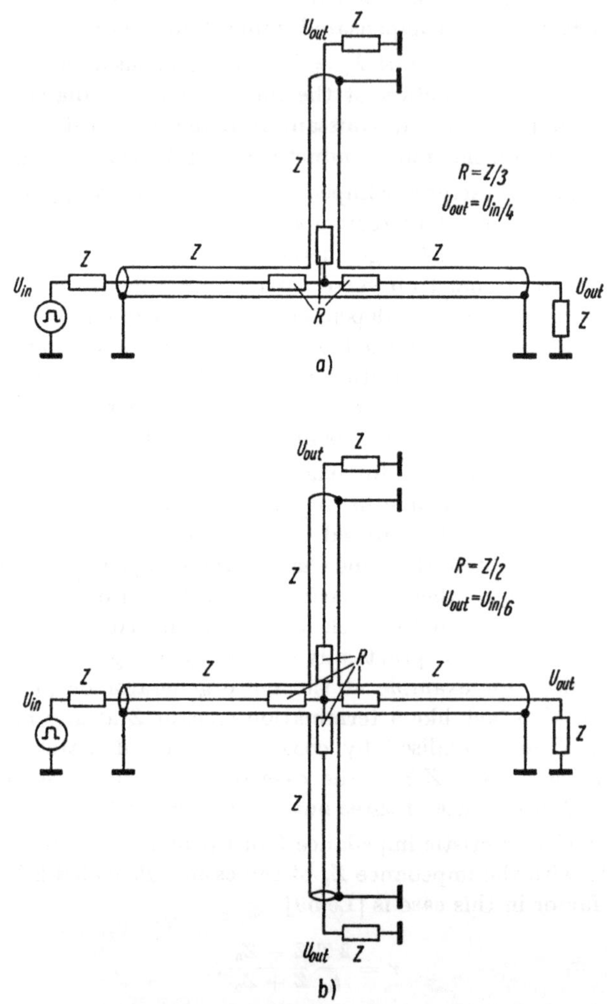

Fig. 62. Reflexion-free branchings of a transmission line having the characteristic impedance Z.
a) For 2 outputs. *b)* For 3 outputs.

3.1.3 ELECTRON VALVES AND THYRATRONS

For use in nanosecond technique electron valves must have a high gain-bandwidth product GB or high S/C ratio. Conventionally, this product is associated with an RC-coupled amplifier consisting of identical stages. Then $GB = S/2\pi(C_i + C_o + 5\,\text{pF})$ applies, with C_i and C_o the input and output capacitance, respectively, of the amplifier stages and 5 pF as additional circuit capacitance. With frame-grid pentodes of steep transconductance values of about 250 MHz are obtained for the gain-bandwidth product, for example, with a type E 810 F valve. With high-frequency transistors in emitter connection, on the other hand, the obtainable current gain-bandwidth product f_T is located above 1 GHz. In this case $f_T = \beta f_\beta$ denotes the frequency level where current gain β has dropped to 1.

For the high-frequency behaviour of electron valves the electron transit time and the lead inductance are of essential importance [Le 59; Me 62, chapter L]. Disk-sealed triodes, klystrons and similar types obtain a considerable expansion of the transmittable frequency band, however, they are ultra-high frequency elements and not suited for the circuits described in the subsequent sections.

Valves intended for pulse circuitry, in general, have to meet the following requirements, viz. a high gain-bandwidth product, a high transconductance to anode current ratio, a low heating power consumption with reference to the anode current, special cathode without intermediate layer, long life time and close-tolerance construction. The last mentioned demands are best met by long-life valves and nuvistors in metal-ceramic design.

For use in pulse circuitry, transistors have practically displaced the electron valve. A detailed review of nanosecond valve circuit technique has been given by *Lewis* and *Wells* [Le 59], and by *Morugin* and *Glebovich* [Mo 64]. Only a few special valves, therefore, will be dealt with here.

a) Secondary-Emission Valves. Among the better known valves, type EFP-60 and its equivalent type 6 B1П from the Soviet Union deserve mention. These valves feature a positive-potential dynode placed before the anode, causing the cathode current to reach the dynode first. Almost all the electrons liberated here are collected by the anode. This makes two output pulses of opposite polarity available. The dynode signal is in phase with the input pulse at the control grid. Due to the high stray capacitance only about 200 MHz is obtained as gain-bandwidth product. For trigger circuits these valves have proved very useful. Feedback can be effected either from the anode to the cathode or from the dynode to the control grid (see Section 3.2.4, [Mo 52; Ma 60]). Secondary-emission valves, however, have no satisfactory long-term stability.

b) Beam Deflexion Tubes [Le 59, chapter 4]. Type 6BN6 (Soviet equivalent type 6A3П) features two control grids having nearly the same characteristic and was initially intended as discriminator of frequency-modulated signals.

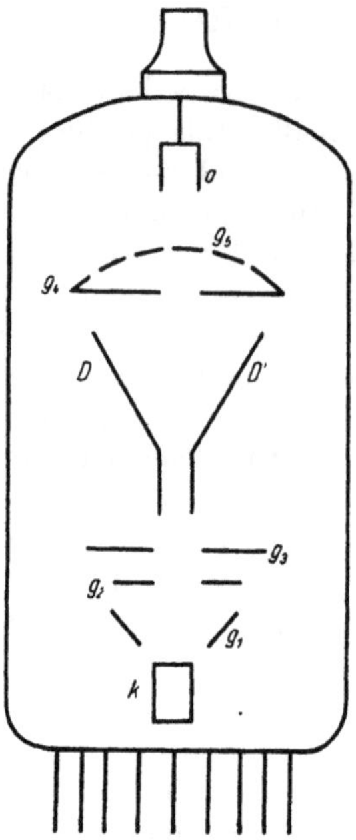

Deflexion of a focussed electron beam is utilised for the control in such a manner that with the one grid the anode current can be turned off irrespective of the potential impressed on the other grid. Referred to saturation state, a pulse height in the order of 1·5 to 3 V is sufficient for the purpose, depending on operating voltage. In the beam deflexion valve E 80 T (Fig. 63) the electron beam passes the focussing electrodes g_2 and g_3 to be deflected by the control electrodes D and D'. The beam is permitted to pass through the slot in g_4 to reach the anode a when the control voltages impressed on D and D' have the same values.

Fig. 63. Cross-section of the beam deflexion tube E 80 T (schematically).

THYRATRONS

Because of the high operating voltage, a thyratron can only be combined with difficulty in transistor circuits. On the other hand, semiconductor components are, in many cases, more favourable for the switching function than a thyratron. For this reason the thyratron has lost much of its importance for nanosecond application. A further drawback is the delay time between trigger and output pulse due to the ionising time required for breakthrough. This time difference depends on the gas filling, the geometry, and the operating voltages, and assumes values ranging from 10 to 200 ns [Kn 55; Le 59; Ge 65]. Statistic time fluctuations of the order of 1 ns are also annoying [Ge 65]. With small hydrogen-filled thyratrons the output pulse has a rise time of a few nanoseconds. Thyratrons have gained some significance for the control of spark chambers (see Chapter 5.3), where an anode voltage of several kilovolt and a switching current of over 100 A is employed.

3.1.4 DIODES AND TUNNEL DIODES

JUNCTION DIODES

In nanosecond pulse technique the semiconductor diodes serve primarily as switching element in circuits. Besides a low forward and a high back resistance a diode should feature brief switching periods. Transition from the blocked state to the conduction state differs considerably from the transition of the conduction state to the blocked state. Fig. 64 illustrates the equivalent circuit of a semiconductor diode [Me 62]. The static resistance R_D of the depletion layer in forward direction ranges between a few Ω to 100 Ω, and in reverse direction this amounts to 10 MΩ. Thickness of the depletion layer increases with impressed reverse voltage U [see equation (25)]. Consequently, capacitance of the depletion layer is $C_D \propto 1/\sqrt[n]{U}$; with $n = 2$ to 3. Point contact diodes especially have very low values amounting to $C_D = 0{\cdot}1$ to 1 pF, where the *Schottky* barrier of a small-area metal—semiconductor contact is utilised.

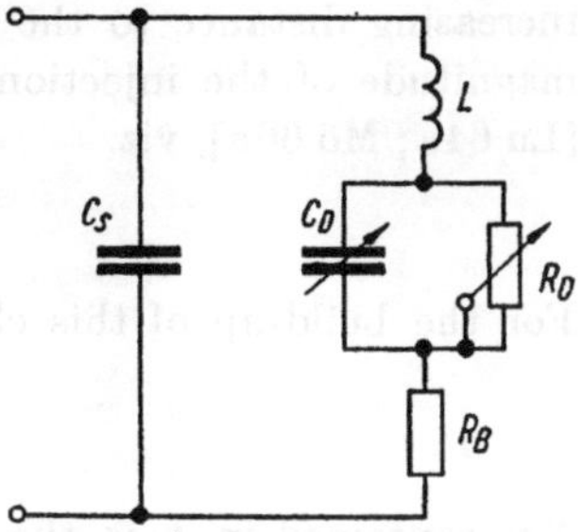

Fig. 64. Equivalent circuit of a semiconductor diode. According to *Meinke* and *Gundlach* [Me 62]. R_D resistance of the depletion layer, R_B resistance of the base, C_D capacitance of the depletion layer, C_s capacitance of case and leads, L inductance of the leads.

Path resistance R_B at the base may range between 10 and 500 Ω. For point contact diodes this is [Me 58]

$$\text{for conventional structure} \quad R_B = \frac{1}{4\,a\,N\,e\,\mu}, \tag{70a}$$

$$\text{for micro-etch structure} \quad R_B = \frac{w}{\pi\,a^2\,N\,e\,\mu}, \tag{70b}$$

where $2\,a$ is the diameter of the point contact, w the thickness of the undepleted bulk, N the concentration and μ the mobility of the majority carriers. A low path resistance can be obtained with highly doped material of high carrier mobility. Thus, III—V compounds are very favourable, for example, GaAs with $\mu_e = 4{,}500$ cm²/Vs (see data for Ge and Si in Table 8) [Je 58; Me 58; Sh 59].

Stray capacitance C_s of the leads and the case usually amounts to 0·3 to 3 pF. Together with the inductance L (≈10 to 30 nH) of the leads its influence can be practically overcome, provided the diode is mounted coaxially.

LIMITING THE SWITCHING TIME THROUGH MINORITY CARRIER STORAGE

Apart from the static characteristic of diodes and transistors the dynamic effects associated with carrier storage determine the switching behaviour of semiconductor layers. When a p-n junction (see Section 2.4.1) is forward biased a forward current I_f flows. Charge carriers are injected into the neighbouring layer of opposite conduction state, which appear as excess minority carriers. There they recombine with the majority carriers present in great number. The average lifetime τ of the minority carriers (reduction to 1/e of their initial number) depends on the number of recombination centres present in the material, and amounts to about $\tau = 0{\cdot}5$ to 500 ns. This causes the charge Q_B to be stored in the bulk material, with a concentration that decreases exponentially with increasing distance to the junction. In stationary state Q_B depends on the magnitude of the injection current I_i and on the average carrier lifetime τ [Lu 64a; Mö 66a], viz.

$$I_i = \frac{Q_B}{\tau}. \tag{71}$$

For the build-up of this charge the so-called charge control equation holds:

$$I_i = \frac{Q_B(t)}{\tau} + \frac{dQ_B}{dt}. \tag{72}$$

Not contained in I_i is the forward current component which passes into the depletion layer for space charge compensation and which decreases the width of the depletion layer.

The voltage U across the diode consists of the voltage drop U_D across the depletion layer and the voltage drop U_B at the bulk resistance, with a time behaviour consistent with the level of the injection current (Fig. 65). With a low injection current the influence of the delay due to the storage charge build-up dominates. The p-n junction shows capacitive behaviour. On the other hand, when the injection current is sufficiently high so that the density of the injected minority carriers approaches that of the majority carriers the conductivity of the path region increases to such an extent through the storage charge as to cause a reduction of the voltage drop U_B with time (inductive behaviour). With increasing injection current the time needed for the voltage to reach final value decreases in any case. Inductive behaviour of p-n junctions under high current load has been dealt with, for instance, by *Spenke* [Sp 58]. General expressions for the numerical calculation of the voltage behaviour and the switching time of very low and very high injection currents are found with *Möschwitzer* and *Landgraf-Dietz* [Mö 66a].

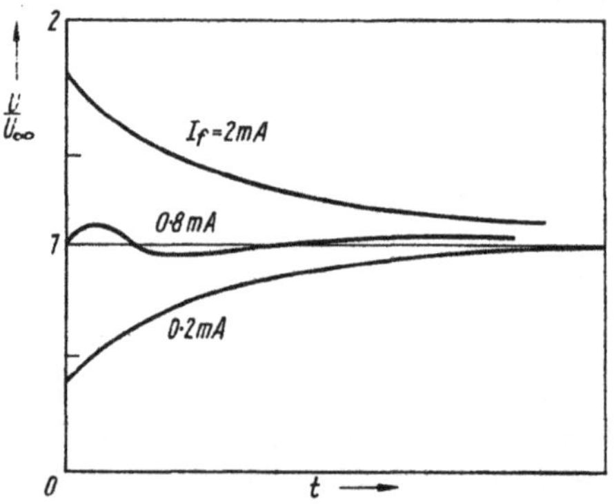

Fig. 65. Diode voltage when turning-on a forward current I_f. According to *Ko* [Ko 61].

Much more significant for the practical application of semiconductor diodes is their behaviour during transition from conduction to blocked state. For measurement the principal arrangement given in Fig. 66*a* is used. Initially the diode passes the current $I_f \approx U_1/R_1$. At the moment $t = 0$ the diode is blocked by the voltage U_2. Minority carriers stored in the path region through the forward current are collected and a reverse current I_r is produced. Behaviour of the latter is given in Fig. 66*b*. The respective diode voltage curve is shown in Fig. 66*c*. From the field of the instantaneous reverse current characteristic of the diode in Fig. 66*d* the magnitude of I_r can be obtained [Hu 56]. Initial minority carrier density depends on the magnitude of the preceding forward current. As long as minority carrier density is high enough a constant saturation current $I_R \approx U_2/R_2$ flows during the storage period t_s. As soon as the carrier density has become zero near the junction the reverse current decreases because the flow of stored carriers from a long distance gradually lessens until finally the static reverse current I_{r0} only is present. The fall time t_f defines the period where the reverse current has decayed to 10% of its maximum. Total reverse recovery time then is

$$t_{rr} = t_s + t_f \, . \tag{73}$$

With a diode consisting of a highly doped p-layer and a low-doped n-layer hole storage in the n-base primarily occurs. From the charge control equation (72) derive the expressions for storage time and fall time [Mö 65]

$$t_s = \tau_v \ln \frac{1 + \dfrac{I_f}{I_r}}{1 + K_0} \, , \tag{74a}$$

$$t_f = \frac{K_0 \tau_v + R_2 \overline{C}_D}{1 + K_0} \ln 10 \, , \tag{74b}$$

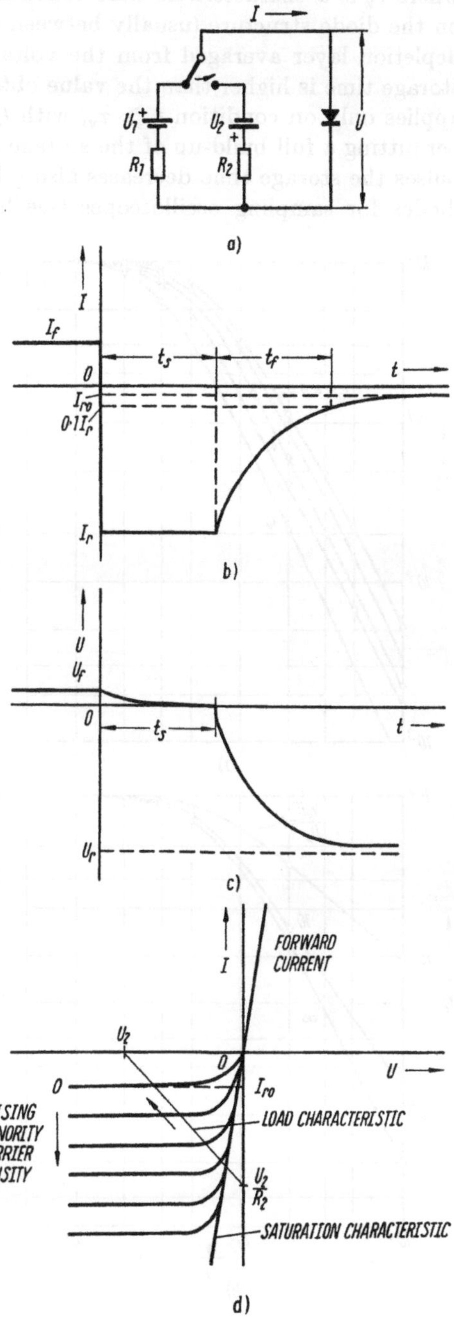

Fig. 66. Diode behaviour when changing from the conduction state to the blocked state. According to *Lunze* [Lu 59; Lu 64a].

a) Principle circuit for measurement. *b)* Current curve. *c)* Voltage curve. *d)* Instantaneous reverse current characteristic.

where τ_v is a characteristic time constant, K_0 the correction factor dependent on the diode structure (usually between 0·2 and 1), $\overline{C}_D$ the capacitance of the depletion layer averaged from the voltage values U_1 and U_2. For $I_f/I_r \lessapprox K_0$ storage time is higher than the value obtained from equation (74a). The latter applies only on condition $t_0 \gg \tau_v$, with t_0 the duration of the forward current permitting a full build-up of the storage charge. With shorter forward current pulses the storage time decreases also [Mö 65], which is of interest with mixer diodes for sampling oscilloscopes (see Section 4.3.3). The time constant τ_v, characterising the time behaviour also for very high injection currents [Mö 66a], depends on the production parameters of the diode, as illustrated in Fig. 67. When bulk thickness w is high compared to diffusion length L_p of the holes,

$$\tau_v \approx \tau_p \qquad (75a)$$

applies, with τ_p representing the mean lifetime of the holes. Small-base diodes with $w \ll L_p$ have a better time behaviour, which is expressed with

$$\tau_v \approx \frac{w^2}{2\,D_p} = t_t\,, \qquad (75b)$$

where D_p is the diffusion constant and t_t the mean transit time of the holes through the base layer. In this case, recombination of the charge carriers occurs primarily at the base contact. Recombination speed V_0 of the latter has therefore considerable influence

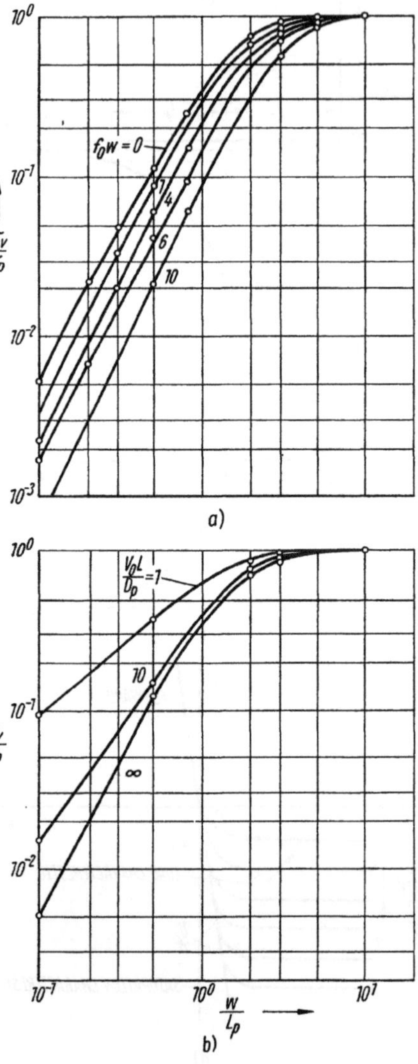

Fig. 67. Dependence of the time constant τ_v on the production parameters of a p^+-n diode. According to *Möschwitzer* [Mö 65].

a) Influence of a drift field.
b) Influence of the base contact.

τ_p mean lifetime of holes,
w width of base layer,
D_p diffusion constant,
$L_p = \sqrt{D_p \tau_p}$ diffusion length,
V_0 recombination speed at base contact,
$f_0 = e\,|E|/k\,T$ field factor,
$L = \dfrac{1}{2\sqrt{(f_0/2)^2 + 1/L_p^2}}$ equivalent diffusion length.

(see Fig. 67*b*). Inhomogeneous doping of the bulk material also is favourable for the time behaviour of the small-base diode. This produces a drift field $\boldsymbol{E}$ which accelerates the charge transport (see Fig. 67*a*).

It is necessary in any case to reduce the lifetime τ_p of holes by incorporating additional recombination centres, for instance, gold or nickel in germanium diodes. Differences in the results of time behaviour calculation, however, depend primarily on the assumptions for the model [Be 65b; Gu 66a].

HOT CARRIER DIODES

Minority carrier storage is avoided when the p-n junction is replaced by an epitaxial metal—semiconductor contact with *Schottky* barrier formation at the boundary [Hu 56; Sp 56]. Electrons passing from the n-layer into the metal, for example, travel with a velocity greater than the thermal velocity of the metal electrons ("hot carriers"), due to the height of the *Schottky* barrier. Through the interaction between the metal lattice and the metal electrons, the "hot carriers" loose their excess energy so rapidly that on applying a reverse voltage the barrier cannot be overcome and no contribution is delivered to the reverse current. Compared with point contact diodes, this type has a somewhat higher capacitance ($\approx$1 pF) but can deliver a higher current. With this arrangement a switching time of a fraction of 1 ns can be expected, although time behaviour is limited by diode impedance. Inductance can be reduced to $L = 3$ nH by respective packaging [Kr 63].

CHARGE-STORAGE DIODES

For utilising the stored charge to produce a steep pulse the development of the charge-storage diode† is of special interest. Against the conventional diode, this type features long-life minority charge carriers which permit the storage of a high charge with a low forward current. Inhomogeneous doping with an impurity concentration that drops sharply in the junction produces an inbuilt field which retards the movement of the minority carriers (see Section 2.4.1). This causes minority carrier storage to concentrate primarily in the immediate vicinity of the junction. Because practically all charge carriers are collected during the storage period, fall time t_f is small, ranging from 0·1 to 0·5 ns. The higher the gradient of concentration, the shorter will be the fall time. A disadvantage of the high concentration of charge carriers stored in a thin layer and on a small area is the modulation of the path resistance and consequent reduction of the retarding field. Thus, with increasing forward current I_f fall time t_f rises rapidly. With diodes of larger junction area this effect is less pronounced because the charge density is lower. Resulting increase of the capacitance, however, affects the time behaviour unfavourably. For the most suitable conditions the optimum junction area must be found. Charge-storage diode theory has been detailed by *Moll* et al. [Mo 62a].

† Sometimes called *Boff*, snap-off, or step recovery diode.

As shown in Fig. 66c the short fall time of the reverse current is concurrent with a voltage pulse which drops sharply at the moment t_s. This may be utilised, for instance, to block a subsequent charge-storage diode to obtain successive pulse steepening [Hu 63].

With charge-storage diodes a current gain I_r/I_f can be effected provided the stored charge from a low forward current — e.g. $I_f = 1$ mA during 10 ns — is collected through a high reverse current, for instance, $I_r = 10$ mA during 1 ns [Ch 63]. This amplification action of the charge-storage diodes in combination with tunnel diodes is useful to improve nanosecond logic circuitry. This has been dealt with by *Sear* [Se 63].

TUNNEL DIODES

Conduction Mechanism

The shape of the volt-ampere characteristic of very highly doped p-n junctions, as has been found by *Esaki* [Es 58], is caused by the tunnel effect, which is responsible also for the *Zener* breakthrough (Fig. 68). In p-n junctions with usually prevailing impurity concentrations of $N < 10^{16}$ to $10^{17}/cm^3$ a sudden current breakthrough occurs only with a high reverse voltage level. With a doping concentration of $N = 10^{18}$ to $10^{20}/cm^3$, however, the *Fermi* energy, i. e. with a 50% occupation probability at temperatures above absolute zero, is not located any more in the energy gap but in the p-region in the valence band and in the n-region in the conduction band (see Fig. 28 and 29, p. 61 and 63). In the absence of the external voltage U both *Fermi* energies in the p-n junction are located on the same level causing the partial overlap of the valence band of the p-layer and the conduction band of the n-layer, in this case. Further, on such a high doping the width w of the depletion layer amounts to only about 10 to 20 nm

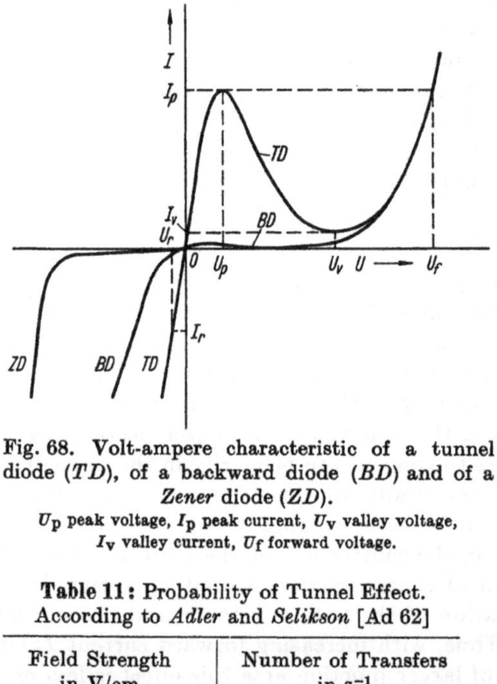

Fig. 68. Volt-ampere characteristic of a tunnel diode (*TD*), of a backward diode (*BD*) and of a *Zener* diode (*ZD*).
U_p peak voltage, I_p peak current, U_V valley voltage, I_V valley current, U_f forward voltage.

Table 11: Probability of Tunnel Effect. According to *Adler* and *Selikson* [Ad 62]

Field Strength in V/cm	Number of Transfers in s^{-1}
1×10^5	10^{-45}
5×10^5	10^{-6}
1×10^6	10^7
5×10^6	10^{11}
1×10^7	10^{13}

and in consequence a field strength of $E = 10^5$ to 10^6 V/cm prevails in the depletion layer. As Table 11 shows, there is a high probability on such field strength for the electrons to transfer to a similar energy state in the adjacent layer, i.e. to "tunnel through" the forbidden zone. Although the *Zener* breakthrough occurs already on $U = 0$ V, the tunnel currents in both directions cancel each other because for the occupied states of one layer an equal number of unoccupied states exist in the other layer. When an external reverse voltage U_r is applied the p-layer *Fermi* energy is raised against the n-layer *Fermi* energy level. This causes the tunnel current of the populated p-layer to prevail and pass into the unoccupied states of the same energy level in the n-layer (Fig. 69). Thus, a tunnel diode has no rectification effect, but shows a reverse current I_r which increases rapidly with rising reverse voltage (Fig. 68). For the same reason, when a low forward voltage is applied the tunnel current from the n-layer into the p-layer dominates. The forward current rises until, on condition $U = U_p$ of the forward voltage, the occupied states of the n-conduction band are balanced by maximum of unoccupied states of the p-valence band, i.e. the peak current I_p is reached. On further increase of the forward voltage, band-overlap and thus the tunnel current I_T decreases, and reaches zero when the overlap is cancelled on $U \approx U_v$. Around this voltage level the normal diffusion current of a p-n junction diode starts, which effects the renewed rise of the tunnel diode characteristic. To the magnitude of the resulting valley current I_v a certain excess current contributes which is caused by the tunnel current passing intermediate levels of impurity in the energy gap.

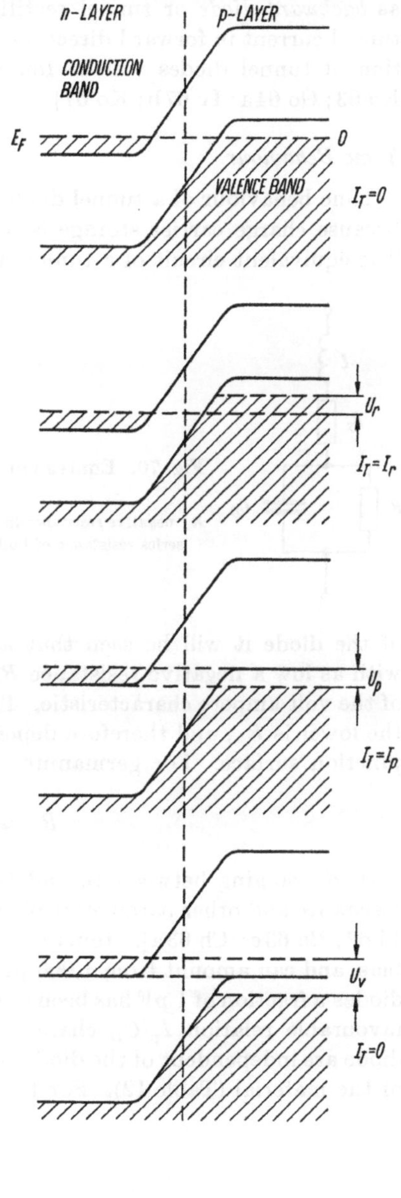

Fig. 69. Location of the energy bands in a tunnel diode and the resulting tunnel current I_T at various external voltages U.

Diodes having a narrow junction but, with no band-overlap at voltage zero, i.e. with *Fermi* energy in the n-region at the bottom of the conduction band, yield a steep tunnel current rise on reverse voltage application and result only in the gradual onset of the diffusion current under forward voltage. As a consequence of the steep onset of the reverse current and the flat initial section of the characteristic curve such tunnel diodes may be used in conventional diode mode when poling is reversed. This type of diode is known as *backward diode* or tunnel rectifier. The circuit symbol is drawn with the tunnel current in forward direction. A review of the properties and the application of tunnel diodes will be found in various papers [Ha 61c; Ad 62; Fi 62; Ko 63; Go 64a; Iv 67b; Ko 67].

Time Behaviour

Time behaviour of a tunnel diode is not limited by the conduction mechanism because charge carrier storage is not associated with the tunnel current flow. The equivalent circuit of a tunnel diode is given in Fig. 70. For fast switching

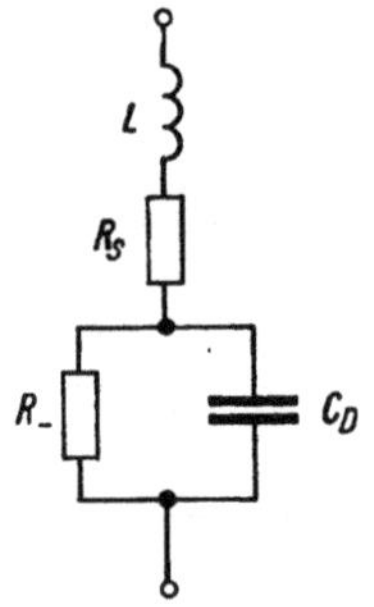

Fig. 70. Equivalent circuit of a tunnel diode. According to *Fistul'* and *Shvarts* [Fi 62].

R_- negative resistance in descending part of curve, C_D depletion layer capacitance, R_s series resistance of bulk material, contacts and leads, L series inductance of leads.

of the diode it will be seen that a low junction capacitance C_D is required, with as low a negative resistance R_- as possible, i.e. a steeply descending part of the volt-ampere characteristic. Thus, the higher the tunnel effect probability the lower is R_-, and therefore depends on the band gap, on doping and on the junction surface. For germanium diodes the approximate relation is [Fi 62]

$$R_- \approx 0{\cdot}4 \frac{U_V - U_P}{I_P - I_V}, \tag{76}$$

with R_- ranging between 10 and 100 Ω. Measuring methods for the negative resistance and other parameters of the diode are found in several papers [Ad 62; Fi 62; Ba 63c; Ch 63a]. Junction capacitance depends on doping, surface and bias, and can amount from 1,000 pF to less than 1 pF [Nu 67]. With backward diodes a fraction of 1 pF has been obtained [Go 63]. Product $R_- C_D$ and the most favourable relation I_p/C_D characterising the switching behaviour of a tunnel diode are independent of the diode surface and specifically given by the property of the material (Table 12). For the parasitic series resistance $R_s \approx 0{\cdot}1$ to 1 Ω applies. Parasitic series inductance should be as low as possible, and with a sophisticated package design $L < 0{\cdot}1$ nH has been obtained [Nu 67].

For a certain working range the tunnel diode represents a voltage-controlled negative resistance, which permits its use as amplifier, oscillator and high-speed switch. Depending on bias and value of the load resistance, according to Fig. 71, the load line has one or three intersections with the tunnel diode characteristic, with the circuit acting either as free-oscillating, monostable or bistable multivibrator (see Sections 3.2.4 and 3.7.2). In approximation, for tunnel diode trigger circuits the output pulse rise time t_r is [Go 63]

$$t_r \approx C_D \frac{U_f - U_p}{I_p - I_v}. \tag{77}$$

It is evident that the ratio I_p/C_D must be large for high-speed switching action. Rise time values underpassing 50 ps have been obtained. The advantages of tunnel diode circuits are a simple design and very high speed, to which must be added power amplification capability, low consumption, compact dimensions, a high degree of reliability and low sensitivity against radiation and temperature. The two-pole character where input and output are completely coupled is a serious drawback of the tunnel diode. Close tolerances of the diode parameters, of the operation voltage and of the other circuit elements are another requirement. That impedance and output signal are relatively low is also unfavourable. By combining with other components these disadvantages can be offset to a certain extent, although a combination with transistors is not as fast [To 60; Am 61; Me 63]. Coupling charge-storage diodes with tunnel diodes has turned out particularly successful because the properties of both supplement each other favourably and a specially high gain-bandwidth product is obtained [Ch 63; Se 63].

Table 12: Properties of Tunnel Diodes Made of Different Semiconductor Materials. According to *Fistul'* and *Shvarts* [Fi 62]

Material	Peak-to-Valley Ratio I_p/I_v	Peak Voltage U_p, in mV	Valley Voltage U_v, in mV	Forward Voltage U_f, in mV	Maximum Permissible Temperature T_{max}, in °C	Time Constant $R_- C_D$, in ps	I_p/C_D in mA/pF
Si	3 to 4	80 to 100	400 to 500	700	400	2000	<0·5
Ge	10 to 15	40 to 70	270 to 350	450	250	500	0·3 to 1
GaAs	40 to 70	90 to 120	450 to 600	1000	600	100	10 to 15
GaSb	15 to 20	30 to 50	200 to 250	450	300	100	—
InSb	7 to 10	—	—	200	25	5	—

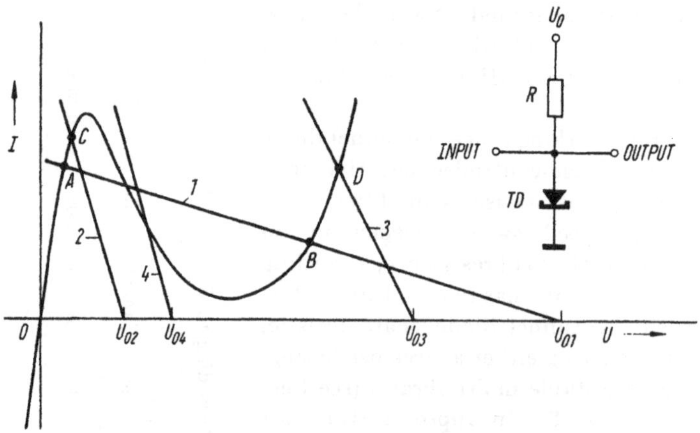

Fig. 71. Several designs of tunnel diode trigger circuits.

Load line 1: bistable multivibrator with stable working points A and B, $R > R_-$. *Load lines 2 or 3:* monostable multivibrator with stable working point C or D, $R < R_-$. *Load line 4:* astable multivibrator, $R < R_-$.

Characteristic of Tunnel Diode Combinations

Combinations of one or several tunnel diodes with other components have interesting characteristics and permit a great number of switching functions [Si 61; Ko 63; Wi 63; Go 64a; Mu 67a]. Some examples of these combinations are given in Fig. 72, p. 124. Such a combination characteristic is formed from the various individual characteristics, i.e. with components in parallel, from the sum of the currents at the same voltage level and with series-circuited components from the sum of the voltages at the same current level. In the latter case, however, two tunnel diodes cannot operate simultaneously in the same part of the descending characteristic. *Earnshaw* and *Tan* [Ea 66] give a graphic method for the construction of the static characteristic. To derive the functional details, for example, of a scaler with series-connected tunnel diodes, it is necessary to regard the dynamic parameters as well as the parameters of the external circuit [Ta 67c].

Connecting a tunnel diode and a resistor $R = R_-$ gives the tunnel resistor [Co 63a] which has the characteristic of a constant current source in a specific voltage range. Connected in series, N tunnel diodes have a characteristic with N descending parts and permit the assembly of a trigger circuit having $N + 1$ stable states, for example, in scaler circuitry. Parameter deviations restrict N to between 10 and 15. By connecting a tunnel diode in series with a diode or a backward diode the tunnel diode characteristic can be shifted to a higher voltage level, which allows, for instance, germanium tunnel diodes to be adapted to silicon transistors. In Fig. 72 *h* the combinations shown in Fig. 72 *e* and 72 *g* are connected in parallel. This set-up has three stable states and is suitable for producing bipolar pulses in memory circuits [Wi 63].

3.1.5 TRANSISTORS

A transistor can be imagined as being assembled from two p-n junctions with either the two n-regions (p-n-p transistor) or the two p-regions (n-p-n transistor) forming a common zone, the base. The forward poled emitter diode injects charge carriers into the base which are collected by the reverse poled collector diode. Principally, the forward current of the emitter diode consists of holes in a p-n-p-type and of electrons in an n-p-n-type transistor (see Section 3.1.4). Extensive information regarding theory and circuitry of transistors is to be found in existing literature, e.g. [Sh 57; Do 59; Fe 60;Le 63; Sa 63a; Ba 64; De 64b;Uz 64;Ro 65]. In order to make the following circuit description clearer only some of the important relations will be dealt with here.

TRANSISTOR EQUIVALENT CIRCUITS

There is a certain analogy between the properties of an electron valve and of a transistor as is evident from the respective family of static characteristics. Due to inherent coupling the equivalent circuit of a transistor is far more complicated than the electron valve equivalent circuit, especially with high frequency. Although the diffusion phenomena should be depicted by transmission lines, representation by lumped components is preferred, and further simplification has to be resorted to for practical handling in calculations. Against the equivalent circuits derived, more or less formally, with the aid of the four-terminal network theory, the equivalent circuit constructed on the basis of physical considerations gives a better insight because the elements of such an equivalent circuit permit direct physical interpretation.

Fig. 73*a* shows the equivalent circuit of a common-base transistor where the forward resistance r_e of the emitter diode amounts to

$$r_e = \frac{kT}{e_0} \frac{1}{I_E} = \frac{26}{I_E/\mathrm{mA}} \,\Omega \tag{78}$$

on emitter current flow I_E. The base resistance r_b consists primarily of the base-spreading resistance. As the back resistance of the collector diode is finite the collector junction resistance r_c has to be located parallel to the current source αI_E. With epitaxial transistors the collector-spreading resistance can, in most cases, be neglected because the high-ohmic collector junction is very thin.

Frequency behaviour of a transistor is primarily determined by the dependence on frequency of the current amplification and by the capacitances of the p-n junctions. On low frequency, in a common-base transistor, this dependency of the short-circuit current gain α is approximated by

$$\alpha \approx \frac{\alpha_0}{1 + j\,\dfrac{\omega}{\omega_\alpha}} \,. \tag{79}$$

No.	Circuit	Volt — Ampere Characteristic
a	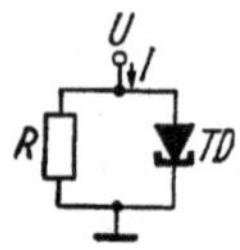	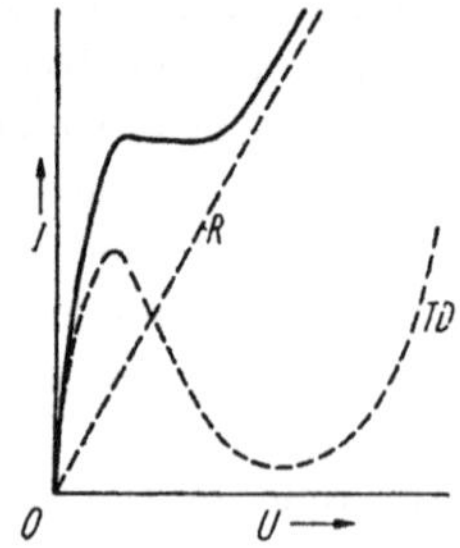
b	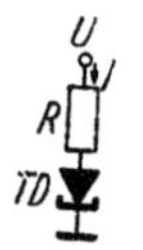	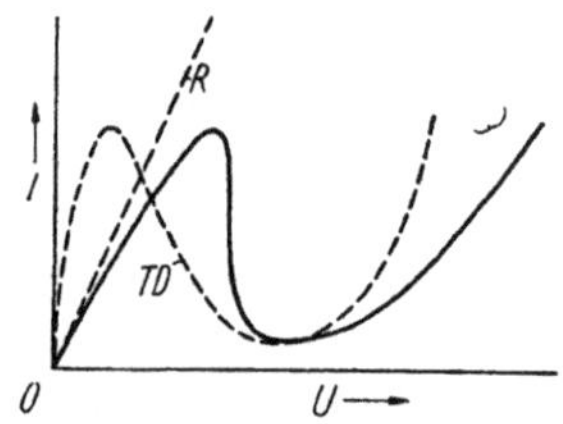
c	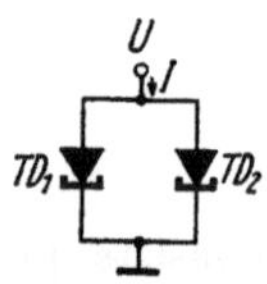	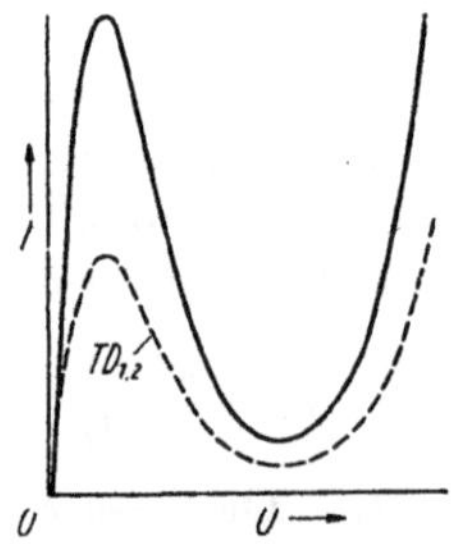
d		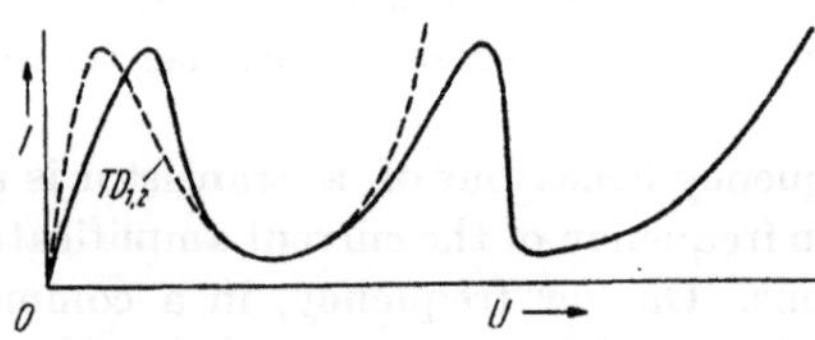

Fig. 72. Characteristics of tunnel diode combinations. According to

No.	Circuit	Volt – Ampere Characteristic
e		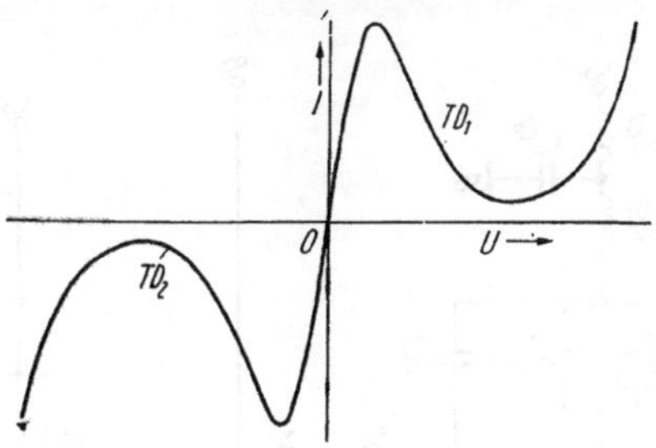
f	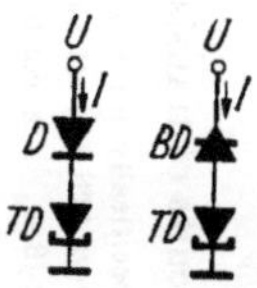	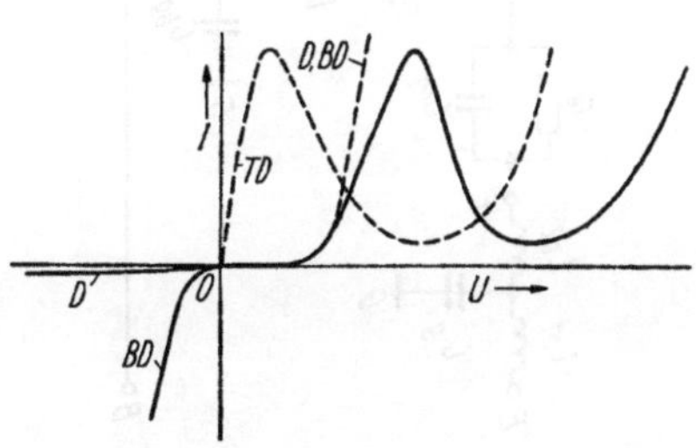
g	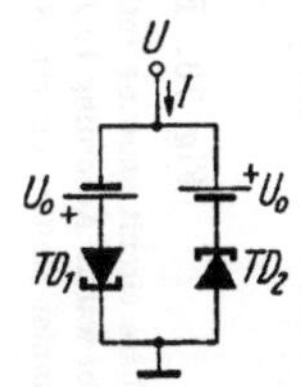	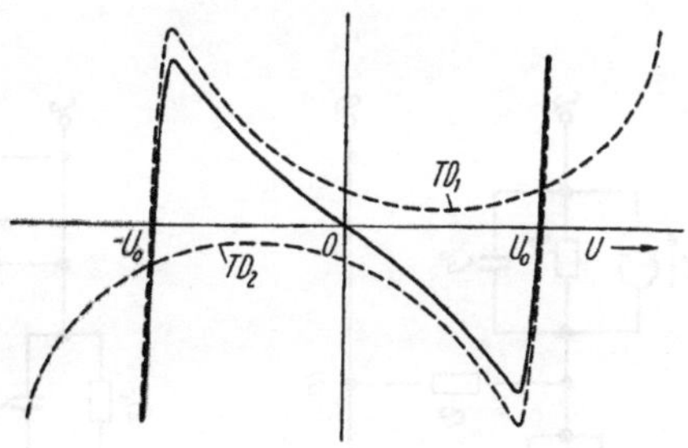
h	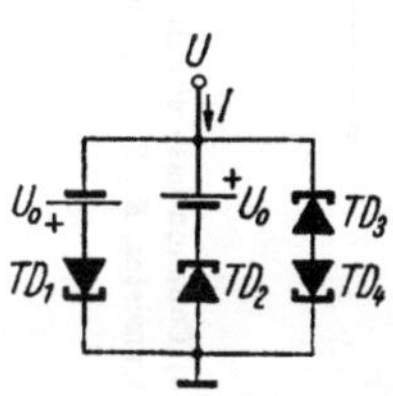	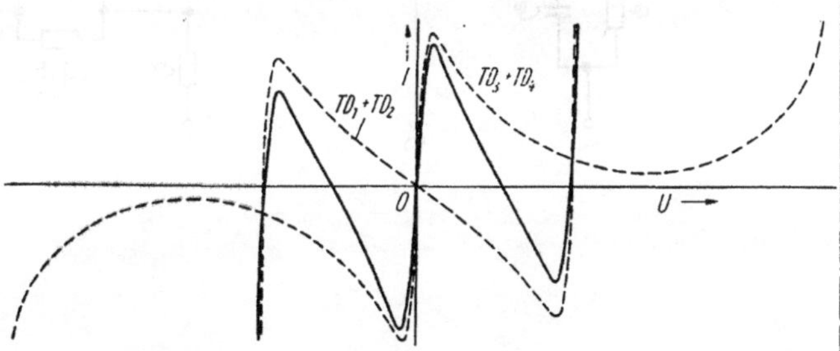

Williams and *Hamilton* [Wi 63] and *Gostrem* and *Zinov'ev* [Go 64 a].

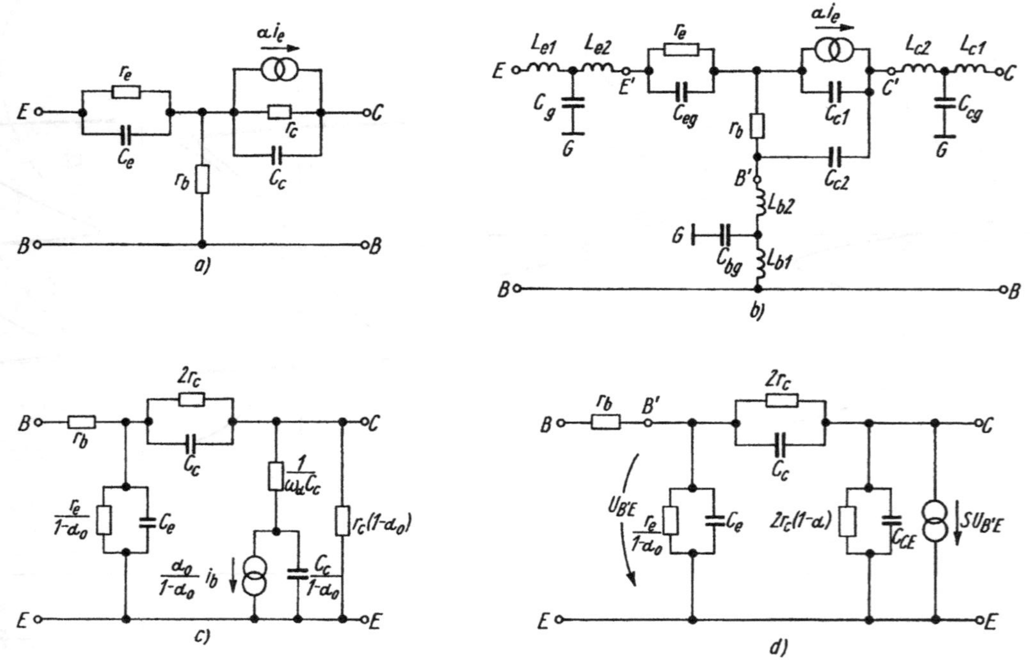

Fig. 73. Equivalent circuits of transistors.

a) Common-base circuit. *b*) Common-base circuit showing influence of can, specifically for v.h.f. transistors; *G* can, connected to base of base-circuited transistor, *E'*, *B'*, *C'* inputs of transistor wafer. According to *Thommen* and *Strutt* [Th 65]. *c*) Common-emitter circuit, according to *Pritchard* [Pr 54]. *d*) Common-emitter circuit, $S = \alpha_0/\left(r_e(1 + j\,\omega/3\,\omega_\alpha)\right)$. According to *Giacoletto* [Gi 54].

When reaching the cut-off frequency ω_α the current gain has decreased to $\alpha_0/\sqrt{2}$; ω_α is expressed by

$$\omega_\alpha = \frac{1}{r_e\, C_e} \quad \text{or} \quad f_\alpha = \frac{\omega_\alpha}{2\,\pi}. \tag{79a}$$

In case of a homogeneously doped base this can be written

$$\omega_\alpha = \frac{2{\cdot}43\, D}{w^2}, \tag{79b}$$

where D is the diffusion constant and w the base width. For frequencies to about 0·1 ω_α equation (79) represents a good approximation, but with higher frequencies the phase response has to be taken into account, and the equation becomes

$$\alpha \approx \frac{\alpha_0}{1 + j\dfrac{\omega}{\omega_\alpha}}\, e^{-j\times 0{\cdot}2\omega/\omega_\alpha} \tag{80}$$

with the factor 0·2 denoting the homogeneous base transistor.

At both p-n junctions a space charge region is formed which appears as capacitance in the equivalent circuit. From the voltage dependency of this space charge capacitance (see Section 3.1.4) it can be concluded that the emitter junction capacitance is higher than the capacitance of the collector diode.

Diffusion capacitance $C_d = \Delta Q/\Delta U$ results from the charge alteration ΔQ in the base (a change in the number of injected holes) at a voltage change ΔU. In general, diffusion capacitance is the main contribution to emitter junction capacitance C_e, while with the collector junction capacitance C_c the space charge capacitance is the main contributor. From the equations (79a, b) the emitter diffusion capacitance derives with

$$C_e = \frac{1}{r_e\, \omega_\alpha} \approx \frac{w^2}{2\, r_e\, D} \tag{81}$$

and reaches 10 to 100 pF with high-frequency transistors. Collector junction capacitance C_c is lower and consists of the components C_{c1} and C_{c2} in mesa- and planar-transistors (Fig. 73*b*).

Besides the behaviour of the transistor wafer, the capacitance between the electrode connections and the influence of the can need consideration with v.h.f. transistors. Fig. 73*b* shows the equivalent circuit of a transistor where can influence is taken into account [Th 65]. Of the capacitance against the can the lead-through of the wires furnishes the main contribution and therefore can be assumed to represent lumped elements (C_{eg}, C_{bg}, C_{cg}). Further, *Thommen* and *Strutt* [Th 65] have shown that the equivalent circuit of the transistor wafer, as drawn in Fig. 73*b*, is also applicable for the GHz region. Due to the influence of the can, however, the cut-off frequencies f_α and f_β are not suitable for characterising the whole transistor.

Figs. 73*c* and 73*d* show frequently used equivalent circuits for common-emitter transistors. The latter, according to *Giacoletto* [Gi 54], has the disadvantage that the voltage between the emitter and the inner point B' is included in calculation. The last two equivalent circuits apply for frequencies amounting to $\omega \leqq \omega_\alpha$, while that in Fig. 73*a*, with α according to equation (79), is suitable for frequencies to $0{\cdot}1\ \omega_\alpha$.

BASIC TRANSISTOR CIRCUITS

The three basic transistor circuits most important in practical work are characterised in Table 13 (see p. 130). Specified equations are based on the equivalent circuits in Fig. 73 with the influence of the transistor-can not taken into account. Equations as well as the circuits apply only for small-signal behaviour. Grounded-collector and grounded-emitter circuits permit current amplification, while grounded-base and grounded-emitter transistors are suitable for voltage amplification. Table 13 contains the most frequently used current gain factors and cut-off frequencies, and the approximations for the input and output resistance [Mi 56; Wi 62; Uz 64]. Resistance values apply for low frequencies and for the conditions $r_c \gg r_b > r_e$ and $(1-\alpha) \ll 1$. For the purpose of demonstration, numerical values for resistance, cut-off frequency and rise-time constants entered in the table have been referred to data of a high-frequency transistor type 2 N 976.

For designing transistorised pulse circuits the knowledge of rise-time constant data and their dependence on the internal resistance of the generator R_G and on the load resistance R_L is of special importance [Mi 56; Lu 61; Wi 62; Sa 63a; Ba 64; Ba 64a]. When a step-type input voltage is applied, the transistor output current increases exponentially with the time constant τ. Rise time amounts to

$$t_r \approx 2{\cdot}2\,\tau\,. \tag{82}$$

For the rise-time constant τ_0 one gets [Lu 61]

$$\tau_0 = \frac{1}{2\pi f_\alpha}\,\frac{R_E + R_B + r_b}{R_E + r_e + (R_B + r_b)(1-\alpha)} \tag{83}$$

provided the load resistance is low and, therefore, capacitance and resistance of the collector junction can be neglected. Consistent with type of transistor circuit, values of R_G, 0 or R_L are used for the external resistance R_E and R_B.

With a high load resistance R_L in the collector lead the rise-time constant in the collector circuit increases due to the *Miller*-effect capacitance, thus

$$\tau \approx \tau_0\,(1 + 2\pi f_\alpha R_L C_c) \tag{84}$$

with $C_c/C_e \ll 1$. The factor $(1 + 2\pi f_\alpha R_L C_c)$ especially affects the already extended rise-time of the common-emitter transistor stage [Iv 67]. If $R_L C_c$ equals $1/2\pi f_\alpha$ and can therefore be neglected with reference to τ_0, the rise-time constant is, nevertheless, increased by the factor 2.

Voltage amplification has been given by *Williams* and *Neiler* [Wi 62] with the expression

$$V_u = V_{u0} \frac{1}{1 + j f/f_{in}} \frac{1}{1 + j f/f_{out}}, \tag{85}$$

where V_{u0} is the voltage amplification on low frequencies, and f_{in} and f_{out} the cut-off frequencies of the input and output circuits respectively. Because $f_{in} \approx 1/2\pi\tau_0$ and $f_{out} = 1/2\pi R_L C_c$ this expression is equivalent to equations (83) and (84). Circuit and electrode capacitance values also contribute to the magnitude of the capacitance C_c.

Table 13 and equation (84) permit the following conclusions:

(i) A minimum rise time of a common-base stage is obtained when driven by a high impedance source. With common-emitter and common-collector circuits a low impedance source will give minimum rise time. According to equation (84) the rise time of common-base and common-emitter stages increases with rising R_L. On the other hand, due to $(R_G + r_b) \gg [r_e + (R_G + r_b)(1 - \alpha)]$, in a common-collector circuit a decrease of the rise time can be expected with rising emitter resistance R_L, as has been experimentally proven [Lu 61; Ba 64a].

(ii) From the rise-time constant τ_0 of a common-emitter circuit it becomes evident that of two transistors having the same cut-off frequency f_α, the transistor with the lower current gain factor α yields the better results.

(iii) Basic circuits yield minimum rise time when mismatched, e.g. for the common-base transistor circuit with its very low input impedance the high impedance pulse generator is most favourable.

(iv) Due to its very low output impedance the common-collector transistor circuit (emitter follower) recommends itself as impedance converter for subsequent very low impedance driven stages, for example, a common-emitter transistor. Even with a relatively high internal resistance R_G of the generator, the common-collector circuit yields a low rise time.

These considerations are taken into account for wide-band amplifiers as illustrated in Section 3.3, where a common-collector transistor connection, for instance, is used as input stage which acts as impedance converter. Thus, a very low impedance supply to the emitter circuit is assured, as is necessary to obtain a low rise time. To keep the influence of the collector resistance R_L low another transistor with an earthed base is inserted in the collector circuit of the common-emitter stage (cascode stage).

If a high internal resistance R_G of the generator applies, the rise-time constant τ_0 of the emitter stage assembly can be reduced to a level applicable for $R_G = 0$ by connecting a capacitance C_G parallel to the resistance R_G. With $\tau_G = R_G C_G = 1/2\pi f_\beta$ and $R_G \gg |r_b (1 + j\omega\tau_G)|$ the external RC-element and the transistor input act as compensated voltage divider [Sa 63a]. This "quasi voltage control" permits the stabilisation of the base current, contrary to operation on $R_G = 0$.

Rise-time values given were calculated for a step-type generator voltage and for the current gain factor α according to equation (79). Using the better approximation of equation (80) the time delay of the output pulse becomes [Ba 64]

$$t_d \approx 0{\cdot}2/\omega_\alpha = 0{\cdot}1\, w^2/D\,. \tag{86}$$

Pulse response delay due to the finite transit time of the minority charge carriers through the base is low against the pulse rise time.

BEHAVIOUR ON LARGE-SIGNAL AND OVERDRIVING CONDITIONS

On small-signal behaviour the relation between input and output data is well defined, while on large-signal switching operation this is not so. Fig. 74 shows a section of the respective family of characteristics. Apart from the avalanche region that has not been entered, three regions are distinguishable, viz. the active small-signal gain region; the residual current region where the switch is "off" and a high voltage and low current obtain; and the residual voltage region, where the switch is "on" and a low voltage but a high current obtain. On a very low voltage U_{CE}, e.g. due to a high voltage drop across the collector resistance R_L, the bias of the collector diode reverses its polarity to forward direction and the collector then acts as emitter and also injects holes into the base. This condition is known as saturation and is characterised by a high charge carrier density in the base. Any further increase of the emitter current only results in a rise of the base current. The residual voltage of the collector-base junction amounts to about 0·1 V and depends on the spreading resistance of the base and of the collector. A minimum collector cut-off current obtains when the emitter diode is also blocked.

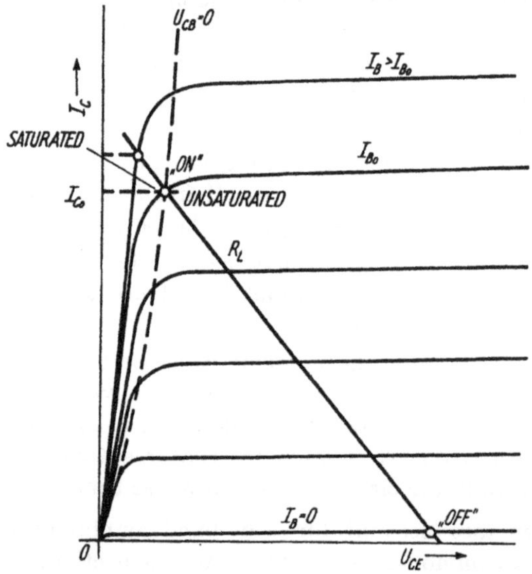

Fig. 74. Section from the family of static characteristics with the load line R_L and the working points "ON" and "OFF" in switching operation.

It is evident that charge carriers additionally stored on saturation of the base will result in a considerably higher delay than with small-signal drive operation (see Section 3.1.4). The collector current is determined by the con-

Table 13: Transistor Circuits

	Common-Base Circuit	
Circuit Diagram	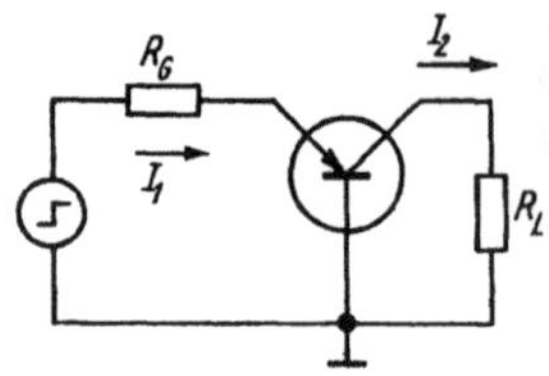	
Short-circuit current Amplification $V_{io} = \frac{I_2}{I_1}\Big\vert\, R_L = 0$	$\alpha \lessapprox 1$	0·9
Cut-off frequency $(V_{io}/\sqrt{2})$	f_α	900 M
Input impedance[b]		
short-circuit $R_L \approx 0$	$r_e + r_b(1-\alpha)$	3
open-circuit $R_L = \infty$	r_b	55
Output impedance[b]		
voltage control $R_G \approx 0$	$r_c\left(\frac{r_e + R_G}{r_b} + 1 - \alpha\right)$	60
current control $R_G = \infty$	r_c	1
Rise-time constant (for $2\pi f_\alpha R_L C_c \ll 1$)	$\tau_0 \approx \frac{1}{2\pi f_\alpha} \quad \frac{R_G + r_b}{R_G + r_e + r_b(1 -}$	
voltage control $R_G = 0$	3·0 ns	
cable connection $R_G = 100\ \Omega$	0·27 ns	
current control $R_G = \infty$	$\frac{1}{2\pi f_\alpha} = 0{\cdot}18$ ns	

[a] Numerical data refer to the high-frequency transistor type 2 N 976, with value $r_c = 1\ \mathrm{M\Omega}$.

[b] For low frequencies with the approximation $r_c \gg r_b > r_e$, according to [Mi

[c] $R_L = 100\ \Omega$.

Ausschlagtafel 5660 Meiling/Stary

es in Small-Signal Operation[a]

-Emitter Circuit		Common-Collector Circuit	
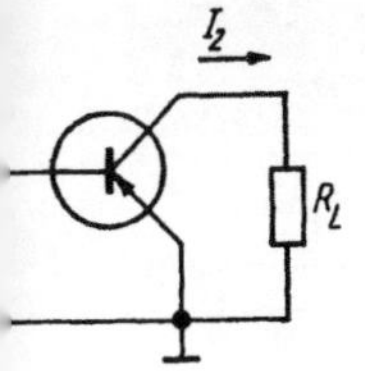		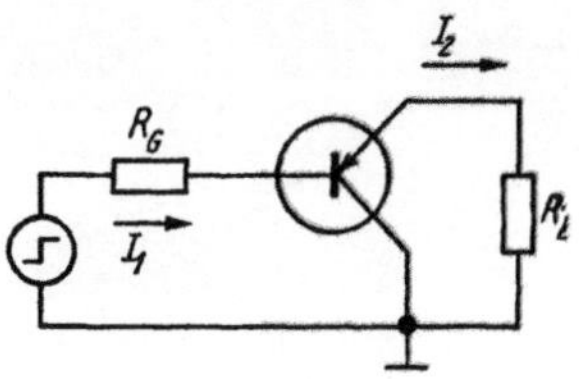 (emitter follower)	
$\frac{\alpha}{-\alpha} \gg 1$	280	$\gamma = \frac{1}{1-\alpha} \gg 1$	
$\cdot\ \alpha) \approx \frac{f_\alpha}{\beta}$	3·6 MHz		
$\frac{r_e}{-\alpha}$	1 kΩ	$r_b + \frac{r_e + R_L}{1-\alpha}$	1 kΩ
	55 Ω	$r_b + r_c$	1 MΩ
$1 - \alpha\Big)$	60 kΩ	$r_e + (r_b + R_G)(1-\alpha)$	3 Ω
$-\ \alpha)$	4 kΩ	$r_c(1-\alpha)$	4 kΩ
$\frac{R_G + r_b}{R_G + r_b)(1-\alpha) + r_e}$		$\tau_0 \approx \frac{1}{2\pi f_\alpha} \quad \frac{R_L + R_G + r_b}{R_L + r_e + (R_G + r_b)(1-\alpha)}$	
3·0 ns		0·27 ns } [c]	
7·5 ns		0·43 ns	
$\frac{}{f_\beta} = 45$ ns		$\frac{1}{2\pi f_\beta} = 45$ ns	

, $r_e = 3{\cdot}3\ \Omega$, $r_b = 55\ \Omega$, $C_e = 55$ pF, $C_c = 1{\cdot}6$ pF [Ba 64a] and the assumed

centration and the distribution of the minority carriers in the base region, which in turn depend on the input current, on the time and on the elements connected to the input and the output.

For the description of the switching behaviour, transistor inherent parameters are used, which must be conveniently measurable and are little influenced by current and voltage changes. Treating the transistor as a charge-controlled device, according to *Sparkes* and *Beaufoy*, has proved to be a better method than using the small-signal equivalent circuit [Sp 57; Kl 62]. The advantage of this method lies in the clearer calculation and in the fact that the occurring parameters can be actually measured during the switching operation [Lu 64a].

In the main, four charge-control parameters are used [En 62; Kl 62; Th 62; Lu 64a; Mö 65; Re 67a; Ha 68]:

(i) The collector time constant τ_c:
In normal operation the injected holes move to the collector in the time τ_c either through diffusion or the action of a drift field. Thus, total charge Q_B in the base is determined by the collector current I_C and the time constant τ_c, with the relation

$$\tau_c = \frac{Q_B}{I_C} . \tag{87}$$

This strictly applies only for the stationary state; where, however, the stationary distribution associated with Q_B is obtained in a time that is smaller than τ_c. This makes equation (87) applicable for frequencies up to $1/2\pi\tau_c$ [Kl 62]. The time constant τ_c further represents the interval where the charge of the base region, with the emitter connection interrupted, has dropped to $1/e$ of its initial value. It is practical to define the parameter τ_c for a working point on the saturation threshold $U_{CB} = 0$ (see Fig. 74) by

$$\tau_{c0} = \frac{Q_{B0}}{I_{C0}} . \tag{87a}$$

For a diffused transistor the following equation holds [En 62; Kl 62]:

$$\tau_c = \frac{1{\cdot}22}{\omega_\alpha} \approx (0{\cdot}5 \text{ to } 1)\,\frac{w^2}{2D} . \tag{88}$$

(ii) Current amplification β_0 at the saturation threshold ($U_{CB} = 0$) in a common-emitter transistor:

$$\beta_0 = \frac{I_{C0}}{I_{B0}} . \tag{89}$$

In stationary state only as many electrons have to be fed-in to maintain the electron charge in the base as are lost through recombination with holes. Therefore

$$\tau_p = \frac{Q_B}{I_B} \tag{90}$$

applies, with τ_p denoting the mean hole lifetime in the base, which together with equations (87a) and (89) becomes

$$\tau_p = \beta_0 \tau_c . \tag{91}$$

(iii) Collector space charge Q_{V0}:

Driving $U_{CB} = -U_C$ to the saturation threshold $U_{CB} = 0$ causes a considerable change of the collector voltage resulting in a change of the space charge Q_{V0} in the collector junction [Kl 62; Lu 64a]

$$Q_{V0} = \int_{-U_C}^{0} C_c(U_{CB})\, dU_{CB} . \tag{92}$$

With a mean collector junction capacitance $\overline{C}_c$ this can be expressed in approximation [Mö 65]

$$Q_{V0} \approx \overline{C}_c U_C \approx \overline{C}_c R_L I_{C0} . \tag{93}$$

(iv) Storage time constant τ_s:

In contrast to the parameters dealt with under (i), (ii) and (iii), the parameter τ_s refers to a case of overdrive, and represents the recombination

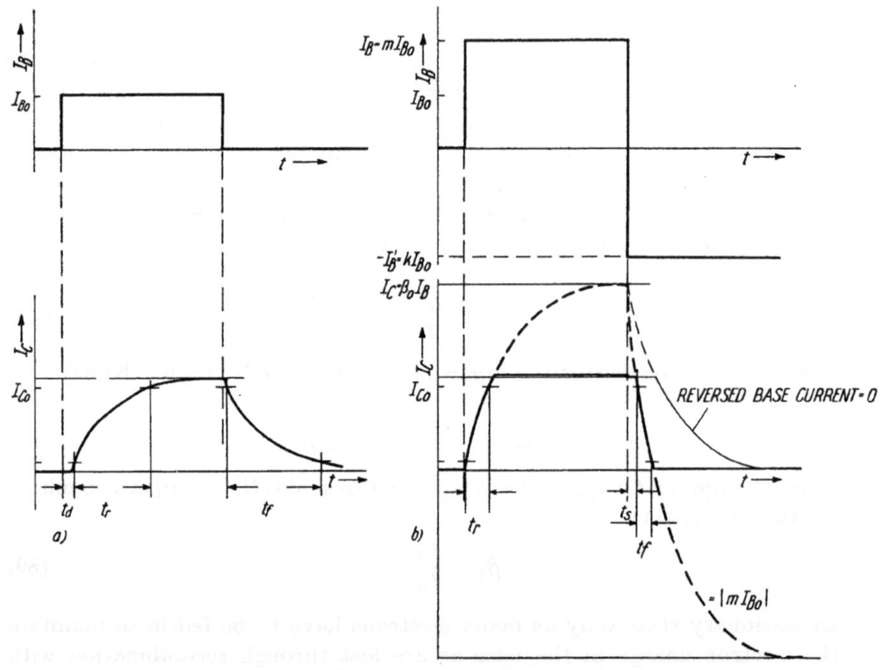

Fig. 75. Driving behaviour of an emitter-circuited transistor.

a) When driven to the threshold where $U_{CB} = 0$. b) When driven into the saturation region; with the collector current pulse sketched in with and without extraction current.

time constant of the excess charge Q_{Bs} which is discharged as excess base current $I_{Bs} = I_B - I_{B0}$ and is expressed with

$$\tau_s = \frac{Q_{Bs}}{I_{Bs}}. \tag{94}$$

As a rule τ_s is somewhat smaller than τ_p because of the greater influence the surface recombination rate exercises in the saturation range [Kl 62]. With modern diffused-base transistors τ_s is in the order of 10 ns.

These four parameters permit the calculation of the switching time on current drive ($R_G = \infty$). When turning-on the base current I_B of an emitter circuited transistor, for example, the rise time can be calculated from the charge balance, thus

$$I_B = \frac{Q_B}{\tau_p} + \frac{d}{dt}(Q_B + Q_V). \tag{95}$$

Calculation of fall time and storage time is handled in an analogue manner. *Baldinger* and *Simmen* [Ba 68] determined storage time with a saturated transistor cut-off by a low impedance generator, e. g. a tunnel diode.

Fig. 75 shows the characteristic time constants of the pulse, while Table 14 contains the results of the common-emitter transistor [Kl 62; Th 62; Mö 65]. Symbol m denotes the overdrive factor of the base current and signifies by how much the base current exceeds the value I_{B0} needed for driving U_{CB} to zero.

From Fig. 75 and Table 14 it will be seen that with rising m the rise time is reduced, but that the pulse is extended by the storage time t_s. Storage time t_s

Table 14: Rise Time, Fall Time and Storage Time of a Switching Transistor for Common-Emitter Circuit and Current Control ($R_G = \infty$). According to [Kl 62; Th 62; Lu 64a; Mö 65]

Common-Emitter Circuit†	Example††		
Overdrive factor $m = \frac{\beta_0 I_B}{I_{C0}} = \frac{I_B}{I_{B0}}$	1	2	2
Extraction factor $k = \frac{\beta_0 I'_B}{I_{C0}} = \frac{I'_B}{I'_{B0}}$	0	0	2
Rise time $t_r = \tau_p \left(1 + \frac{\overline{C}_c R_L}{\tau_c}\right) \ln \frac{m - 0{\cdot}1}{m - 0{\cdot}9}$	$2{\cdot}2\, t_0$	$0{\cdot}55\, t_0$	$0{\cdot}55\, t_0$
Fall time $t_f = \tau_p \left(1 + \frac{\overline{C}_c R_L}{\tau_c}\right) \ln \frac{k + 0{\cdot}9}{k + 0{\cdot}1}$	$2{\cdot}2\, t_0$	$2{\cdot}2\, t_0$	$0{\cdot}32\, t_0$
Storage time $t_s = \tau_s \ln \frac{k + m}{k + 1}$	0	$0{\cdot}7\, \tau_s$	$0{\cdot}29\, \tau_s$

† Analogue equations [Mö 65] apply for the base circuited transistor with $\alpha_0 I_E$ instead of $\beta_0 I_B$ and τ_c in place of τ_p.

†† Approximately corresponds to Fig. 75. For simplification $t_0 = \tau_p \left(1 + \frac{\overline{C}_c R_L}{\tau_c}\right)$.

and the fall time t_f can be reduced by reversing the polarity of the base current (see Fig. 75*b*). The extraction factor k is $\beta_0 I'_B/I_{CO}$, with I'_B representing the extraction current from the base [Th 62; Lu 64a].

Reduction of the space charge in the emitter junction is accompanied by a delay of the turn-on time when changing from blocked state to forward state. With the emitter junction voltage U_e [Th 62] the delay time is

$$t_d \approx C_e U_e/I_B . \tag{96}$$

Thus, with rising overdrive of the base current I_B the delay time t_d decreases, and in many cases the latter receives no consideration in overdrive behaviour analysis.

PROTECTING THE TRANSISTORS AGAINST SATURATION

With a saturated transistor, maximum switching speed from "on" to "off" is limited by the storage time. This often requires that provision be made to restrict or to prevent driving the transistor to saturation state, of which some examples are illustrated in Fig. 76. In the circuit of Fig. 76*a* the diode conducts as soon as the collector current exceeds a value determined by $|U - U_V|$

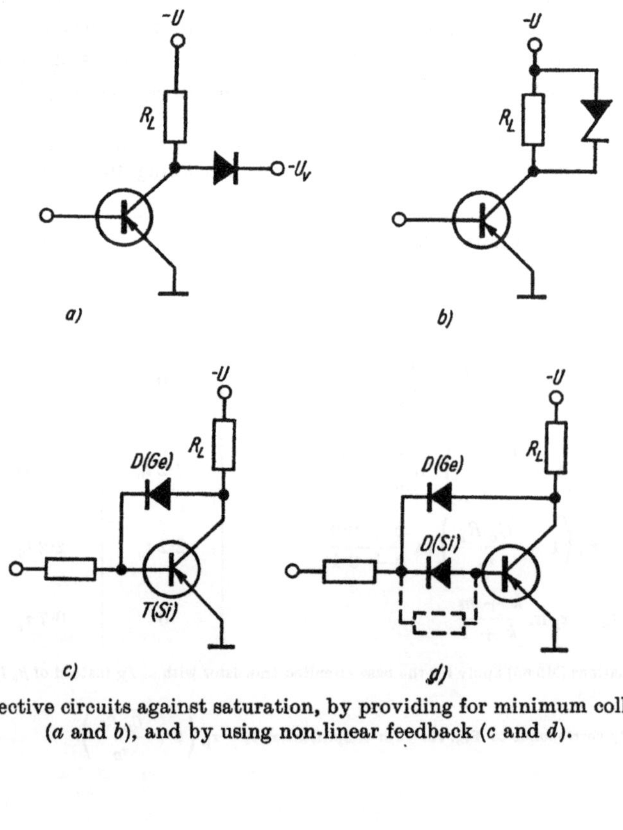

Fig. 76. Protective circuits against saturation, by providing for minimum collector voltage (*a* and *b*), and by using non-linear feedback (*c* and *d*).

and R_L, and in Fig. 76*b* the *Zener* diode serves to stabilise the voltage drop across R_L. Of advantage here is the insensitivity against spread of the transistor parameters, a drawback is the increased parasitic capacitance and therefore the effective capacitance C_c. In the circuit shown in Fig. 76*c* the diode conducts when U_{CB} reaches zero, by taking over part of the input current and partially by-passing the transistor. Saturation cannot be prevented entirely. Most satisfactory results are obtained with a germanium diode and a silicon transistor [Ba 64]. With the germanium diode replaced by a hot carrier diode (cp. Section 3.1.4) the effective storage time constant reduces to 10 to 50% of that of the original transistor and is 1 ns at best [Ta 67a]. If a resistor or a silicon diode is interposed in the transistor base lead, as shown in Fig. 76*d*, the germanium diode between collector and input already conducts when the collector-base junction is still blocked by a negative voltage.

Demands made on transistors, which have to be met for small-signal and large-signal operation and a short rise time, cannot be technologically solved without contradictions and force one to make compromises.

On the one hand, a high cut-off frequency f_α can be obtained through a narrow width w of the base [see equation (79b)], on the other hand it is desirable for the current gain factor α to noticeably underpass unity to provide for a high cut-off frequency f_β of emitter circuited transistors. For this reason gold doping, or a similar measure, is resorted to with switching transistors to obtain a reduction of the time constant τ_p.

Spreading resistance r_b of the base must be low because it determines the rise-time constant (see Table 13) as well as the cut-off frequency of the power gain. The frequency f_{max} where the power gain has dropped to unity represents the maximum obtainable oscillating frequency of an oscillator circuit for the respective transistor, and is expressed with [Me 62; Sa 63a]

$$f_{max} \approx \sqrt{\frac{f_\alpha}{30\, r_b\, C_c}}\,. \tag{97}$$

The collector junction capacitance C_c exerts an enormous influence on the high-frequency behaviour of a transistor. According to equation (97) the collector junction capacitance C_c is included in f_{max} and through the factor $(1 + 2\pi f_\alpha R_L C_c)$ in the rise-time constant. When selecting a transistor the required electric strength has to be taken into consideration along with the previously mentioned parameters f_α, f_β, r_b and C_c. It can be stated that in general the micro-alloyed transistors and the epitaxial transistors have the most favourable properties for high-speed switching behaviour. This also applies with reference to saturation.

AVALANCHE TRANSISTORS

This type of transistor differs from the normal transistor in so far as it is operated with a much higher collector voltage. This results in a breakthrough between emitter and collector and causes an avalanche-type current rise. A hole (minority charge carrier) injected from the emitter of a p-n-p transistor into the base is accelerated to such an extent that further holes are produced by collision ionisation which move rapidly to the collector. Electrons liberated through the ionisation move as majority charge carriers in direction of the emitter and can also give rise to further ionisation. The ionising power of electrons in germanium amounts to about 50% of the holes, while in silicon the electrons cause a 4·5-fold stronger ionisation. For germanium, neglecting the differing ionising power, the multiplication factor M can be written

$$M = \frac{1}{1 - (U_{CB}/U_{CBe})^n}, \tag{98}$$

where for abrupt p-n junctions (alloy transistors) $n \approx 3$ and for a graded base (diffused transistor) $n \approx 6$ applies [Ba 64], and U_{CBe} denotes the breakthrough voltage ($M \to \infty$) when the emitter lead is interrupted ($I_E = 0$).

Fig. 77 shows the family of static characteristics of an avalanche transistor, with the avalanche region located between the breakthrough voltages U_{CBb} and U_{CBe} with $I_B = 0$ and $I_E = 0$, respectively. Here U_{CBb} with $I_B = 0$ signifies an unconnected base and an operating voltage between emitter and collector, which blocks the collector junction and renders the emitter junction conducting. U_{CBe} with $I_E = 0$ denotes an open emitter and the collector-base junction biased in reverse direction. Between these voltages the relation [Ba 64; Lu 64a]

$$U_{CBb} = \sqrt[n]{1 - \alpha}\, U_{CBe} \tag{99}$$

is valid, with α as the current gain factor of the base-circuited transistor. On U_{CBb} the multiplication factor M amounts to $1/\alpha$. In a transistor the collector current increases with the collector voltage. Maximum collector voltage, where the collector current tends to infinity, equals the breakthrough voltage U_{CBe}.

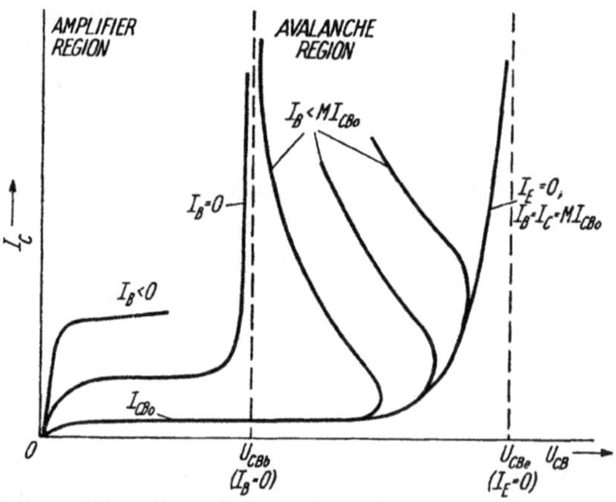

Fig. 77. Family of static characteristics of an avalanche transistor. According to *Lunze* [Lu 64a].

The family of characteristics, as shown in Fig. 77, has parts with a negative resistance which permits the use of avalanche transistors for trigger circuits. The principle of a monostable trigger circuit, as it is frequently used, is illustrated in Fig. 78*a*. Circuit function is based on the discharge of the capacitor C across the avalanche transistor. On an operating voltage $-U_C$, and with a collector load R_C, the working point A of the avalanche transistor is located on curve $I_E = 0$ when the emitter is blocked by a positive bias $+U_B$ (see Fig. 78*b*). This causes I_C to become equal to I_B, making a bias of

$$U_B > I_A (R_B + r_b) \tag{100}$$

necessary for blocking the emitter. The collector current at the working point A is represented by $I_A = M_A I_{CBO}$, while M_A denotes the assigned multiplication factor. If the condition expressed by equation (100) does not apply, an astable trigger circuit would result.

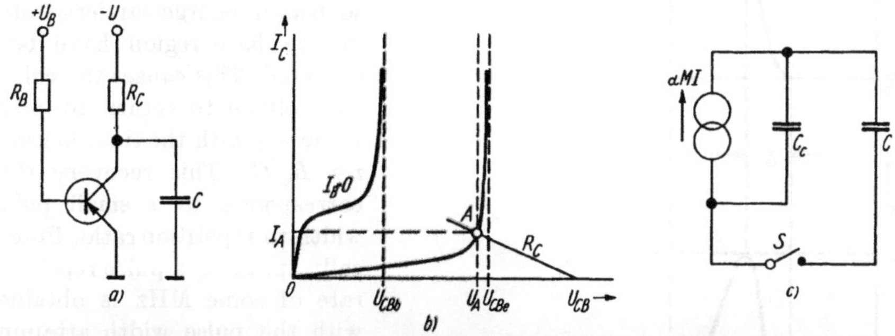

Fig. 78. Trigger circuit with an avalanche transistor. According to *Baranowski* and *Jankowski* [Ba 64].

a) Principle circuitry. *b*) Location of working point. *c*) Equivalent circuit.

Fig. 78*c* shows the equivalent circuit of the trigger stage where, with the emitter-base junction conducting through a negative pulse, the feedback loop is closed with switch S and the discharge of C initiated. Triggering occurs when the following condition is met:

$$\frac{\alpha M C}{C + C_c} \geq 1, \tag{101}$$

that is when αM reaches such a level that the total gain exceeds unity despite the current flow through the collector junction capacitance C_c. Using a high capacitance C enables the trigger event to operate. Curves of the discharge current of capacitance C and of the collector voltage are given in Fig. 79. Initially the discharge current increases rapidly and then linearly because the reduction of the collector voltage leads to a decrease of the multiplication factor M. The peak value of the discharge current I_p is nearly proportional to the value of the capacitance C and may rise to some amperes. Rise time of the current pulse, however, is independent of C and underpasses the level of the cut-off frequency f_α of a normally operated transistor. The reason for this is a high collector voltage, which narrows the base width, and the majority charge carriers which enter the base due to the multiplication effect and cause the appearance of an accelerating field [Ba 64]. With a capacitance C of some picofarad, the collector voltage drops on discharge to about the level of U_{CBb}, which generally gives a symmetric current pulse. In the case of a high capacitance, the voltage across the transistor falls below the value U_{CBb} and nearly becomes zero. This process is slower than the preceding trigger event down to the voltage U_{CBb} and is also independent of C, resulting in an asymmetric current pulse. A detailed analysis has been made by *Baranowski* and *Jankowski* [Ba 64].

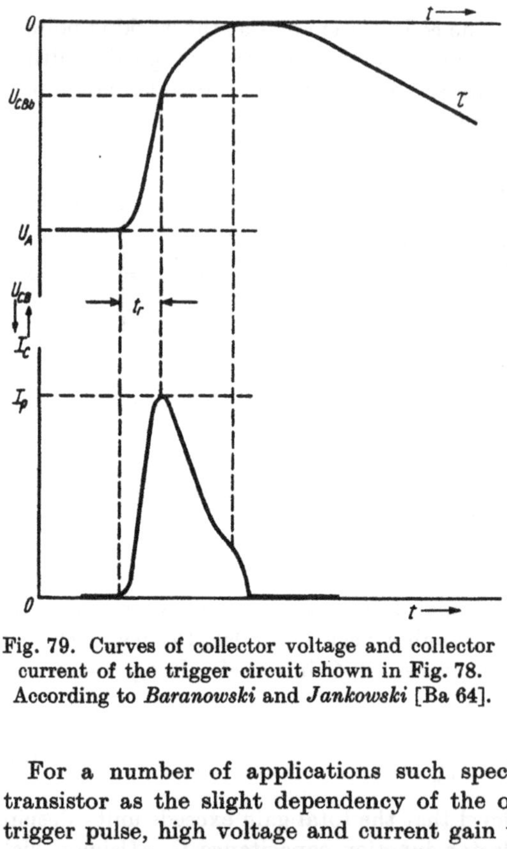

Fig. 79. Curves of collector voltage and collector current of the trigger circuit shown in Fig. 78. According to *Baranowski* and *Jankowski* [Ba 64].

The current pulse ceases as soon as all charge carriers stored in the base region have been collected. This causes the collector voltage to return to initial value U_A with the time constant $\tau \approx R_C\,C$. This recovery time corresponds to a small pulse-width-to-repetition ratio. Practically speaking a pulse repetition rate of some MHz is obtained with the pulse width attaining some nanoseconds. In general, avalanche transistors operate better the higher M is at the working point A, i.e. on high quiescent current I_A. Here, as well as with the maximum pulse repetition rate the permissible power dissipation $U_A\,I_A$ has to be taken into consideration.

For a number of applications such specific properties of the avalanche transistor as the slight dependency of the output pulse on the shape of the trigger pulse, high voltage and current gain with high input sensitivity, short rise time and simple circuitry have been found very useful. Maximum pulse repetition rate restriction to about 10 MHz and the time delay dependent on trigger pulse height prove to be a drawback at times [Mi 62a; Gy 63]. Most suitable for use as avalanche transistors are types with a thin base and radially symmetric structure, while mesa and similar transistors are not as favourable because a direct breakthrough path is absent [Fu 60; He 61a]. Current rise-time values obtained with selected transistors of various types differ less than the fall time values [Fu 60]. With several types of micro-alloyed p-n-p transistors, *Henebry* [He 61a] obtained minimum trigger voltages of 0·1 to 1 V, an avalanche range between 15 and 100 V, and delay and rise-time values from 0·3 to 4 ns. Apart from these, special avalanche transistor types are available, e.g. ASZ 23 or NS 1110.

3.2 PULSE SHAPING

3.2.1 GENERAL

Pulse shaping, generally understood to be alteration of the pulse shape in accordance with specific considerations, represents one of the main objects of pulse technique. Circuitry employed for this purpose cannot be classified without resorting to a certain random selection. For the present context the division into linear and non-linear pulse shaping seems most feasible, consistent with the linear or non-linear dependence of the output pulse height on the maximum height of the input pulse. Table 15 (see next page) gives the principal methods.

Various combinations of individual pulse shaping modes are employed for pulse circuits, a fact which complicates uniform presentation. Thus it is necessary to describe specific circuitry and special properties together.

3.2.2 PULSE SHAPING WITH TRANSMISSION LINES

A line of finite length terminated with an ohmic resistance equal to the characteristic impedance Z acts like a transmission line of infinite length and represents a load for a pulse generator similar to an ohmic resistance Z. If R_G is the internal resistance of the generator (Fig. 80*a*), the fraction $U_i(0) = U_G\, Z/(R_G + Z)$ of the generator voltage U_G is delivered to the line. With a non-terminated line the same holds for the period $t < 2\,T$, where T is the transit time of a pulse in the line. When the line is terminated by the resistance $R_L = R$ the fraction

$$r = \frac{R - Z}{R + Z} \tag{102}$$

of the voltage wave travelling in the line is reflected at the line output and returned to the line input. If $R < Z$, r and, therefore, the reflected voltage is negative. At the input, the line is terminated with the internal resistance $R = R_G$ of the generator which may give rise to further reflexions. At any point of the line the total voltage is equivalent to the sum of all partial waves travelling back and forth. Depending on the relative length $T_0/2\,T$ of the input pulses and the magnitude of the terminating resistances R_L and R_G very different pulse shapes result at the input and output of the line [Gr 60]. This will be all the more pronounced whenever complex terminating networks (*RLC* elements) are also taken into consideration [Le 59].

Seitzer [Se 62b] investigated the influence of imperfections, with known characteristic impedance and location, in homogeneous lines on the transmission of a step function and developed a calculation method suitable for computer processing. A gas-filled coaxial cable was used by *Rosset* [Ro 68a] to locate and determine the intensity of ionising radiation which changes the cable impedance through gas ionisation and causes partial reflexion. This facilitates the beam adjustment in high-energy accelerators.

Table 15: Methods of Pulse Shaping

Type of Shaping	Type of Circuit	Main Components	Description
A) Linear pulse shaping			
1. attenuation	resistance networks	passive elements	Section 3.1.1
2. differentiation, integration	*RC* and *RL* networks	passive elements	
3. shaping with transmission lines	mismatched cables	passive elements	Section 3.2.2
4. pulse stretching	stretchers	valves transistors	Section 3.2.3
5. amplification	amplifiers	valves transistors	Section 3.3
B) Non-linear pulse shaping pulse standardising (trigger circuits)	a) multivibrator circuits monostable: univibrator *Schmitt*-trigger bistable: binary (flip-flop stage)	valves, transistors tunnel diodes	for standardising, general: Section 3.2.4 for timing: Section 3.4.2 for pulse height discrimination: Section 3.6 for scaling circuits: Section 3.7 for pulse generation: Section 4.1
	b) blocking oscillators	valves, transistors	
	c) trigger circuits with secondary-emission valves; ditto avalanche transistors		

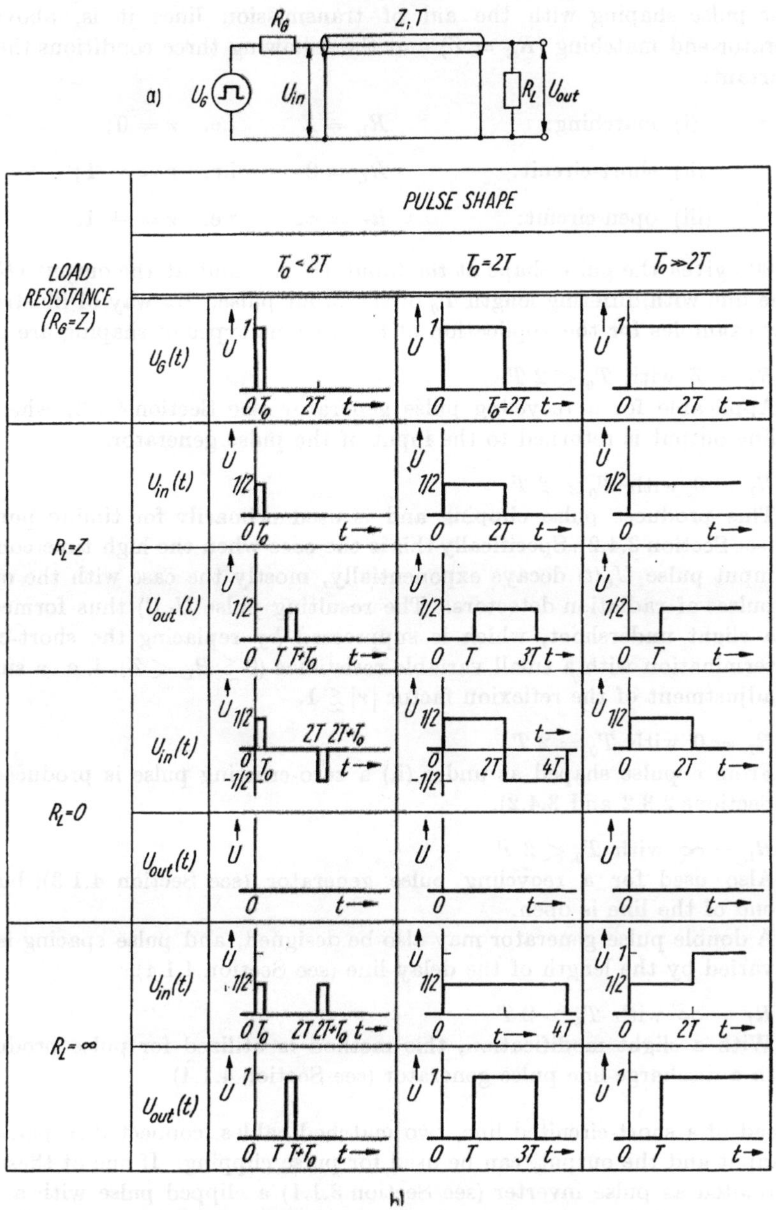

Fig. 80. Pulse shaping with transmission lines.

a) Principle circuit diagram: R_G internal resistance of generator, R_L load resistance, Z characteristic impedance, T delay time of transmission line.

b) U_{in} pulse shape at input and U_{oub} pulse shape at output of transmission line for various widths T_0 of pulse U_G.

For pulse shaping with the aid of transmission lines it is, above all, generator-end matching ($R_G = Z$) and the following three conditions that are important:

(i) matching: $R_L = Z$ i.e. $r = 0$;

(ii) short-circuit: $R_L = 0$ i.e. $r = -1$;

(iii) open circuit: $R_L = \infty$ i.e. $r = +1$.

Fig. 80*b* gives the pulse shape at the input ($U_{in}(t)$) and at the output ($U_{out}(t)$) of the line with differing length T_0 of the input pulse. By way of illustration, a few examples for the application of the various types of shaping are given:

(i) $R_L = Z$ with $T_0 < 2\,T$
Applicable for a recycling pulse generator (see Section 4.1.3) where the line output is returned to the input of the pulse generator.

(ii) $R_L = 0$ with $T_0 \gg 2\,T$
This produces pulse clipping and is used primarily for timing purposes (see Section 3.4.2). Specifically this is the case when the high time constant input pulse $U_G(t)$ decays exponentially, mostly the case with the output pulses of radiation detectors. The resulting pulse $U_{in}(t)$ thus formed has a slight undershoot, which is suppressed by replacing the short-circuit termination with a small variable resistance ($0 \lessapprox R_L < Z$), i. e. a suitable adjustment of the reflexion factor $|r| \lessapprox 1$.

(iii) $R_L = 0$ with $T_0 = 2\,T$
From a pulse shaped as under (ii) a zero-crossing pulse is produced (see Sections 2.3.2 and 3.4.2).

(iv) $R_L = \infty$ with $T_0 < 2\,T$
Also used for a recycling pulse generator (see Section 4.1.3), but the end of the line is open.
A double-pulse generator may also be designed, and pulse spacing is then varied by the length of the delay line (see Section 4.1.4).

(v) $R_L = \infty$ with $T_0 \gg 2\,T$
With a slight modification, this method is utilized for pulse production in a discharge-line pulse generator (see Section 4.1.4).

Instead of a short-circuited line, two matched cables, connected in parallel at the input and the output, can be used for pulse clipping. If one of these lines is circuited as pulse inverter (see Section 3.1.1) a clipped pulse with a width equal to the difference of the delay times in both cables is produced at the output. Against the short-circuited line, this has the advantage that by varying the delay in one line, not only the width but also the polarity of the shaped pulse can be changed [Me 65b].

3.2.3 PULSE STRETCHER

Pulse stretcher devices are intended to change a short needle pulse to a pulse with a flat top or a pulse with a gradually descending trailing edge. When the output pulse height in a stretcher circuit is proportional to the charge of the input pulse, it is known as integrating stretcher. In a non-integrating pulse stretcher, the output pulse height is proportional to the height of the input pulse. A simple RC element is sufficient to achieve integration. Non-integrating pulse stretchers which serve to produce a practically rectangular output pulse will now be described. This type of stretcher circuit is needed, for instance, when measuring the height of nanosecond pulses with a pulse height analyser.

Fig. 81*a* illustrates the principle of a pulse stretcher. With switch S open, the capacitor C is charged to the level of the peak value of the input pulse through the diode path D. As soon as the input signal has passed maximum value the diode blocks. The capacitor voltage remains constant until the switch S is closed causing the capacitor C to discharge. This mode of charge and discharge of a storage capacitor is at times also used with a time analyser (start-stop converter — see Section 3.4.4). Fig. 81*b* shows a design sketch of a pulse stretcher. Detailed information regarding properties and various circuitries of valve equipped pulse stretchers are found with *Maeder* [Ma 58] and *Gruhle* [Gr 60]. *Sarazin* et al. [Sa 60] mention a pulse stretcher which is linear for a pulse duration of 5 to 100 ns across 0·8 to 14 V and delivers output pulses of about 100 ns width. *Goyot* et al. [Go 67c] describe a transistorised version for pulses of 0·5 to 14 V height and more than 10 ns length.

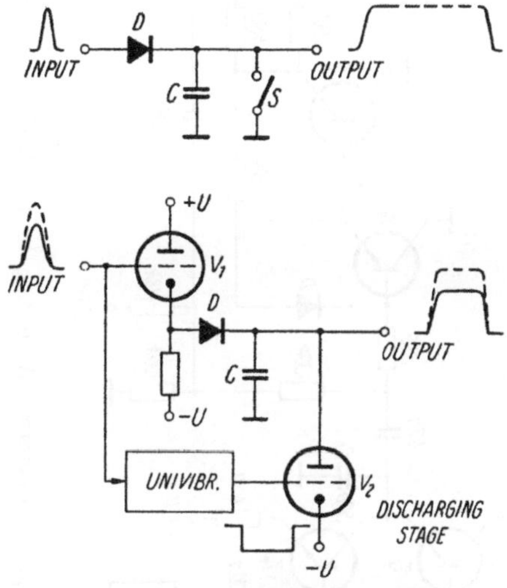

Fig. 81. Pulse stretcher system.
a) Principle circuit diagram. *b*) Block diagram.

The transistorised pulse stretcher shown in simplified presentation in Fig. 82 has been developed by *Weddigen* and *Haase* [We 65]. The negative input pulse is amplified by transistor T_1, passed through the emitter-follower stage T_2, and is delivered simultaneously to the charging stage T_3 and to the generator G, which is a tunnel-diode univibrator. Capacitor C_1 serves for storing. In quiescent state a current of 2 mA flows through T_3 and T_4, which is interrupted by the generator G when the input signal arrives. This causes the capacitor C_1 to be charged through T_3 until the voltage is equal to the peak value of the input signal at the charging stage T_3. The voltage is constant and is tapped by a cascaded emitter follower with high input impedance ($\approx 4\ M\Omega$). After the generator pulse has decayed, transistor T_4 returns to conduction state and the voltage across C_1 assumes initial value. An assembly of this type serves to stretch input pulses of a height between 20 mV and 2·5 V and a length ranging from 6 to 300 ns to flat-top pulses of a width of 1 μs. Proportionality deviation between the heights of the input pulse and the output pulse is around 1%.

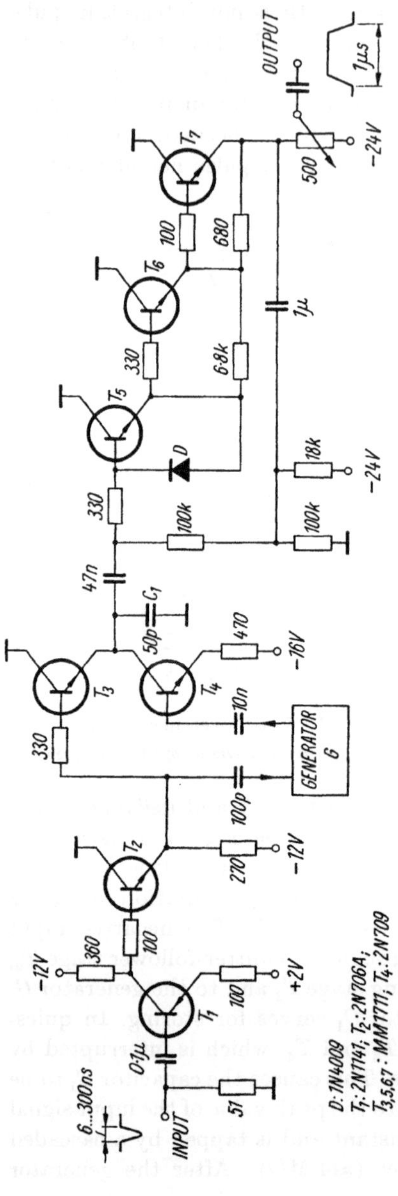

Fig. 82. Pulse stretcher circuit, simplified diagram. According to *Weddigen* and *Haase* [We 65].

Integrating pulse stretcher designs for nanosecond pulses have been described by *Hayashi* and *Yoshimura* [Ha 62], by *Anderson* and *Porat* [An 67], and by *Keller* [Ke 64]. *Keller* uses the input circuit of a field-effect transistor for integration.

3.2.4 TRIGGER CIRCUITS

Generally, an output signal is delivered by the trigger circuit when the input signal (trigger pulse) exceeds a predetermined threshold. Trigger circuits where the output signal is released on zero-crossing of the input signal are described in connection with timing methods in Section 3.4.2.

Trigger circuits dealt with here should meet the following requirements:

(i) delivery of an output pulse of constant shape, irrespective of the location of the threshold and of the part of the input pulse which exceeds threshold level;
(ii) an adjustable threshold level, that is stable and can be conveniently reproduced;
(iii) low threshold uncertainty, especially for pulse height measurements;
(iv) small time-jitter between input pulse and output pulse, especially for time measurements;
(v) short dead time and recovery time values.

The non-linear transmission characteristic of a trigger circuit, i.e. the switching process, is produced through a negative resistance caused either by the positive feedback in valve or transistor equipped amplifier circuitry or by a specialised component with a respective characteristic, such as a tunnel diode.

Monostable Trigger Circuits

This type of circuit has only one stable state of equilibrium. Capacitive or inductive storage elements are used to make the output pulse longer than the triggering signal. After the output pulse has been delivered, a certain period of recovery is necessary to restore the initial state of the storage elements and, therefore, of the trigger circuit.

Bistable Trigger Circuits

Two stable states are characteristic for these circuits, where initial state is assumed only after the second input signal has been received. The output pulses consist of current or voltage steps of constant height and differing polarity. Small storage elements serve as "memory" in the switching process. Due to short recovery time a bistable trigger stage is able to handle pulse repetition rates up to one order of magnitude higher than is possible with monostable circuits of similar design.

Schmitt-Trigger Circuits

The *Schmitt*-trigger circuit represents a modified version of the two mentioned circuits. In general, it operates without storage elements, so that the length of the rectangular output pulse is nearly as long as the length of the input pulse above the response threshold.

Monostable circuits and the *Schmitt*-trigger circuits are described here, while the discussion of bistable stages as the basic element of scaler circuitry is reserved for Section 3.7.

As is known, multivibrator circuits operating with valves or transistors contain at least two stages. With the *Schmitt*-trigger both stages are d.c. feedback coupled or operate with a very high time constant feedback. The monostable or one-shot multivibrator (univibrator) has an *RC* coupling element the time constant of which has a deciding influence on the duration of the unstable state. Free-running circuits (*astable* multivibrators) have two *RC* coupling elements for determining duty cycle. Trigger circuits with secondary-emission valves, avalanche transistors or tunnel diodes, and blocking oscillators on the other hand, generally have only one active stage.

TRIGGER CIRCUITS WITH ELECTRON VALVES

In valve circuitry, trigger stages using *secondary-emission valves* have been developed which can be considered relatively fast for their kind. In 1952

Moody et al. [Mo 52] presented a discriminator circuit which consisted of a Ge-diode and the secondary-emission valve EFP-60 in the succeeding trigger stage (Fig. 83), with D_1 the biased discriminator diode. In quiescent state valve V_1 conducts, but the feedback branch from the dynode to the grid is interrupted

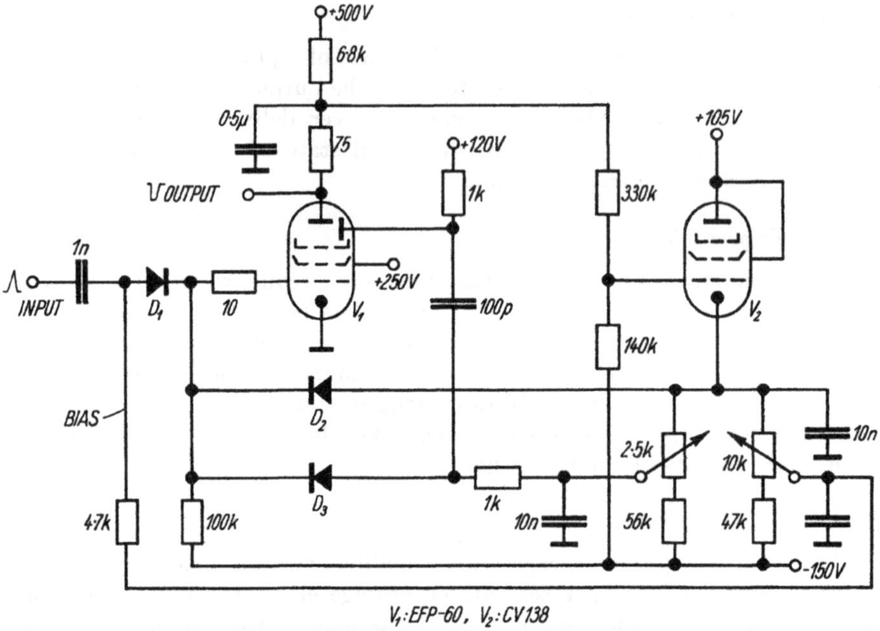

Fig. 83. Fast discriminator circuit, simplified diagram. According to *Moody* et al. [Mo 52].

by the diode D_3 which is biased in reverse direction with about 1 V. Positive feedback becomes effective only above a certain input pulse height ($\approx$0·1 V). Required bias voltages are tapped at the valve V_2. The trigger stage responds to a pulse of some nanoseconds width, and an output pulse of about 0·3 μs width is delivered. *Hayashi* and *Yoshimura* [Ha 62] have described a trigger stage where feedback is effected from the anode to the cathode of the type EFP-60 valve which is blocked in quiescent state. Monostable operation can also be obtained by blocking the control grid of the EFP-60, this, however, reduces sensitivity [Sw 59]. Other circuitry examples are mentioned in Section 4.1.2.

Trigger Circuits with Two Systems of Triodes or Pentodes

They give reliable performance for the delivery of output pulses down to some tenths of a microsecond in width. *Farley* [Fa 58] developed a modified *Schmitt*-trigger circuit using a third valve system in place of the common cathode resistance (Fig. 84). Here, a positive input pulse opens the valve V_1 which is blocked in quiescent state. This causes the current through valve V_2 to drop. The voltage rise at the anode of V_2 is coupled to the valve V_3 and serves, similar to a *White* cathode follower, to aid the decrease of the cathode potential of V_1 and V_2. This results in an accelerated opening of valve V_1 and a faster switching. Hysteresis remains low because the diodes limit the anode voltage changes. Together with a biased amplifier this trigger stage is employed as discriminator and permits the separate analysis of two 15 ns in width pulses at 40 ns spacing between their peaks. This circuit can also be modified for negative input pulses [Sa 62].

In the univibrator as described by *Jung* et al. [Ju 65a] a cathode follower is used for coupling. Output pulses can be set ranging from 30 to 300 ns. *Rethmeier* et al. [Re 62] designed a univibrator circuit where the positive feedback is effected through the common cathode of both valve systems of the trigger circuit and through an additional amplifier valve.

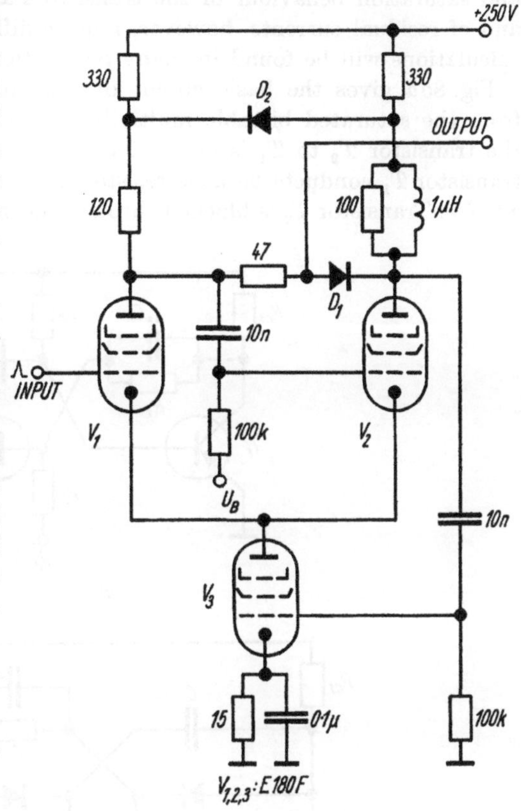

Fig. 84. Fast *Schmitt*-trigger circuit, simplified diagram. According to *Farley* [Fa 58].

Blocking Oscillators

With blocking oscillators triggering is effected by phase displacing the anode signal by 180° with the aid of a pulse transformer and coupling it to the control grid. This causes the anode current to rise to the level of grid current limitation. In this manner an intense pulse up to 10 μs width is obtained, with capacitance and leakage inductance of the transformer exerting a material influence on width and rise time of the pulse (see Section 3.1.1). Although pulse width down to 10 ns and a rise time of 5 ns have been obtained [Ma 57], blocking oscillator circuits have gained only little importance in nanosecond technique.

Theory and circuits of blocking oscillators have been reviewed by *Morugin* and *Glebovich* [Mo 64].

TRANSISTORISED TRIGGER CIRCUITS

Transistorised univibrator and *Schmitt*-trigger circuits can be constructed according to the same principle used for electron valve circuits. Overdrive and saturation behaviour of the transistors and the influence of temperature and of residual currents, however, require different dimensioning. Respective calculations will be found in many publications, e. g. [Ne 61; Ba 64; Re 65a].

Fig. 85*a* gives the basic circuit of a symmetric univibrator which differs from the saturated bistable multivibrator mainly in that the feedback from the transistor T_2 to T_1 is effected through capacitor C_B. In quiescent state transistor T_1 conducts because resistor R_B is connected to the operating voltage U_C. Transistor T_2 is blocked through the voltage divider R_1 R_2. If, in order

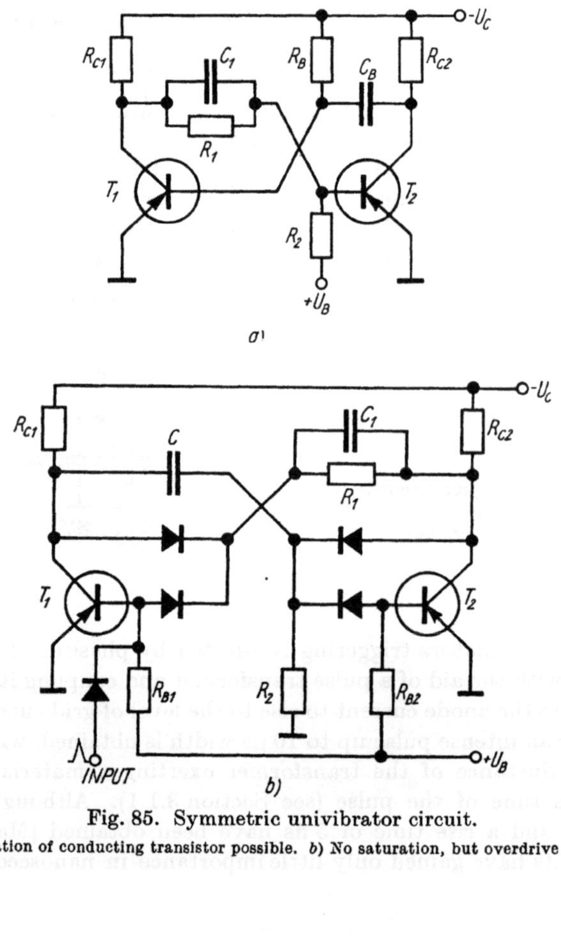

Fig. 85. Symmetric univibrator circuit.
a) Saturation of conducting transistor possible. *b*) No saturation, but overdrive possible.

to prevent a storage effect, the conducting transistors have to be operated at the boundary level between amplification and saturation (see Fig. 74, p. 130), the maximum collector current $I_{C_{max}}$ and the base current I_B of the transistor T_1 become

$$I_{C_{max}} \approx \frac{U_C}{R_{C1}} \approx \beta I_B \approx \frac{\beta U_C}{R_B}$$

therefore

$$R_B \approx \beta R_{C1} .$$

After a positive trigger pulse has been applied to the base of T_1, or a negative signal to the base of T_2, the system shifts to unstable state, i.e. T_1 is blocked and T_2 conducts. It is likewise possible to select R_1, R_2 and U_B in such a manner that transistor T_2 is also kept at the saturation threshold. In the case under consideration, pulse width is about $0{\cdot}7\ R_B C_B$, but becomes shorter with rising temperature. Smaller R_B values give a pulse length that is less sensitive to temperature fluctuations [Ba 64].

After resetting, C_B must be recharged to initial level. The recovery period is dependent on R_{C2} and may be longer than the pulse width. Adding diodes or transistors to the system will materially reduce this recharging time [Ri 64; Ha 66a]. The parallel circuited elements R_1 and C_1, and the input impedance of T_2 can be viewed as compensated voltage divider. If T_2 is not subjected to overdriving, $C_1 \approx 1/2\,\pi f_\beta R_1$ can be assumed as an approximation. The use of higher capacitance values for C_1 will accelerate switching, but the recharging time-constant $C_1 [R_1 \| (R_{C1} + R_2)]$ should be kept small against pulse length.

The rise time of the pulse at the collector of T_2 corresponds to the rise time of a common-emitter transistor, provided that no overdriving occurs. In this case, the decay time constant of the pulse becomes about $R_{C2} C_B$ when T_2 blocks.

Fig. 75, p. 132 illustrates that overdriving the base reduces the rise time and decay time when switching on and off. The principle circuit of a univibrator is given in Fig. 85*b*, where overdriving occurs during the opening event of the transistors [Su 58]. With diode-combination systems as shown in Fig. 76 saturation is prevented. In the one-shot trigger circuit shown, the transistor T_1 conducts and the transistor T_2 is blocked in quiescent state. In the main the time constant $R_{C1}C$ determines the pulse width. It will be an advantage to include an inductance in series with R_{C1} (or with R_{C2} in Fig. 85*a*), to serve as supplementary storage element. This gives, on the one hand, a steeper discharge function resulting in less time-jitter on resetting, and on the other a greater portion of the collector current is available during switch-over for driving the other transistor [Su 58].

By introducing a common-emitter resistance for both transistors the additional base supply $+U_B$ is not necessary. This offers some advantages. An unbypassed emitter resistor limits the emitter current and, on overdrive, prevents the transistors from being strongly drawn into the saturation range. The

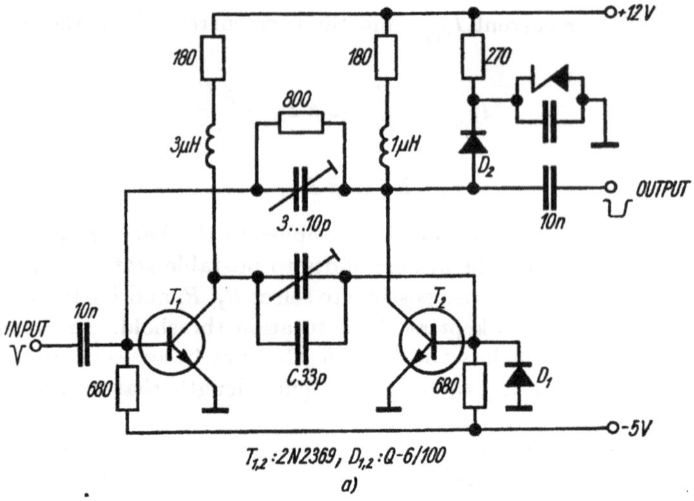

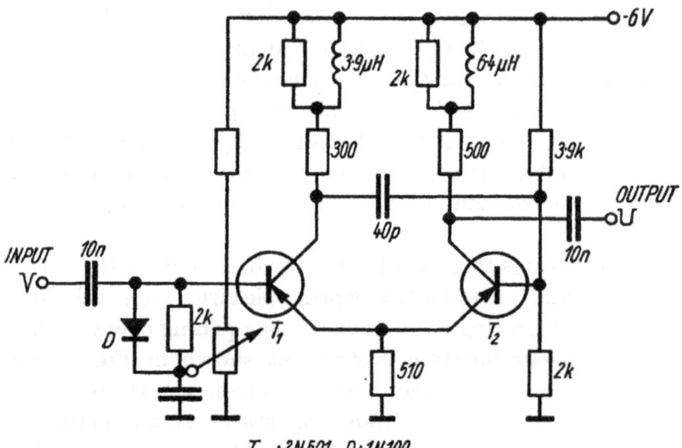

Fig. 86. Univibrator circuits.

a) Pulse width ≧28 ns, rise time 1·8 ns, pulse height 9·5 V, repetition rate 10 MHz. According to *Righini* [Ri 64].
b) Pulse width 35 ns, rise and fall time 10 ns, pulse height 2 V, repetition rateupto 15 MHz, height of input pulse 40 mV to 1·7 V. According to *Fitch* [Fi 59].

feedback across the common-emitter resistance permits the passing of the trigger pulse to the "free" base, without subjecting the pulse source and the feedback path to an additional load.

Two univibrator systems are given in Fig. 86, both of which are suitable for pulse repetition rates in the order of 10 MHz. Fig. 86*a* shows a univibrator according to *Righini* [Ri 64], where transistor T_1 is saturated in quiescent state and transistor T_2 is blocked. Recovery time is short because the capacitor C is rapidly recharged through the low saturation resistance of transistor T_1 and the diode D_1. Overshoot is eliminated by the diode D_2. The univibrator circuit according to *Fitch* [Fi 59] features a common emitter resistance (Fig. 86*b*). In quiescent state transistor T_1 is blocked. Other univibrator circuitry is found in various papers, e.g. [Ba 61; Ve 61]. These are d.c. coupled, with the pulse width determined by the inductances, allowing a maximum pulse repetition rate of some 10 MHz.

Univibrators with Capacitive Emitter Coupling

This is another group of univibrator circuits, of which Fig. 87 gives an example of the many possible modifications [Ba 64]. In the system, transistor T_1 is blocked in quiescent state and T_2 conducts due to the impressed current I_e. The current source for I_e may, for example, consist in a common-base transistor. On arrival of a negative pulse a portion of the current I_e is passed through C to the transistor T_1 until transistor T_2 is blocked, due to the voltage drop across the resistance R_g. Diode D is also in non-conduction state. After the current flow through C ceases transistor T_2 becomes conducting, i.e. the circuit changes over. Pulse width amounts to about CR_g. Capacitor C is subsequently recharged through diode D and resistance R_d. In the described circuit the transistors do not reach the saturation region. Another advantage is that the collector of transistor T_2 is not subjected to load by the feedback. Rise time and decay time of the output pulses are about equal and with low R_C ($R_C C_c < 1/2\pi f_\alpha$) are in the order of $1/f_\alpha$. Thus, a steep pulse slope is obtainable with high-speed transistors. The volume of *Baranowski* and *Jankowski* [Ba 64] contains a circuit modification with a capacitor C of constant current discharge, which delivers a pulse without noticeable undershoot at the collector of transistor T_2.

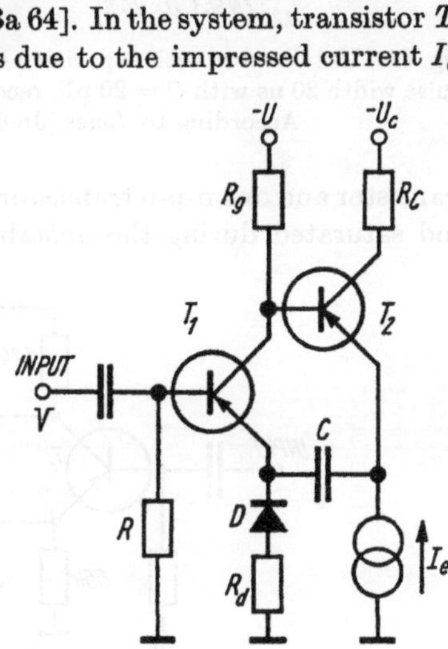

Fig. 87. Asymmetric univibrator with capacitive emitter coupling.

Univibrator circuitry with capacitive emitter coupling is also included in publications by *Baker* et al. [Ba 60] and *Jones* [Jo 63]. Fig. 88 shows a univibrator circuit according to *Jones*, where the transistor T_1 is conducting in quiescent state. Function of the system, however, is similar in principle to that shown in Fig. 87. Recovery time is determined from the product of the forward resistance of the diode D and of the capacitor C.

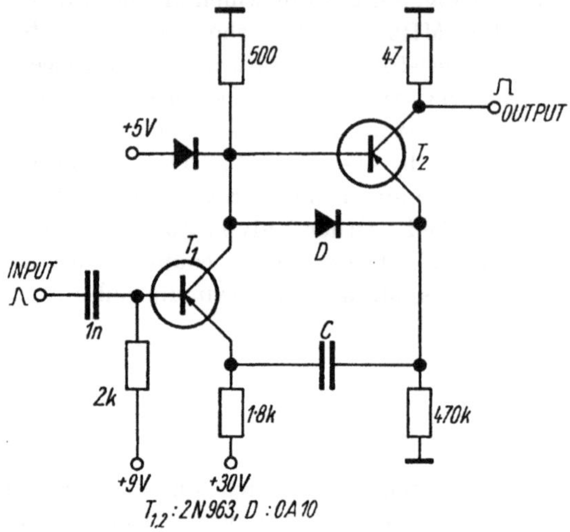

Fig. 88. Univibrator with capacitive emitter coupling. Pulse width 20 ns with $C = 20$ pF, recovery time 30 ns. According to *Jones* [Jo 63].

It is evident from the above, that in univibrator circuitry the triggered transistor can be conducting or blocked in quiescent state. The first instance produces a higher trigger sensitivity, while the second provides for a more sharply defined threshold, which is bias adjustable. *Wieber* [Wi 63a] developed a univibrator with one p-n-p transistor and one n-p-n transistor (Fig. 89), both conducting in quiescent state and saturated during the unstable phase. In quiescent state the diode D is

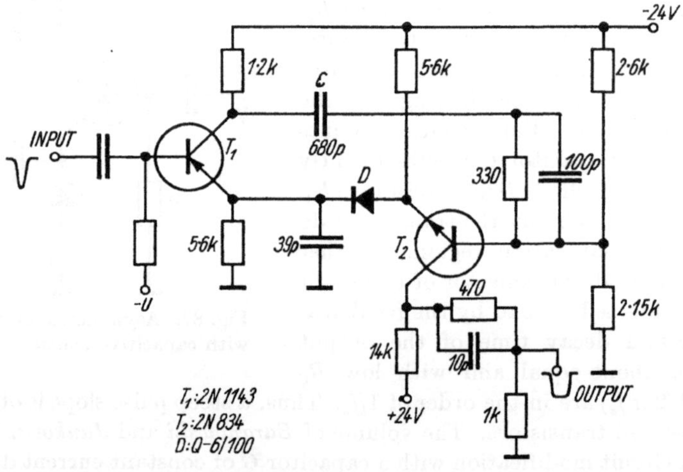

Fig. 89. Univibrator with conducting transistors. Pulse width 0·5 μs to 1 ms (T/ns ≈ 1·47 C/pF), triggering delay <1 ns. According to *Wieber* [Wi 63a].

blocked and the feedback interrupted. A negative trigger pulse eliminates the interruption, permitting both transistors to transfer to saturation state. For this reason the circuit is not suitable for very short pulses, but features very little trigger delay, amountingless than 1 ns.

Transistorised Blocking Oscillators

Against similar valve systems, these have the advantage of delivering a flat topped pulse. Pulse transformer construction poses the main problem in the design of blocking oscillators, when, short and steep pulses are desired (see Section 3.1.1). In the nanosecond technique this system has not gained much importance because avalanche transistors provide a more convenient method of producing a short high-intensity pulse. *Feldman* [Fe 60 a] describes a blocking oscillator system which delivers pulses of a minimum width of 15 ns with a dead time of 25 ns. With circuits given by *Moody* et al. [Mo 63] and *Mineev* [Mi 66 b], pulses with a rise time of 2 to 3 ns and a width of some hundred nanoseconds are obtained. Approximate data calculations of transistorised blocking oscillators have been reviewed by *Baranowski* and *Jankowski* [Ba 64].

Trigger Circuits Containing Avalanche Transistors

Here, the carrier multiplication in the transistor on high operating voltage is utilised. As the principle function of this type of transistor circuit has been detailed in Section 3.1.5, the following description will, therefore, be restricted to some practical examples. A simple trigger circuit is shown in Fig. 90*a* [Al 63 a]. The high-ohmic resistance R determines the quiescent current of the transistor. To reduce recovery time, R is subdivided into R_1 and R_2. Capacitor C is charged with the time constant $R_2\,(C + C_c)$, where C_c represents the collector capacitance of the transistor. This permits the triggering of the system with pulses spaced less than 100 ns apart. Although, due to the discharge of C_1 and its very high recharging constant R_1C_1, only a pulse train of some microseconds duration can be obtained. A low resistance R or a recharging circuit, as mentioned by *Henebry* [He 61 a], will provide for a high pulse repetition rate on continuous duty, however, the permissible collector dissipation has the final word with respect to maximum repetition rate. The circuit shown in Fig. 90*a* gave a repetition rate up to 3 MHz. Systems of similar design deliver a pulse train of maximum 40 pulses in 4 μs [Fu 60; Ch 63 c; De 66 a]. Further circuitry is mentioned in [He 61 a] and [Gy 63].

To avoid undesirable charge storage the base resistance R_3 should not be too high. Generally, maximum obtainable trigger sensitivity is around 0·1 to 0·2 V. *Pardies* et al. [Pa 65], however, achieved a sensitivity of some millivolts. Improved temperature stability of the threshold can be attained with temperature-dependent resistors in the collector or in the voltage divider chain for the base bias [Al 63 a].

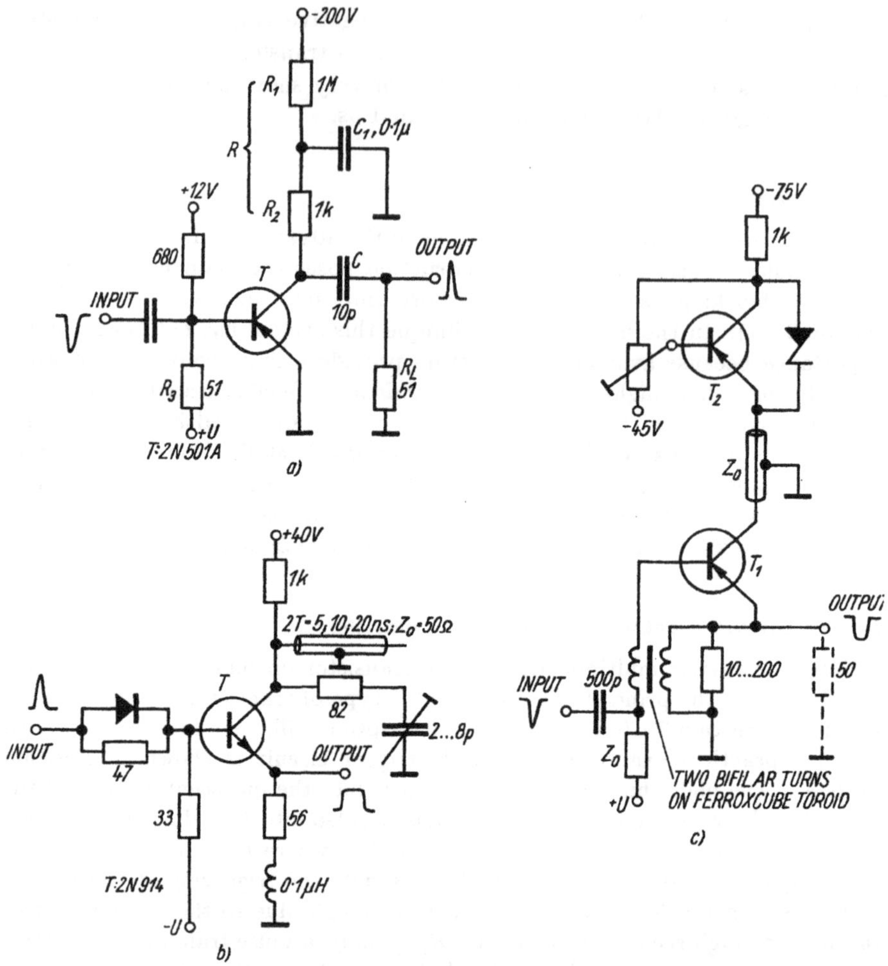

Fig. 90. Trigger circuitry with avalanche transistors.

a) Height of output pulse 15 V, rise time 0·3 ns, pulse duration at base 4 ns. According to *Alon* and *Rausch* [Al 63a]. *b)* Height of output pulse 8 V (on 50 Ω), rise time 0·5 ns, pulse width 5 to 20 ns, repetition rate max. 5 MHz. Simplified diagram according to *Samueli* et al. [Sa 64a]. *c)* Trigger circuit incorporating transformer. According to *Henebry* [He 61a].

With a p-n-p transistor system, the load resistance R_L causes a positive pulse shaped by the voltage behaviour at the collector and by the time constant $R_L C$ [St 62a]. Negative pulses are obtained at an inserted emitter resistor.

Incorporating a transmission line, operating either open-ended [Ha 62a; Ta 63a; Sa 64a; Pr 65] or short-circuited [He 61a; Br 62; Sa 64a], a steep pulse is obtained with a width of twice the transit time in the line. Fig. 90*b* gives an example of this type of circuit [Sa 64a], where a constant current is produced in the open-ended line, which makes a rectangular pulse available at the emitter for the duration of double the delay time.

In the described trigger circuits operating with avalanche transistors the output signal is delayed by an interval dependent on width and rise time of the input pulse. *Henebry* [He 61a] has designed the trigger stage shown in Fig. 90*c*, where a transformer is incorporated offering two possibilities to use the sense of winding. In negative feedback mode, current passing through the emitter circuit induces a pulse which opposes the input pulse. Due to the delay in the transistor and the transformer this effect becomes noticeable only about 1 or 2 ns after the start of the input pulse. Because of this, the transistor "sees" only the first part of this pulse, irrespective of its width. Reversing the poling of one transformer winding causes positive feedback which boosts the triggering event and, with slow rising input pulses, reduces threshold uncertainty. The working point of the avalanche transistor T_1 is set by the transistor T_2, while the nearly short-circuited transmission line determines pulse width. On continuous duty a maximum repetition rate of 2 MHz has been obtained, and on intermittent duty this rose to 20 MHz. Another version of this circuit [Al 63a] provides for a line with 1 ns transit time causing part of the collector pulse to counteract the input pulse. At a maximum repetition rate of 55 MHz this obtains pulses of 1·5 ns width and 15 V height.

In place of the base, the emitter-end may also be employed for triggering an avalanche transistor. Fig. 91 gives the principle of the system designed by *Bramson* [Br 62], where a negative pulse serves to trigger the avalanche transistor T_1. This causes the capacitor C to discharge and produces an intense pulse which drives the normal transistor T_2 into the saturation region. Formation of the collector pulse, coming from T_2, with the aid of a short-circuited transmission line and interposed common-base transistor stage, is not shown in the sketch. Transistor T_2 must be maintained in saturation state by a high-ohmic base resistor for a sufficient length of time. Thus, transistor T_2 stretches the avalanche transistor pulse, so to speak. With this circuit output pulses of 50 mA and a repetition rate exceeding 10 MHz were obtained. A slightly modified version of this design has been used as pulse shaper for nanosecond coincidence measurements by *Michaelis* et al. [Mi 64].

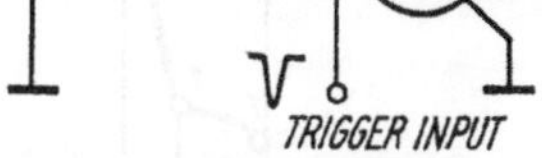

Fig. 91. Trigger system showing the principle according to *Bramson* [Br 62].

TRIGGER CIRCUITS WITH TUNNEL DIODES

Repetition rates of 100 MHz and more could be obtained with trigger circuitry only after the appearance of the tunnel diode. The principle of the system illustrated in Fig. 92 gives the basis for the construction of a monostable trigger

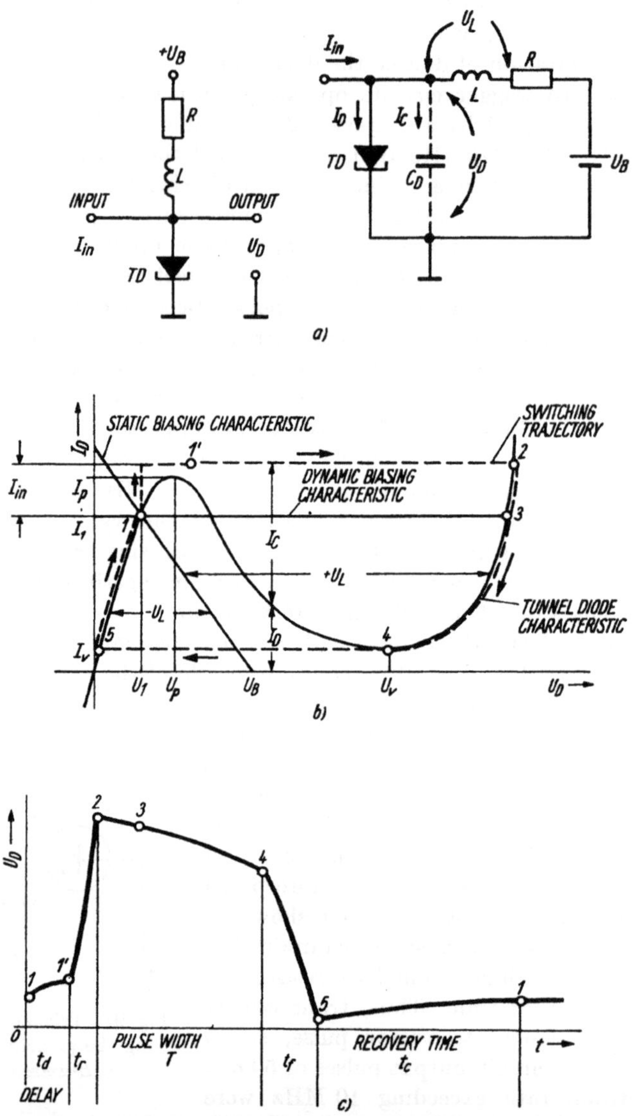

Fig. 92. Schematic diagram of a tunnel diode univibrator. According to *Bergman* et al. [Be 62].

a) Principle diagram and simplified equivalent circuit. *b*) Characteristic of tunnel diode with ideal switching trajectory (broken line), *1* working point. *c*) Shape of output pulse.

stage using a tunnel diode [Be 62; Iv 67b]. In this case, an ideal tunnel diode with the diode capacitance C_D in parallel is assumed for simplification. The load characteristic determined by U_B and the series resistor R intersects the tunnel diode characteristic at the stable working point *1*. Inductance L must be selected to result in a horizontal dynamic load characteristic. This causes the entire input current I_{in} to be delivered to the tunnel diode during the switching process. If $(I_1 + I_{in})$ exceeds the peak current I_p the tunnel diode is shifted to high-voltage state, drawn as broken line trajectory in the illustration. For the transfer from point *1* to *2*, the switching speed depends on the recharging of the capacitance C_D. Because

$$\frac{dU_D}{dt} = \frac{I_C}{C_D} \tag{103}$$

switching speed increases with rising current $I_C = I_1 + I_{in} - I_D$ (see Fig. 92*a* and *b*). For this reason speed is low around the peak region of the tunnel diode characteristic, which practically corresponds to a delay of the output pulse (see Fig. 92*b* and *c* – point $1 \to 1'$). With rising overdrive condition $\Delta I = I_{in} + I_1 - I_p$ this delay decreases rapidly.

When the current I_{in} is cut-off after reaching the working point *2* the latter shifts to *3*, which is also unstable because of the voltage U_L across the inductance L. Therefore the current changes with

$$\frac{U_L}{L} = \frac{dI_L}{dt} \tag{104}$$

until U_L disappears. This causes the working point to shift from *3* to *4* and then fall to *5*. During the subsequent recovery period, now with reverse-poled voltage U_L, the current through L rises until the stable working point *1* with $U_L = 0$ is reached.

This type of univibrator works as current amplifier because the available output current nearly equals I_1 and may exceed the input current pulse I_{in} by several times.

Many publications have appeared dealing with the switching behaviour of tunnel diode systems. In theoretical analysis the tunnel diode characteristic [Be 64b] has either been approximated in sections, represented by straight lines [La 62a; St 62b] or by parabolic curves [In 62; Ab 66], or regarded numerically in its true form [Be 62; Ko 63a; Pa 64; Wi 64]. In some cases, the series inductance and the series resistance of the tunnel diode have also been included [Or 66]. For trigger signals in univibrator systems and *Schmitt*-trigger circuits, needle pulses [Wi 64], rectangular pulses [La 62a; Ko 63b], long step-pulses [In 62] and pulses having a finite rise and decay time similar to the signals of scintillation counters, were used [Ko 63a; Pa 64; Ab 66]. Diagrams for the dynamic behaviour of tunnel-diode univibrators are given by *Abbattista* et al. [Ab 66] representing pulse length, resolving time, threshold behaviour and trigger delay in dependence of circuit parameters. Calculation of time delay and time jitter for small signals in the presence of amplitude noise have been variously mentioned [Ab 66; Ba 67a; Nu 67; Ma 68]. Within the scope of this book mention can only be made of a few results to detail the qualitative consideration derived from Fig. 92.

For the calculation of delay time and pulse rise and decay the influence of inductance can be neglected if the time constant L/R is high against the switching time. Analoguously, the capacitance C_D can be ignored when determining the duration of the unstable phase.

For rectangular trigger pulses where the rise time is short compared to switching time, the following approximation applies for the leading edge of the univibrator pulse [La 62a; St 62b; Ko 63b]

$$t_r \approx C_D \frac{U_f - U_p}{I_p - I_v} \approx \frac{C_D}{I_p} \Delta U \,. \tag{105}$$

Germanium tunnel diodes ($\Delta U \approx 0{\cdot}4$ V) have a rise time of $t_r \approx 2$ ns and $t_r \approx 0{\cdot}1$ ns with $C_D/I_p = 5$ pF/mA and $C_D/I_p = 0{\cdot}2$ pF/mA, respectively. By placing a tunnel diode in a coaxial housing, *Cummings* [Cu 65] obtained a value of $t_r \approx 70$ ps.

For the estimation of the decay time t_d the assumed shape of the peak of the tunnel diode characteristic has a pronounced influence on calculation, which explains the considerable difference of the equations derived by *Lacour* [La 62a] and by *Kononov* et al. [Ko 63b]. With an overdrive of $\Delta I = I_{in} + I_1 - I_p$ *Kononov* expressed the delay time for the rectangular trigger pulse with

$$t_d \approx \frac{\pi}{6} \frac{C_D}{\sqrt{I_p \Delta I}} \Delta U \,. \tag{106}$$

According to *Lacour* the main contribution for t_d is proportional to $C_D/\Delta I$ when the threshold is slightly exceeded. With rising overdrive ΔI the delay time and rise time of the output pulses become shorter when an input pulse is present with a finite rise time [Ko 63a]. With $\Delta I > 0{\cdot}5\, I_p$ this reduction is only very slight.

Provided the dynamic characteristic is known, the leading edge can be graphically obtained [Sk 62; Bl 63a]. Equation (103) can also be written $\Delta U_D \approx (\Delta t/C_D)\, I_C$, where, with given time interval Δt and the diode capacitance C_D assumed as constant, $\Delta U_D \approx \text{const.} I_D$ is then obtained. The respective design and a related example are given in Fig. 93. This method could be applied also for assessing the trailing edge. However, the actual switching trajectory *4* → *5* deviates too much from the ideal version shown in Fig. 92 in the region of the valley (see [Be 62; Wi 64]).

A rising load causes the dynamic characteristic to shift downward, and rise time increases. With load resistors down to $4\,|R_-|$ switching time is usually 20% higher at the most than with no-load operation [St 62b]. R_- represents the mean negative resistance of the falling portion of the tunnel diode characteristic.

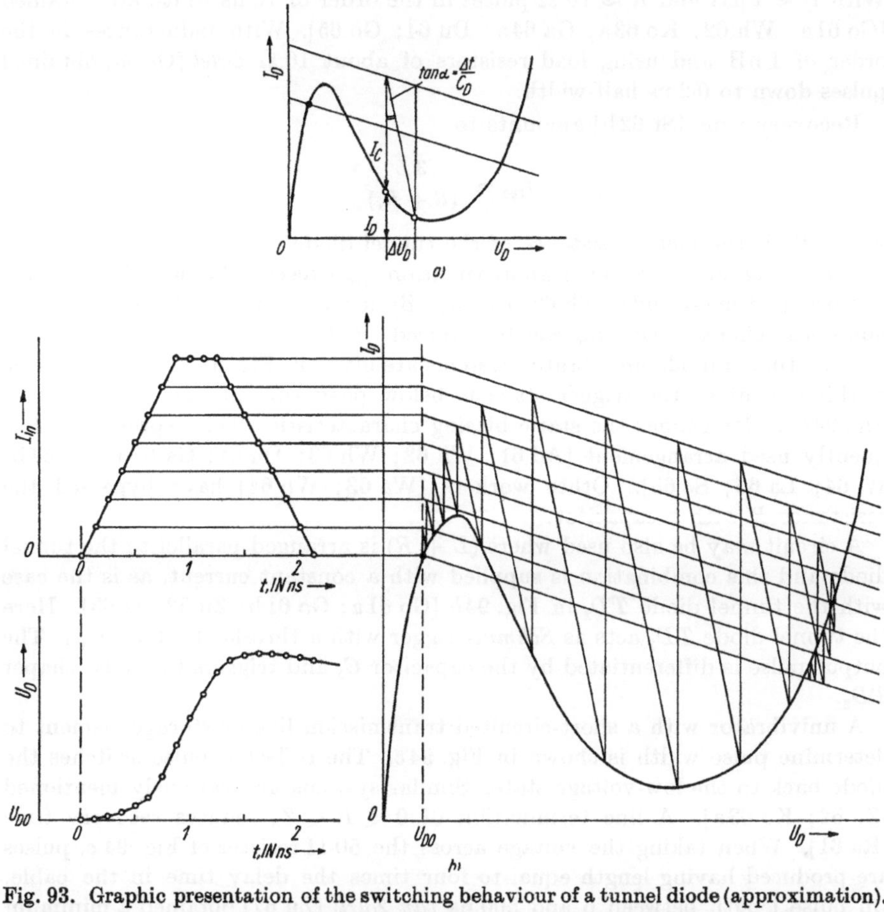

Fig. 93. Graphic presentation of the switching behaviour of a tunnel diode (approximation). *a)* Principle diagram. According to *Skilling* [Sk 62]. *b)* Construction of the leading edge when transferring to high-voltage state; time scale is referred to the condition where $I_p \approx 20$ mA, $C_D \approx 20$ pF, $\Delta t \approx 0{\cdot}12$ ns.

Minimum width of a rectangular trigger pulse is given by $t_d + t_r$, according to *Kononov* [Ko 63b]. Provided the working point *1* is located above the valley current I_v [Wi 64], however, signals of a width to switch the tunnel diode to just "above peak" suffice (see [La 62a]).

For the univibrator as shown in Fig. 92 the pulse width, according to *Lacour* [La 62a], is approximated with

$$T = \frac{L}{R + R_-} \ln\left(1 + \frac{I_p - I_v}{I' - (I_p - I_v)}\right), \tag{107}$$

where

$$I' = \frac{U_f - U_p}{R + R_-}.$$

With $L \approx 1\ \mu H$ and $R \approx 10\ \Omega$ pulses in the order of 10 ns width are obtained [Go 61a; Wh 62; Ko 63a; Ga 64a; Du 64; Go 65]. With inductances in the order of 1 nH and using load resistors of about 10 Ω *Ortel* [Or 66] obtained pulses down to 0·2 ns half-width.

Recovery time [St 62b] amounts to

$$t_{rec} \approx \frac{3\,L}{(R + R_+)}\,, \tag{108}$$

where R_+ is the mean resistance of the tunnel diode in the low-voltage region $5 \to 1$ of Fig. 92 *b*. In rough approximation t_{rec} exceeds T in length by nearly one order of magnitude [Wh 62; Go 65]. By using a load resistor with a static non-linear characteristic t_{rec} can be reduced [Be 62; Wh 63].

Some tunnel-diode univibrator systems are given in Fig. 94, where the stable working point of the trigger stage is below peak current. In the circuit of Fig. 94*a* R determines the static biasing characteristic. This represents a frequently used arrangement [Ad 61; Wh 62; Wh 63; Du 64; Ga 64a; Go 64b; Wi 64; La 65; Sc 65]. Other workers [Wa 63; Wh 64] have bypassed the resistance R with a capacitor.

A circuit may be also used where $(L + R)$ is arranged parallel to the tunnel diode and this combination is supplied with a constant current, as is the case with the tunnel diode TD_2 in Fig. 94*b* [Go 61a; Go 61b; Zu 62; Iz 63]. Here the tunnel diode TD_1 acts as *Schmitt*-trigger with a threshold set by U_{B1}. The output pulse is differentiated by the capacitor C_1 and triggers the pulse shaper TD_2.

A univibrator with a short-circuited transmission line as storage element to determine pulse width is shown in Fig. 94*c*. The reflected pulse switches the diode back to the low-voltage state. Similar systems are variously mentioned [Zu 62; Ko 63a]. A line termination of $0 < R < Z_0$ reduces recovery time [Ra 61]. When taking the voltage across the 50 Ω resistor of Fig. 94 *c*, pulses are produced having length equal to four times the delay time in the cable. On pulse length between 6 and 500 ns *van Zurk* [Zu 67] obtained a minimum resolving time of 10 ns.

An arrangement with two series-circuited univibrators is given in Fig. 94*d* [Ad 61; Wh 62], where the tunnel diode TD_1 serves as discriminator. Height of the output pulse and its width are still slightly influenced by the shape and the size of the trigger pulse and by the trigger threshold. The preformed pulses are standardised in the succeeding univibrator circuits TD_2 and TD_3. The incorporated coupling resistor R_k reduces the load of the tunnel diodes and decreases the static and dynamic interaction of the two stages. Through the voltage dividing action of R_k and TD_2 the influence of pulses, which are not sufficient to trigger TD_1, is nearly completely suppressed. If a constant resolving time is desired with the trigger circuit, the resolving time of TD_2 must exceed that of TD_1. It is of advantage to insert a separating diode in the coupling system, where the diode D_1 separates the tunnel diode TD_3 from the

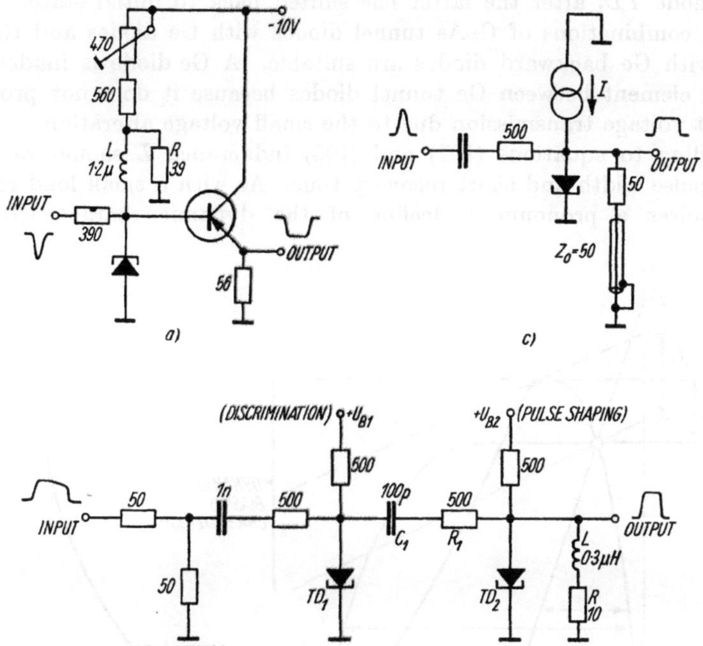

Fig. 94. Univibrator systems with tunnel diodes.

a) Width of output pulse 15 ns, height of output pulse 0·7 V, threshold instability <10 mV, maximum delay 2 ns. According to *Golutvin* and *Zanevskii* [Go 64b]. b) Width of output pulse 10 ns. According to *Gorodetzky* et al. [Go 61a]. c) Width of output pulse 4 ns, determined by short-circuited cable, cycle time 12 ns. According to *Lacour* [La 62a]. d) Width of output pulse 5 ns, inductance $L \approx 0{\cdot}5$ μH, pulse height 0·33 V (on 50 Ω); through D_2 or D_3 pulses are delivered showing no undershoot. According to *Whetstone* and *Kounosu* [Wh 62].

tunnel diode TD_1 after the latter has shifted back to initial state. For this purpose, combinations of GaAs tunnel diodes with Ge diodes and Ge tunnel diodes with Ge backward diodes are suitable. A Ge diode is inadequate as coupling element between Ge tunnel diodes because it does not provide for sufficient voltage transmission due to the small voltage alteration.

According to equations (107) and (108) inductance L should be low for narrow pulse width and short recovery time. As with a small load resistance this requires a pronounced decline of the dynamic characteristic which

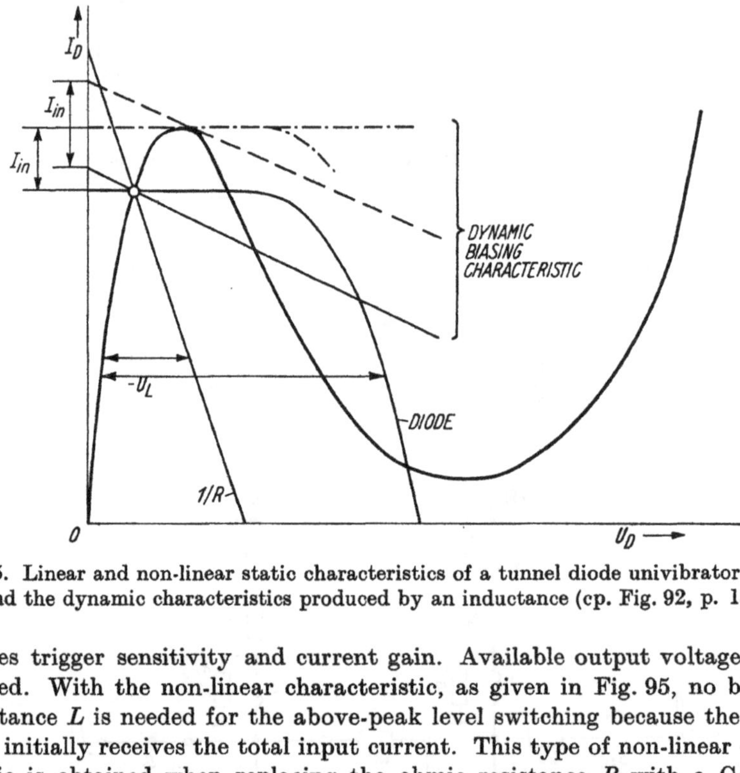

Fig. 95. Linear and non-linear static characteristics of a tunnel diode univibrator system, and the dynamic characteristics produced by an inductance (cp. Fig. 92, p. 156).

reduces trigger sensitivity and current gain. Available output voltage is also reduced. With the non-linear characteristic, as given in Fig. 95, no blocking inductance L is needed for the above-peak level switching because the tunnel diode initially receives the total input current. This type of non-linear characteristic is obtained when replacing the ohmic resistance R with a Ge diode [Wh 63], a backward diode [Be 62; Go 63; Rh 63] or a tunnel resistor [Co63a]. The use of backward diodes has attained wide preference. Compared with the described circuits a lower inductance is sufficient to obtain the same output pulse width due to the lower voltage $+U_L$. For this reason and in consequence of the higher voltage $-U_L$ recovery time is considerably shortened.

A trigger circuit assembled from two tunnel diode univibrator systems and backward diodes is given in Fig. 96 [Be 62] — this circuit operates at a repetition rate of more than 200 MHz. For working point setting the backward diodes BD_1 and BD_2 are biased with +90 mV. This circuit is used as logic

"OR"-gate. To obtain a higher current gain a tunnel diode with a high peak current was selected for TD_2. The use of the backward diode BD_3 for coupling the two univibrator systems has proved to be favourable because connection to TD_2 remains interrupted until TD_1 has switched to high-voltage state. This provides the tunnel diode TD_1 with the total input current for initial switching. Connection with the latter is interrupted after TD_2 has changed over, which prevents TD_1 from influencing TD_2 during the further operating cycle. Besides this, mutual static interaction of the two univibrators is small because, in

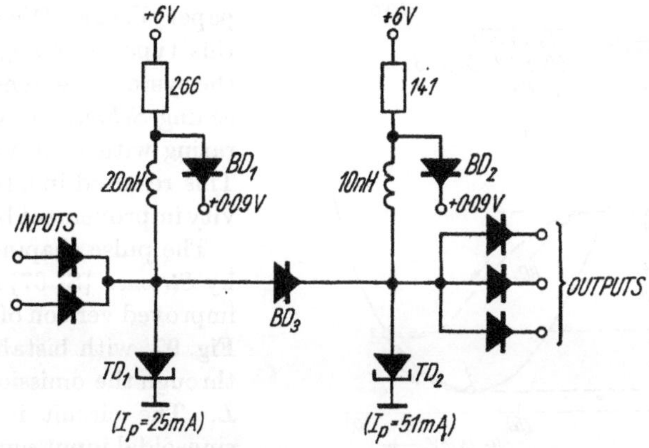

Fig. 96. "OR"-gate with tunnel diode univibrators. According to *Bergman* et al. [Be 62]. Height of input pulse 7·4 mA, delay 0·7 ns, height of output pulse 10 mA, width of output pulse 2 ns.

initial state, BD_3 has a high resistance. Only parasitic pulse transmission occurs due to the inherent capacitance of this backward diode. The backward diodes at the input and the output ends serve to isolate several systems of this type, especially when using coaxial cable elements. *Bergman* et al. [Be 62] have calculated the switching trajectory of the two coupled univibrator systems and taken the parasitic components into consideration. The resulting trajectory differs considerably from the schematic presentation given in Fig. 92. Similar calculation for tunnel diodes with two stable working points were made by *Cooperman* [Co 67e].

A trigger circuit featuring a short recovery time has been designed by *Verweij* [Ve 63a]. It is shown in Fig. 97 together with its characteristic. Because the transistor T is not conducting, in quiescent state (*1*) the tunnel diode TD receives the total current of $7{\cdot}5 + 10 = 17{\cdot}5$ mA. When the input current I_{in} exceeds 2·5 mA the tunnel diode shifts to high-voltage state (*2*). The backward diode BD_1 is blocked and the transistor T receives the current of 10 mA, which passes through the resistor R_2. Under the influence of the inductance L and of the backward diode BD_2 the working point shifts from *2* to *3*. It then switches to *4* and immediately moves to *5* because the backward diode BD_1 becomes conducting as soon as the tunnel diode is in low-voltage state. The recovery time resulting from the transfer of the working point from *5* → *1* is therefore only about three times as long as the pulse width during the working point shift from *2* → *3*. The output pulse shows no droop if the backward diode BD_1 remains blocked during the transfer from *2* → *3*. In a succeeding paper *Verweij* [Ve 66] described this type of arrangement where the system is extended by a preceding *Schmitt*-trigger stage operating with a 5 mA tunnel diode. This resulted in a trigger sensitivity improvement by the factor 5.

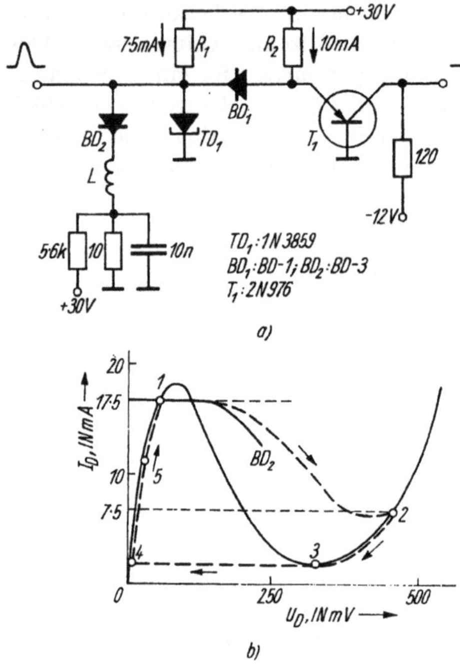

Fig. 97. Univibrator system. According to *Verweij* [Ve 63a].

a) Circuit diagram: width of output pulse 2 ns at $L = 0$, width of output pulse 10 ns when $L = 0{\cdot}35$ μH, repetition rate max. 100 MHz.
b) Characteristic of tunnel diode showing switching trajectory.

The pulse shaping stage given by *Stromer* [St 67] represents an improved version of the circuit in Fig. 97, with bistable behaviour through the omission of BD_2 and L. The circuit is suitable for sinusoidal input signals of repetition rates up to 700 MHz. Output pulses have a rise and fall time of 0·5 ns which is largely determined by the switching behaviour of the transistor.

SCHMITT-TRIGGER SYSTEMS WITH TUNNEL DIODES

In this arrangement the input pulse switches the tunnel diode to the high-voltage state, but in contrast with univibrator systems, the tunnel diode retains this state until the end of the input pulse or until a supplementary circuit resets the tunnel diode to initial state.

Back-switching at the end of the input pulse presupposes that the load characteristic has only a single stable working point in initial state. Fig. 98*a* illustrates the principle of this *Schmitt*-trigger version. Triggering takes place when $U_{in} \geqq U_p + RI_p$ applies for the input voltage. This circuit is characterised by high speed and short recovery time because, with the exception of parasitic inductances and capacitances, no other storage elements are present. Width of the output pulse approximately equals the duration of the input pulse. In quiescent state a current of $I_D = I_0$ flows. Above a maximum current I', which is dependent on the load resistance R, the tunnel diode does not return to initial state when the input signal approaches zero. Fig. 98*b* gives the dependence of the tunnel diode output voltage U_{out} with reference to the input voltage U_{in} of the *Schmitt*-trigger, with $I_0 = 0$. In the diagram, $I_0 > 0$ corresponds to an input voltage $U_{in} \approx RI_0$, while resetting occurs approximately at $U_{in} \approx RI'$. It will be noticed further that only a relatively

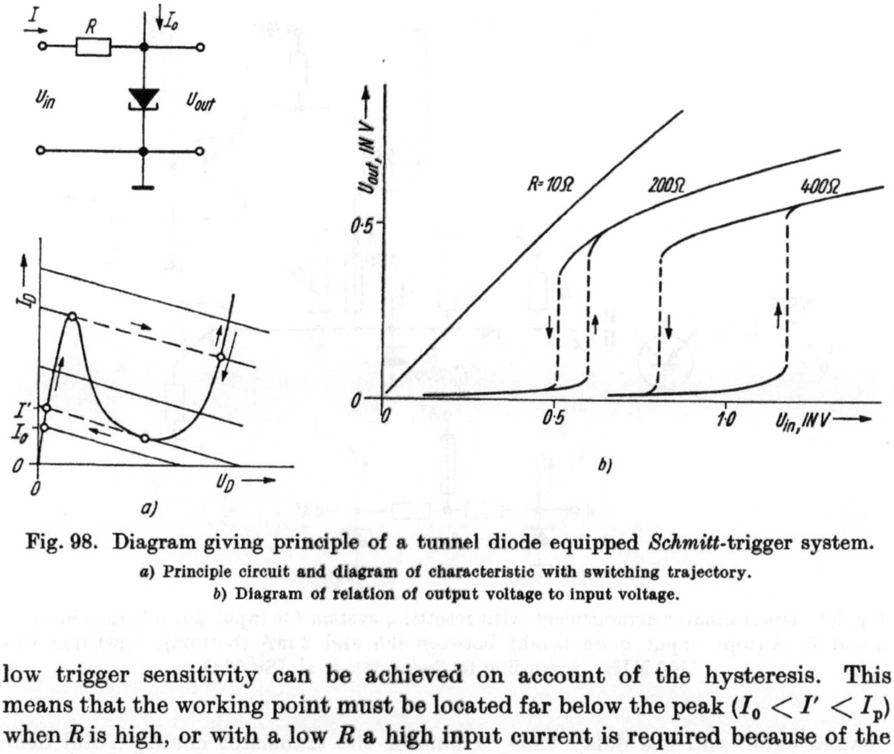

Fig. 98. Diagram giving principle of a tunnel diode equipped *Schmitt*-trigger system.
a) Principle circuit and diagram of characteristic with switching trajectory.
b) Diagram of relation of output voltage to input voltage.

low trigger sensitivity can be achieved on account of the hysteresis. This means that the working point must be located far below the peak ($I_0 < I' < I_p$) when R is high, or with a low R a high input current is required because of the steep load characteristic.

In the arrangement given in Fig. 94*b* the tunnel diode TD_1 operates as *Schmitt*-trigger [Go 61 a], and with $R = 500\ \Omega$ the trigger threshold is located between 5 and 50 mA. The *Schmitt*-trigger system of *Infante* and *Pandarese* [In 62] responds to a pulse that has a width of less than 10 ns and a height which exceeds 2 V.

Resistance R and current I_0 can also be selected to obtain a load characteristic intersecting the tunnel diode characteristic twice, resulting in two stable points ($I_0 > I'$). This achieves a higher trigger sensitivity but requires a resetting circuit. Delay of the latter, however, must be longer than the input pulse width, otherwise multiple triggering will occur. The use of a resetting circuit is recommended if the width of the output pulse has to be independent of the shape and size of the trigger signal. A high-speed discriminator with resetting circuit [Su 62a] is given in Fig. 99. In this version, the negative input pulse switches the tunnel diode to high-voltage state, which in this case is also stable. The negative voltage step passes through the transistors T_2, T_3, T_4, and the coaxial cable to emerge as a positive pulse. The resulting positive voltage step resets the tunnel diode to the low-voltage state. This yields an output pulse of a width which is equal to the transit time in the

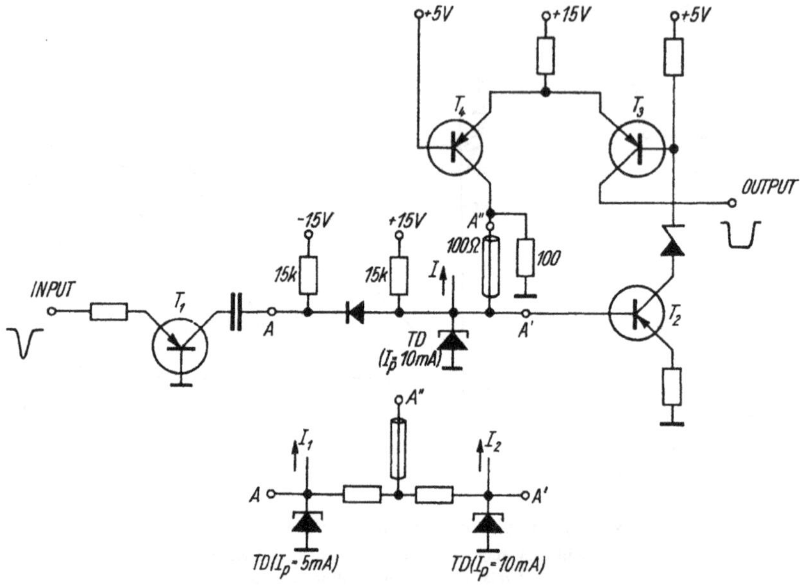

Fig. 99. Discriminator arrangement with resetting system for input pulse height between 2 and 5 mA (top), input pulse height between 0·5 and 2 mA (bottom), repetition rate 100 MHz. According to *Sugarman* et al. [Su 62a].

coaxial cable and the delay time of 3 ns of the transistor chain. Total dead time amounts to about twice the delay of the coaxial cable element. A discriminator of this type can handle pulse repetition rates of 100 MHz. To increase trigger sensitivity two tunnel diodes with different peak current levels may also be employed. These are connected in series and are reset by the same pulse (see bottom sketch in Fig. 99). The same principle of resetting action through a delay line has also been used by other workers [Bl 63a; Ve 64; Jo 65a; Nu 67]. A discriminator arrangement where the moment of resetting is determined by an *RC* element has been developed by *Welter* [We 65a]; and *Verweij* [Ve 63a] gives a similar setup. Another version is a chain of several selected trigger stages with successively higher peak current values, where the last stage is a univibrator for resetting the stages to initial state [Ve 65; Ve 66].

3.3 PULSE AMPLIFIERS

High-frequency transistors lend themselves admirably for assembling simple and extremely reliable pulse amplifiers with rise times in the nanosecond range. In most cases gain factors of between 5 and 100 and a maximum output pulse height up to 10 V are sufficient. Transistor circuits are superior to electron valve equipped amplifiers because complete d.c. coupling is more easily achieved due to the lower operating voltages and the availability of complementary p-n-p and n-p-n transistors. General requirements for "fast" amplifiers differ only little from those for "slow" amplifiers. In combination with high-speed discriminators it is primarily a short rise time, a low overshoot, satisfactory linearity and temperature stability that are necessary. Usually true to shape amplification of the input pulse, particularly the trailing edge is not necessary.

VOLTAGE- AND CURRENT-MODE SYSTEMS

Most radiation detectors deliver current pulses with a total charge proportional to the energy loss of the registered particles in the detector. If the time constant of the detector output is high against current pulse width, the current pulse is integrated, i.e. a voltage-mode signal is received. In current-mode signal operation, however, current and voltage pulses at the detector output have nearly the same time behaviour (see Fig. 17, p. 35). *Fairstein* and *Hahn* [Fa 65b] differentiate between three distinct amplifier systems:

(i) Voltage-Mode System

In this traditional and most widely used system the voltage-mode signal is amplified during which process it is suitably differentiated and integrated to avoid pile-up and other effects.

(ii) Current-Mode System

The current-mode signal is amplified practically without integration. Only when it reaches the output of the system the signal is integrated to form a voltage pulse with a height proportional to the charge of the initial current-mode signal. Of advantage here is that gating and other logical operations can be performed on short current-mode signal operation with a minimum interference between adjacent pulses. This method fails if the energy resolution is determined by the signal-to-noise ratio of the preamplifier circuit. In such a case the following system is recommended:

(iii) Voltage- and Current-Mode System

In this circuit the voltage-mode signal is first amplified in the preamplifier, then differentiated and further amplified as current-mode signal as under (ii).

With the current-mode system the rise time in the amplifier should not exceed the current pulse width. Therefore, an amplifier can be used in the mode of operation described under (i) or (ii), depending on the width of the detector pulse.

The above terms should not be confused with the designations voltage and current amplifier. A voltage amplifier essentially corresponds to the vacuum tube amplification principle and is widely used. A current amplifier always has a low impedance in the forward path, resulting in a high current pulse with only little voltage swing [Ru 64; Ar 65]. This mode of operation is best suited for current-controlled transistors and permits the complete utilisation of the gain-bandwidth product.

NEGATIVE FEEDBACK

As detailed in Section 3.1.5, the common-emitter transistor stage which permits current as well as voltage amplification, shows unsatisfactory frequency behaviour compared with the two other basic transistor circuits. Negative feedback improves cut-off frequency and stability of amplifier systems. The stabilising action becomes effective within a period of 2 to 5 times of the rise time of the negative feedback amplifier, i.e. a period that is short against the rise time of the open-loop amplifier [Fr 64; Ha 66; Fr 67]. Considerable difficulties impede accurate negative feedback dimensioning. On high frequency the current gain factor is complex [see equations (79) and (80)], and the input and output impedances depend on frequency and drive. For this reason, negative feedback is employed only for a maximum of two successive stages

Table 16: Single-Stage Series and Shunt Feedback for Emitter Circuited Transistors

	Series Feedback	Shunt Feedback
	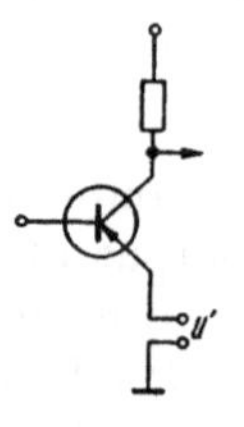	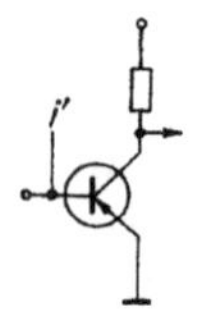
Negative feedback through	voltage u'	current i'
current amplification†	unchanged	lower
voltage amplification†	lower	unchanged
input and output impedance†	higher	lower
preferable source impedance	$R_G \approx 0$	$R_G \approx \infty$
preferable load impedance	$R_L \approx 0$ (short-circuit)	$R_L \approx \infty$ (open-circuit)
Negative feedback inactive on	$R_G = \infty$ (current control not influenced by u')	$R_G = 0$ (short-circuited current i')

† Compared to the common-emitter stage without feedback.

in multi-stage pulse amplifiers with a rise time in the nanosecond region. Frequently it will be necessary to carry out experimental balancing for the most favourable input step response.

Series-feedback and shunt-feedback are the two basic negative feedback systems, where either the voltage u' is in series or the current i' is in parallel to the input. Voltage u' and current i' are proportional to voltage or current at the output. In transistor circuits the negative feedback current is fed-in at the base with the shunt-mode, while, with a series feedback system, the negative feedback voltage is fed-in at the emitter. This negative feedback results in a change of the gain factor and of the input and output impedance. For a single-stage negative feedback circuit qualitative behaviour is given in Table 16. Respective details are found in various papers [Ch 63b; Le 63; Sa 63a; Uz 64; Er 66]. If both negative feedback modes are utilised in the same stage, current and voltage gain decrease, but the input impedance can be widely varied [Sa 63a].

SINGLE-STAGE NEGATIVE FEEDBACK

This is widely preferred in common-emitter circuits. With a common-base stage the rise time of the collector current is extended through an external base resistance; while in the common-emitter circuit the emitter resistance R_E (Fig. 100a) yields a better frequency behaviour, as can be seen from equation

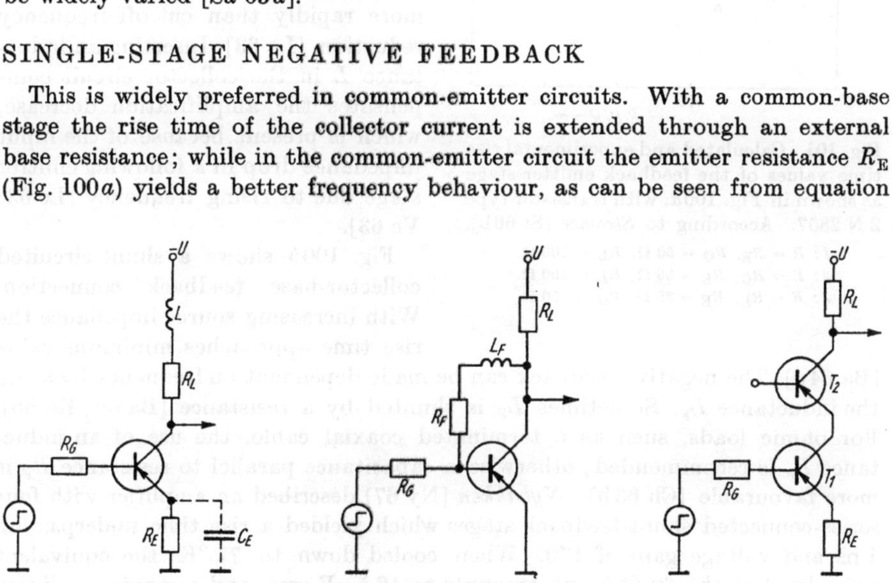

Fig. 100. Single-stage amplifier circuits.
a) **Series-feedback stage.** *b)* **Shunt-feedback stage.** *c)* **Cascode stage.**

(83) in Section 3.1.5. *Stromer* [St 66b] investigated the dependency of the rise time of negative feedback connected common-emitter circuits with reference to various parameters, of which an example is given in Fig. 101. Generally, for R_E a resistance not exceeding 50 to 60 Ω is selected. Voltage gain amounts approximately to $-R_L/R_E$. Connecting a capacitor C_E in the order of 10 pF parallel to the resistance R_E permits the balancing of the output pulse to

a minimum rise time without overshoot (aperiodic limiting condition). Sometimes it may be feasible to switch-in C_E only above a specific pulse height [Al 63]. High-frequency peaking with C_E results in a steeper pulse rise only if the rise time is primarily determined by the capacitance in the collector circuit [Le 63]. In any other case C_E does not bring any marked improvement because the improved gain through reduced negative feedback is compensated by the decrease of the cut-off frequency of the common-emitter circuit. Replacing the capacitor C_E with a series-connected resonant circuit allows amplification to increase more rapidly than cut-off frequency reduction [Le 63]. Inserting an inductance L in the collector circuit compensates the amplification decrease, which is present because of the input impedance drop in a following emitter stage due to rising frequency [Le 63; Ve 63].

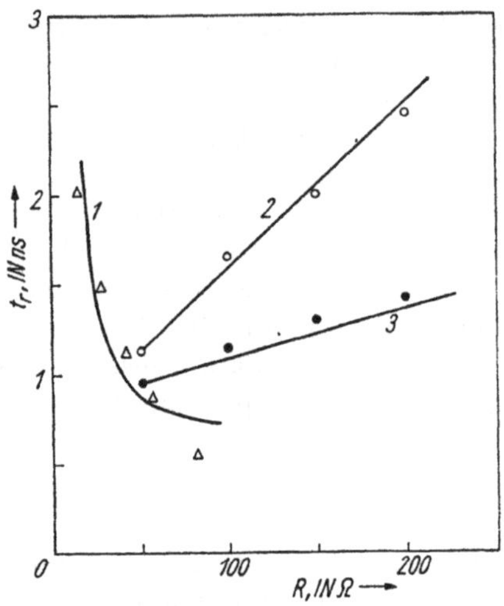

Fig. 101. Calculated and experimental rise time values of the feedback emitter stage as shown in Fig. 100*a*, with transistor type 2 N 2857. According to *Stromer* [St 66b].

1: $R = R_E$, $R_G = 50\ \Omega$, $R_L = 100\ \Omega$.
2: $R = R_G$, $R_E = 39\ \Omega$, $R_L = 100\ \Omega$.
3: $R = R_L$, $R_E = 39\ \Omega$, $R_G = 50\ \Omega$.

Fig. 100*b* shows a shunt-circuited collector-base feedback connection. With increasing source impedance the rise time approaches minimum value [Ba 64a]. The negative feedback can be made dependent on frequency by using the inductance L_F. Sometimes L_F is shunted by a resistance [Ba 59; Ro 65]. For ohmic loads, such as a terminated coaxial cable, the use of an inductance L_F is recommended; otherwise a capacitance parallel to resistance R_F is more favourable [Ch 63b]. *Nybakken* [Ny 67] described an amplifier with four series-connected shunt-feedback stages which yielded a rise time underpassing 1 ns and voltage gain of 170. When cooled down to 77 °K the equivalent noise level at the 50 Ω input amounts to 12·5 μV rms, and a maximum linear output voltage on 50 Ω of about 0·3 V was obtained.

Fig. 100*c* shows the cascode stage — a further basic amplifier connection. It consists of the common-emitter transistor T_1 and the common-base transistor T_2, and combines the advantage of a short rise time and a high voltage swing [Iv 67a]. Rise time of T_1 is short due to the feedback connection with R_E and because of the low collector impedance represented by the input impedance of T_2. A collector resistance R_L in the order of some 100 Ω can be used due to the good frequency behaviour of the common-base stage T_2. Voltage amplification amounts to approximately $-R_L/R_E$. As with the series-feedback stage, a correcting capacitance can be employed at the emitter end of T_1 or T_2 and an inductance in series with R_L [Wi 62; Al 63; Sa 63b; Iv 67a; La 67a].

The rise time increase observed with some basic circuits on increasing output pulse height has been investigated by *Baldinger* and *Simmen* [Ba 64a]. Results obtained with a transistor 2 N 976 are given in Fig. 102. In conjunction with

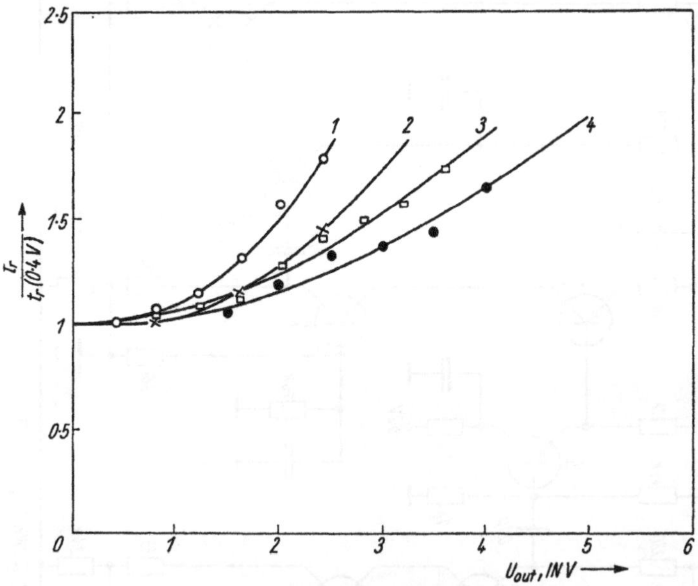

Fig. 102. Relative rise time in dependence on output signal height, with $R_L = 50\ \Omega$, transistor type 2 N 976. According to *Baldinger* and *Simmen* [Ba 64a].

1: Shunt-feedback stage (Fig. 100 *b*).
2: Series-feedback stage (Fig. 100 *a*).
3: Common-collector stage (emitter follower).
4: Common-base stage.

the data compiled in Tables 13 and 16, this demonstrates that an amplifier having a common-collector input stage, followed by a cascode stage, will yield good results for rise time and maximum output amplitude. An amplifier circuit according to *Epstein* [Ep 63] is shown in Fig. 103 which includes two such combinations (T_1 to T_3 and T_4 to T_6). Through the use of p-n-p transistors and n-p-n transistors it was made possible that, on positive input pulses, every transistor is shifted to the region of higher emitter current [Ep 63; Pi 66]. Further examples, some with d.c. coupling, are given by other workers [Wi 62; Al 63; La 63a; Ve 63; Iv 67 a]. The rise time of all amplifiers cited in this section is between 1 and 5 ns. *Dubrovskii* has compiled details on a large number of amplifier circuits [Du 66].

In cases where only series-feedback and shunt-feedback stages are employed optimum gain and rise time can only be obtained when the stages are arranged

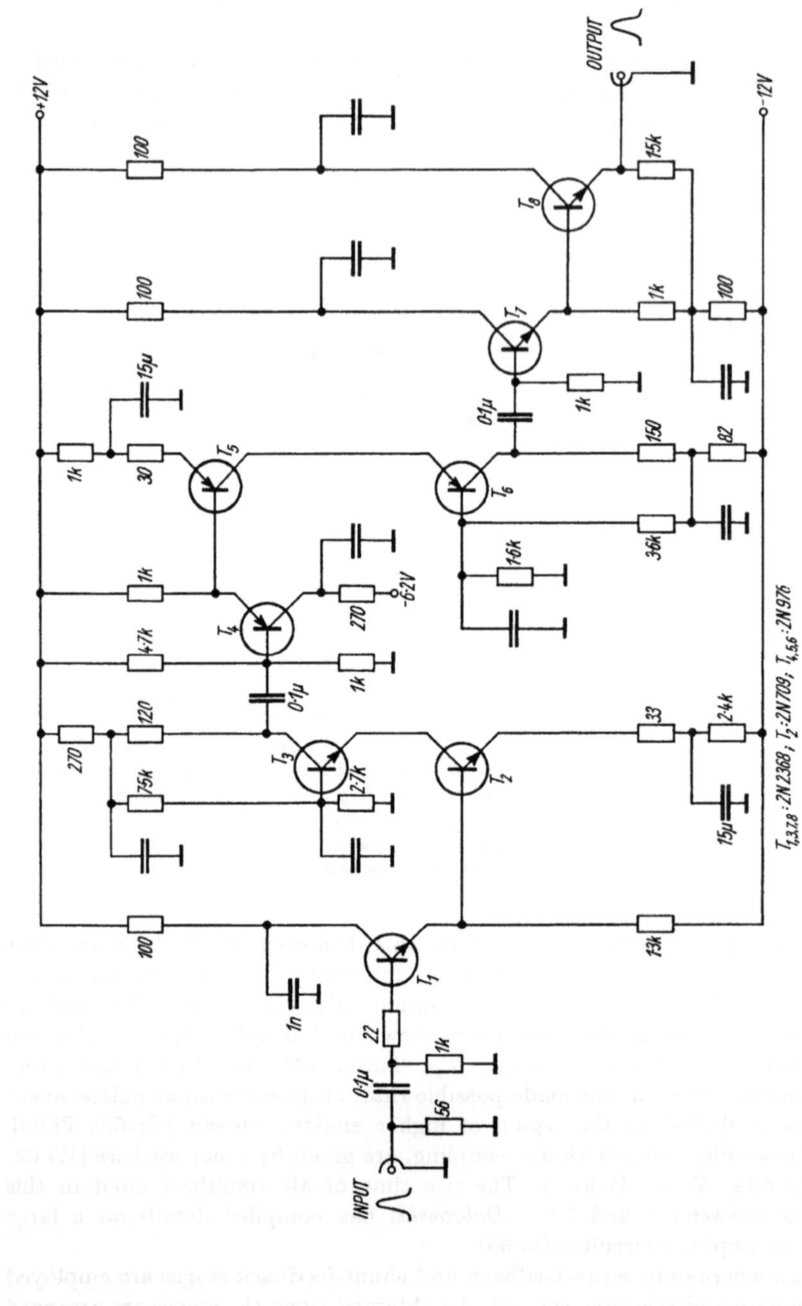

Fig. 103. Simplified circuit of a fast amplifier. According to *Epstein* [Ep 63].
Rise time less than 2·0 ns, gain 10, output pulse height 5 V with a load resistance of 125 Ω.

alternately in the cascade [Ch 63b]. Then the series-feedback stage as a high-ohmic source feeds the shunt-feedback stage which has a low input impedance (mismatched feedback design). Amplifiers mentioned by *Baldinger* and *Simmen* [Ba 64a], *Beneteau* and *MacIntosh* [Be 61] and *Millard* [Mi 67] are of this design and consist of a series-feedback stage which is followed by a shunt-feedback stage. *Coli* et al. [Co 65] give a pulse amplifier circuit where voltage amplification is effected by two common-base stages which are coupled through a common-collector stage for current amplification.

For use with semiconductor detectors *Lunsford* [Lu 64b] has developed very-wide band amplifiers, where the basic amplifier stage shown in Fig. 104 consists of the common-base stage T_1 and the series-feedback stage T_2. The resistance R_s strongly reduces the output impedance of T_1. Connecting of several such basic stages in series will always produce small voltage signals only, which makes this circuitry similar to a current amplifier. In an assembly containing five basic stages in cascade formation a current gain of 100 at a rise time of 1 ns was achieved. At the output, pulses of a maximum height of 1 V with a load resistance of 50 Ω were obtained. The amplifier of *Tojo* [To 67] comprises three shunt-feedback stages, similar to Fig. 100 *b*, but with $L_F = 0$ and the output on the emitter end, with interposed common-base stages for impedance transformation. At a rise time of 3·2 ns the current gain is about 50.

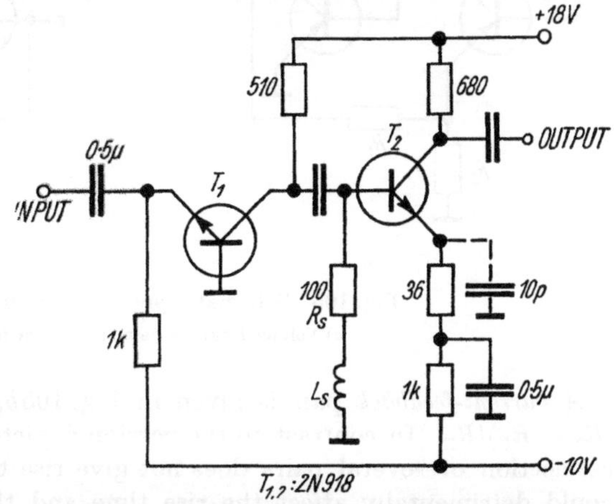

Fig. 104. Basic amplifier stage with a rise time of less than 1 ns. According to *Lunsford* [Lu 64b].

TWO-STAGE NEGATIVE FEEDBACK

Compared to the single-stage feedback version, in this mode a second transistor stage is added in the feedback loop. Fig. 105 gives two of the four possible basic circuits where the transistor base serves as input electrode [Ch 63b].

A voltage-feedback pair is shown in Fig. 105*a*, with the voltage amplification amounting to approximately $(R_1 + R_E)/R_E$. This feedback design is very sensitive to both the input and the output terminations. A practical example is shown in Fig. 106 [Ja 65]. *Schapper* [Sc 64b] describes an amplifier circuit which combines a p-n-p transistor and an n-p-n transistor and makes the *Zener*-diode coupling unnecessary.

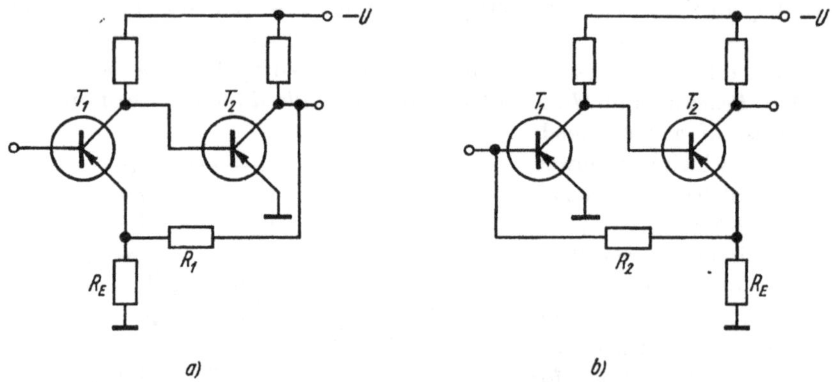

Fig. 105. Schematic presentation of feedback pairs.
a) Voltage feedback pair. *b*) Current feedback pair.

A current-feedback pair is given in Fig. 105*b*, where current gain is about $(R_2 + R_E)/R_E$. In contrast to the version depicted in Fig. 105*a*, here the series connection of several pairs does not give rise to any mutual influence which could detrimentally affect the rise time and the gain. *Bowes* and *Granville* [Bo 64b] employed thin-film technique to develop a multi-stage amplifier.

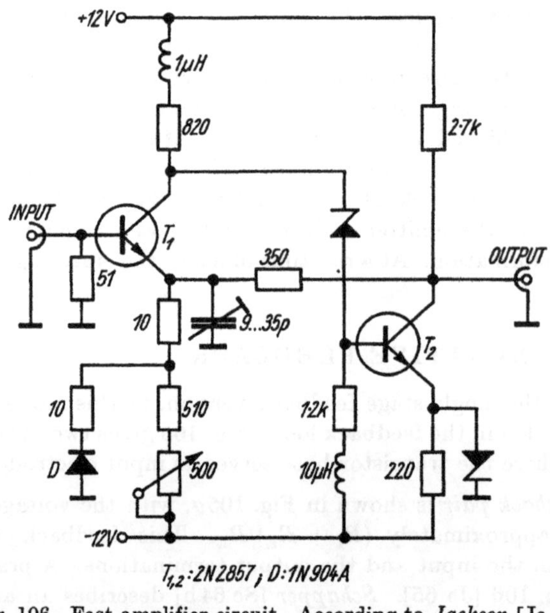

Fig. 106. Fast amplifier circuit. According to *Jackson* [Ja 65].
Rise time 1 ns, gain 10, maximum output amplitude 1 V at a load of 50 Ω, delay 3 ns.

Manfredi and *Rimini* [Ma 67] designed current-feedback pairs where, in contrast to Fig. 105*b*, the output signal is taken from the emitter end of the transistor T_2 and delivered to the load resistance via an emitter follower. With a 10-fold gain of one pair a rise time between 1 and 2 ns could be obtained.

Cherry and *Hooper* [Ch 63b] demonstrated that in wideband amplifiers a maximum gain-bandwidth product is already obtainable with a cascade design of alternate series-feedback and shunt-feedback stages. At the best, a feedback pair can achieve only the same end at the expense of greater design complexity.

The speed of transistor pulse amplifiers is ultimately limited by the current gain-bandwidth capability of the transistors used. In voltage amplifiers the speed is reduced by various unavoidable circuit capacitances. In the current

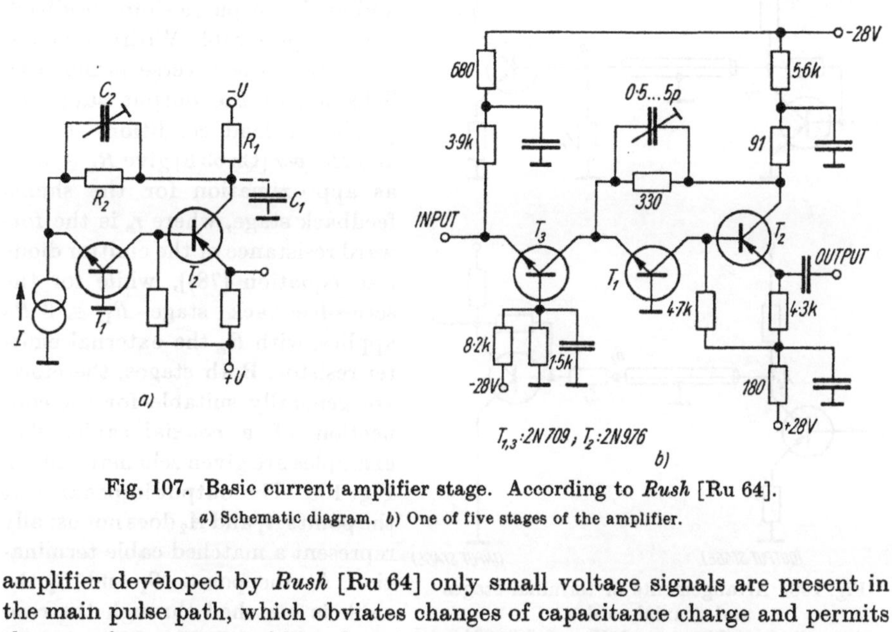

Fig. 107. Basic current amplifier stage. According to *Rush* [Ru 64].
a) Schematic diagram. *b*) One of five stages of the amplifier.

amplifier developed by *Rush* [Ru 64] only small voltage signals are present in the main pulse path, which obviates changes of capacitance charge and permits the use of transistors of low electric strength. A schematic illustration of one stage of this current amplifier is given in Fig. 107*a*. Most satisfactory operation of the circuit is achieved with current control ($R_G \to \infty$) and a short-circuited output. On the aperiodic limiting condition with the current control mode the minimum rise time is $t_r \approx 1/f_{T1}$ with f_{T1} the gain-bandwidth product of transistor T_1. Gain is $A = (R_1 + R_2)/R_1$ which obtains a minimum rise time without overshoot at a gain of $A = 4 f_{T2}/f_{T1}$. A single stage with $t_r \approx 1{\cdot}25$ ns is shown in Fig. 107*b*. For the coupling of several stages the input common-base transistor serves as impedance transformer. The complete amplifier has five such stages and produces a current gain of 1,580 with a rise time of $t_r = 3{\cdot}05$ ns. An amplifier of similar design as in Fig. 107 is mentioned by *Pigneret* et al. [Pi 66], and *Goyot* et al. [Go67 b], where the output pulses are taken from the collector of transistor T_2 with polarity reversed. With the series-connected feedback pairs the common-base transistor T_3 of Fig. 107 is replaced by a low-ohmic resistance which is in series with the emitter of transistor T_1.

INPUT AND OUTPUT STAGES OF AMPLIFIERS

For the terminal stages at the input and output of amplifiers, series-feedback or shunt-feedback stages are most suitable because their impedance is less frequency-dependent than is the case with a common-collector or a common-base stage [Ch 63 b]. Instead of the common-collector stage at the amplifier input a series-feedback stage, and at the output, a shunt-feedback stage are preferable. With a common-base stage the reverse is the case. Selection of the output stage depends on load condition. *Cherry* and *Hooper* [Ch 63 b] give $R_L \geqq 30\, r_e$ as approximation for the shunt-feedback stage, where r_e is the forward resistance of the emitter diode [see equation (78)], while for the series-feedback stage $R_L \leq 3\, R_E$ applies, with R_E the external emitter resistor. Both stages, therefore, are generally suitable for the connection of a coaxial cable. Two examples are given schematically in Fig. 108. The output impedance at the points A_1 and A_2 does not usually represent a matched cable termination. At the points B_1 and B_2 the cable is matched through the resistances Z_1 and Z_2. Matching at point B_1 can be better managed than at point B_2 because of the high input impedance of the circuit. *Lennartz* and *Taeger* [Le 63] have described examples of the generally complex termination networks Z_1 and Z_2.

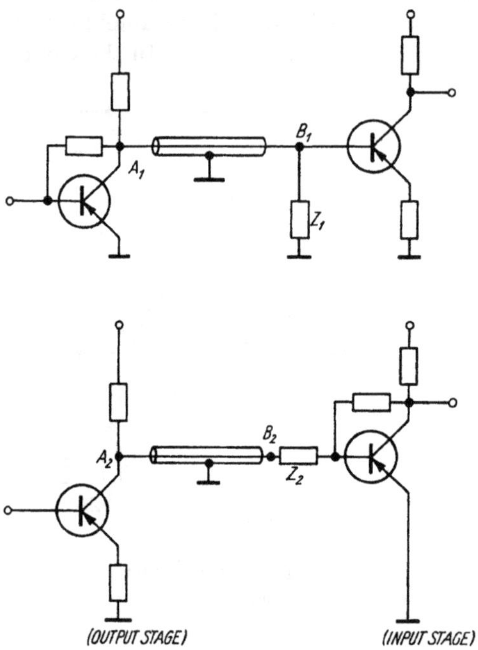

Fig. 108. Arrangements of terminal stages and coaxial cables.
According to *Cherry* and *Hooper* [Ch 63 b].

OTHER TYPES OF AMPLIFIERS

Pulse amplifiers with passive amplification elements have also been developed. In the circuit described by *Ageno* and *Felici* [Ag 62 a] voltage amplification is effected with pulse transformers which are connected through common-collector stages as impedance converters (current amplification) . It is also possible to use pulse transformers [Ko 62] or coaxial cable inverters [Ha 62] for current amplification, and common-base transistors for matching the various stages and for voltage amplification.

D.c. coupling gives good overshoot behaviour and count-rate independent zero level. On overdrive condition the transistors should not be forced into the saturation region but should be cut off. Non-blocking amplifiers which, however, have a rise time of more than 10 ns, are mentioned by *Hahn* and *Guiragossian* [Ha 63], *Conrad* and *Friese* [Co 64], and *Blankenship* and *Nowlin* [Bl 66]. Problems relating to non-blocking are discussed by *Fairstein* and *Hahn* [Fa 65b].

Amplification of pulses from semiconductor detectors (see Section 2.4.1) is effected with charge-sensitive amplifiers. Their rise time is $t_r \approx t_0/\gamma K$, where t_0 is the rise time with open feedback loop, K the open loop gain and γ the ratio of the feedback capacitance to the input capacitance [Ch 61]. The obtainable small rise time, however, cannot be utilised in practice. According to equation (43) voltage rise occurs at the amplifier input with the time constant $\tau \approx R_b C_A C_B/(C_A + C_B)$, where R_b is the series resistance, C_B the capacitance of the depleted detector region, and C_A the effective input capacitance of the amplifier. With $C_A > 1$ nF for the charge-sensitive amplifier, τ is about $R_b\, C_B$ and generally is large compared to the charge collection time in surface barrier detectors and similar types. In that case it is advisable to use voltage-sensitive preamplifiers with $C_A \ll C_B$. *Katzenstein* and *Alexander* [Ka 63] developed a low-noise high-impedance amplifier for semiconductor detectors which consists of a common-collector input stage and four series-feedback stages. Its input capacitance is ≈ 2 pF and its input impedance ≈ 100 kΩ. With this amplifier a voltage gain of 35 and a rise time of $t_r < 2$ ns were obtained. In the arrangement described by *Wahl* [Wa 64a] the detector pulse is passed to a cathode follower and then separately amplified in a fast and a slow channel. Further examples are mentioned in Section 3.4.2.

In preamplifiers for semiconductor detectors field-effect transistors are being increasingly used. Up till end 1967, however, fast amplifiers operating with field-effect transistors have only seldom been described [Ho 63; Go 66b]. *Pigneret* et al. [Pi 66] and *Goyot* et al. [Go 67b] give the circuit of a low-noise preamplifier with an input cascode stage using a field-effect transistor of high cut-off frequency. It has a rise time of around 1 ns.

For the sake of completeness, the *distributed amplifier* must be mentioned. Although the electron valve equipped distributed amplifier technique brought a considerable increase of the bandwidth compared to conventional pulse amplifiers [Mi 56; Le 59; Mc 62; Mo 64; Ch 67a], this could not be achieved with transistorised distributed amplifiers. One of the first versions of this type of amplifier was presented by *McMullen* [Mc 59], where four stages at a bandwidth of 290 MHz resulted in a voltage gain of 1·8. *Dubrovskii* [Du 66] and *Chen* [Ch 67a] compiled further references. Fig. 109 gives the system of one

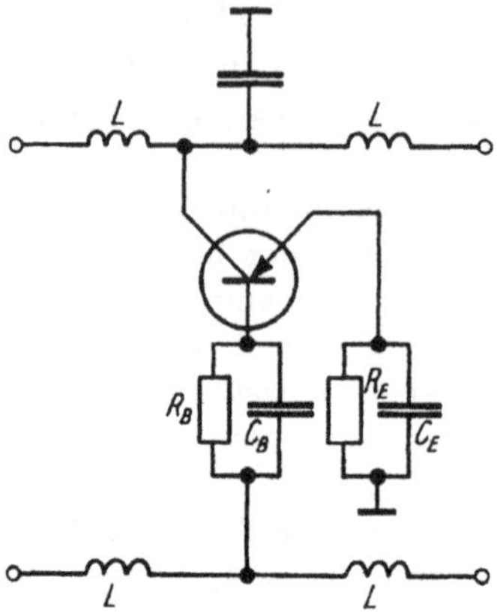

stage of a distributed amplifier. For the time constant condition $R_B C_B = R_E C_E = 1/2\,\pi f_\beta$ applies [Ro 65], because the RC elements together with the transistor input represent a compensated voltage divider. *Moser* [Mo 67a] developed a distributed amplifier with 8 field-effect transistors and obtained a rise time of 6·5 ns at a voltage gain of 3.

Fig. 109. Basic system of a distributed amplifier stage.

3.4 TIME ANALYSERS

3.4.1 GENERAL

Time measuring techniques differ for the various time ranges and, in general, electronic time analysers are used for the measurement of time differences in the 10^{-11} to 10^{-3} s range, for instance, to determine the lifetime of excited states of atomic nuclei or for measuring the time-of-flight of elementary particles (see Sections 5.1 to 5.3). For determining time differences of less than 10^{-11} s electron-optical methods are employed (cp. Section 2.3.3), provided that the involved event is associated with the emission of light.

In the nanosecond range time analysers are utilised for two main purposes [Ch 61a; Fi 65; Am 64a; Ga 65]:

(i) Measuring the *time correlation* of two or more pulses, e.g. the pulses from several radiation detectors, with the aid of *coincidence circuits* (see Section 3.4.3). This permits picking-off a specific event from among other events not correlated in the desired manner.

(ii) Measurement of *time interval distribution*, as is necessary for lifetime and time-of-flight investigations. *Time-to-pulse-height converters* (Section 3.4.4) and *time-to-digital converters* (Section 3.4.6) are mostly used for this purpose.

In general, response and output of coincidence and converter circuits are influenced by the shape and the size of the examined pulses. For the time measurements in the nanosecond range, therefore, the detector output pulses, at least those exhibiting height fluctuations, must be standardised. The timing circuits employed to obtain the desired information are discussed in the following section.

3.4.2 TIMING

The problem of timing involves the use of a non-linear circuit to convert the detector output pulses into standard pulses for the input of the time analyser. Shaping stages for this purpose have to come close to fulfilling two important conditions for a high-accuracy time measurement, i.e. to allow high time resolution:

(i) The moment of the appearance of the shaped pulse, usually its leading edge, must be determined by the part of the detector signal which, with maximum statistic accuracy, represents the moment of the involved event.

(ii) During shaping no additional time fluctuations may be introduced, for instance, fluctuations caused by a differing height of the detector output pulses.

Leaving supplementary compensation circuits aside, only one of these conditions can usually be met satisfactorily. *Pisarevskii* and *Soshin* [Pi 60] and *Gatti* and *Svelto* [Ga 65] give a survey of the various timing methods. The following description deals with scintillation counter pulse shaping in the first instance, while details of timing methods involving semiconductor detectors are dealt with in conclusion.

LIMITING

Pulses of uniform height are delivered by the valve or the transistor of the limiter circuit when the detector signals cut off the quiescent current flow. Pulses limited in this manner are formed to the same length by a short-circuited coaxial cable. Two basic arrangements are given in Figs. 110*a* and *b*, showing a valve in grounded cathode circuit and a p-n-p transistor in common-base connection. Fig. 110*c* gives a qualitative presentation of the pulse shape at the input and at the output, with polarity reversal neglected. It will be seen that rise time and decay time of the limited pulse depend on the height of the input pulse. The differing slope steepness has the effect of an additional time shift ("walk") in the time analyser, so that the measured time difference does not occur with the true value. This characteristic of limiting circuits is of particular disadvantage when the pulses to be limited have a long rise time (for instance, when integrated pulses of a NaI(Tl) scintillation counter are involved), or when a wide pulse-height spectrum has to be allowed for time measurement to obtain sufficient detection efficiency. With a lower limiting threshold this additional time shift remains small. This, however, presupposes the high gain of a photomultiplier because the cut-off of a current $I = 10$ mA requires a current of the same magnitude with common-base connected transistors and, in case of a valve, a voltage in the order of 1 V.

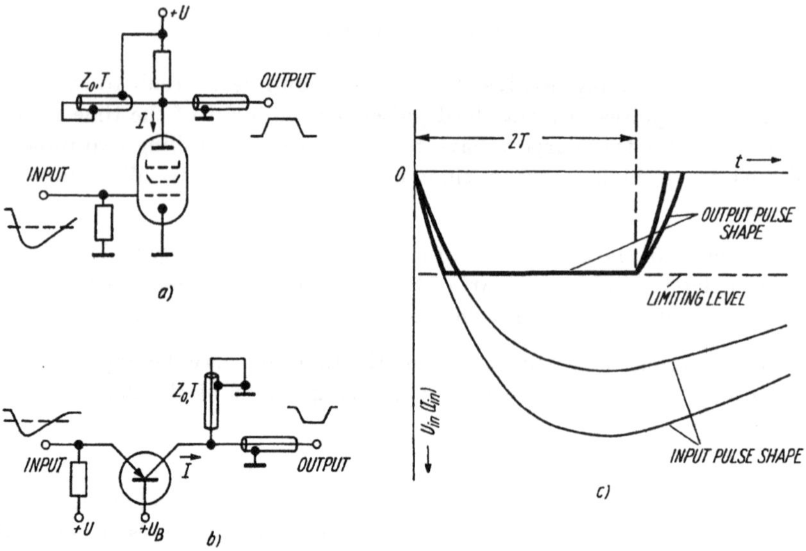

Fig. 110. Principle of a limiter.
a) In cathode-base circuit. b) With common-base transistor. c) Pulse shape diagram.

Valve-Equipped Limiter Circuits

If the limiting valve shown in Fig. 110*a* is blocked by a negative voltage pulse the output pulse height amounts to $Z_0\, I/2$, where Z_0 is the characteristic impedance and I the cut-off anode current. Output pulse width is $2\,T$, with T the one-way transit time in the pulse shaping cable. Integration of the detector signal is usually effected at the input of the limiter circuit. If the limited pulses are also integrated in the time analyser (time-to-pulse-height converter in overlap principle), machine time at least approaches the centroid of the detector current pulse up to the moment when the limiting valve is blocked (see Fig. 18, p. 36). As has been described in Section 2.3.2, minimum statistic time fluctuation is obtained if only the first part of the detector pulse is utilised, i.e., if, depending on the decay time constant of the scintillator, approximately the initial 1 to 25 per cent of the integrated detector signal serves to cut off the limiter current. Therefore, a low limiting level permits the time information to be picked off from the detector pulses with optimum statistic accuracy and, at the same time, keeps the undesirable walk from differing slope steepness small.

Limiting integrated detector pulses in this manner has two disadvantages that cannot remain unmentioned:

(i) Pulses to be limited must decay slowly and must be longer than $2\,T$ to permit well-formed output signals without obscuring undershoot, otherwise pile-up effects can occur even on a low count rate. If a second pulse arrives before the first has decayed sufficiently for the limiter current to have regained full value, a limited pulse with a reduced height is delivered. If the limiter circuit is coupled to a detector through a capacitor of moderate value, the negative detector pulse at the limiter input causes a positive undershoot. During this period an increased limiter current may result and the limited pulses become too high.

(ii) All the detector pulses the height of which underpasses the limiting threshold, e.g. photomultiplier noise, also produce output signals which may interfere with the time analyser accuracy.

These disadvantages affect the obtainable time resolution by causing a flattening of the coincidence curve slopes and producing slowly decaying tails. On the other hand, on full width at half-maximum coincidence curves are less sensitive.

Despite its serious drawbacks limiting with a valve circuit for a long time represented the most important method in practical timing work in the nanosecond region. Some examples of these limiter circuits are given in Fig. 111. Limiter valves require a high gain-bandwidth product *GB*, and the anode-current/grid-voltage characteristic must show a sharp cut-off bias without tail. The quiescent current through the limiting valve can be stabilised by a d.c. feedback through a cathode resistor R_k (Fig. 111*a*). To block the limiting valve with minimum voltage the cathode potential must be kept constant during the duration of the pulse, for instance, by a high capacitance in parallel to R_k. This, however, allows the stabilisation of the mean cathode current only, which still depends on pulse rate. Removing the current, flowing through R_k during the duration of a pulse, in a diode (Fig. 111*a* — [Le 59, Si 61a, Fo 63]) or through a transistor connection [St 66c] is more satisfactory. As a compromise between stabilisation and outlay (negative voltage source), a resistor R_k with shunted capacitor is frequently employed in the cathode lead, with a capacitance of some microfarads and a voltage drop across R_k in the order of 10 V.

In the circuit developed by *Jones* [Jo 60] the limiter valve is operated with a constant current for the control grid and screen grid for stabilisation (Fig. 111*b*). Additional stabilisation is given by the feedback connection through the resistor R_1 common to both grids. Diode *D* is blocked in quiescent state and conducts during the duration of the negative input pulse. This makes the stabilising action ineffective until the input pulse is past. Another advantage of this limiter circuit is that, because of the constant current through R_2, the integrated pulse from the photomultiplier shows a linear fall and does not decay exponentially. With the dimensioning given in the presented case, fall time amounts to 1 μs.

Limiter circuits with secondary emission valves [Jo 56; Sa 58] also necessitate stabilising measures. A circuit with d.c. feedback from the anode to the control grid is mentioned in reference [Vy 63].

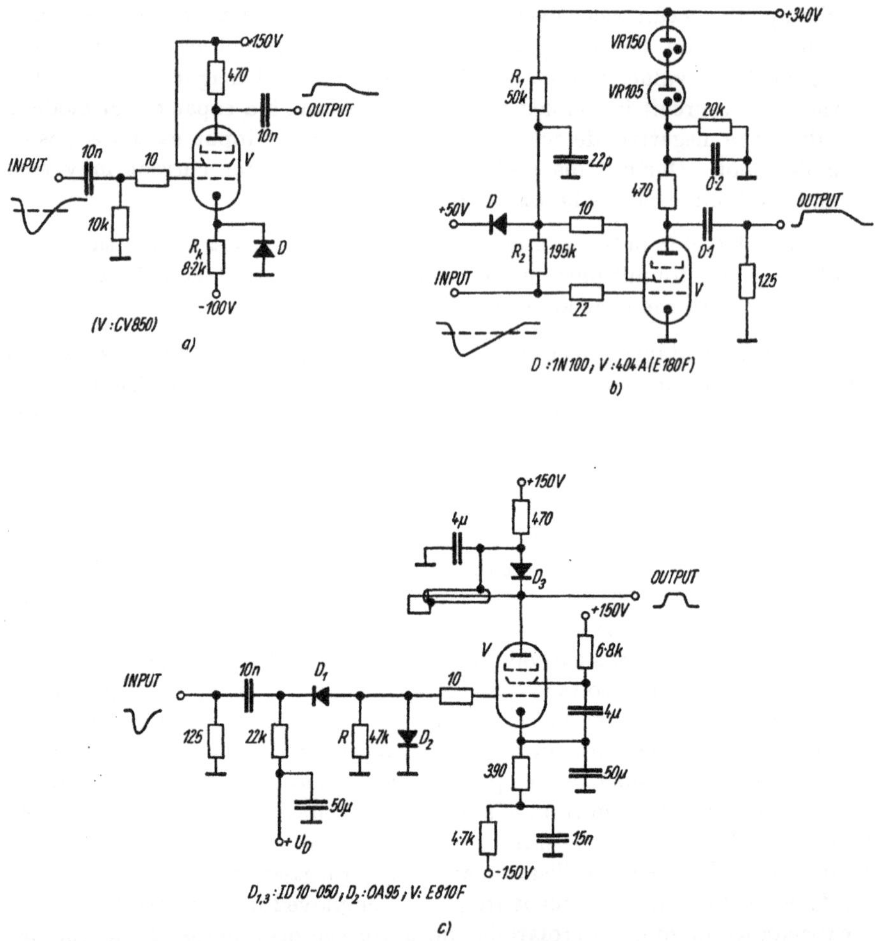

Fig. 111. Limiter circuits for scintillation counter pulses (slightly simplified).
a) Stabilised by R_k. According to *Lewis* and *Wells* [Le 59]. b) Stabilised through the control grid and screen grid currents, pulse height 0·94 V at a resistance of 100 Ω. According to *Jones* [Jo 60]. c) With discriminator diode (D_1), output pulse 2·8 V, pulse width 100 ns. According to *Aspelund* [As 63].

To avoid count-rate-dependent effects, the limiter should be d.c. coupled to the detector output. The positive swing after an RC coupling element will produce only a slight increase of the limiter current if a high cathode resistance is used, as in Fig. 111*a*.

Sometimes a discriminator diode is inserted in the pulse path to keep the background noise of the photomultiplier away from the limiter circuit [Ne 59; As 63; Du 64a]. In the arrangement according to *Aspelund* [As 63], given in Fig. 111*c*, the biased diode D_1 works as discriminator, while D_2 and R stretch

the short non-integrated pulses of the scintillation counter. The shaping cable is not matched at the anode-end, resulting in a pulse height $Z_0 I$. This causes overshoot which is largely suppressed by the diode D_3 [Ga 60a].

***Bartl* and *Weinzierl* [Ba 63b] developed a limiter where the positive dynode pulses are first delivered to a cut-off valve (E 810 F) which has a discriminating function. The negative output signal then blocks the following limiter valve. For this an adjustable interval of about 0·2 V width of the dynode pulse is sufficient. In contrast to the described limiters, one point of the leading edge of the detector pulse is utilised only, as is also the case with a trigger circuit.**

Because integrated multiplier pulses are generally handled in limiter circuits, the direct mounting of a limiter to the socket of the photomultiplier is recommended. This may, however, give rise to additional measuring error when limited pulses are transmitted to the time analyser. Such an error may be caused by reflexions in a short cable from unsatisfactory cable terminations and by pulse distortion due to attenuation in a long cable. With time analysers operating on the overlap principle, the limiter is often an essential component of the circuit arrangement and can, therefore, not be incorporated in the scintillation counter assembly. For such cases, it has been found advantageous to connect a cathode follower to the photomultiplier which then serves as a so-called "prelimiter", and which is loaded with the characteristic impedance of the connecting cable to the time analyser. Output pulses of the cathode follower, in this instance, must have sufficient height to assure the blocking of the original limiter. Arrangements of this type have severally been employed [Ga 60b; Go 60; Fo 63; Bo 64a].

Transistorised Limiter Circuits

Transistors, representing current controlled components, give minimum rise time in common-base connection. Fig. 112 gives the circuit of a transistorised

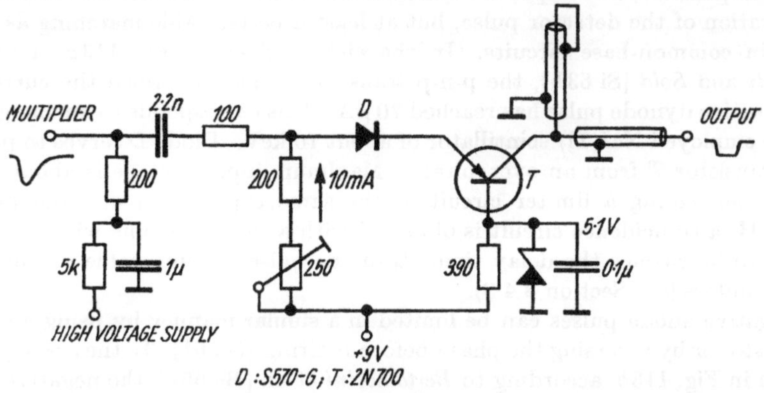

Fig. 112. Limiter circuit with common-base transistor. According to *Sugarman* et al. [Su 62b].

limiter according to *Sugarman* et al. [Su 62b]. With a common-base transistor the input impedance does not exceed 50 Ω. For this reason the effective anode resistance of the photomultiplier for the circuit shown is less than 100 Ω and the detector pulses are not integrated. Thus the leading edge of the limited pulse reproduces the current rise of the detector signal. Machine time is determined by the succeeding circuit. In contrast to the limitation of the integrated detector pulses, this current-mode arrangement can accept higher pulse rates. With the short current pulses from organic scintillators the width of the output pulse of the limiter is restricted to about 10 ns if a short-circuited coaxial cable is used for pulse shaping. Limiting with a common-base connected transistor is less suitable for scintillation counters incorporating slow-decay inorganic scintillators, because the delivered current pulses are far smaller than from an organic scintillator. With increasing reverse voltage the input impedance of the limiting transistor also increases and makes the non-reflecting cable termination maybe difficult. For this reason it is preferable to mount the limiter directly to the multiplier output without interposing a cable. The diode D in Fig. 112 is blocked after transistor T has cut off the current. On excessively high detector signals this prevents breakthrough on the emitter-base junction of transistor T. Various descriptions of similar limiting arrangements, as shown in Fig. 112, are available [Su 60; Ro 62; Du 63; Op 64; Ve 64; Ma 65; Ve 66]. Frequently, d.c. coupling to the photomultiplier is employed. The work of *Vergezac* and *Kahane* [Ve 64] describes a circuit where the pulses are shaped to 6 ns width by a short-circuited cable and further processed in the following trigger stages with tunnel diodes. Positive dynode pulses can be limited with an n-p-n transistor [Ma 65]. Resorting to two series-connected transistors in common-base circuit, the capacitive through-coupling of the input pulse is practically avoided.

Common-emitter transistors result in a high input sensitivity. Such limiter arrangements are suitable for pulses from a slow-decay scintillator. The higher input impedance, as compared to the common-base transistor, allows a certain integration of the detector pulse, but at least a better cable matching as is the case in common-base circuits. In the circuit given in Fig. 113*a*, according to *Sidi* and *Sold* [Si 63b], the p-n-p transistor is blocked when the current of the positive dynode pulse has reached 70 μA. This corresponds to an energy loss in the employed NaI(Tl) scintillator of about 10 keV. Diode D serves to protect the transistor T from an overvoltage. Maximum input voltage is about 0·6 V. When connecting a limiter circuit of the same type and a discriminator to point II, a coincidence circuit is obtained with a resolving time which is about equal to four-times the delay time T_d of the pulse shaping cable common for both limiters (see Section 3.4.3).

Negative anode pulses can be limited in a similar manner by using an n-p-n transistor or by reversing the phase before limiting [Ga 61]. In the arrangement shown in Fig. 113*b*, according to *Bertolaccini* et al. [Be 65c], the negative input pulse drives the p-n-p transistor T_1 to saturation state. In the collector circuit

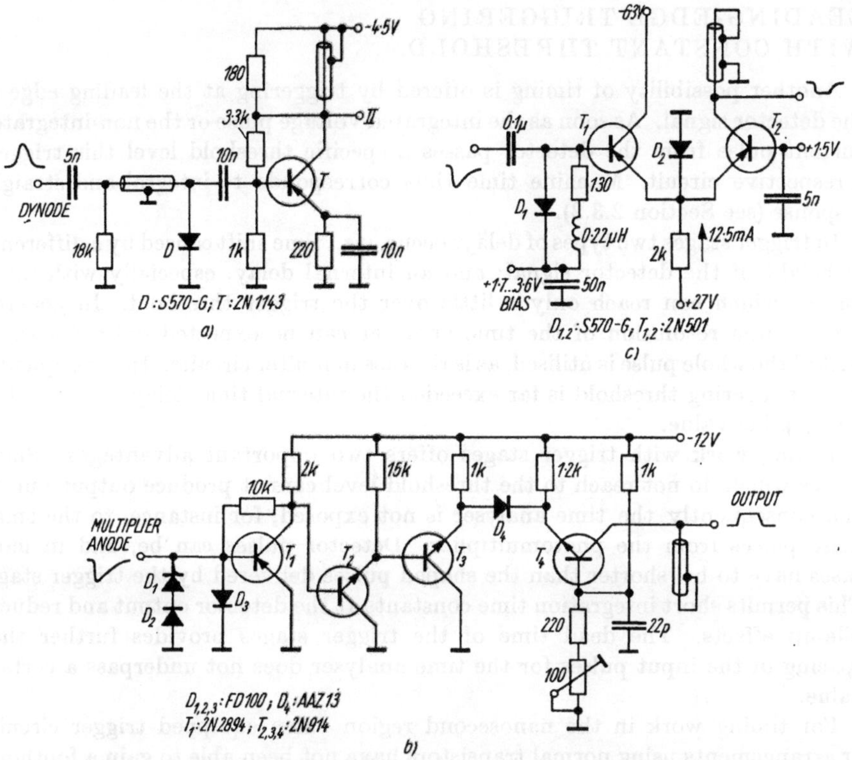

Fig. 113. Limiter circuits of high sensitivity, for slow-decay scintillators.
a) Common-emitter transistor. According to *Sidi* and *Sold* [Si 63b]. *b)* Double limiting (in T_1 and T_4). According to *Bertolaccini* et al. [Be 65c]. *c)* Long-tailed emitter pair. According to *Barna* et al. [Ba 60a].

of T_1 the pulse is integrated with a time constant of about 50 ns. The two emitter followers T_2 and T_3 produce a positive high current pulse which cuts off the current of the second limiter system T_4.

Another limiter version according to *Barna* et al. [Ba 60a] is seen in Fig. 113*c*. Here, the transistor T_1 of the long-tailed emitter pair is blocked in quiescent state, and a negative input signal switches the emitter current from T_2 to T_1. On complete current change-over this arrangement obtains a well limited negative pulse without direct capacitive through-coupling. This limiter is insensitive to small input pulses, because the transistor T_1 is blocked by a variable bias and acts as discriminator. Positive undershoot is reduced by the diode D_1, while the diode D_2 serves to prevent an excessive negative emitter potential. The long-tailed pair represents an emitter follower with attached common-base stage and consequently short inherent rise time. This amounts to about 2 ns if the input signal exceeds the threshold by 0·5 to 1 V. The use of a similar circuit is described by *Dunaitsev* [Du 64b].

LEADING-EDGE TRIGGERING WITH CONSTANT THRESHOLD

Another possibility of timing is offered by triggering at the leading edge of the detector signal. As soon as the integrated voltage pulse or the non-integrated current pulse from the detector passes a specific threshold level this triggers a respective circuit. Machine time thus corresponds to integral and straight response (see Section 2.3.2).

In trigger stages two types of delays occur, viz., time shift caused by a difference in height of the detector signal; and an internal delay, especially with input pulses, which can reach only a little over the trigger threshold. In general, a maximum resolution of the time analyser can be expected only if a small part of the whole pulse is utilised, as is the case in limiter circuits. In consequence, if the triggering threshold is far exceeded the internal time delay is reduced to a negligible value.

Timing work with trigger stages offers two important advantages. Small pulses which do not reach to the threshold level cannot produce output pulses, and consequently the time analyser is not exposed, for instance, to the small noise pulses from the photomultiplier. Detector pulses can be, and in most cases have to be, shorter than the shaped pulses delivered by the trigger stage. This permits short integration time constants at the detector output and reduces pile-up effects. The dead time of the trigger stages provides further that spacing of the input pulses for the time analyser does not underpass a certain value.

For timing work in the nanosecond region valve equipped trigger circuits or arrangements using normal transistors have not been able to gain a foothold. On the other hand, circuits equipped with avalanche transistors or tunnel diodes can satisfy the requirements with reference to sensitivity and fast response. Design and operation of these trigger stages have been described in Section 3.2.4. Here, only some of the points will be touched on which are of particular importance for timing with trigger circuits in the nanosecond and the picosecond regions.

It must be pointed out that output pulses of trigger circuits incorporating avalanche transistors or tunnel diodes are not entirely unaffected by the shape of the trigger signal from the detector. For this reason repeated shaping is nearly always resorted to, with the first stage of the circuit usually acting as limiter or discriminator, and the second stage producing the standardised signals.

Trigger Circuits with Avalanche Transistors

For the connection of an avalanche transistor to the photomultiplier an intermediate stage is used. A transistorised limiter circuit or a transistorised shaping circuit [Sa 64b; Gr 67b], a tunnel diode system [La 65] or another avalanche transistor circuit [Al 63a; Mi 64] have been employed for the purpose. Arrangements of this type are also suitable for slow-decay scintillators [Mi 64; La 65]. Fig. 114 gives the trigger stage developed by *Samueli* et al. [Sa 64 b], where the transistors T_1 and T_2 represent a cascode amplifier. The pulse is limited by the transistor system T_3 and T_4, shortened to 7 ns through the short-circuited cable and transferred through T_5 to the avalanche transistor T_6 (cp. Fig. 90 *b*, p. 154). The resulting output pulse is delivered to a time-to-pulse-height converter. This arrangement features good long-term stability (better than 40 ps/day) and

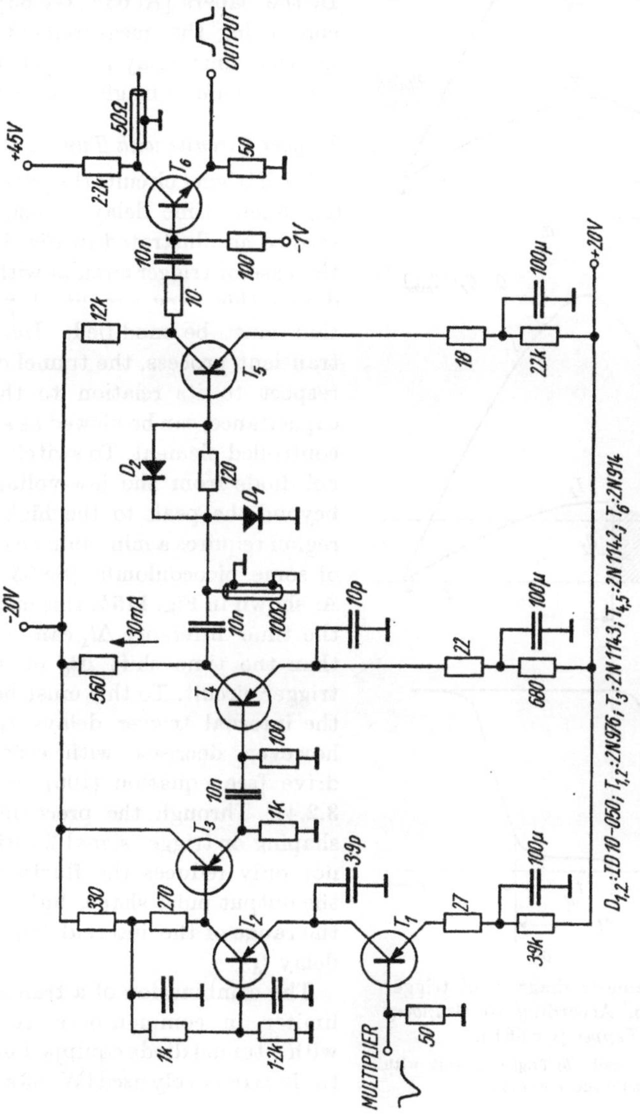

Fig. 114. Shaping and limiting of pulses (in T_1 to T_5) and avalanche transistor trigger stage (T_6). According to *Samueli* et al. [Sa 64 b].

Width of output pulse 25 ns, height of output pulse 8 V, rise time <0·7 ns.

a high time resolution. With two photomultipliers, type XP 1020, and plastic scintillators, type NE 102 A, with ^{60}Co radiation a full width at half-maximum of $2\tau_0 = 0{\cdot}29$ ns of the coincidence curve was obtained. This result also indicates that the statistic time fluctuation of avalanche breakthrough may be far less than 1 ns [Gy 63; Sh 66]. In the papers [Al 63a; Gy 63] details concerning the measurement of the internal time delay in dependency on the input pulse height are found.

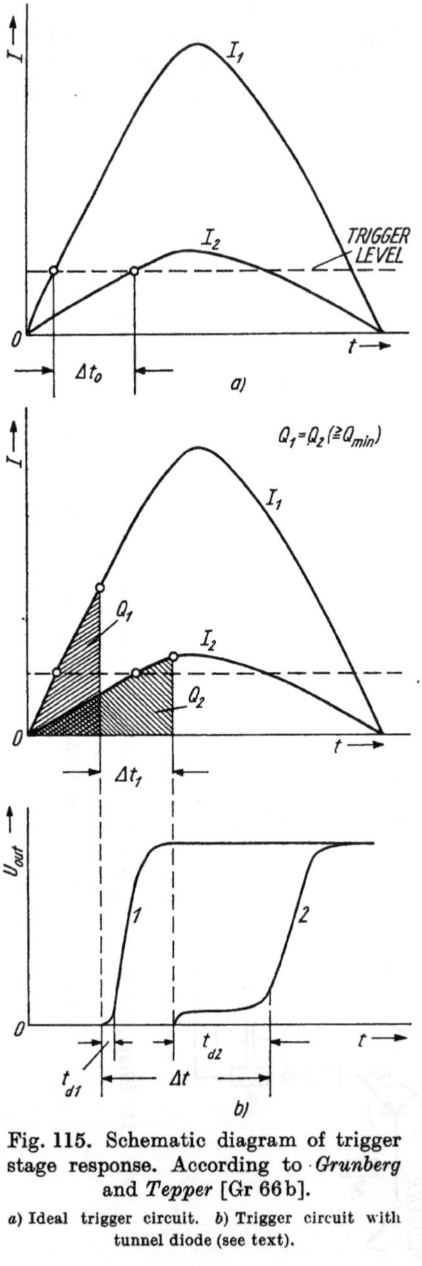

Fig. 115. Schematic diagram of trigger stage response. According to *Grunberg* and *Tepper* [Gr 66b].

a) Ideal trigger circuit. *b*) Trigger circuit with tunnel diode (see text).

Trigger Circuits with Tunnel Diodes

With trigger circuits the pulse-height dependent time delay is usually described as illustrated in Fig. 115*a*. In the case of trigger circuits with tunnel diodes this very simplified presentation must be modified. During the transient process, the tunnel diode, in respect to its relation to the diode capacitance, can be viewed as a charge-controlled element. To switch the tunnel diode from the low-voltage state beyond the peak to the high-voltage region requires a minimum charge $Q_{\min}$ of some picocoulombs [Se 63; Co 67]. As shown in Fig. 115*b*, this means that the time difference Δt_1 can be shorter than the time shift Δt_0 of an ideal trigger circuit. To this must be added the internal trigger delays t_d which, however, decrease with rising overdrive [see equation (106) in Section 3.2.4]. Through the preceding pulse shaping or trigger signal limiting, this not only reduces the fluctuations of the output pulse shape, but decreases the range of the internal trigger time delay t_d.

The combination of a transistorised limiter in common-base connection with a tunnel diode equipped univibrator is extensively used [Wi 63a; Op 64].

Sometimes two tunnel diode univibrators are arranged in series [Ad 61; Wh 62; Go 65] or a germanium diode is employed for pre-limiting [Du 64; Cl 67]. *Ophir* [Op 64] developed the circuit shown in Fig. 116, where the transistor T_1 delivers a pulse of 3 mA on blocking which switches the univibrator tunnel diode *TD* to high-voltage state. Input impedance of the circuit is about 90 Ω but does not increase appreciably when T_1 is blocked, because the diodes D_1 and D_2 then conduct. In the design presented the part of the input

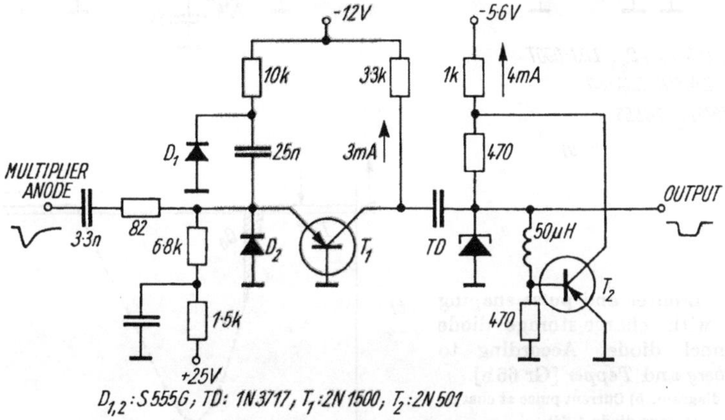

Fig. 116. Pulse shaper circuit with tunnel-diode trigger stage. Width of output pulse 100 ns. According to *Ophir* [Op 64].

pulse located between 1 and 3 mA determines the output pulse. A time-to-pulse-height converter circuit is connected to the pulse shaping stage. With ^{60}Co radiation, a full width at half-maximum of the coincidence curve $2\tau_0 = 0{\cdot}46$ ns was obtained using photomultipliers, type 56 AVP, and plastic scintillators, type Naton 136. Slope steepness of the coincidence curve was $T_{1/2\,\mathrm{instr}} = 67$ ps. In this circuit, and in the version developed by *Wieber* [Wi 63a], tunnel diodes were employed with a peak current of 5 mA. Such tunnel diodes represent a certain compromise between the temperature drift, which is proportional to the peak current, and the switching behaviour, which is generally more favourable with a higher peak current.

Grunberg and *Tepper* [Gr 66b] used a *charge-storage diode* as pulse shaping element. Fig. 117 gives the shaping stage. Pulses limited by the transistor T_1 open the transistor T_2, which is blocked in quiescent state. Small pulses, such as the noise pulses from the photomultiplier, are removed through the diode D_4. The negative pulse coming from the transistor T_2 interrupts the quiescent current I_f through the charge-storage diode *CSD*, which produces a current pulse of 3 mA height and a width in the order of 1 ns. The total charge Q_0 of the standardised pulse, which is proportional to I_f, must be high enough to switch the following tunnel diode *TD* to high-voltage state. On the other hand, the desired charge Q_0 must be low enough to obtain a low response threshold,

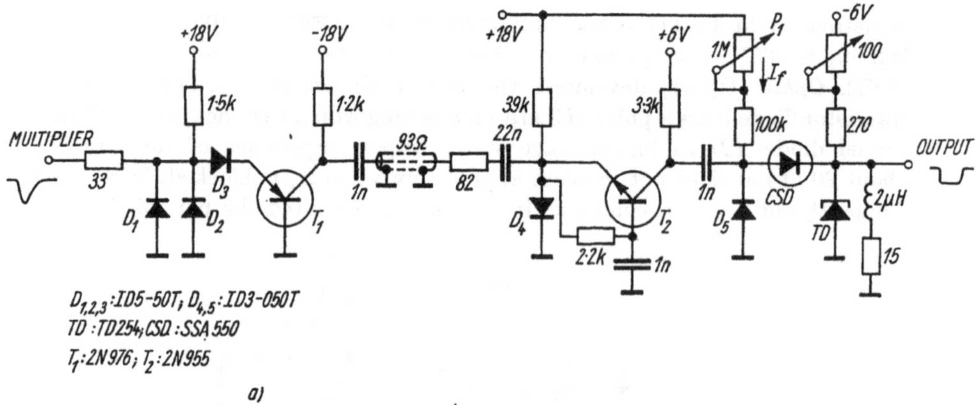

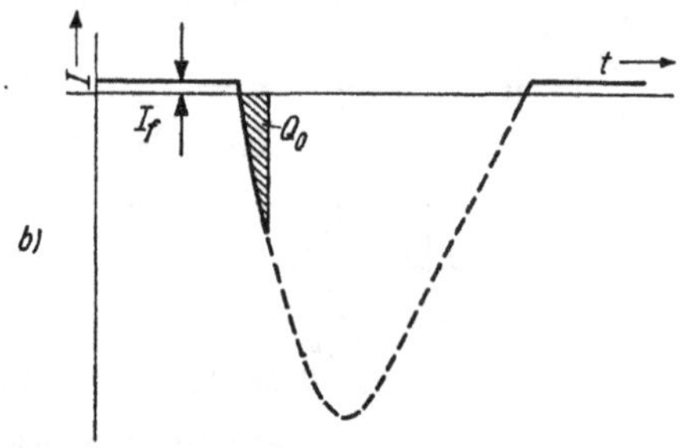

Fig. 117. Limiter and pulse shaping circuits with charge-storage diode and tunnel diode. According to *Grunberg* and *Tepper* [Gr 66b].
a) Circuit diagram. b) Current pulse at charge-storage diode *CSD*.

as shown in Fig. 115*b*, and thus results in a small time shift Δt_1. Utilising plastic scintillators and photomultipliers of type 56 AVP, with ^{60}Co radiation, a full width at half-maximum of the coincidence curve of $2\tau_0 = 0{\cdot}94$ ns was obtained when every detector pulse of a height between 10 and 200 mA was admitted; or $2\tau_0 = 0{\cdot}54$ ns resulted in the range of 10 to 65 mA.

TRIGGERING ON REACHING A CONSTANT PULSE FRACTION (ZERO-CROSSING TIME PICK-OFF)

In trigger circuits having a constant response threshold a pulse-height dependent time shift occurs (cp. Fig. 115*a*, p. 188). This is avoided if the trigger stage is actuated by a constant fraction of the total pulse. For this purpose several possibilities are available. If the base line of the pulse is reduced by a specific fraction of total height (Fig. 118 *a*) the trigger level, located above or on the base line (zero-cross), is always reached at the same moment irrespective of pulse height [We 56; Jo 57; Dr 60; Ge 67a; Ju 67; Ge 68; Jo 68].

The crossover point of a doubly differentiated detector signal is also independent of pulse height (Fig. 118*b* — [Ma 57a; Gr 59]). Zero-crossing time pick-off is, in most cases, achieved with a trigger circuit where the trigger stage is switched back to initial state on the crossover point of the input signal. In such cases the trailing edge of the output pulse of the trigger circuit contains the desired time information. Another possibility is shown in Fig. 118*c* where a pedestal of constant height is superimposed on the doubly differentiated detector pulse. With the response threshold located at the height of the pedestal the trigger circuit also responds on zero-crossing of the input signal [Ar 59; Sh 66; Wi 66; Bu 67].

Zero-crossing time pick-off is widely used because it is a relatively uncomplicated method and suitable trigger circuits with tunnel diodes can be easily set up. A special characteristic of zero-cross timing has to be mentioned in this connection. The moment of crossover is determined by two points of the pulse (cp. Fig, 19, p. 37), for example, 50% and 100% of its final height, if the integrated pulse is shaped by two short-circuited cables with transit times longer than the pulse rise time (see Fig. 19*a*). If transit time is shorter or a lower differentiating time constant applies, the fixed points shift to smaller pulse fractions. This, however, makes triggering more difficult because pulse height as well as the steepness of the crossover is reduced. In the end effect, this means that the first part of the pulse is not utilised for triggering on zero-crossing and, therefore, no optimum timing accuracy is achieved (cp. Sections 2.3.2 and 2.4.1). The question is either to tolerate a certain pulse-height dependent time shift with consequent introduction of an additional compensation circuit, or

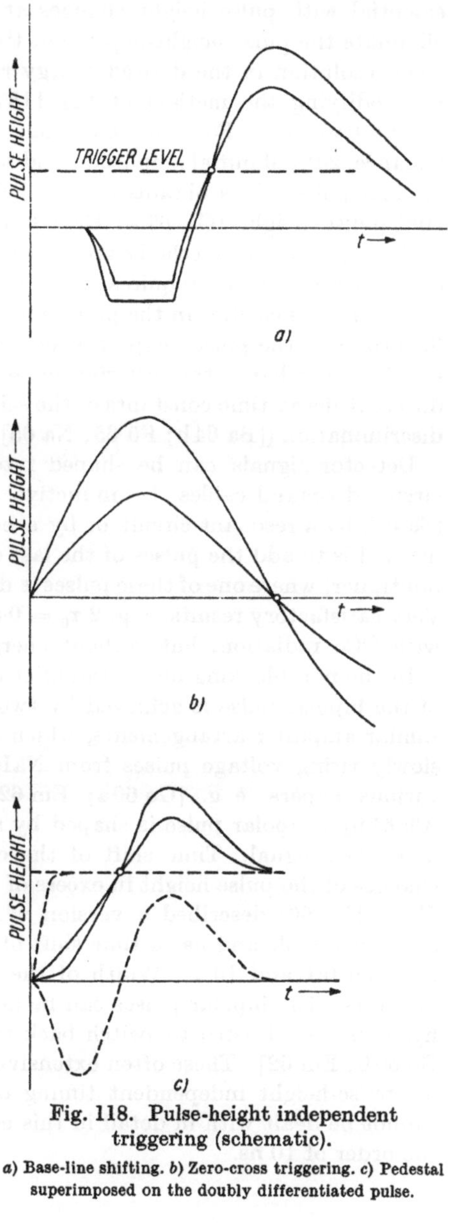

Fig. 118. Pulse-height independent triggering (schematic).
a) Base-line shifting. *b*) Zero-cross triggering. *c*) Pedestal superimposed on the doubly differentiated pulse.

to eliminate the pulse-height dependency over a wide pulse range through the zero-crossing method to the detriment of the time resolution [Be 66c]. In coincidence spectroscopy operating with slow-decay scintillators, such as NaI(Tl), the second possibility has gained considerable favour. The reason is that maximum timing accuracy is not so important, while it is essential with pulse height changes at a ratio of 20:1 and more to largely eliminate the pulse-height-dependent time shift and to obtain a closely constant time resolution in the desired energy range. Good time resolution is achieved by modifying the method of Fig. 118 *a* that the trigger threshold is located on the base line. For this the anode current pulse is first attenuated to, for instance, 20% of initial amplitude and then added to the inverted and delayed original pulse. This obtains a pulse-height independent zero-cross at the first 20% pulse height [Ge 67 a; Ge 68; Jo 68].

The zero-crossing method has one principal difficulty; the pulse-height independent crossover point requires a constant shape of the detector pulse. For this reason, non-linearities in the photomultiplier, e.g. space charge effects, and non-linearities in the pulse shaping stage may give rise to undesirable time shift. On the other hand, the differing moments of zero-crossing, resulting from the different decay time constants of the scintillators can be made use of for particle discrimination ([Ba 64b; Fü 65; Na 65] etc.).

Detector signals can be shaped into zero-crossing pulses either by short-circuited coaxial cables, by inductive and capacitive differentiating networks [Sc 45], by a resonant circuit or by a combination of these elements. Another method is to add the pulses of the last dynode and of the anode of the photomultiplier, where one of these pulses is delayed by a cable [Ne 66, chapter 11.3]. Very satisfactory results, e. g. $2\,\tau_0 = 0{\cdot}53$ ns, were obtained by *Conrad* [Co 67d] with ^{60}Co radiation, but without energy selection.

In the non-blocking linear amplifier developed by *Fairstein* [Fa 56], shaping of the bipolar pulse is achieved by two short-circuited cables. Descriptions of similar amplifier arrangements, which have been specially constructed for the slowly rising voltage pulses from NaI(Tl) scintillation counters are found in various papers, e. g. [Go 60a; Em 62]. In the *Chase* and *Svelto* amplifier [Ch 61b] a bipolar pulse is shaped by subtracting the delayed signal from the undelayed signal. Time shift of the crossover point of the bipolar pulse on changes of the pulse height in excess of 10:1 usually amounts to less than 10 ns. *Hatch* [Ha 66] described a version which selectively uses an *RC* network or a shaping cable and has a time shift of only ± 2 ns with output pulses ranging between 0·1 and 10 V. Width of the primary pulse is generally between 0·5 and 1 μs. The bipolar pulses can be delivered to a *Schmitt*-trigger stage where hysteresis is adjusted to switch back the stage on zero-crossing [Gr 59; Ch 60; Go 60b; Em 62]. These often extensive circuits, which are principally intended for pulse-height independent timing of signals from slow-decay scintillators, cannot be dealt with in detail in this context. Coincidence resolving time is in the order of 10 ns.

Zero-Crossing Trigger Stages with Tunnel Diodes

When the detector signal excites a resonant circuit the phase of the induced oscillation is associated with the centroid of the total pulse and, therefore, independent of pulse height [Co 57]. *Arecchi* [Ar 59] describes a circuit where zero-crossing of the oscillation is registered as shown in Fig. 118*c*. The version of *Bjerke* et al. [Bj 62; Wi 63b] given in Fig. 119, uses a critically damped resonant circuit consisting of the transformer inductance and the capacitance between the anode and the last dynode of the photomultiplier. Oscillating frequency is about 80 MHz and, therefore, cycle duration approximately equals current pulse width obtained with organic scintillators. The tunnel diode TD_1 is biased near the peak current. Shortly *after* zero-crossing of the oscillation the tunnel diode univibrator switches to high-voltage state and delivers pulses of a width of 3 ns and a rise time of 0·15 ns. These pulses are differentiated with a time constant of 0·2 ns by the capacitor C and the characteristic impedance Z of the coaxial cable, and then transmitted to the time analyser. After another pulse shaping through TD_2 the resulting pulses are brought in coincidence with the output pulses of a pulse shaping stage of analogue design, and trigger the univibrator system TD_3. The tunnel diodes TD_1 and TD_2 are mounted together with the capacitor C in a coaxial housing. With 10-fold changes of the pulse height of short light pulses a time shift of less than 0·3 ns could be observed.

Switching speed of a tunnel diode is higher when shifting to the high-voltage state than vice versa. For the working point of the tunnel diode in zero-cross discriminators two choices are available, as illustrated in Fig. 120. With the first version the stable working point A is located below the peak, while the second version shows point A on the high-voltage branch. The zero-cross discriminators, as described in a number of papers [Bj 62; Ga 64a; Ef 65; Ve 66;

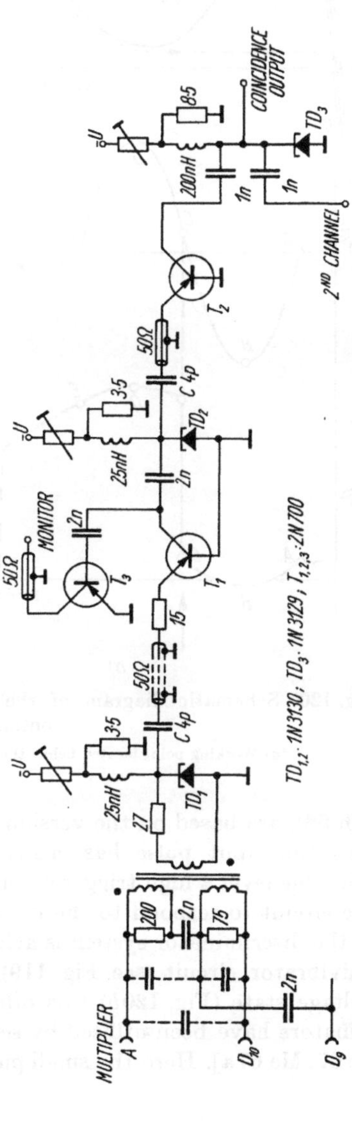

Fig. 119. Pulse shaping circuit with zero-cross triggering. According to *Bjerke* et al. [Bj 62].

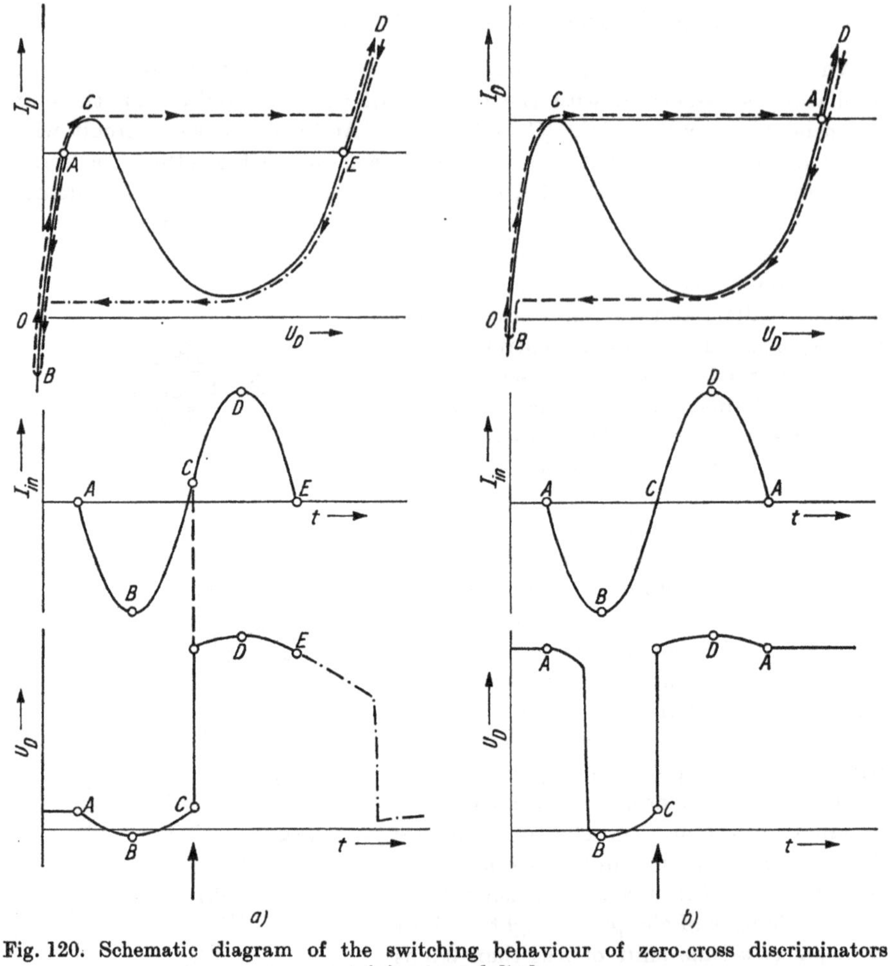

Fig. 120. Schematic diagram of the switching behaviour of zero-cross discriminators containing tunnel diodes.

a) Working point located below the peak. *b*) Working point on the high-voltage branch.

Wh 66], are based on the version given in Fig. 120*a*. Triggering always occurs *after* the input pulse has crossed the base line. To restrict this fault to a tolerable level a high trigger sensitivity is necessary, which will, however, cause the circuit to respond to the noise pulses from the photomultiplier. Resetting of the discriminator system is achieved by incorporating the tunnel diode in a univibrator circuit (see Fig. 119). With the working point located on high-voltage state (Fig. 120*b*) this offers some advantages; these zero-cross discriminators have been utilised by several workers [Or 63; Al 65b; Br 65b; Ab 67; Be 67; Me 67a]. Here the small pulses have no effect. The moment of switching

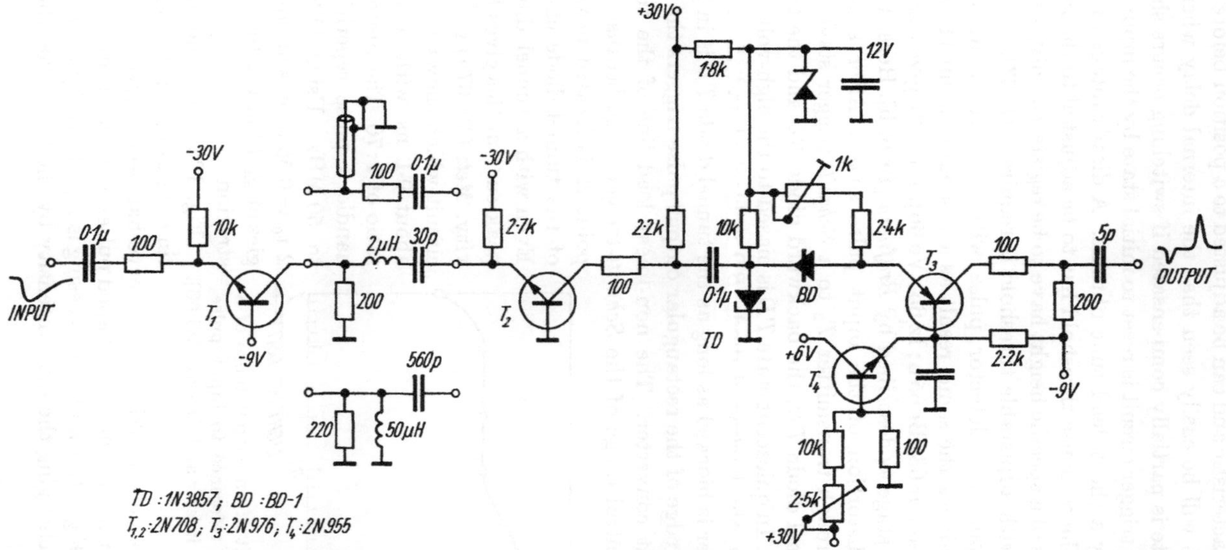

Fig. 121. Crossover pulse shaping and triggering, short-circuited cable and series-connected resonant circuit for pulses with $t_r \approx 5$ ns, $t_f \approx 10$ ns, time shift 0·1 ns with input pulses of 2 to 20 mA, parallel-connected resonant circuit for pulses with $t_r \approx 250$ ns, $t_f > 1$ μs, time shift 2 ns with input pulses of 2 to 20 mA. According to *Brafman* [Br 65b].

back to high-voltage state can be varied by respectively locating the working point A on the characteristic and can be adjusted to a position before or after the crossover point. It will be easily seen that the internal delay when switching to high-voltage state is partially compensated if switching occurs shortly *before* zero-crossing. This trigger circuit is reset to initial state by the input signal itself and, therefore, only a short dead time obtains. A disadvantage of the version in Fig. 120 is that the response threshold has to be adjusted to the most favourable level. If pulses of a specific height have to be registered only, a supplementary discriminator with adjustable threshold is required [Or 63].

Essentially, shaping the detector pulse with a short-circuited cable (see Fig. 19c, p. 37) produces the same result as a resonant circuit [Bj 62] and has been frequently used [Or 63; Br 65b; Iz 65; Ve 66]. Fig. 121 gives an example of a zero-cross trigger stage as described by *Brafman* [Br 65b]. Here, transistor T_1 must not cause a distortion of the input pulse shape. After shaping pulses are transmitted through transmitter T_2 to a *Schmitt*-trigger stage. The latter consists of the tunnel diode TD, the backward diode BD and the transistor T_3 (cp. Fig. 97, p. 164). In quiescent state TD is biased to the high-voltage region as shown in Fig. 120b, the backward diode BD is blocked and the transistor T_3 conducts. The latter is blocked as long as the tunnel diode TD is in low-voltage state. The trailing edge of the rectangular output pulse triggers the subsequent time-to-pulse-height converter. The non-linear load line of the tunnel diode TD spreads the control range of the *Schmitt*-trigger stage because its working point A is located near the valley of the tunnel diode characteristic. Even with a tunnel diode of high peak current this gives high sensitivity. *Metz* [Me 67a] gives a similar circuit which shows a time shift of about 0·1 ns with a pulse-height ratio of 1:70. The timing stage can handle a pulse repetition rate up to 50 MHz. Using ^{60}Co radiation $2\,\tau_0 = 0{\cdot}55$ ns was obtained on energies of $\geqq$ 10 keV admitted for registration.

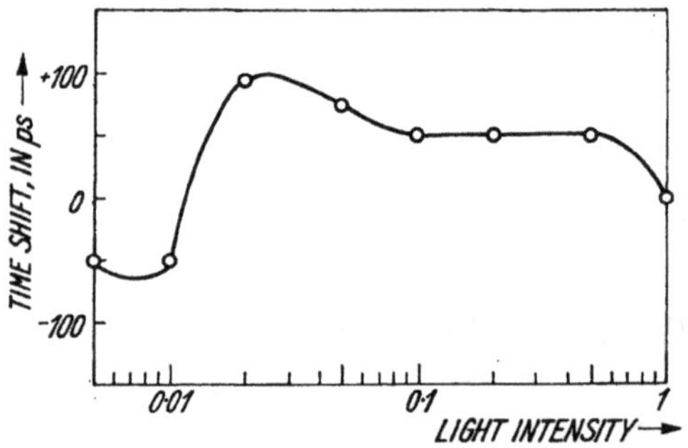

Fig. 122. Time shift of an output pulse of a shaping stage with reference to input pulse height. According to *Wieber* and *Lefevre* [Wi 66].

Fig. 122 shows the output pulse time shift of a trigger stage in dependence of input pulse height [Wi 66]. The curve has been measured with a light-pulse generator. In this instance, the shaped pulses were superimposed by a pedestal as indicated in Fig. 118c, and zero-crossing is registered by tunnel diode discriminator. When changing the light intensity by the factor 200 the time shift is less than $\pm$ 0·1 ns.

Whittaker [Wh 66] used a zero-cross discriminator with a tunnel diode biased below the peak. This circuit has been tested in coincidence mode with α-particles and neutrons at an energy of 14 MeV. For detection photomultipliers type 56 AVP and organic scintillators were used. With recoil protons of an energy level between 0·1 and 14 MeV, corresponding to a light intensity ratio of 1:500, a resolving time of $2\tau_0 = 1{\cdot}7$ ns was obtained. Taking the neutron transit time in the scintillator into account, the resolving time of the entire arrangement is equivalent to $2\tau_0 = 1{\cdot}4$ ns.

COMPENSATING PULSE-HEIGHT DEPENDENT TIME SHIFTS

To obtain maximum statistic timing accuracy the first part of the detector pulse has to be utilised. This can be achieved through limiting or triggering by the leading edge of the pulse. Occurring pulse-height dependent time shift phenomena may be compensated, more or less satisfactorily, by suitable compensation circuits.

The simplest method is to pick off a specific fraction of the detector signal, superimpose it on the output signal of the time analyser and thus compensate the influence of the pulse height fluctuations. This method is suitable for time-

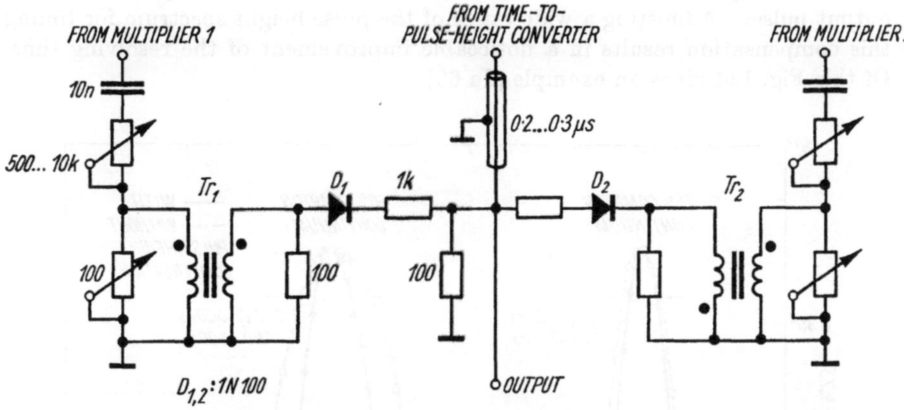

Fig. 123. Pulse-height compensator. According to *Rodda* et al. [Ro 63].

to-pulse-height converters and coincidence stages derived from converter circuits, and has been employed by several workers [Be 60; Wa 61 a; Ro 63; Sc 63; Ad 64; Av 68; Se 68]. In Fig. 123 the circuit of a simple and stable compensating stage is shown. It is assumed that, at the input of the time-to-pulse-height converter, the pulses from the scintillation counter *2* are delayed against those from the counter *1*. If the converter circuit operates on the overlap principle (see Section 3.4.4) a higher pulse from the counter *1* causes a lower output pulse of the converter. On the other hand, a higher pulse from counter *2* results in an increased output signal of the converter. For this reason the compensating signals in the circuit of Fig. 123 must have different polarity, which is achieved by the reversing transformer Tr_2. Thus, in this arrangement the output signal of the converter undergoes all the more correction the higher the detector pulse. *Wagner* et al. [Wa 61a] use a compensating arrangement where the correction quantity is produced from the difference between the detector signal and a standard pulse. In this case correction increases with decreasing height of the detector pulse.

Complete compensation of pulse-height dependent time shift cannot be obtained with linear compensation methods. Better results can be expected when using non-linear methods [Dr 67].

Thieberger [Th 66] developed an arrangement where scintillation counter pulses are delivered to two time-to-pulse-height converters and the resulting output pulses are added. Pulse-height dependency could be eliminated by reversing the sequence of the input pulses at the two converters. At a variation of the pulse height of about 5:1 the time shift is about 30 ps. *Lapshin* et al. [La 66] also used two time-to-pulse-height converters.

Abou-Leila et al. [Ab 65] and *Jastrzebski* et al. [Ja 66] employed a method where compensation is effected before limiting. Through a capacitor of some picofarads, connected between the control grid and the anode of the limiter valve, strongly differentiated detector pulses are added to the positive limiter output pulses. Admitting a wide range of the pulse height spectrum for timing this compensation results in a noticeable improvement of the resolving time. Of this Fig. 124 gives an example [Ja 66].

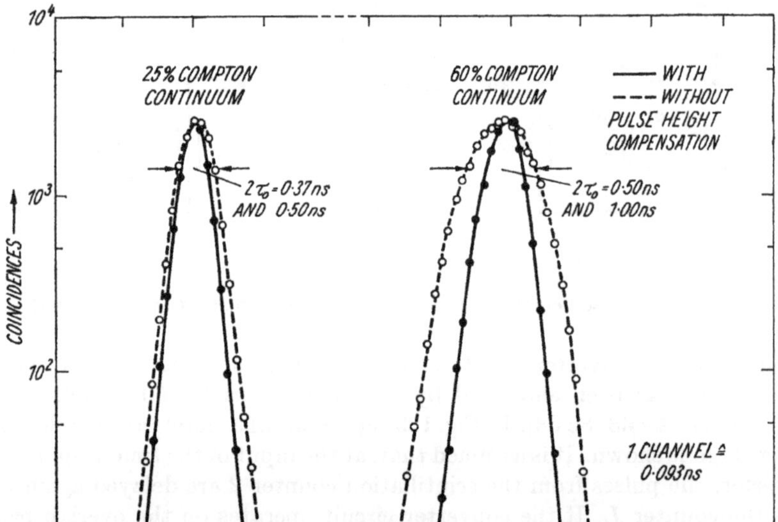

Fig. 124. Influence of pulse height compensation on time resolution of a coincidence assembly, measured with ^{60}Co radiation. According to *Jastrzebski* [Ja 66].

TIMING WITH SEMICONDUCTOR DETECTORS

In comparison with scintillation counters the timing with semiconductor detectors is difficult, principally for the following reasons:

(i) The charge delivered from semiconductor detectors is very small. In consequence, the resulting pulses have to be amplified before they are suitable for shaping in a limiter circuit or a trigger stage. For amplification the influence of noise has to be taken into consideration.

(ii) Because of the different charge collection time, pulses, especially those from thick p-i-n detectors, have no constant shape. This causes time fluctuations if zero-crossing time pick-off is employed for the reduction of the pulse-height-dependent time shift.

iii) As compared with photomultipliers, only one output is available. Under certain conditions, measured energy resolution can be less satisfactory when a "fast" channel is connected for timing.

According to the details presented in Section 2.4.1 the best time resolution can be expected with semiconductor detectors if the first part of the detector signal is used for triggering. Arrangements utilised by the various authors differ, essentially in the mode of coupling a fast channel to the detector. Timing is effected with a limiter or a trigger circuit.

Surface barrier detectors and *p-n junction detectors* yield the best time resolution and are utilised for protons and α-particles with energies of some MeV or for fission products. In most cases the additional noise contributions that may be caused by the coupled fast channel, are of no importance. With these detectors the charge collection time is in the order of 1 ns. Rise time of the amplified pulse corresponds to collection time if a voltage sensitive amplifier with a low input capacitance is used, as has been discussed at the end of Section 3.3 [Ka 63; Me 66a].

In an arrangement developed by *Cobb* [Co 63b], the signal from the surface-barrier detector is passed to a cathode follower and delivered to two separate channels for timing and for the measurement of energy (Fig. 125*a*). In the fast channel the signal experiences amplification through several distributed amplifiers connected in series to be finally shaped, as with a scintillation counter, by a valve limiter and a short-circuited coaxial cable. Resolving time on coincidences of 6 MeV α-particles and γ-quanta, detected by a scintillation counter, amounted to 9 ns. A similar circuit design has been used by *Wahl* [Wa 64a], where α-particles with an energy of 10 MeV gave a resolving time of less than 0·5 ns. On a lower energy level the time resolution is impaired by noise. *Neal* and *Kraner* [Ne 65] have used a nuvistor valve in a cathode follower for coupling the fast channel.

All of the constructions mentioned so far show a certain adverse influence on the energy resolution, whereas with the two following modifications no detrimental phenomena could be observed. *Williams* and *Biggerstaff* [Wi 64a]

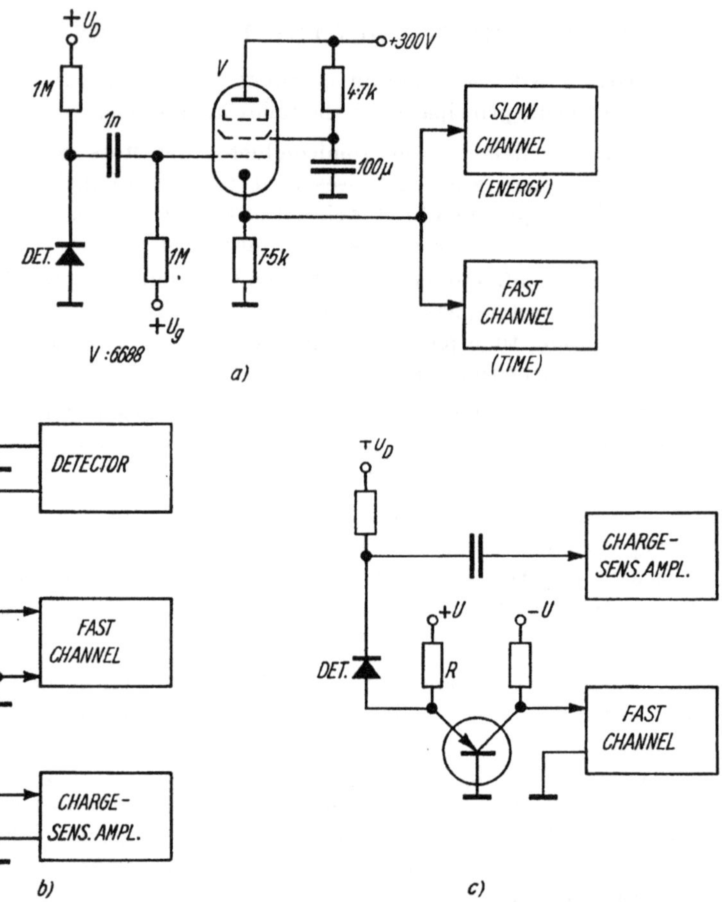

Fig. 125. Connection of a p-n detector for timing (schematically).
a) With cathode-follower. According to *Cobb* [Co 63b]. *b*) With transformer coupling. According to *Williams* and *Biggerstaff* [Wi 64a]. *c*) Current pulse pick-off through a fast transistor. According to *Papadopoulos* [Pa 66].

used a transformer *Tr* for the coupling of a fast channel (Fig. 125*b*). Consistent with the winding ratio, the noise voltage of the fast amplifier is reduced and divided between the detector capacitance and the high input capacitance of the charge-sensitive amplifier. The fast channel consists of a voltage-sensitive transistorised amplifier (gain 250, rise time 3·5 ns) followed by a tunnel-diode trigger stage. Depending on the admitted energy range, with fission products, surface-barrier detectors yielded a time resolution in the order of 1 ns. A second version was given by *Papadopoulos* [Pa 66], where a transistor is connected in series with the detector, with current from the latter passing through the transistor (Fig. 125*c*). In the attached fast amplifier the current pulse is amplified. Properties of the detector are not influenced by the transistor because the emitter voltage is very low and the emitter current, which is far stronger than the detector reverse current, is removed through resistor R. As rise time of the current pulses is little influenced by the attached charge-sensitive amplifier, this method is recommended for fast timing [Al 67a].

Utilising the current pulse from a Ge(Li) detector yields steep leading pulse edges. *Bertrand* et al. [Be 66b] achieved a rise time of the current pulse of approximately 4 to 5 ns with a Ge(Li) detector of 10 mm thickness and protons of 60 MeV energy. A terminated cable with a characteristic impedance of 125 Ω served as load resistance.

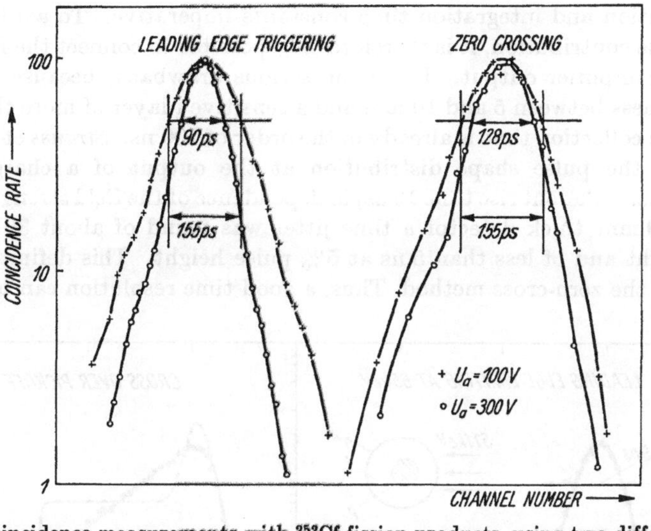

Fig. 126. Coincidence measurements with ^{252}Cf fission products, using two different timing methods and two different voltages U_a of the silicon surface-barrier detectors. Width of the admitted energy range 12 MeV. According to *Meyer* [Me 66a].

Investigations with reference to the response behaviour of silicon surface barrier detectors to fission products were undertaken by *Meyer* [Me 66a]. He found that a little better time resolution could be attained with a low trigger threshold than is possible with the zero-crossing method, as Fig. 126 indicates.

Detection of electrons and γ-quanta with energies up to a few MeV reveals different conditions because of the relatively high noise contribution. *Walthard* et al. [Wa 65] investigated the influence of the noise on the time resolution of fast detectors, such as silicon surface-barrier detectors. It was found that, depending on the type of the lower frequency limitation, the rise time of the amplifier should be selected about 2 to 10 times higher than the rise time of the detector signal. Coincidence measurements between 316 keV electrons, registered by a silicon surface-barrier detector with a layer of less than 1 mm thickness, and γ-quanta with an energy of 468 keV, which were detected by a scintillation counter, yielded a resolving time of $2\tau_0 \approx 12$ ns. *Sakai* [Sa 63] employed a 2 mm Si(Li) detector for coincidence measurements of 1 MeV electrons and γ-quanta, utilising the zero-cross method, and obtained a resolving time of about 10 ns.

Timing problems with *p-i-n detectors*, for example, of lithium-drifted germanium, could not yet be solved to satisfaction. Of special annoyance is the often asymmetric shape of the coincidence curves and the presence of slowly decaying tails (cp. Fig. 36*b*, p. 74). For the measurement of γ-quanta with Ge(Li) detectors maximum energy resolution usually represents the predominating requirement. This makes the use of charge-sensitive amplifiers and the selection of optimum differentiation and integration time constants imperative. To avoid the additional noise contribution, it is therefore only possible to connect the fast channel to the preamplifier output. This is no serious drawback, because with layers of a thickness between 5 and 10 mm and a sensitive i-layer of more than 10 cm^3, the charge collection time is already in the order of 100 ns. *Strauss* et al. [St 66a] examined the pulse shape distribution at the output of a charge-sensitive preamplifier (inherent rise time 10 ns) in dependence of the field strength (Fig. 36). With a 10 mm thick detector a time jitter was found of about 30 ns on 50% pulse height and of less than 5 ns at 5% pulse height. This definitely negates the use of the zero-cross method. Thus, a good time resolution can be achieved,

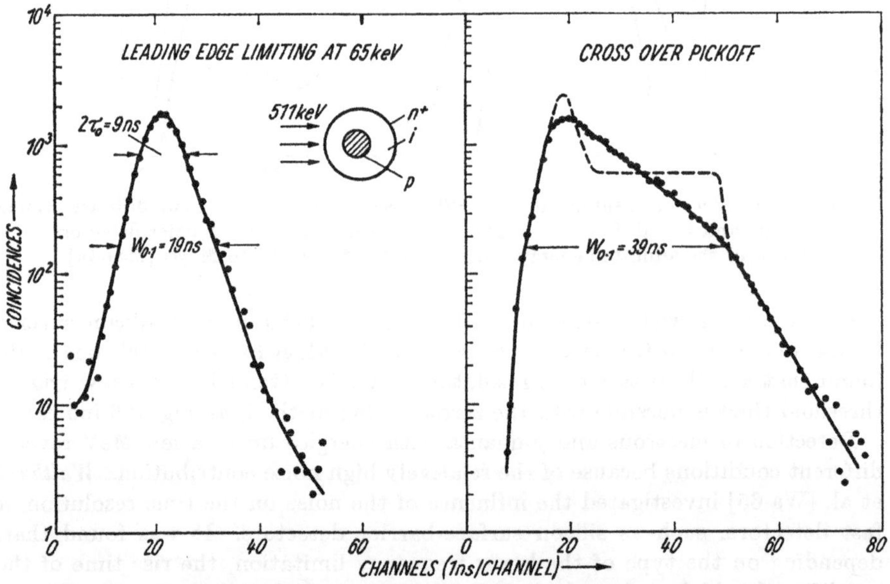

Fig. 127. Time distribution of leading-edge triggering and zero-cross triggering, with a coaxial Ge(Li) detector, at an operating voltage of 1,000 V and ^{22}Na radiation; second detector: NaI(Tl) scintillation counter. According to *Malm* [Ma 66b].

if a fast low-noise preamplifier is combined with a trigger stage, which is triggered by the pulse base and with the lower limit of the trigger threshold given by noise. Charge-sensitive preamplifiers with a rise time of 30 ns were used by *Ewan* et al. [Ew 66], *Graham* et al. [Gr 66a] and *Malm* [Ma 66b]. Two coincidence curves measured by *Malm* [Ma 66b] are reproduced in Fig. 127. These make the superiority of leading-edge timing on low threshold evident. *Pigneret* et al. [Pi 66] achieved a three times better coincidence resolving time, where, however, the Ge(Li) detector was only 2 mm thick. It is of interest, that a field-effect transistor is employed for the input stage of the fast channel, which is connected directly to the detector output. *Gorni* et al. [Go 67e] and *Michaelis* [Mi 68] also achieved good time resolution. Here, the timing signal has been derived from the output of the charge-sensitive amplifier which had a rise time of 20 ns and 150 ns, respectively. In the arrangement of *Gorni* et al. a limiter forms a signal proportional to the detector pulse rise time which is utilised for time-shift compensation.

3.4.3 COINCIDENCE CIRCUITS

For the measurement of events correlated in time, coincidence circuits which deliver an output signal only on simultaneous arrival of n signals at their n inputs are frequently used. A coincidence circuit is characterised by a *resolving time* $2\tau_0$, i.e. the period where the arrival of the input signals can be considered as "simultaneous". Time resolution in the nanosecond region had already been achieved with these circuits in the late 1940's, and today resolving time values of a few picoseconds can be attained with jitter-free input pulses. In most coincidence assemblies resolving time is, in the main, determined by the length of the input pulses, but, with very short intervals, the properties of the components and the design of the circuit assume greater importance. In addition, statistical fluctuations of the time shift between the event to be measured and the input pulse of the coincidence circuit adversely affect the resolving time (cp. Sections 2.3.2, 2.4.1 and 3.4.2). When the input pulses occasioned by prompt events are mutually delayed by a successively varied transmission line this produces the so-called prompt coincidence curve $N(T)$; i.e. the dependency of the number N of the measured coincidences on the relative time delay T. Full width at half-maximum of this prompt coincidence curve is generally defined as resolving time $2\tau_0$.

Sensitivity is described by the minimum height of the coincident input pulses causing a response of the coincidence circuit. Depending on type and dimensioning of the coincidence circuit a single input pulse may also produce an output pulse, often caused by parasitic coupling. The *singles rejection ratio*, which denotes the relation of the output pulse height of a coincidence stage of coincident pulse input to single pulse input is, with fast coincidence circuits, at times, not higher than 2. Single pulses can be suppressed entirely if a discriminator is connected to follow the coincidence circuit.

Utilising suitable operating conditions and respective input pulse polarity every coincidence stage may also act as *anticoincidence* circuit, which delivers an output signal only when no input signal arrives at the anticoincidence input during the resolving time. To avoid interference with high singles count rate, dead time of the anticoincidence circuit must be kept to a minimum, for example, by d.c. coupling and pulse shaping with a partially reflecting transmission line [Ak 66].

When uncorrelated events occur during resolving time they are also accepted as coincidence. This is called *chance coincidence* and for a double-coincidence circuit this rate is

$$N_c = 2\,\tau_0\,N_1\,N_2\,, \tag{109}$$

with N_1 and N_2 representing the singles count rates. In practice the ratio to the rate of genuine coincidences N_g must be low.

$$\frac{N_c}{N_g} = 2\,\tau_0\,N \tag{109a}$$

is independent of counter efficiency, but is determined by the source activity N and the resolving time $2\,\tau_0$. Coincidence measurements involving a strong source, therefore, require a low resolving time. As the reduction of counter efficiency by, for instance, enlarging the detector spacing, however, will not improve the relation. This requirement was one of the reasons for the advance into nanosecond pulse technique. Further chance coincidence contributions may be produced when a fast-slow system, as shown in Fig. 1, p. 2, is utilised for the simultaneous energy selection of coincident events (see Section 5.1.3).

A number of authors have furnished reviews of coincidence circuits [Ba 56a; Le 59, chapter 7; Ku 60; Gr 60, chapter 4; Ch 61a, chapter 7; Go 64c; Ne 66, chapter 11; Re 67, chapter 5].

MULTIPLICATIVE MIXING

A double coincidence stage with multiplicative mixing is, in principle, represented by two series-connected open switches S_1 and S_2 which are closed by the two input pulses and thus produce an output signal at a load resistance R_L (Fig. 128*a*). A *multi-grid valve* with two of the grids biased below the cut-off point may, for example, be used as switching element. A version was tried using several valves connected in parallel, as with a distributed amplifier [Wi 50]. The very high singles rejection ratio is limited only by parasitic coupling across the grid-anode capacitance. Because of the nearly identical control behaviour of the two control grids the gated beam tube, type 6 BN 6 (Soviet equivalent type 6 A 3 П), is usually used (Fig. 128*b*). For obtaining high sensitivity it is driven with low voltage. Electron transit time between the two control grids amounts to approximately 2 ns and must be balanced by a supplementary signal delay before the second control grid. With pulses from a mercury-switch pulse generator *Fischer* and *Marshall* [Fi 52] measured a resolving time of $2\tau_0 \approx 0{\cdot}3$ ns.

Utilising a beam-deflexion tube, type E 80 T (Fig. 63, p. 112), a resolving time of less than 10 ps was obtained by *Christiansen* [Ch 59]. In this circuit one pulse opens the control grid g_1, while the other arrives at the deflexion plate D and causes the electron beam to sweep across the slotted diaphragm g_4 situated in front of the anode. Several circuit versions have been designed for this type of tube [Bl 63b], which are suitable for use as time-to-pulse-height converter (cp. Section 3.4.4), the same as the 6 BN 6 tube equipped circuits [Ga 60a].

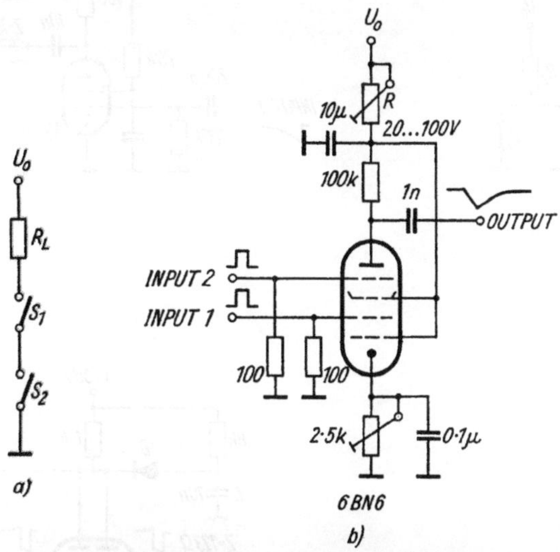

Fig. 128. Multiplicative mixer stage.
a) Principle. *b*) With gated-beam tube 6 BN 6.
According to *Fischer* and *Marshall* [Fi 52].

With pulsed accelerators in particular endeavours are being made to utilise the *photomultiplier* itself as coincidence stage. For this purpose one of the initial dynode stages is blocked and a positive pulse derived from beam pulsing used to open the stage. *Vysochansky* et al. [Vy 63; Vy 65] employed a sinusoidal control voltage with a frequency of 70 MHz and an amplitude of some 10 V to achieve a resolving time of $2\tau_0 \approx 2$ ns. Grid-controlled photomultipliers require a far lower voltage, in the order of 1 V (cp. Section 2.3.1). The low conductivity of the photocathode layer limits the control speed. This can be overcome by a second grid which is either attached on the face plate at the outside [Vy 66] or placed inside the cathode region [Gö 64; Mu 65]. In this manner a resolving time of $2\tau_0 = 1{\cdot}4$ ns could be obtained with 5 GeV/c-π^--mesons. Values of 0·5 ns were measured with a generator delivering simultaneous electrical and light pulses of 1 ns width. It is expected to be able to reduce the time still further [Gö 66].

One of the oldest fast coincidence arrangements, the *oscilloscopic method* [Ho 50], is also based on multiplicative mixing. Here one input pulse serves to actuate the X-sweep and the intensifier of the oscilloscope, while the other input pulse is delivered to the Y deflexion plate. A slotted mask is placed before the screen where the pulse image produced on coincidence appears and is then registered by the attached photomultiplier. In measuring triple coincidences the third pulse is employed as bright-up pulse. Several slots in the mask

Fig. 129. Additive mixer stages with diodes.

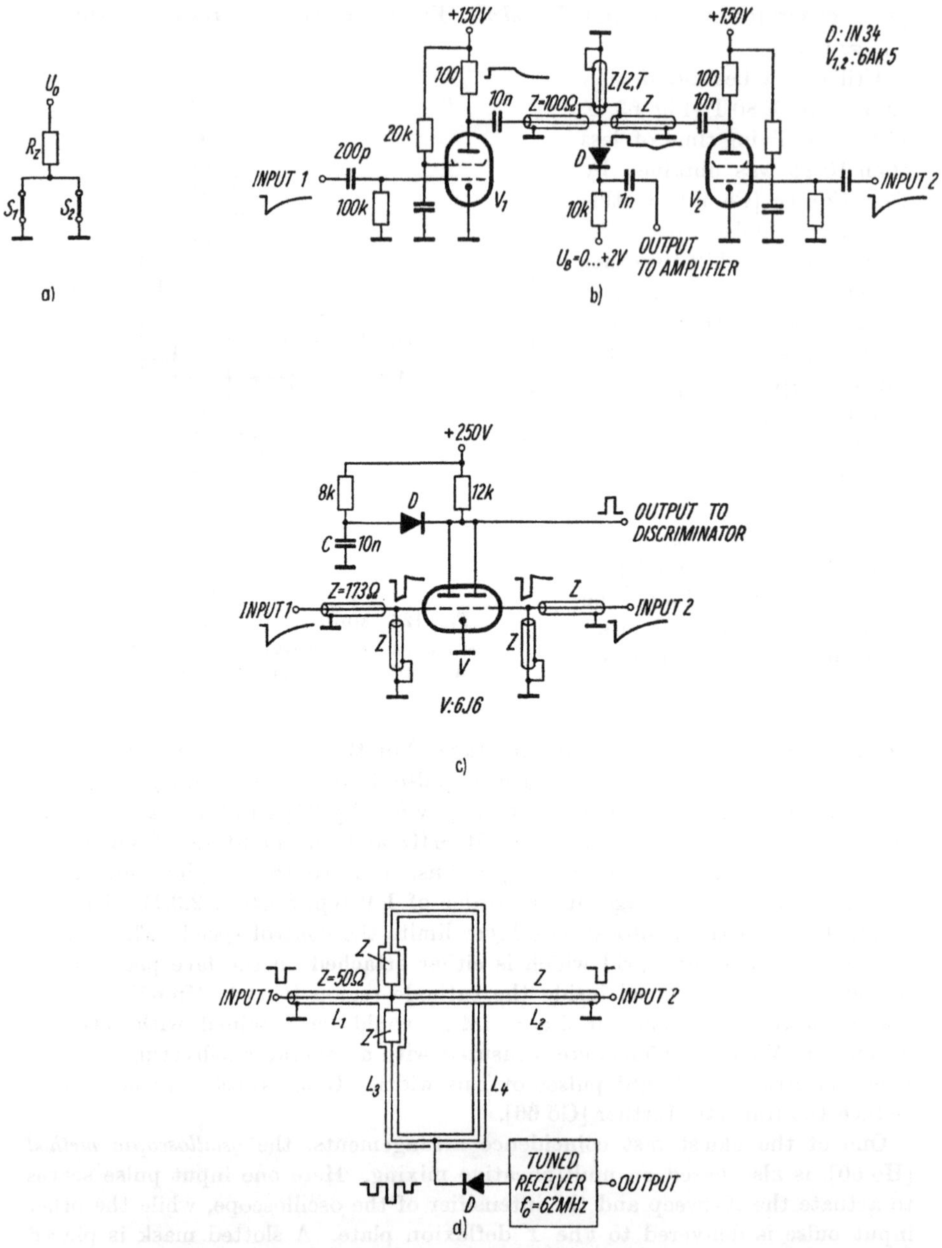

Fig. 129. Additive mixer stages with diodes.

a) Principle. *b)* Partial cable reflexion. According to *Bell* et al. [Be 52]. *c)* Use of a clamping diode. According to *Garwin* [Ga 50]. *d)* Selection of a *Fourier* component. According to *Dicke* [Di 47].

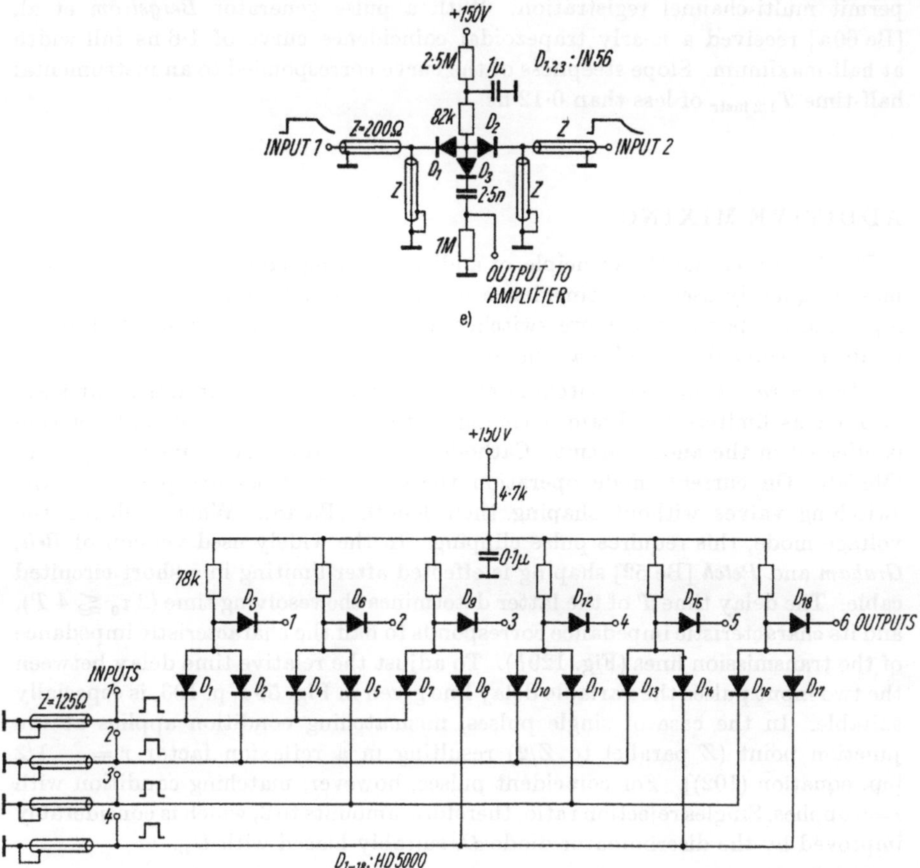

Fig. 129. Additive mixer stages with diodes.

e) Diode mixer stage. According to *de Benedetti* and *Richings* [Be 52a]. *f)* Diode matrix system. According to *Caini* and *Tove* [Ca 64]. *g)* Diode coincidence stage. According to *Rudenko* [Ru 58].

permit multi-channel registration. With a pulse generator *Bergström* et al. [Be 60a] received a nearly trapezoidal coincidence curve of 1·6 ns full width at half-maximum. Slope steepness of the curve corresponded to an instrumental half-time $T_{1/2\,\mathrm{instr}}$ of less than 0·12 ns.

ADDITIVE MIXING

Fig. 129*a* presents the principle of additive mixing (*Rossi* circuit), which is most frequently used for coincidence circuits, especially for $n > 2$. Here, the input pulses open two or more switches, that are closed in quiescent state, to produce a coincidence pulse at the load resistance.

Electron valves used as switching elements which conduct in quiescent state also act as limiters to obtain a timing signal [Ba 48; Ga 50; Be 52]. Mixing is effected in the anode circuit. Cathode circuit adding has gained no ground [Me 50]. On current mode operation the detector pulses are passed to the switching valves without shaping their length [Ba 48]. When utilising the voltage mode, this requires pulse clipping. In the widely used version of *Bell*, *Graham* and *Petch* [Be 52] shaping is effected after limiting in a short-circuited cable. The delay time T of the latter determines the resolving time ($2\,\tau_0 \lessapprox 4\,T$), and its characteristic impedance corresponds to half the characteristic impedance of the transmission lines (Fig. 129*b*). To adjust the relative time delay between the two input pulses the variable delay line given in Fig. 59*a*, p. 106, is especially suitable. In the case of single pulses, mismatching condition applies at the junction point (Z parallel to $Z/2$) resulting in a reflexion factor $r = -1/2$ [cp. equation (102)]. For coincident pulses, however, matching condition with $r = 0$ applies. Singles rejection ratio, therefore, amounts to 2, which is considerably improved by the discriminator diode D, suitably biased with U_B.

Garwin [Ga 50] gives the version shown in Fig. 129*c*, where clipping is effected before the pulses reach the limiting valves. In quiescent state more than half the total current of both triode systems flows through the clamping diode D. As a result part of the current continues to flow through the diode because of the action of the capacitor C, when only one triode system is cut off. Therefore, only an output pulse height of $\approx$0·3 V is produced, while on coincidence an output pulse height of $\approx$20 V is obtained. *Varlamov* et al. [Va 63] describe a circuit with four coincidence and two anticoincidence channels which yielded a resolving time of 10 ns and had a dead time of 30 ns. *Budyashov* et al. [Bu 67b] also used this type of circuit.

Additive mixing with *semiconductor diodes* enjoys wide favour. *Dicke* [Di 47] was probably the first to develop a fast coincidence stage ($2\,\tau_0 \approx 3$ ns). It is presented in Fig. 129*d*. Here branching causes two pulses for every pulse appearing at the inputs *1* or *2*, which are then added at the diode D. Cable length has been selected to quench that component of the *Fourier* spectrum of the pulse-pair which has the frequency f_0 to which the narrow-band amplifier is tuned. This is expressed by

$$L_4 - L_3 = L_2 - L_1 = \frac{v}{2 f_0}, \tag{110}$$

where v is the velocity of pulse propagation in the cable. On coincidence, the amplitude of the partial pulses that have passed through $L_1 + L_4$ and $L_2 + L_3$ add, because they have the same delay time. Due to the non-linearity of the diode characteristic complete quenching with the *Fourier* components of the other two partial pulses does not occur any more. Without limiting a pulse, height variations of 10 : 1 could be tolerated. In place of the diode an overdriven distributed mixer amplifier has been used by *Hornyak* and *Yoshiki* [Ho 59].

In the diode mixer stage from *de Benedetti* and *Richings* [Be 52a] a quiescent current flowing through the diodes D_1 and D_2 is turned off by the limited and length-shaped positive pulses (Fig. 129*e*). On appearance of a single pulse the still open diode takes the current which yields a very good singles rejection ratio and a high sensitivity. Limiting the input pulse height is not absolutely necessary because the height of the coincidence pulse is determined by the lower of the input pulses. Diode D_3 serves as pulse stretcher and particularly limits the obtainable minimum resolving time. Such diodes must have a very low forward resistance, a high reverse resistance and a fast switching behaviour. Furthermore, their capacitance must be small to avoid the occurrence of parasitic coupling. *De Vries* [Vr 62] estimated the obtainable minimum resolving time with 10 ps.

A circuit using a transistor as discriminator was employed by *Miller* and others [Mi 59; Em 62; Wa 63a]. With fast switching diodes *Michaelis* et al. [Mi 64] obtained a singles rejection ratio of 12 before the negative biased diode D_3. A four-fold coincidence arrangement with four anticoincidence systems as part of a counter telescope is described by *Barna* et al. [Ba 61].

To measure all six possible double coincidence combinations in four detectors, *Caini* and *Tove* [Ca 64] utilised, for example, the diode matrix arrangement illustrated in Fig. 129*f*. With limited pulses shaped to a width of 2 ns they obtained a resolving time of 1 ns and a coincidence ratio of <15.

Chaplin and *Candy* [Ch 59a] used pulse transformers for clipping. In a four-fold coincidence system *Madey* [Ma 55] employed an additional clamping diode, as did *Garwin* in the version given in Fig. 129*c*.

Rudenko [Ru 58; Vy 61] found an interesting solution which can also operate without height limitation and without current flow through the diodes on quiescent state (Fig. 129*g*). Single pulses appearing, for example, at the input *1*, are sharply differentiated by the network consisting of the capacitor C_1 and the low forward impedance of diode D_2, capacitor C_2 and the impedance $Z/2$. Coincident input pulses block diode D_1 and D_2, and the pulse stretcher diode D_3 charges the input capacitance C_i of the amplifier to the level of the input pulse height. In a slightly modified design, this principle has also been utilised for a triple coincidence — single anticoincidence arrangement [Ru 62].

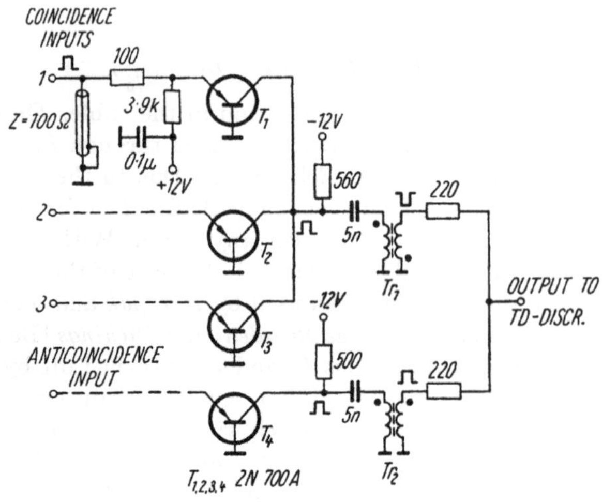

a)

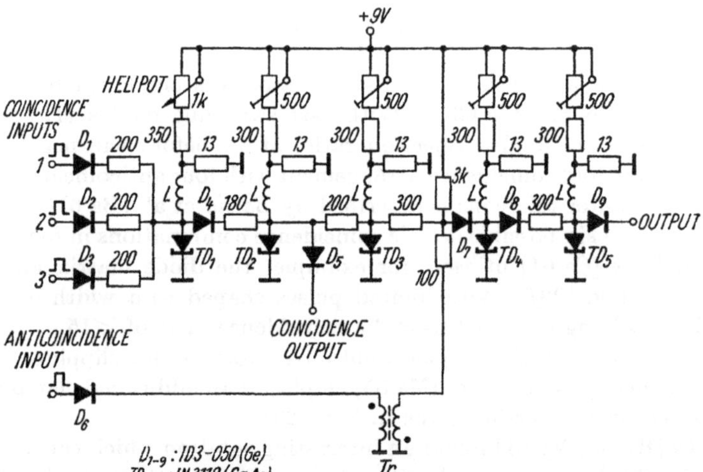

b)

Fig. 130. Additive mixer stages.

a) With common-base transistors. According to *Banner* and *Teiger* [Ba 64c]. b) Tunnel-diode coincidence circuit. According to *Whetstone* and *Kounosu* [Wh 62].

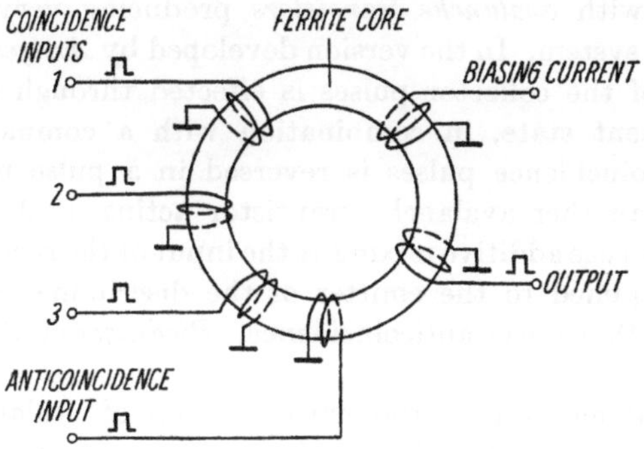

c)

Fig. 130. Additive mixer stages.

c) With pulse transformer. According to *Ispiryan* and *Oganesyan* [Is 62].

In the circuit of *Bell* et al. (Fig. 129*b*) *transistors* can replace the switching valves. *Barna* et al. [Ba 60a; Ak 66] developed a circuit where long-tailed transistor pairs are used as limiters (Fig. 113*c*, p. 185) and adding takes place at the collector of the transistors open in quiescent state. *Sidi* and *Sold* [Si 63b] employed simple common-emitter stages (Fig. 113*a*). Higher switching speed, however, is obtained with common-base stages working either as limiter [Ro 62] or which are opened by the input pulse [Ga 64a]. With the arrangement given in Fig. 130*a*, length shaping of the height-limited pulse is effected as in Fig. 129*e*, before it enters the mixing stage [St 63a; Ba 64c]. The anticoincidence channel T_4 is designed identically, but the pulse transformer Tr_2 effects no polarity reversal. As a result the anticoincidence pulse and the coincidence pulse cancel each other. Negative pulses serve to trigger the tunnel diode discriminator. This circuit has a resolving time of $2\,\tau_0 = 2$ ns at 100% coincidence efficiency and can handle a maximum pulse repetition rate of 35 MHz. Including limiter transistors sensitivity is 250 mV. With a four-fold coincidence — single anticoincidence circuit *Kolpakov* [Ko 64a] obtained a sensitivity of 150 mV. *Infante* and *Pandarese* [In 62] used pulse transformers in front of the clipping cables.

Analogue to Fig. 129*c*, point contact clamping diodes with a switching time of 0·2 ns were employed in the collector circuit of the switching transistors by *Sugarman* et al. [Su 60; Su 62a]. With d.c. coupling this four-fold coincidence — single anticoincidence arrangement permitted a maximum pulse repetition rate of 100 MHz. The circuit arrangement of *Whittaker* [Wh 66b] is designed to operate at the same speed and delivers an output signal as soon as two of three input signals are present.

Utilising additive mixing in the emitter circuit of switching transistors working as emitter followers, a transistor may serve as common emitter resistor and perform the function of the anticoincidence channel [Du 63].

Trigger stages with *avalanche transistors* producing very high pulses can replace the limiter system. In the version developed by *Artiges* and *Brun* [Ar 61] additive mixing of the collector pulses is effected through diodes, which are blocked in quiescent state, in combination with a common shaping cable. Polarity of the coincidence pulses is reversed in a pulse transformer which serves to trigger another avalanche transistor acting as discriminator. *Alon* and *Rausch* [Al 63 a] use additive mixing at the input of the reversing transformer. Negative pulses applied to the emitter of the discriminator transistor avoid breakthrough and thus effect anticoincidence. *Riedinger* et al. [Ri 67 c] describe a similar circuit.

The additive mixing mode in the emitter circuit of avalanche transistors is utilised by *Gygi* and *Schneider* [Gy 63]. As soon as a transistor fires, the other transistor of the system becomes biased through the common emitter resistor and prevents it from also being triggered. Only when breakthrough condition (cp. Section 3.1.5) applies for both transistors simultaneously they will fire together. Despite the wide input pulses, resolving time amounts to only 0·2 ns.

With *Pardies* et al. [Pa 65] the negative emitter pulses of the p-n-p avalanche transistors are passed to a diode mixer stage which permits triggering of the following avalanche-transistor discriminator system without having to resort to polarity reversal. Here, trigger sensitivity is 3 mV and the maximum repetition rate 10 MHz.

When *tunnel-diode trigger circuits* are used to produce shaped pulses this results in a recovery time down to some nanoseconds, which is of particular interest for operation as anticoincidence system. Coincidence circuits equipped with tunnel diodes permit pulse repetition rates in the order of 100 MHz and have a maximum sensitivity of approximately 10 mV or 1 mA [Fr 61; Go 61 b; Wh 62; Du 64; Go 64 c; Bu 65 a; Ef 65; Go 65; Ba 66 e; Iv 67 b; Sa 67]. In most cases, discrimination against the output of single pulses is also effected with a tunnel-diode trigger circuit. In the circuit of Fig. 129*e* the diodes can be replaced by tunnel diodes [In 61].

Whetstone [Wh 62] used the circuit given in Fig. 130*b*, where the input pulses are transmitted from the tunnel-diode discriminator (Fig. 94*d*, p. 161) with an input sensitivity of 50 mV. Buffer diodes and series resistors reduce the reaction on the respective preceding trigger stage. A helical potentiometer is used to adjust the threshold of the discriminator TD_1 to single pulse, or double- and triple-coincidence operation. For anticoincidence the input pulse is inverted with transformer *Tr* and added to the output pulse of the buffer stage TD_3, preventing trigger action of the anticoincidence stage TD_4. Output pulses are supplied by TD_2 and TD_5. *Bushnin* et al. [Bu 67 a] use a circuit where additive mixing is effected in a similar manner.

The short rise time of the tunnel-diode pulses allows a sharp differentiation, for example, with a capacitor $C = 4$ pF placed before a 50 Ω cable ($\tau = 200$ ps) [Bj 62]. This reduces the feed-through of the slower trigger pulses and a very short coïncidence input pulse is produced (cp. Fig. 119, p. 193). With this arrangement coincidence curves were obtained by measurement having slopes as steep as 10 ps per decade of count rate drop, and a temperature drift of less than 3 ps/°C.

Additive mixing of current pulses with a *pulse transformer* is another interesting modification (Fig. 130*c*). Here, a biasing direct current opposed to the input currents, magnetically biases the rectangular ferrite core in such a manner that it requires the sum of all the three input currents to bring the ferrite core to high flux density state. This causes a large signal to be produced in the output winding only on coincidence. A reverse poled input winding effects anticoincidence. Time resolution and singles rejection ratio depend on the figure of merit of the rectangular ferrite core [Is 62].

DIODE BRIDGE SYSTEMS

Balanced diode bridge systems are sometimes employed as coincidence mixer stages (Fig. 131) which have a sensitivity of approximately 100 mV and a singles rejection ratio in the order of 100. Irrespective of height, the non-limited single input pulses produce no output signal. On coincidence, the balance of the bridge system becomes disturbed.

In the version given in Fig. 131*a* this is caused by the non-linear characteristic of the diode D_3, through which the summated current of both bridge branches flows on coincidence [Ba 48a; Me 48; Mo 49]. In the analogue circuit of *Lundby* [Lu 51] only the anode pulses are used, which, however, requires a difference amplifier, as in the versions of *Bay* [Ba 51] and *Strauch* [St 53].

The principle of the modification by *Bay* [Ba 51; Ba 56a] is given in Fig. 131*b*, where a negative single pulse at the input *1* rapidly charges the capacitors C_1 and C_2 through the resistors R_1 and R_2. On discharge, long pulses of equal polarity and height develop at the resistors R_3 and R_4. By exchanging D_2 and C_2, which causes pulses of opposite polarity to appear at R_3 and R_4, the use of a difference amplifier can be omitted. Between the potentiometer slides a resistor chain is interposed, with a slow amplifier connected at the centre. Negative pulses at the input *2* block D_2 and cannot charge C_2, which prevents production of a slow output signal. On coincidence C_2 is also not charged due to the voltage applied to both ends. This seriously disturbs the bridge balance and leads to a large output pulse.

An analogue circuit has been developed by *Strauch* [St 53] which contains no capacitors and requires a fast amplifier. In Fig. 131*c* a four-fold coincidence — single anticoincidence assembly of this type is given [Du 64a; Du 64b]. Here, the single pulses arriving at the inputs *2* to *4* produce a small output signal due to the finite reverse resistance of the diodes D_2 to D_4. This output signal is compensated by the resistors R. Coincident pulses at the inputs *2* to *4* are, therefore, not registered. When pulses coincide at all four inputs, a larger pulse is produced at the left edge of the bridge than at the right because the

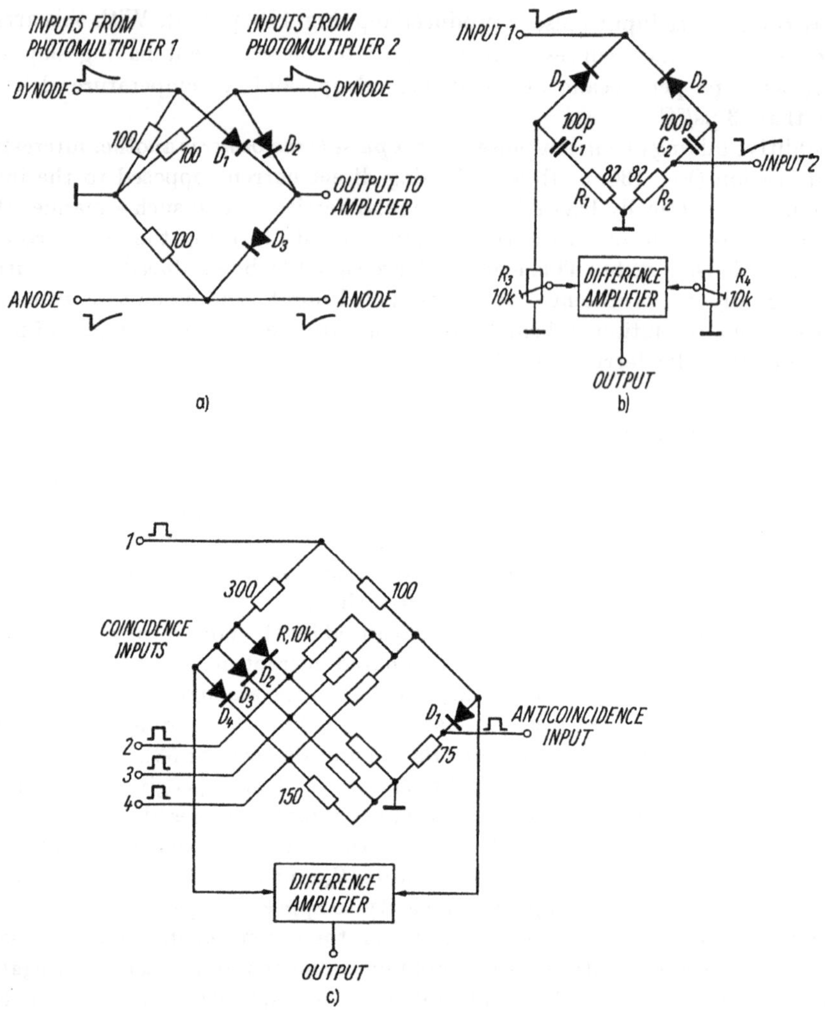

Fig. 131. Diode bridge systems.

a) According to *Baldinger* et al. [Ba 48a]. *b*) According to *Bay* [Ba 51]. *c*) According to *Strauch* [St 53; Du 64a].

diodes D_2 to D_4 are blocked. This disturbs the balance of the bridge and causes a large output pulse. A simultaneously arriving anticoincidence pulse also blocks diode D_1 which restores the balance of the bridge system, and no output signal is produced. Several anticoincidence branches may also be used. With limited pulses shaped to a width of approximately 2 ns *Dunaitsev* and *Prokoshkin* [Du 64a] obtained a resolving time of about the same magnitude.

DIFFERENTIAL-COINCIDENCE CIRCUITS

The differential-coincidence principle of *Bay* [Ba 56a] yields a high sensitivity, amounting to 100 mV, a singles rejection ratio of 100 to 200, and a resolving time which is only about 1 per cent of the length of the input pulse. Here, two coincidence circuits which produce a maximum output pulse height on input pulse coincidence are connected together in such a manner that they operate with a mutual delay of $2\,\Delta T$. Fig. 132*a* shows the dependence of the output pulse height on the relative time delay T of the input pulses. The output pulses of the two coincidence circuits are subtracted from each other which yields the diagram given in Fig. 132*b*. If the difference $U_{II} - U_{I}$ is negative its polarity is reversed by a phase rectifier as suggested in Fig. 132*c*. The attached discriminator does not deliver an output signal if the delay time of the input pulses is less than $2\,\tau_0$. The dashed-line curve illustrates the reduced influence of changes of input pulse height or discriminator threshold on time resolution. To obtain an output signal on coincidence, a slow coincidence circuit is arranged in anticoincidence with the differential coincidence output (Fig. 132*d*). It must be noted that the resolving time of the slow coincidence circuit must be less than $2\,T_d$ (Fig. 132*c*) to prevent an output signal on $|T| > T_d$. In deviation of the initial arrangement of *Bay*, Fig. 132*e* gives a differential coincidence stage with two diode adding systems [Me 61], which are symmetrically delayed in relation to each other [Ak 57].

If, instead of the difference in amplitude, the count rate difference is formed at the output of the two delayed coincidence circuits, this method looses many of its advantages [Mi 56a; Bo 57].

Noise and component instability determine the lower limit of the resolving time of differential coincidence circuits. Without resorting to specialised switching diodes, *Brunner* [Br 64] obtained a resolving time of $2\,\tau_0 = 1$ ps by measurement with constant pulses of 15 ns width and approximately 50 V height. A strong dependence, however, could be observed of resolving time on pulse height and, when semiconductor diodes were employed, on temperature.

Input pulse difference may also be utilised by replacing the differential coincidence stage of the circuit in Fig. 132*d* with a tunnel-diode anticoincidence stage incorporating a transformer to reverse the polarity of one of the input pulses [Ku 63].

In the arrangement of *Meunier* and *Teiger* [Me 59a; Me 60], which has been derived from a chronotron circuit [Me 57], the input pulses have opposite polarity and, on coincidence, cancel each other at the diode D_2 in the middle of the delay line (Fig. 133*a*). On coincidence, the negative input pulse *1* charges the capacitors C_1 and C_3 and actuates the coincidence stage after difference amplification. Positive single pulses block the diodes while negative singles also charge the capacitor C_2, causing the difference amplifiers not to give an output signal. Singles rejection ratio amounts to 80. Resolving time is practically independent of width and height of the input pulses and is determined by the delay between the diodes D_1, D_3 and D_2.

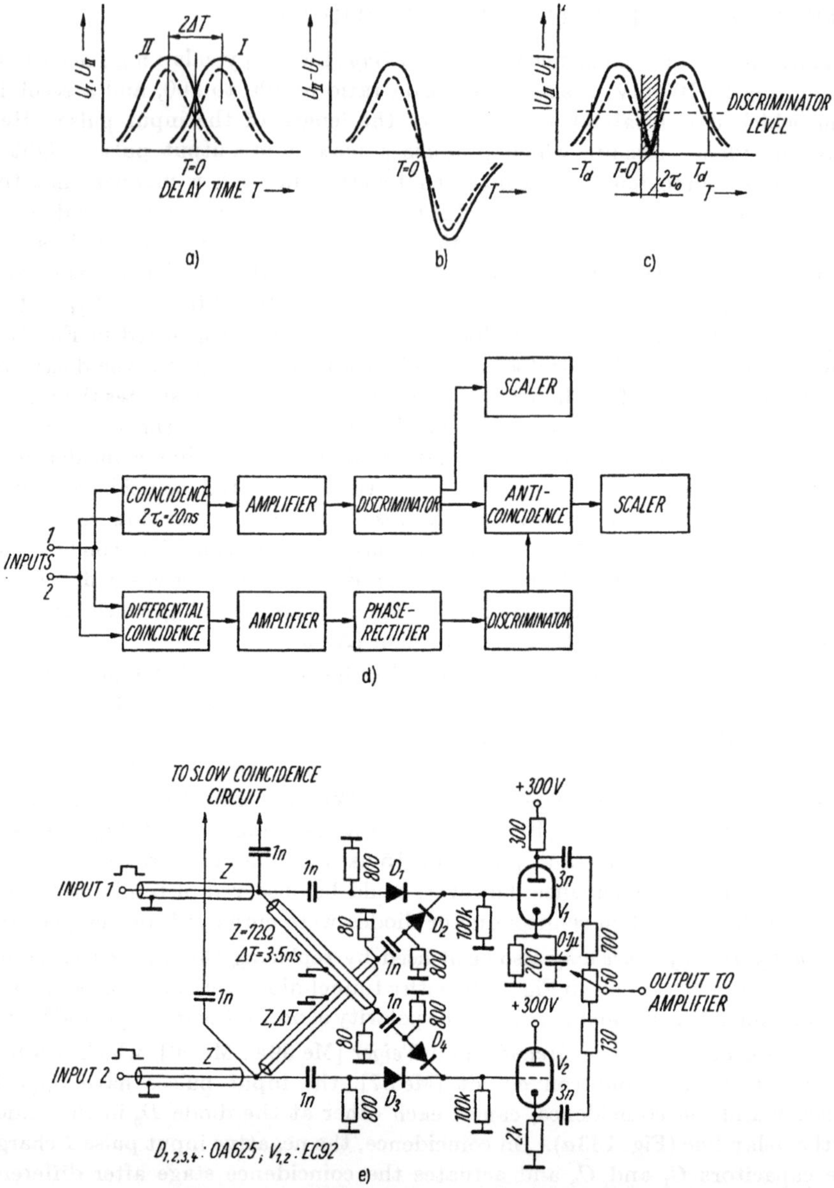

Fig. 132. Differential coincidence circuit. According to *Meiling* et al. [Me 61].

a) Output pulse height dependence of two coincidence circuits, mutually delayed by $2\,\Delta T$, on the relative time delay T of the input pulses. b) Output pulse height after difference formation. c) Output pulse height after phase rectification. d) Block diagram of the circuit. e) Differential coincidence circuit: input pulse width 15 ns, mutual time delay $2\,\Delta T = 7$ ns.

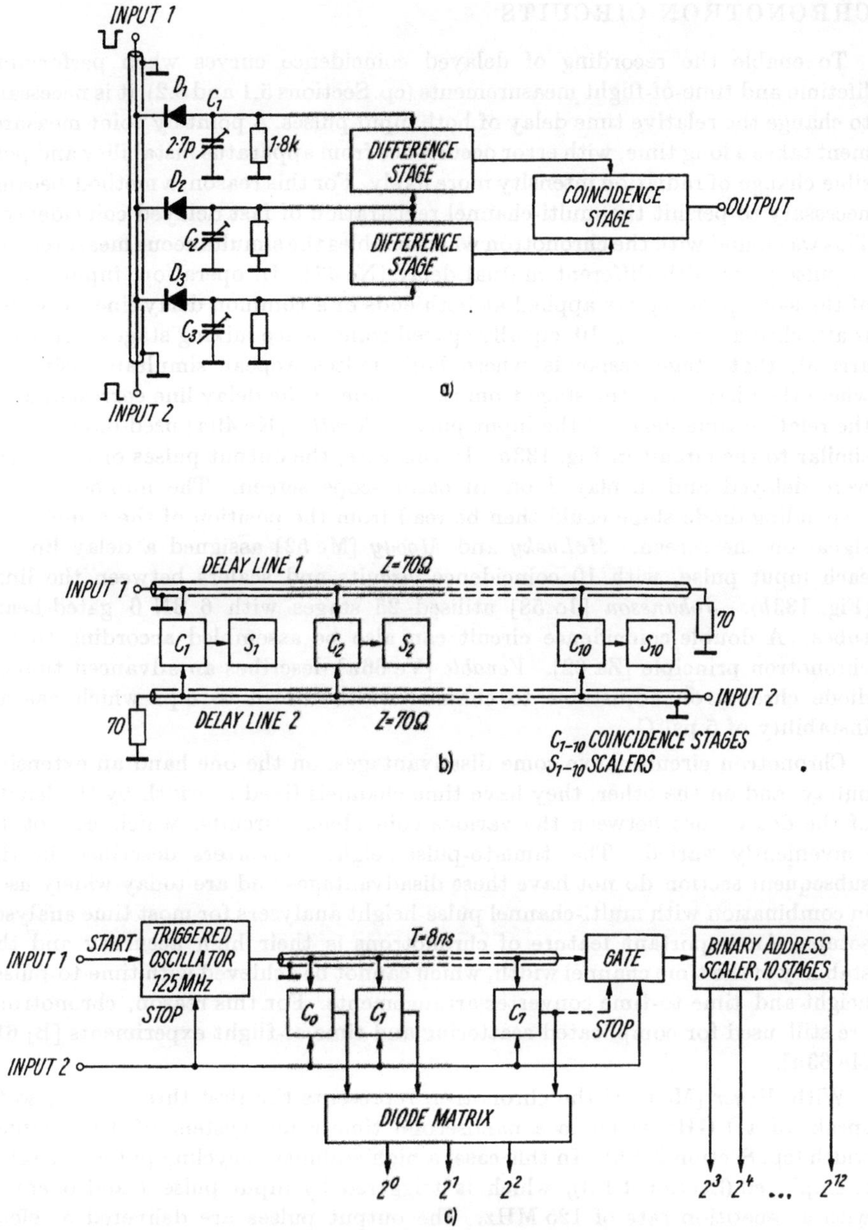

Fig. 133. Use of the chronotron principle.

a) Coincidence circuit. According to *Meunier* and *Teiger* [Me 59a; Me 60]. b) Chronotron system. According to *McLusky* and *Moody* [Mo 52]. c) Simplified diagram of 2^{13}-channel time analyser of 1 ns minimum channel width. According to *Meyer* [Me 63a].

CHRONOTRON CIRCUITS

To enable the recording of delayed coincidence curves when performing lifetime and time-of-flight measurements (cp. Sections 5.1 and 5.2), it is necessary to change the relative time delay of both input pulses. A point by point measurement takes a long time, with error occurrence from apparatus instability and possible change of radiation intensity more likely. For this reason, a method became necessary to permit the multi-channel registration of fast delayed coincidences. This was found with the chronotron which enables the simultaneous measurement of pulse pairs with different mutual delay [Ne 47]. In operation, input pulses of the same polarity are applied at both ends of a common delay line, to which is attached a series, e.g. 10, equally spaced coincidence mixing stages. On pulse arrival, that stage responds where both pulses appear simultaneously, i.e. where the distance of the stage from the middle of the delay line corresponds to the relative time delay of the input pulses. *Keuffel* [Ke 49a] used diode stages similar to the circuit in Fig. 133*a*. In this case, the output pulses of the stages were delayed and displayed on an oscilloscope screen. The number of the responding diode stage could then be read from the position of the coincidence signal on the screen. *McLusky* and *Moody* [Mc 52] assigned a delay line to each input pulse, with 10 coincidence circuits and scalers between the lines (Fig. 133*b*). *Johansson* [Jo 58] utilised 25 stages with 6 BN 6 gated-beam tubes. A double-coincidence circuit can also be assembled according to the chronotron principle [Za 62]. *Venable* [Ve 66a] describes an advanced tunnel-diode chronotron adjustable to channel widths down to 5 ps which has an instability of 5 ps/°C.

Chronotron circuits have some disadvantages, on the one hand an extensive outlay, and on the other, they have time channels fixed in width by the length of the delay lines between the various coincidence circuits, which cannot be conveniently varied. The time-to-pulse-height converters described in the subsequent section do not have these disadvantages and are today widely used in combination with multi-channel pulse-height analysers for most time analyser setups. An important feature of chronotrons is their high accuracy and the stability of the time channel width, which cannot be achieved with time-to-pulse-height and time-to-time converter arrangements. For this reason, chronotrons are still used for complicated scattering and time-of-flight experiments [Bj 61; Me 63a].

With *Meyer* [Me 63a] the chronotron represents the first three stages, so to speak, of a 1 GHz scaler in a nanosecond time coder system of 1 ns channel width (cp. Section 3.4.6). In this case, a high stability recycling pulse generator is employed (Section 4.1.3), which is triggered by input pulse *1* and operates with a repetition rate of 125 MHz. The output pulses are delivered as clock pulses through a delay line to a binary address scaler, equipped with tunnel diodes, which has a maximum counting rate of 300 MHz (Fig. 133*c*). The time interval to arrival of input pulse *2*, which blocks the gate, is registered by the address scaler, with the minimum interval amounting to 8 ns. The interval of the pulse from input *2* to the last preceding clock pulse, located between 0 and 8 ns, is registered by a chronotron system consisting of eight coincidence stages C_0 to C_7. The latter are spaced 1 ns apart and connected via transmission line transformers. The number of the responding coincidence stage is binary-coded in a diode-matrix system. Coincidence resolving time amounts to 1·5 ns. Thus, one, and sometimes two, coincidence stages are always actuated, with the later activated stage instantly reset by the first responding stage. To avoid error counts of the clock pulses when stage C_0 or C_7 responds, the gate is at times also blocked by the coincidence outputs.

Hansen and *Sedlmeyer* [Ha 67 c] describe a commercial chronotron with strip-line delay and 10 coincidence circuits for measuring the rate of rise of fast single transient waveforms.

3.4.4 TIME-TO-PULSE-HEIGHT CONVERTERS

Time interval measurements in the nanosecond region can be based on pulse height measurements by converting the time interval between the signals of two pulse sources into a proportional pulse height and measuring this with a multi-channel pulse-height analyser. Compared with single-channel coincidence curve measurements with the aid of a coincidence circuit (delayed coincidences) multi-channel time analysers with time-to-pulse-height converters have significant advantages:

(i) Multi-channel measuring requires less time than the single-channel mode. Time gain is not proportional to the number of channels, because for the same statistical accuracy of the measuring points on a coincidence curve in a singlechannel mode the measuring time can be adapted to prevailing intensity, for example, by preselecting the number of counts.

(ii) The effect of apparatus instability is less because all parts of the coincidence curve are affected to the same extent, and apparatus faults can be more easily detected.

(iii) In general, better time resolution obtains than with coincidence circuits, because, generally, channel width can be selected small compared to resolving time of the detectors, and in consequence the influence of the finite channel width on the measuring result is negligible.

(iv) Correction of the measuring result due to changing or rapidly decreasing intensity of the source becomes unnecessary, because these changes affect all channels, and thus all points on the coincidence curve, at the same time.

The multiplicity of the circuit versions of time-to-pulse-height converters has been variously described and reviewed [Gr 60; Ch 61 a; Li 61 a; Bo 63; Dr 65; Fi 65; Ma 67 a; Re 67, chapter 7; Og 68]. For assessing the utility of a circuit the time resolution obtained with pulses from a generator is not as important, because this is often better by one or two orders of magnitude than with radiation detectors. Far more significant in this respect are the processed singles count rate, the temperature and long-term stability and the simplicity of the design. To a large extent this depends on the components and type of circuit used in the arrangement. The following data give an approximate outline of the sensible demands that can at present be made on time-to-pulse-height converters [Dr 65].

Dynamic range	5 to 100 ns
Resolving time,	
electronic, dependent on pulse generator	5 to 50 ps
physical, dependent on detectors and energy of radiation	200 to 500 ps
Conversion factor, dependent on post-amplification	50 mV/ns
Non-linearity	0·1 to 5%
Input sensitivity	10 mV to 1 V
Instability,	
temperature	2 ps/°C
long-term	15 ps to 1 ns/10 h
Count rate capability,	
singles, sometimes different for the two inputs	10^4 to 10^6 pps
coincidences	10^3 pps

Compared to time-to-digital converters (see Section 3.4.6), the main disadvantage is the unsatisfactory differential linearity. A marked improvement can be obtained through the method of statistical equalisation described in Section 3.4.6. *Abbattista* et al. [Ab 68a] suggest two versions for the purpose. In principle, instead of the time spacing between start and stop pulse, the time spacing of these pulses to a random clock pulse is measured with two time-to-pulse-height converters. If the clock pulse arrives between a start and a stop pulse the output pulses of the time-to-pulse-height converters are added, but otherwise subtracted. In the case of many events of the same time spacing all parts of the converter characteristic contribute in the statistical mean and differences average out. In combination with a bidirectional register in a *Wilkinson*-type multi-channel analyser the time for analysis can be reduced, e. g. to 1/3, which is another benefit. This method was first applied to analog-to-digital converters [Ab 68].

Function principle divides the time-to-pulse-height converters into two main groups, viz. the overlap type and the start-stop type. As is the case with coincidence circuits, two converter circuits can be combined to form a differential converter.

PULSE OVERLAP TYPE

This type of circuit has been evolved from coincidence circuitry (cp. Section 3.4.3) so that there are many similarities in function and characteristics. Fig. 134 gives the function principle of the overlap type. During the time interval between the appearance of the leading edge of the pulse at input *2* and the trailing edge of the pulse at input *1* the output pulse rises in a linear manner. Because the final height of the latter is $U_0 \propto (L - T)$, the length L of the input pulses must be constant; a demand that is met with pulse shaping in short-circuited transmission lines. The usually exponentially rising and falling edges of the input pulses, on condition $T \approx 0$ and $T \approx L$, respectively, cause a linearity deviation of the converter characteristic (Fig. 134*b*). With

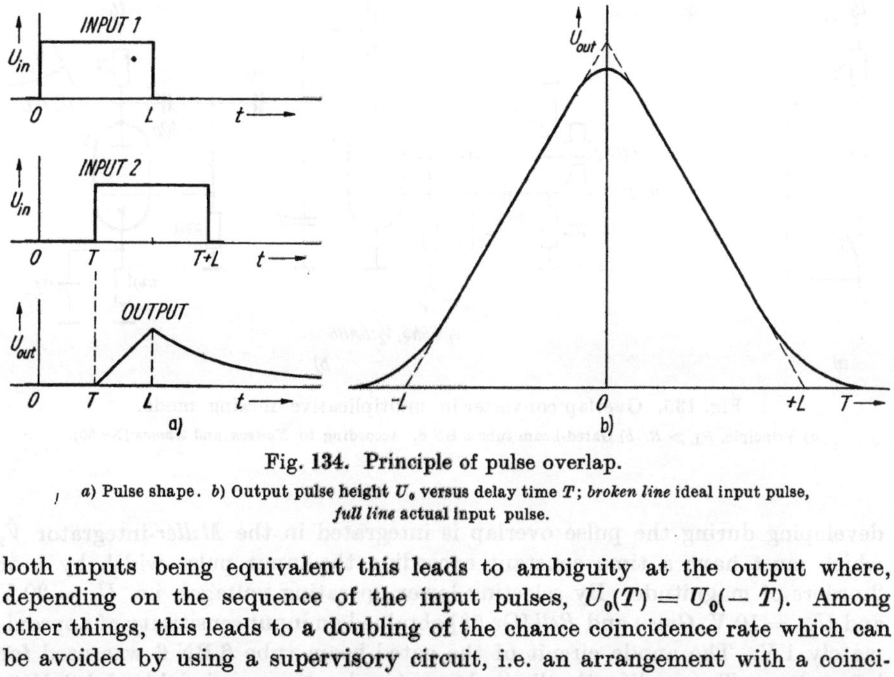

Fig. 134. Principle of pulse overlap.

a) Pulse shape. *b)* Output pulse height U_0 versus delay time T; *broken line* ideal input pulse, *full line* actual input pulse.

both inputs being equivalent this leads to ambiguity at the output where, depending on the sequence of the input pulses, $U_0(T) = U_0(-T)$. Among other things, this leads to a doubling of the chance coincidence rate which can be avoided by using a supervisory circuit, i.e. an arrangement with a coincidence circuit of long resolving time which operates with a time delay L [Gr 58; Ne 59; Bo 60].

One of the significant features of the overlap principle is the high admissible single-pulse count rate, because an output signal is produced only during the interval when pulses are present at both inputs. Simple construction is another benefit. The at times low stability is a disadvantage, which is caused by switching elements, such as diodes, or by valves and transistors that are blocked in quiescent state. Length L of the input pulse restricts the dynamic range. As with coincidence circuits, pulse overlap is either multiplicative or additive.

Multiplicative Mixing Mode

Fig. 135*a* gives the simplified diagram of this converter principle. Only when both switches S_1 and S_2 are closed capacitor C, charged to U_0 level, does discharge through the resistor R. To obtain linear behaviour the time constant RC must be high against the maximum overlapping time. When one of the switches is opened, the capacitor C is charged again through the large resistor R_L. Fig. 135*b* shows a circuit with a gated-beam tube (6 BN 6) which is blocked through both grids in quiescent state [Ne 55; Jo 58; Ne 59]. The output pulse

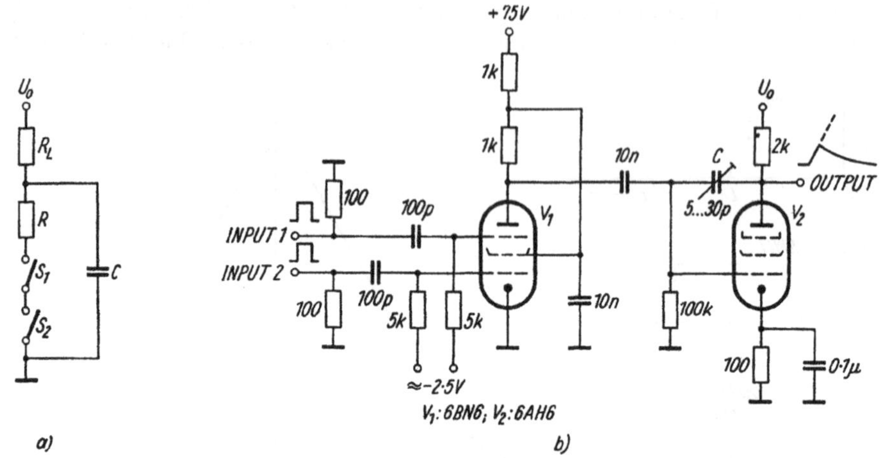

Fig. 135. Overlap converter in multiplicative mixing mode.
a) Principle, $R_L \gg R$. *b*) Gated-beam tube 6 BN 6. According to *Neilson* and *James* [Ne 55].

developing during the pulse overlap is integrated in the *Miller*-integrator V_2 which must have a time constant exceeding the input pulse width by 1 or 2 orders of magnitude. By selecting lower operating voltages, i.e. $U_a = 20$ V and $U_{g2} = 10$ V, *Green* and *Bell* [Gr 58] obtained an input sensitivity of approximately 1 V. The anode circuit of the gated-beam tube 6 BN 6 was used for integration. To amplify the limited input pulses from a height of 1·2 V to a height of 3·5 V, *Božek* et al. [Bo 62a] employed common-base transistor stages before the control grids.

With the 6 BN 6 circuits, an interference sometimes arises because of the parasitic single-pulse coupling due to the capacitance between the unscreened grid *3* and the anode. With measurements at pulsed accelerators, the pulse coming from the latter is, for this reason, transmitted to grid *1* [As 63]. Parasitic pulse transmission can also be compensated [De 63]. Drift is reduced to 20 ps per 5 h operating time by means of good stabilisation, especially of the filament voltage [Ga 62b].

For balancing the differing characteristic of the two control grids, *Samueli* and *Sarazin* [Sa 58] utilised two seven-grid valves, type 6 BE 7, with a common load resistor. In this case, the input pulses are simultaneously applied to grid *3* of one valve and grid *5* of the other valve.

Additive Mixing Mode

In principle, the input pulses cause the two switches, which are closed in quiescent state, to open and during the overlap period of the two input pulses the capacitor C is charged through the resistor R (Fig. 136*a*). *Jones* and *Warren* [Jo 56] modified the coincidence circuit developed by *Bell* et al. (cp. Fig. 129*b*, p. 206), by inserting an integrating network after the discriminator diode. Compared with circuits using 6 BN 6 valves, this frequently employed arrangement features a higher input sensitivity, although temperature sensitivity of the semiconductor diode is a disadvantage. Various modifications to improve linearity and stability have been described [Bo 63]. In place of the diode, an n-p-n transistor, blocked in quiescent state, can be utilised in common-emitter connection, which is opened only during overlap. A bootstrap arrangement in the collector circuit may improve linearity [Ro 63].

Fig. 136*b* gives the converter circuit developed by *Gorodetzky* et al. [Go 60]. After the negative input pulses have been limited in the valves V_1 and V_2 and shaped in the short-circuited cable, the overlap part, which exceeds the threshold *1*, opens the discriminator valve V_3. That part of the output pulses from the latter which is located above the level *2*, blocks the second limiting valve V_4. The shaded portion of the output pulses drawn on the circuit diagram is integrated in the anode system of valve V_4. Double limiting reduces the influence of the instability of the limiting valves V_1 and V_2. A certain unlinearity of the converter characteristic is a disadvantage, and occurs as soon as the pulse overlap is effected between the input pulse edges [So 61; St 66c]. A noticeable improvement of stability is achieved by the current feedback in the screen-grid circuit of the three limiting valves, as shown in Fig. 136*b*, and by stabilising the filament voltage to 4×10^{-4}, particularly of the discriminator valve V_3 [Bo 64a]. By stabilising the screen grid current of the valves V_1 and V_2 (Type E 180 F) and of the charging current of the integrating capacitance by means of pentodes an instability was obtained of $< 0{\cdot}2\%$ through 64 hours [Do 63]. Together with other improvements, *Jung* [Ju 63] introduced a current stabiliser for charging, and a resetting stage for discharging the integrating capacitance. In this manner a very small linearity deviation of less than 0·3% was achieved in the 100 ns dynamic range, concomitant with a maximum count rate capability of 1 MHz on periodic repetition. Utilisation of a selected mercury pulse generator and a biased amplifier resulted in a minimum time scale of 0·37 ps/channel and a resolving time of $2\,\tau_0 = 4{\cdot}8$ ps. With particular care in assembling, the circuit in Fig. 136*b* will yield a resolving time of 7 ps and a singles rejection ratio of 500 [St 66c].

In a similar arrangement proposed by *Samueli* and *Sarazin* [Sa 60a] pulses limited and shaped to suitable width in the preceding stages are added at the common load resistance of two pentode valves. A biased semiconductor diode serves as discriminator with integration effected in the subsequent valve stage. Employing a difference stage for integration will result in less capacitive feed-through of single pulses [Fo 63].

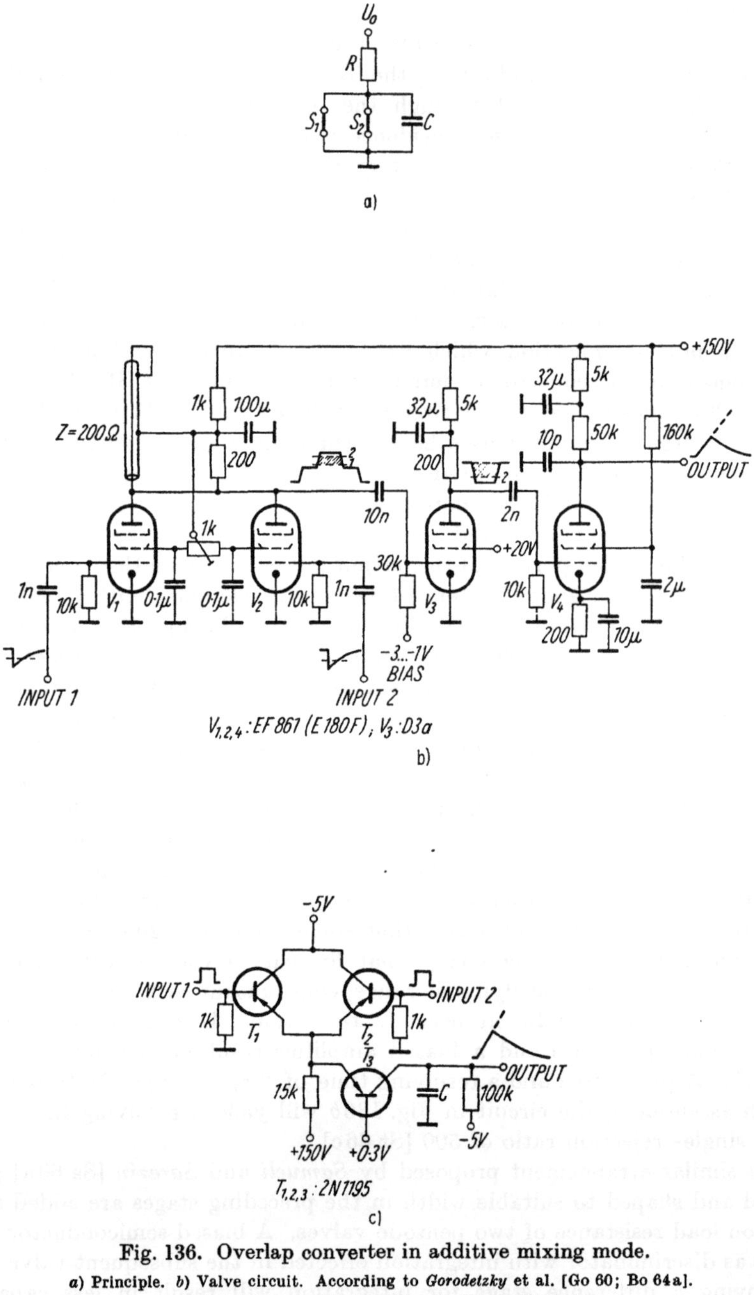

Fig. 136. Overlap converter in additive mixing mode.

a) Principle. *b)* Valve circuit. According to *Gorodetzky* et al. [Go 60; Bo 64a]. *c)* Transistor circuit. According to *Simms* [Si 61a].

The version of *Samueli* et al. [Sa 64b] uses transistorised limiter stages and collector circuit integration of a transistor inserted after the discriminator diode. This circuit can handle 7×10^5 pps and shows an instability of 3·5 to 7 ps/°C and 40 ps per day.

Vergezac and *Kahane* [Ve 64] employ the coincidence system of *Garwin* (cp. Fig. 129*c*). *Grin* et al. [Gr 61] modified the circuit of *de Benedetti* and *Richings* (cp. Fig. 129*e*) by supplementing it with an integrating network inserted after the diode D_3. Good linearity and singles rejection are obtainable if the operating voltage and the characteristic impedance of the shaping cable are kept as high as possible, with the diode quiescent current and the maximum height of the output pulse low [Gr 63]. Diodes should have a short recovery time and a low capacitance. Thus, it can be achieved that output pulse height is practically independent of the height of the input pulse. Stability is, in the main, dependent on the operating voltage and also on the load resistor R and the integrating capacitance, and less on the bias of the diode D_3. With germanium diodes, type S 570 G, drift was approximately 10 ps per 10 h without temperature stabilisation, while the linear range amounted to 65 ns ($L = 70$ ns), non-linearity was $< 1\%$ and electronic resolving time < 40 ps. Instead of the diode D_3, a valve [Be 65d] or an avalanche transistor [Be 65e] may be used. Avalanche transistors can also be utilised for pulse shaping [Wa 64b].

In the circuit of *Simms* [Si 61a; Jo 65a] transistors T_1 and T_2, that conduct in quiescent state, are employed with a large common resistor to stabilise the emitter current (Fig. 136*c*). With the appearance of single pulses, current is taken up by the still open transistor. When a shaped positive pulse blocks both transistors, the blocked transistor T_3 becomes conducting during the pulse overlap period. The collector stray capacitance C serves to integrate the total current flow. Input sensitivity is less than 1 V. In the version of *Mantakas* [Ma 64] a transistor is used as common emitter resistance, through which the system can be blocked.

Bloess et al. [Bl 63b] utilised a beam-deflexion tube, type E 80 T (cp. Fig. 63, p. 112), in an overlap converter design. Here, the input pulses are added and, with the voltage maintained at g_1, delivered to the deflexion plate D to guide the electron beam to the anode. Single pulses are not sufficient for this purpose. Integration is effected in the anode circuit of tube E 80 T. Instability amounts to ≤ 50 ps per 24 h, due to stabilisation of the cathode current flowing to g_4 in quiescent state.

START-STOP TYPE

From the function illustrated in principle in Fig. 137*a* it is evident that after the arrival of a starting pulse the voltage at a capacitor rises linearly until a stop pulse arrives. Compared to the overlap converter system, function is independent of input pulse width and therefore also suitable for the conversion of time differences which exceed the pulse width. If both inputs are equivalent (cp. [Me 63b]) this results in a double-valued output (Fig. 137*b*, curve *1*), analogue to overlap converters. This can be eliminated through the use to a supervisory circuit. Most of the circuits, however, are designed in such a manner that no output pulse is produced if the stop pulse arrives before the start pulse (Fig. 137*b*, curve *2*). Single start pulses will then result in maximum

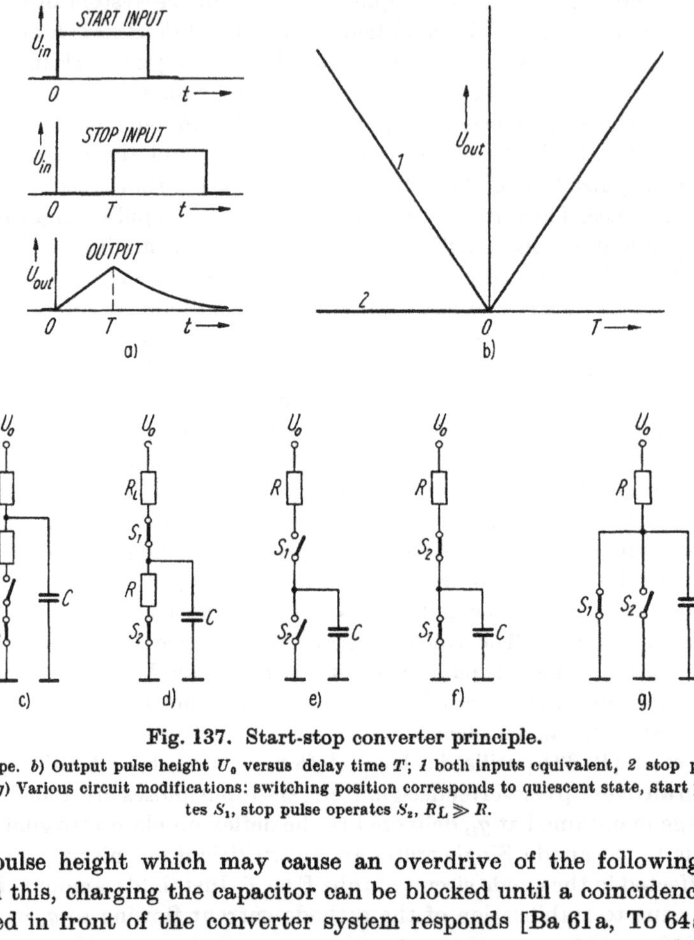

Fig. 137. Start-stop converter principle.

a) Pulse shape. *b*) Output pulse height U_0 versus delay time T; *1* both inputs equivalent, *2* stop pulse cannot start. *c*) to *g*) Various circuit modifications: switching position corresponds to quiescent state, start pulse operates S_1, stop pulse operates S_2, $R_L \gg R$.

output pulse height which may cause an overdrive of the following stages. To avoid this, charging the capacitor can be blocked until a coincidence circuit interposed in front of the converter system responds [Ba 61a, To 64a]. This supervisory circuit becomes unnecessary, for instance, when working with a pulsed accelerator, with the delayed signal extracted from the accelerator used as stop pulse. In this case, the detector signal serves as start pulse and the latter is always accompanied by a stop pulse, apart from the noise pulses from the detector. Thus, the serious disadvantage of the start-stop converter is its low single-pulse capability at least of the start pulse input.

To put the start-stop principle into practical use a number of designs have been developed which are schematically given in Fig. 137*c* to *g*. Converters constructed according to Fig. 137*c* are similar to the overlap-type using multiplicative mixing (cp. Fig. 135*a*). In quiescent state, the capacitor C is charged to U_0-level and discharged through the resistor R after the switch S_1 has closed. As soon as the stop pulse opens switch S_2 the capacitor C recharges through the large resistor R_L.

In the circuit of *Weber* et al. [We 56a] the gated-beam tube V_2 (type 6 BN 6) is opened through grid *1* and blocked through grid *3* (Fig. 138*a*). Stop pulses arriving at a high repetition rate (3·7 MHz) do not result in output signals because they open only the grid *1* of valve V_1. Hence, a previously arriving start pulse opens the grids *3* of V_1 and V_2, actuating the discharge of capacitor C, which stops as soon as a stop pulse fully opens valve V_1, causing the grid *1* of tube V_2 to be blocked by the negative anode pulse of V_1. If the stop pulses arriving at a high repetition rate are applied to grid *3* of the gated-beam tube 6 BN 6, spurious coupling may cause interference and must be compensated [Fr 58]. In a similar circuit [Cu 60] with transistors used as switching elements, after switch S_1 closes, a linearly rising current at first flows through an inductance which charges the capacitor C through a diode, after switch S_2 opens. *Beghian* et al. [Be 58] developed an arrangement where a valve blocked in quiescent state takes up the function of both switches and acts as constant current source during discharge of the capacitor C. The positive start pulse which opens this valve is compensated when the negative stop pulse arrives, thus causing the switching valve to become blocked again.

In the modification according to Fig. 137*d* both switches are closed in quiescent state. The start pulse opens switch S_1 and initiates the discharge of capacitor C through resistor R. As soon as switch S_2 is opened by the stop pulse the voltage applied to capacitor C remains constant and the output pulse becomes flat-topped which facilitates registration. As soon as switch S_1 has closed again capacitor C recharges through the large resistor R_L. A circuit operating according to this principle [Le 55] is given in Fig. 138*b*, which incorporates a secondary-emission valve V_1 as switch S_1 and the pentode V_2 as switch S_2 through the internal resistance R of which the discharge of capacitor C is effected. The neon bulb V_3 serves to stabilise the working range of the circuit.

Fig. 137*e* shows a version where the capacitor C is charged through the resistor R after the start pulse has arrived and is discharged rapidly by the stop pulse. In the circuit given in Fig. 138*c* both valves are blocked in quiescent state [Pf 63]. The start pulse causes the capacitor C to be rapidly charged through valve V_1 and, after a delay time T, to be rapidly discharged through valve V_2 by the stop pulse. This produces a rectangular pulse of width T which is integrated in the subsequent stage. In an analogue circuit *Brun* et al. [Br 62b] employed a diode in place of the valve V_1, and instead of valve V_2 a diode in series with an n-p-n transistor. The arrangements of *Ivanov* and *Nesterenko* [Iv 62] and *Basire* et al. [Ba 63d] are entirely transistorised. In the version of *Oliver* et al. [Ol 67] a diode functions as switch S_1 in Fig. 137*e*, which

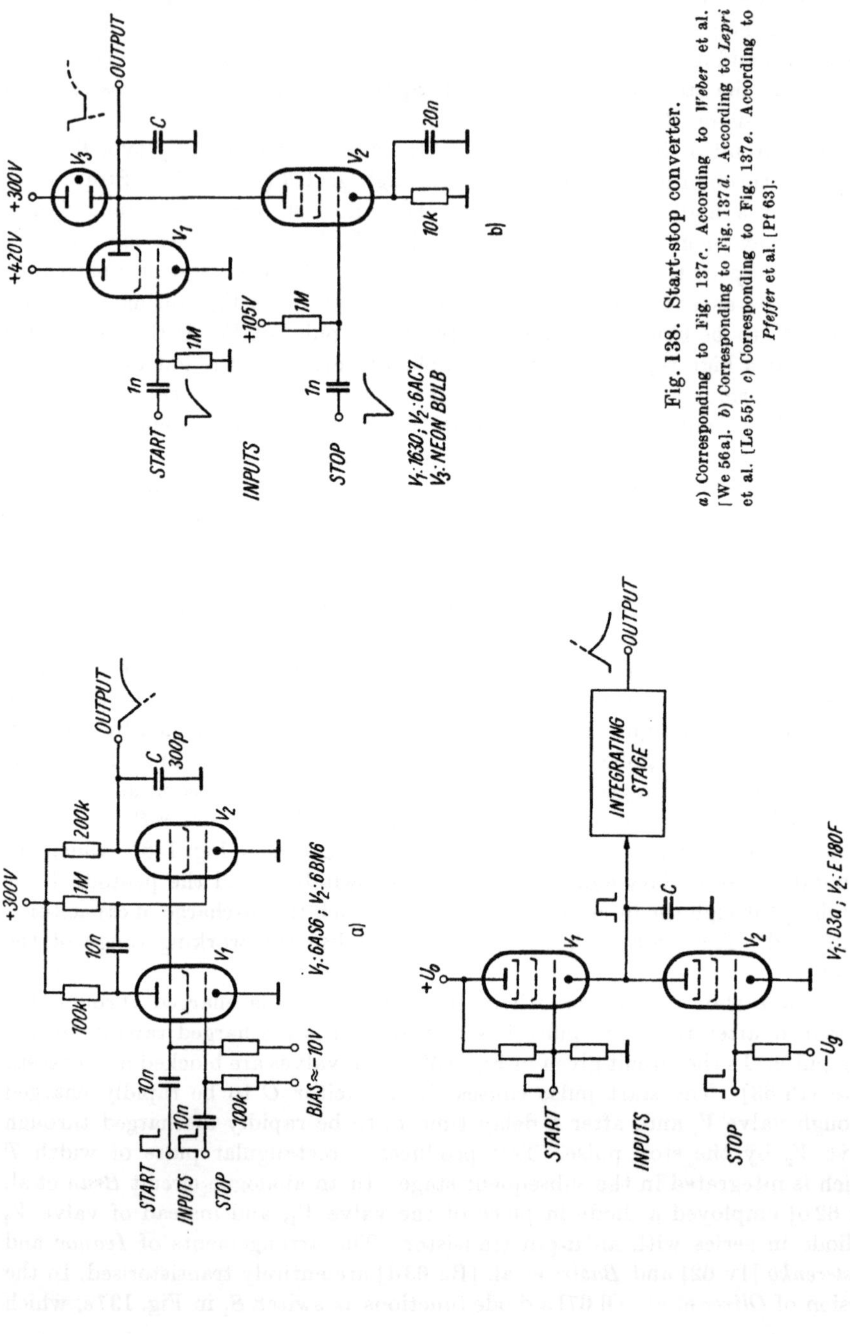

Fig. 138. Start-stop converter.

a) Corresponding to Fig. 137*c*. According to *Weber* et al. [We 56a]. *b*) Corresponding to Fig. 137*d*. According to *Lepri* et al. [Le 55]. *c*) Corresponding to Fig. 137*e*. According to *Pfeffer* et al. [Pf 63].

is turned on by the start pulse and is turned off by the stop pulse. In place of switch S_2 a diode is used which becomes conducting 300 ns after the arrival of the start pulse and discharges the capacitor C.

Fig. 137*f* gives a modification where both switches are closed in quiescent state. Charging of the capacitor C begins after switch S_1 is opened and stops when switch S_2 opens, which produces a flat-topped pulse, similar to the version shown in Fig. 137*d*. Capacitor C discharges only after switch S_1 closes. A transistor circuit [Ga 62c] based on this principle is given in Fig. 139*a*, where the negative start pulse blocks the n-p-n transistor T_1 and the positive stop pulse blocks the p-n-p transistor T_2. The start pulse is shorter than the stop pulse, so that the capacitor C discharges before T_2 again becomes conducting. Thus, the width of the output pulse is determined by the width of the start pulse. The circuit of *Efimchik* et al. [Ef 68] operates on the same principle. *Dardini* et al. [Da 67a] give a circuit where the stop pulse is delivered to the emitter of T_2. The width of the output pulse is determined by a slow-section controlled gate which prepares the start and stop pulse shaper for triggering. With *Maeder* [Ma 64c] a stretching circuit determines the length of the output pulse which keeps switch S_1 in Fig. 137*f* open. Suitable gating rejects undesirable events and permits an input count rate exceeding 10^6 pps.

The converter circuit by *Wieber* [Wi 63a] uses fast switching diodes, type HD 5000, in place of the transistors, where a constant-current source supplies the charging current for the mica-capacitor C. This gives an integral non-linearity well within ± 1 per cent. The instability of the transistorised constant-current source is less than $2 \times 10^{-4}/°C$. Measurements with a mercury pulse-generator yielded a resolving time of $2\tau_0 = 6$ ps and a temperature drift of less than 2 ps/°C. A further development of this type of converter is found with *Henebry* [He 66b].

Brafman [Br 65b] developed the arrangement shown in Fig. 139*b* which operates on the same principle. Here, the tunnel diodes TD_1 and TD_2 are in low-voltage state when switch S is open and TD_3 is in high-voltage state. The backward diodes BD_1 and BD_2 conduct and the transistors T_1 to T_5 are blocked. The start pulse switches the tunnel diode TD_1 to stable high-voltage state, blocks the backward diode BD_1, and opens the transistors T_1 and T_2. Diode D_1, conducting in quiescent state, is blocked and the collector current from transistor T_4 linearly dicharges the storage capacitor C_1. The stop pulse also switches the tunnel diode TD_2 to stable high-voltage state, opens the transistors T_3 and T_5, and blocks transistor T_4, interrupting the discharge of capacitor C_1. Instead, capacitor C_2 is linearly discharged, transistor T_6 transmits increasingly more current until the tunnel diode TD_3, with the tunnel diodes TD_1 and TD_2 connected at point A, switch to the low-voltage state. This restores initial condition, which causes the tunnel diode TD_3 then to switch to the high-voltage state, making the tunnel diodes TD_1 and TD_2 ready for the next switching action. Stretching time is determined by the magnitude of C_2 and R. Closing of switch S keeps TD_3 on low-voltage state, inhibiting both start and stop channels until a positive trigger pulse switches the tunnel diode TD_3. If no stop pulse arrives, the diode D_4 becomes conducting through the increasing negative voltage across C_1, transistor T_7 is blocked and a fictitious stop pulse is fed-in through D_5. The gated output appears only during stretch

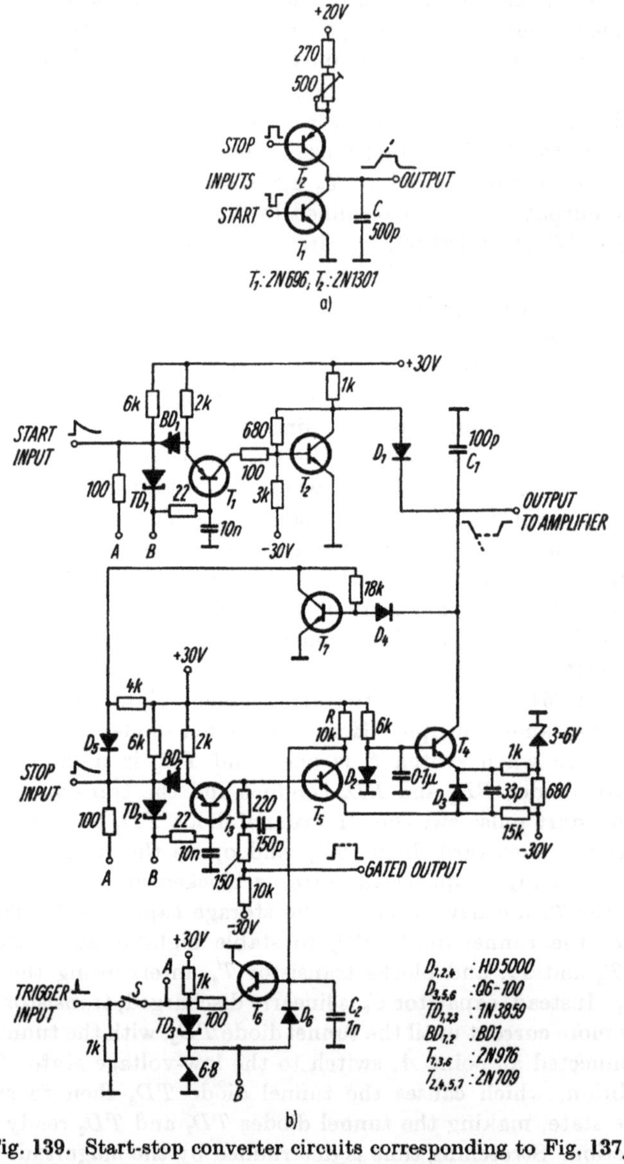

Fig. 139. Start-stop converter circuits corresponding to Fig. 137*f*.

a) According to *Gatti* et al. [Ga 62c]. *b*) According to *Brafman* [Br 65b].

time. In this circuit the conversion factor is variable between 100 mV/ns and 1 V/µs, with the resolving time amounting to $2\tau_0 = 5$ ps and an instability of 2·5 ps/°C. With the timing circuit given in Fig. 121, p. 195, a time shift of 100 ps was measured for pulse height variations of 2 to 20 mA at a pulse rise of 5 ns.

The version of *Tomlinson* and *Brown* [To 64a] shown in Fig. 140*a* operates analogue to the circuit in Fig. 137*g*, where the capacitor is charged as soon as switch S_1 opens and discharges when switch S_2 closes. In quiescent state transistor T_1 conducts and transistor T_2 is blocked. The bootstrap stage stabilises the charging current for capacitor C which results in better linearity.

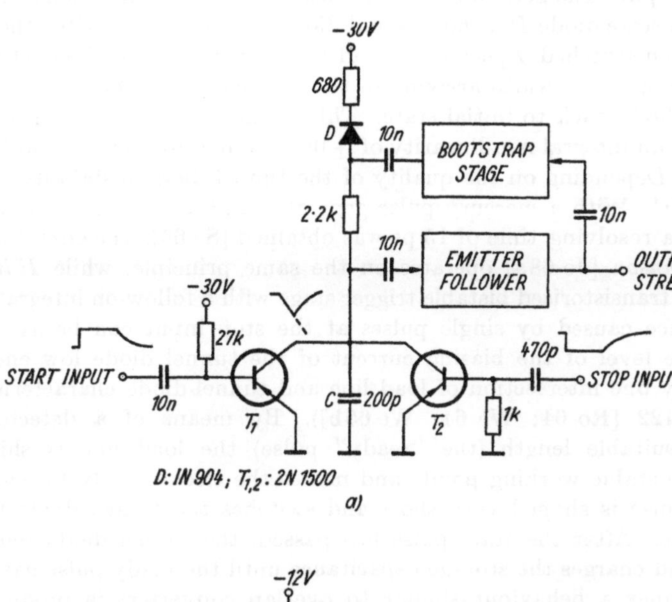

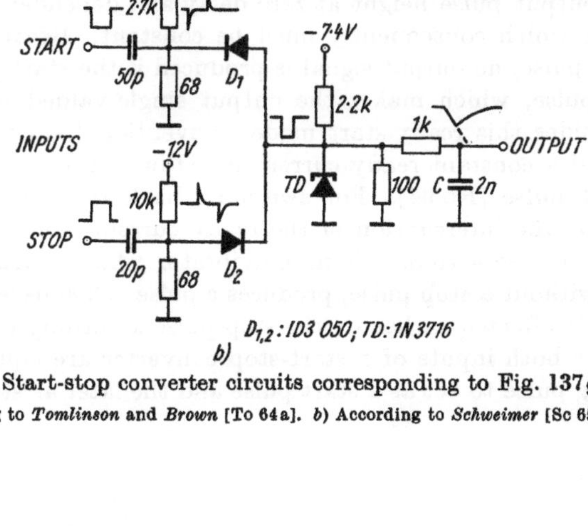

Fig. 140. Start-stop converter circuits corresponding to Fig. 137*g*.
a) According to *Tomlinson* and *Brown* [To 64a]. *b*) According to *Schweimer* [Sc 65].

The principle illustrated in Fig. 137*g* may be utilised by using a single switching element where the capacitor is charged between the opening and the closing of the switch S_1. This modification is particularly adapted to tunnel diode circuits, where a negative start pulse and a positive stop pulse switch the bistable tunnel diode trigger system back and forth (Fig. 140*b*). Proper function can be assured only when every start pulse is followed by a stop pulse, as is the case with pulsed accelerators. Further, trigger pulses must be kept short and have a definite height because they are included in integration [Br 61a; Zu 63; Ma 65; Sc 65]. With the fast gating system of the diodes D_1 and D_2 a high single-pulse repetition rate, up to 40 MHz, is possible at the stop pulse input. The germanium diodes used have a switching time of 0·1 ns. In quiescent state diode D_1 conducts and diode D_2 is blocked. After the tunnel diode has been switched, D_1 is blocked and D_2 conducts, to pass the next positive needle pulse of 1·3 ns width arriving in the stop channel, which then switches the tunnel diode back to initial state. This circuit has an input sensitivity of 50 mV, with an integral non-linearity of $\pm 0{\cdot}5\%$ and a differential non-linearity of $\pm 6{\cdot}5\%$. Depending on the quality of the tunnel diode instability amounts to $\leqq 0{\cdot}7\%/°C$. With a mercury pulse generator and a capacitance reduced to $C = 0{\cdot}2$ nF a resolving time of 15 ps was obtained [Sc 65]. The circuit arrangement of *Johnson* [Jo 68a] operates on the same principle, while *Il'in* et al. [Il 68] use a transistorised bistable trigger stage with a follow-on integrator.

Interference caused by single pulses at the stop input can be avoided by selecting the level of the biasing current of the tunnel diode low enough to produce only one intersection of load line and tunnel-diode characteristic (cp. Fig. 71, p. 122 [Ro 64; We 64; We 65b]). By means of a detector pulse shaped to suitable length (the "ready" pulse) the load line is shifted to produce two stable working points and makes the circuit ready for switching. The start pulse is shaped very short and switches the tunnel diode to high-voltage state. After the start pulse has passed, the tunnel diode remains in this state and charges the storage capacitance until the ready pulse has ceased. In this manner a behaviour similar to overlap converters is produced, i.e. maximum output pulse height at zero delay and dependence on length of the ready pulse, which consequently must be constant. Because of the shortness of the start pulse, no output signal is produced if the start pulse arrives before the ready pulse, which makes the output single-valued without supervisory circuit. Besides this ready-start mode, conventional start-stop operation can be effected if a constant ready current is fed-in at the start input, in addition to the start pulse [Ro 64]. For switching back the tunnel diode either the stop pulse or the interruption of the ready current can be utilised. *Meleshko* et al. [Me 67b] use a tunnel diode univibrator which is triggered by the start pulse and, without a stop pulse, produces a pulse of 300 ns length. Normally the switchback is effected earlier by the stop pulse according to the time spacing.

Sometimes both inputs of a start-stop converter are equivalent causing the first arriving pulse to act as a start pulse and the later as stop pulse. Fig. 141*a*

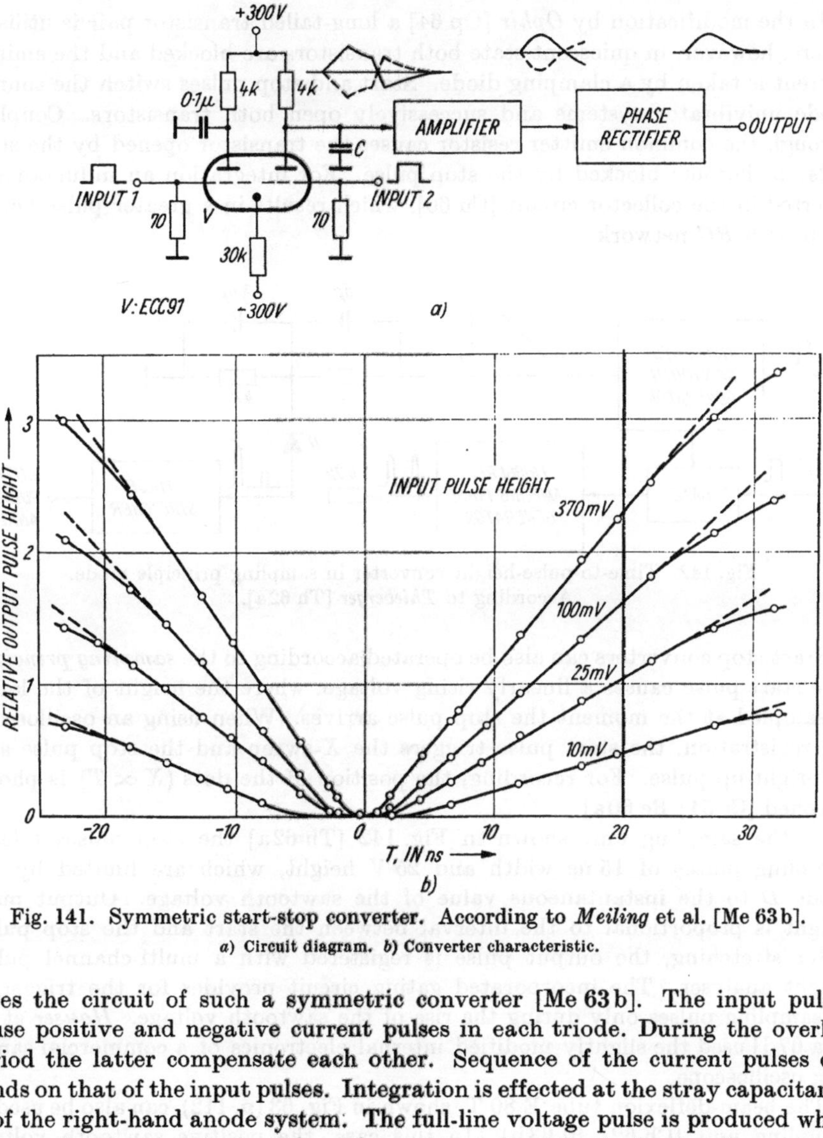

Fig. 141. Symmetric start-stop converter. According to *Meiling* et al. [Me 63b].
***a)* Circuit diagram. *b)* Converter characteristic.**

gives the circuit of such a symmetric converter [Me 63b]. The input pulses cause positive and negative current pulses in each triode. During the overlap period the latter compensate each other. Sequence of the current pulses depends on that of the input pulses. Integration is effected at the stray capacitance *C* of the right-hand anode system. The full-line voltage pulse is produced when input pulse *1* arrives before the pulse at input *2*, while the broken-line pulse is caused on reversed condition. After passing through a phase rectifier the output pulse height only depends on the absolute value of the delay time (Fig. 141*b*). The common cathode resistor has a feedback effect and serves to stabilize the working points. Because both valves operate in the steep part of the characteristic a high input sensitivity results, down to 10 mV. Curvature of the converter characteristic on $T \approx 0$ is caused by the rectifier.

In the modification by *Ophir* [Op 64] a long-tailed transistor pair is utilised, where, however, in quiescent state both transistors are blocked and the emitter current is taken by a clamping diode. Start and stop pulses switch the tunnel-diode univibrator systems and successively open both transistors. Coupling through the common emitter resistor causes the transistor opened by the start pulse to become blocked by the stop pulse. For integration an inductance is inserted in the collector circuit [Cu 60], which results in a greater pulse height than with *RC* network.

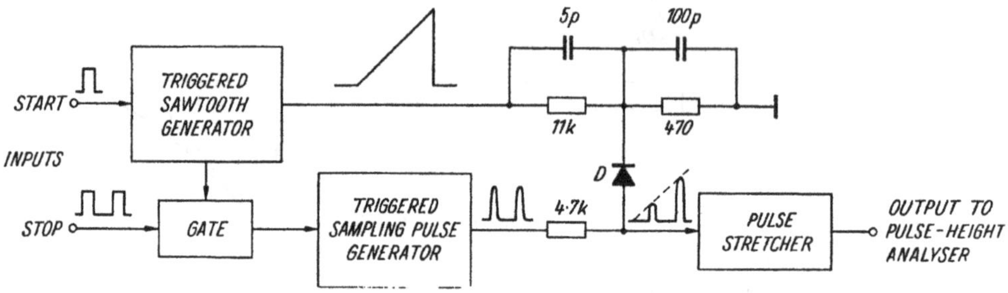

Fig. 142. Time-to-pulse-height converter in sampling principle mode. According to *Thieberger* [Th 62a].

Start-stop converters can also be operated according to the *sampling principle*. The start pulse causes a linearly rising voltage, where the height of the latter is sampled at the moment the stop pulse arrives. When using an oscilloscope for registration, the start pulse triggers the X-sweep and the stop pulse acts as bright-up pulse. For recording, the position of the dots ($X \propto T$) is photographed [Cr 54; Be 60a].

In the sampling unit shown in Fig. 142 [Th 62a] the stop pulses release sampling pulses of 15 ns width and 25 V height, which are limited by the diode D to the instantaneous value of the sawtooth voltage. Output pulse height is proportional to the interval between the start and the stop pulse. After stretching, the output pulse is registered with a multi-channel pulse-height analyser. The incorporated gating circuit provides for the triggering of sampling pulses only during the rise of the sawtooth voltage. *Hauser* et al. [Ha 67d] used the slightly modified internal electronics of a commercial sampling oscilloscope.

The beam-deflexion tube E 80 T, shown in Fig. 63 (p. 112), can also be used as sampling unit [Ch 62; Bl 63b]. In this case, the positive sawtooth voltage produced by the start pulse is applied to the control grid g_1 of the tube blocked in quiescent state. This causes the linear rise of the beam current. Sampling pulses are then applied to the deflexion plate D causing a brief incidence of the beam on the anode. The resulting current pulse is proportional to the interval between start and stop pulses, and is integrated in the anode circuit. Single pulses produce no output signal in this arrangement either. Using pulses from a generator a resolving time of $2\tau_0 = 7$ ps could be measured. This type of tube may also be operated in conventional start-stop mode [Bl 63b], where with the equally biased deflexion plates D and D' the linearly rising beam current always reaches the anode and is cut off by one of the deflexion plates on arrival of the stop pulse. Without integration, a linear-rise voltage pulse is produced at the load resistor with a height corresponding to the interval between the input pulses.

Bloess and *Münnich* [Bl 64] also give a description of a sampling-principle operated transistorised time-to-pulse-height converter which features a linearity range variable between 70 ns and 1 ms. Here, a sampling pulse is only produced on coincidence which avoids output ambiguity. With long-tailed pairs an instability amounting to $< \pm 10$ ps/18 h and < 20 ps/°C was achieved.

The use of the sampling principle has the advantage of covering a wide time range, amounting to approximately 12 orders of magnitude [Th 62a], and the possibility of registering more than one stop signal for every start pulse [Fi 60a].

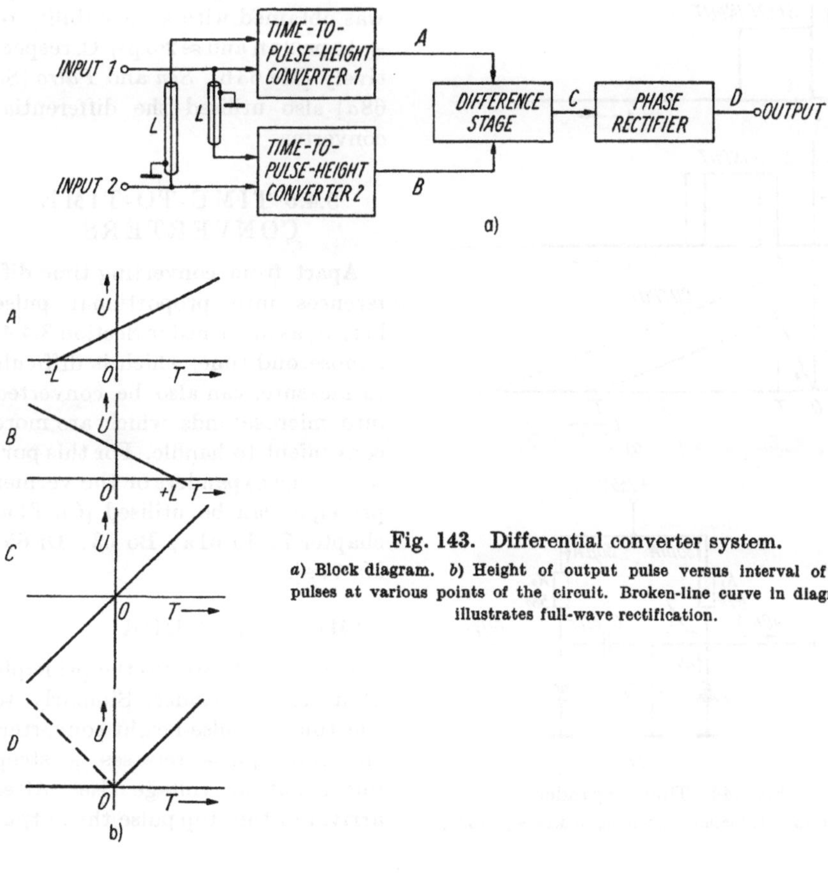

Fig. 143. Differential converter system.

a) Block diagram. *b)* Height of output pulse versus interval of input pulses at various points of the circuit. Broken-line curve in diagram *D* illustrates full-wave rectification.

The seldom occurring signals are, therefore, used for triggering the sawtooth voltage and the frequently arriving signals for releasing the sampling pulses after a suitable dalay. When using such high-speed multistop systems the after-pulses of the photomultiplier can cause interference, as *White* [Wh 67] has shown.

DIFFERENTIAL CONVERTER SYSTEM

As is the case with differential coincidence circuits (cp. Section 3.4.3), a differential converter arrangement can also be assembled from any two time-to-pulse-height converters. In operation, the input pulses are alternately delayed by the interval L (Fig. 143*a*). After difference formation and phase rectifying, the converter characteristic shown in Fig. 143*b* is obtained. *Bay* and *McLernon* [Ba 59a] used a bridge-circuit, while *Chery* [Ch 60a] and *Tintă* [Ti 62] employ 6 BN 6 — overlap converter stages for their arrangements. With two symmetrical start-stop converters, according to Fig. 141*a*, employed in differential connection, a resolving time of $2\tau_0 = 50$ ps was obtained with an instability of ± 15 ps/24 h and $\approx 2{\cdot}5$ ps/°C, respectively [Me 63b]. *Sen* and *Patro* [Se 68*a*] also utilised the differential converter.

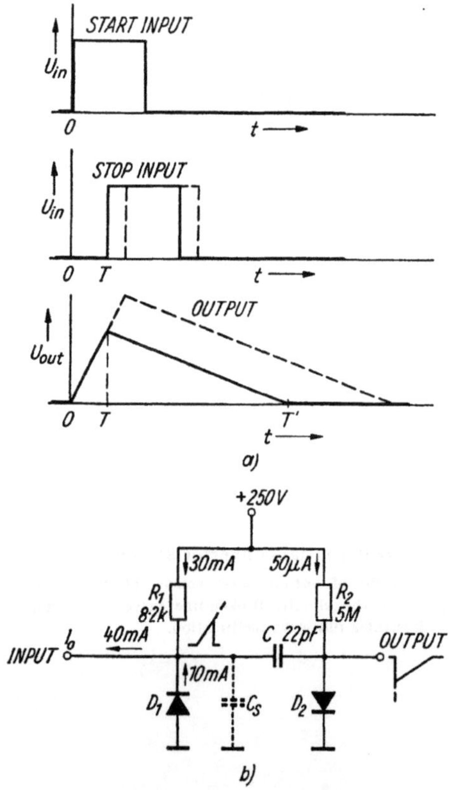

Fig. 144. Time Expander.
a) Principle. *b*) Circuit. According to *Moody* [Mo 52a].

3.4.5 TIME-TO-TIME CONVERTERS

Apart from converting time differences into proportional pulse height, as described in Section 3.4.4, nanosecond time, which is difficult to measure, can also be converted into microseconds which are more convenient to handle. For this purpose, time expanders or the vernier principle can be utilised [Ch 61a, chapter 7; Li 61a; Bo 63; Dr 65; Ma 67a].

TIME EXPANDER

Fig. 144*a* illustrates the principle of a time expander. Similarly to the time-to-pulse-height converter, the start pulse releases a steep linear output voltage rise. After arrival of the stop pulse the output voltage shows a slow linear drop until it reaches zero after the time $T' \propto T$. Thus, time-to-height and height-to-time conversion is effected in a single step. The result is a triangular pulse which is further processed as in a *Wilkinson* type multi-channel pulse-height analyser (cp. Section 3.6.3).

In the circuit by *Moody* [Mo 52a] the initially flowing current I_0 is cut off by the start pulse (Fig. 144*b*). This blocks the diode D_1, and the voltage at capacitor C and capacitance C_s at first rises linearly against the operating voltage. Diode D_2 conducts. On arrival of the stop pulse, the latter turns on the current I_0 again which causes the diode D_1 to become conducting, rapidly discharging the stray capacitance C_s. The storage capacitor C remains charged and diode D_2 is blocked by the negative potential step. Capacitor C_2 discharges with the high time constant R_2C in direction to operating voltage. This stops as soon as the diode D_2 conducts again.

Ferguson et al. [Fe 62] describe a time expander where the storage capacitor is charged with a 20 mA current and discharges with 80 μA, resulting in an expansion factor of 250. An expansion factor of 1,000 is used by *Ettinger* [Et 67] to measure cloud height with a laser radar (cp. Section 5.5). *Thenard* and *Victor* [Th 67] count the time, expanded by the factor 100, with a 100 MHz oscillator and attain with 2,048 channels a channel width of 100 ps with an instability of less than 10^{-4}/°C.

Expansion factors up to 10^4 and better stability can be achieved if, between the arrival of the start and the stop pulses, at first a low capacitance and then a high capacitance is charged to the same voltage level across the same resistor. With this mode, the conversion factor is determined only by the ratio of the two capacitances [Go 62c].

VERNIER PRINCIPLE

In Fig. 145 the basic principle is given [Ba 57]. Input pulses arriving at the interval T actuate two h.f. oscillators or repetitive pulse generators having a slightly differing period $T_1 > T_2$, e.g. by 1%. An output signal is produced when the phases of the two oscillators coincide:

$$T' = n\,T_1 = \frac{T_1}{\Delta T}\,T = k\,T\,, \tag{111}$$

where k is the conversion factor and $\Delta T = T_1 - T_2$. Thus, the interval T' from the output pulse, released on arrival of the input pulse *1*, is proportional to the interval T and can be measured with a time-to-pulse-height converter suitable for microseconds [Co 56a]. Another way is to count the number of periods n to phase coincidence [Le 57] and get the time T digitally in units of ΔT (cp. Section 3.4.6). A review of vernier principle measuring setups is also found with *Rekhin* et al. [Re 67, chapter 8].

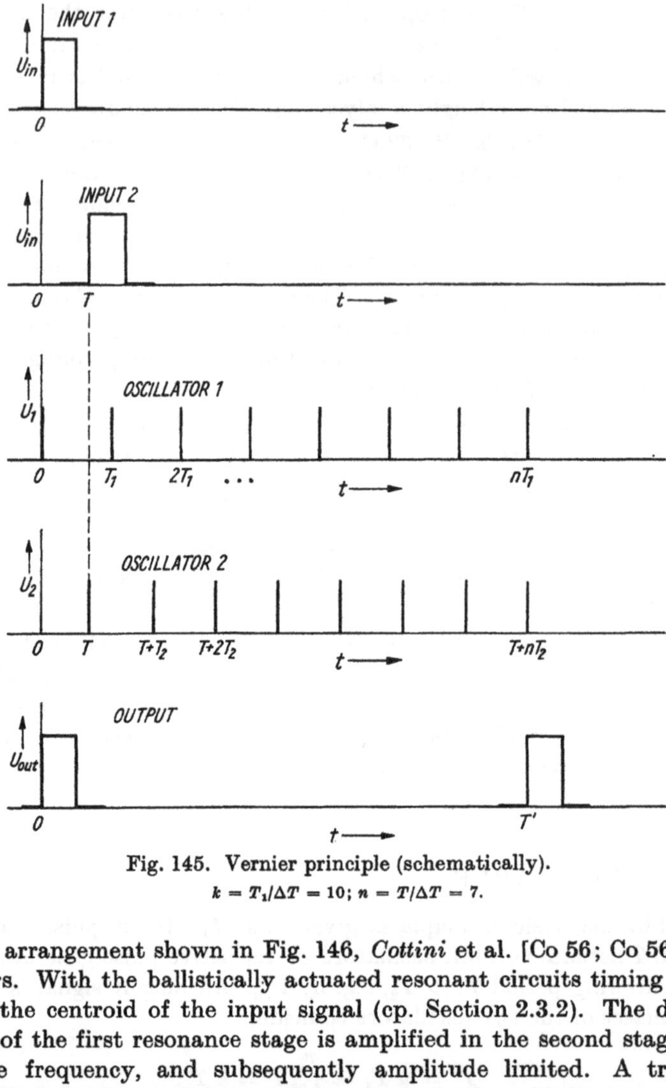

Fig. 145. Vernier principle (schematically).
$k = T_1/\Delta T = 10$; $n = T/\Delta T = 7$.

In the arrangement shown in Fig. 146, *Cottini* et al. [Co 56; Co 56a] use h.f. oscillators. With the ballistically actuated resonant circuits timing is effected through the centroid of the input signal (cp. Section 2.3.2). The damped oscillation of the first resonance stage is amplified in the second stage tuned to the same frequency, and subsequently amplitude limited. A transformer-coupled diode bridge modulator follows, where both oscillations are combined. Of the resulting low-frequency oscillation the first zero-crossing, which is independent of oscillation amplitude, is passed through a low-pass filter for separating the h.f. components and subsequently measured with a millivolt-sensitive discriminator.

In pulsed accelerator measurements, the h.f. voltage of the accelerator pulsation is fed directly to the diode bridge modulator, and only one resonant circuit is necessary for the detector pulse signal [Ch 57; An 63]. Input sensitivity of this setup is approximately 100 mV, and a resolving time of 50 ps was obtained with generator pulses. The conversion factor k depends on the frequency difference of the two resonant circuits and, because of their instability, is limited to a maximum of $k \approx 100$. The whole duration of the period T_1 is available as measuring range. Registration of input pulses with an interval of $T > T_1$ is prevented by a gate-circuit actuated by a supervisory circuit, which are not shown in Fig. 146. A disadvantage is the long dead time and the malfunction of the circuit if a single pulse, for instance, an after-pulse from the photomultiplier, arrives during conversion time.

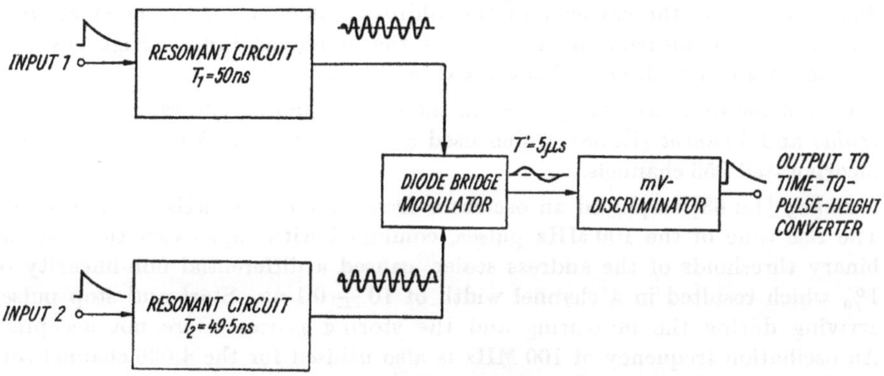

Fig. 146. Time-to-time converter, according to vernier principle. According to *Cottini* and *Gatti* [Co 56a].

Suppressor diodes in the resonant circuits can prevent the single-pulse response until the supervisory circuit is actuated and cancels the resonant circuit block, e.g. for 4 μs [Co 58a; Co 59; Co 59a]. On the other hand, the photomultipliers are blocked during this period. Instability was reduced to <50 ps per day by incorporating a regulating system. Generator pulses fed-in at the two photomultipliers every tenth of a second produce a regulating voltage at the discriminator which is arranged behind the time-to-pulse-height converter. In one of the resonant circuits a diode is used which acts as variable capacitance. The diode is controlled by the regulating voltage and thus stabilises the oscillator frequency.

Artemenkov and *Babykin* [Ar 60a] introduced a temperature stabilisation to 0·5 °C in place of the regulating circuit.

3.4.6 TIME-TO-DIGITAL CONVERTERS

The simplest method of direct time measurement is to produce time markers of a fixed interval T_0 and compare the interval to be measured [Ma 67a; Re 67, chapter 6]. For time marking a specific point of a pulse or a specific phase of a sine wave to which the time analyser responds ("machine time") can be used. The number of accommodated time markers represents time in T_0-units in digital form, which can be conventionally registered with a scaler, a magnetic tape or a ferrite core matrix. Accuracy of the time measurement is determined by T_0 and may be improved to approximately $T_0/100$ by applying the vernier principle. Time-to-digital converters are primarily utilised for time-of-flight measurements (cp. Section 5.2) [Hi 56; Sc 56]. In principle, the converter consists of an oscillator, either free-running or triggered by the start pulse, and a fast address scaler for counting the number of cycles between the start and the stop pulse (cp. Fig. 133*c*, p. 217).

This method has the significant advantages of a wide dynamic range, which depends only on the capacity of the address scaler and of the memory unit, and of a good linearity independent of the employed range. High stability can be obtained with crystal-stabilised oscillators.

One of the first circuits operable in the nanosecond region was developed by *Hillas* and *Tennent* [Hi 58], which used a free-running 20 MHz oscillator and incorporated 255 channels.

Lundy [Lu 63] employed an oscillator frequency of 100 MHz in his version. The rise time of the 100 MHz pulses, combined with slight variations of the binary thresholds of the address scaler, caused a differential non-linearity of 1% which resulted in a channel width of $10 \pm 0{\cdot}1$ ns. Start and stop pulses arriving during the measuring and the storing process were not accepted. An oscillation frequency of 100 MHz is also utilised for the 4,096 channel sets of *Whittaker* [Wh 64], *Grebenyuk* and *Kolpakov* [Gr 67], and the commercial set described by *Durand* and *Giraud* [Du 63b] which operates with a maximum of 65,536 (2^{16}) channels. A repetition frequency of 200 MHz was used by *Ivanov* [Iv 66].

To achieve a channel width of 1 ns, an oscillator frequency of 1 GHz and a respectively fast scaling circuit would be necessary. There are several ways of bypassing this difficulty. *Meyer* [Me 63a] employed for this time-of-flight analyser of 8,192 (2^{13}) channels the arrangement shown in simplified presentation in Fig. 133*c*, p. 217. Interpolation between two oscillator pulses arriving at an interval of 8 ns is handled with a chronotron consisting of the coincidence circuits C_0 to C_7. By means of a zero time delay switch, not shown in Fig. 133*c*, the whole dynamic range can be displaced against the start pulse in steps of 8 ns up to a time limit of 2^{13} ns.

An improved modification of this analyser [Me 66b] with a maximum range of 2^{17} ns and a preset delay of 2^{18} ns, contains an equalising system which reduces the deviation of the channel width from its mean value to ± 4 ps, against the former ± 70 ps without the system.

Durand et al. [Du 63c] describe a transistorised set with four stop pulse inputs and 262,144 (2^{18}) channels of $\geqq 1$ ns width. It has an analysis dead time of 200 to 300 ns, and the start delay is adjustable to $2^{12} \times 64$ ns (262 μs). Interpolation between the start pulse arrival and the following pulse from the free-running oscillator is effected with the circuit shown in Fig. 147*a*. An analogue interpolation arrangement is employed to measure the interval from the stop pulse to the preceding clock pulse. To obtain better stability a 62·5 MHz oscillator was used. With a delay time differing by 8 ns, the clock pulses are passed through the two transmission lines to the interpolation circuit and the mixer stage. At the output of the latter the pulse spacing is, therefore, only 8 ns. The two 62·5 MHz pulse trains are delivered with successive differences of 1 ns to the gating stages G_1 to G_{16}, which are open in quiescent state. Every 8 ns the binary stages B_1 to B_8 are switched backward or forward by the pulses of the two trains. This results in the voltage curves given in Fig. 147*b*. On arrival of the start pulse the gates G_1 to G_{16} are closed and the momentary state of the binary stages is binary coded and stored. Time relation for the opening of the gating circuits G_{17} to G_{19} is selected to exclude a coincidence between the opening of the gating circuit G_{19} and a clock pulse on any position of the start pulse. Otherwise ambiguity of the coarse address counting of 8 ns would result. Total address can be read between two clock pulses without having to stop the address scaler. This "read in flight" possibility permits the measurement of more than one stop pulse for every start pulse, which considerably enhances the effectivity of the system.

Engelmann [En 64] utilised a 500 MHz oscillator and two 500 MHz quinary scalers operating with a mutual delay of 1 ns, where one scaler counts the positive half-cycles and the other the negative half-cycles. A 100 MHz pulse train is delivered at the scaler output, which is then further processed. Reading of the quinary scalers is decade coded and supplies a fine-coding to 1 ns. Counting ambiguity, that may be caused by random phase relations, is prevented by synchronising gates in front of the quinary scalers. By means of frequency doubling and of two current-mode switches the channel width can be reduced to $0{\cdot}5 \pm 0{\cdot}25$ ns.

Instead of the scalers, *De Lotto* et al. [Lo 64; Co 67f] used a series of phase controlled oscillators, with frequencies and phase positions that form a binary code with redundancy to obtain unambiguous indication also on the channel boundaries. Time reading is obtained by sampling the oscillator voltages, with positive and negative voltage representing L and 0, respectively. It is shown that the use of a free-running master oscillator that is not correlated to the start and stop pulses, has the decided advantage of statistical equalisation of the channel width [Ga 63]. A relative standard deviation of the channel width $\sigma = \varepsilon^2/\sqrt{n}$ is obtained if n is the number of time intervals in a cycle and ε their relative standard deviation.

With a cycle duration of 0·8 ns, rather unequally divided in $n = 16$ intervals by a phase measuring device, the channel width amounts to 50 ps, at $\sigma = 0{\cdot}25\%$ with $\varepsilon = 10\%$. Crosstalk between the start and stop channels, however, can cause a periodic non-linearity up to 3%. The measuring range may be up to $2^{18} \times 50$ ps (13 μs) [Co 67f].

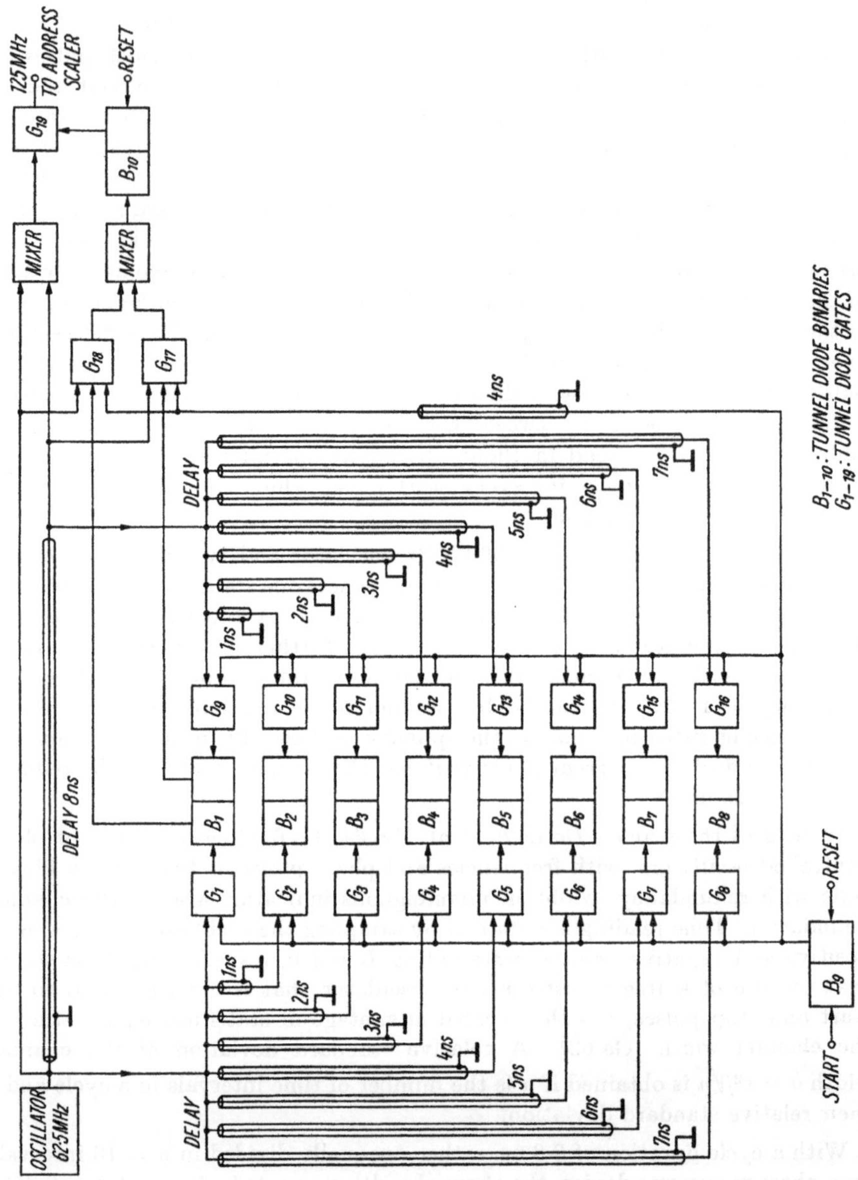
OSCILLATOR 62.5 MHz
DELAY 8ns
DELAY
1ns
2ns
3ns
4ns
5ns
6ns
7ns
MIXER
MIXER
125 MHz TO ADDRESS SCALER
RESET
RESET
START
B_{1-10}: TUNNEL DIODE BINARIES
G_{1-19}: TUNNEL DIODE GATES

a)

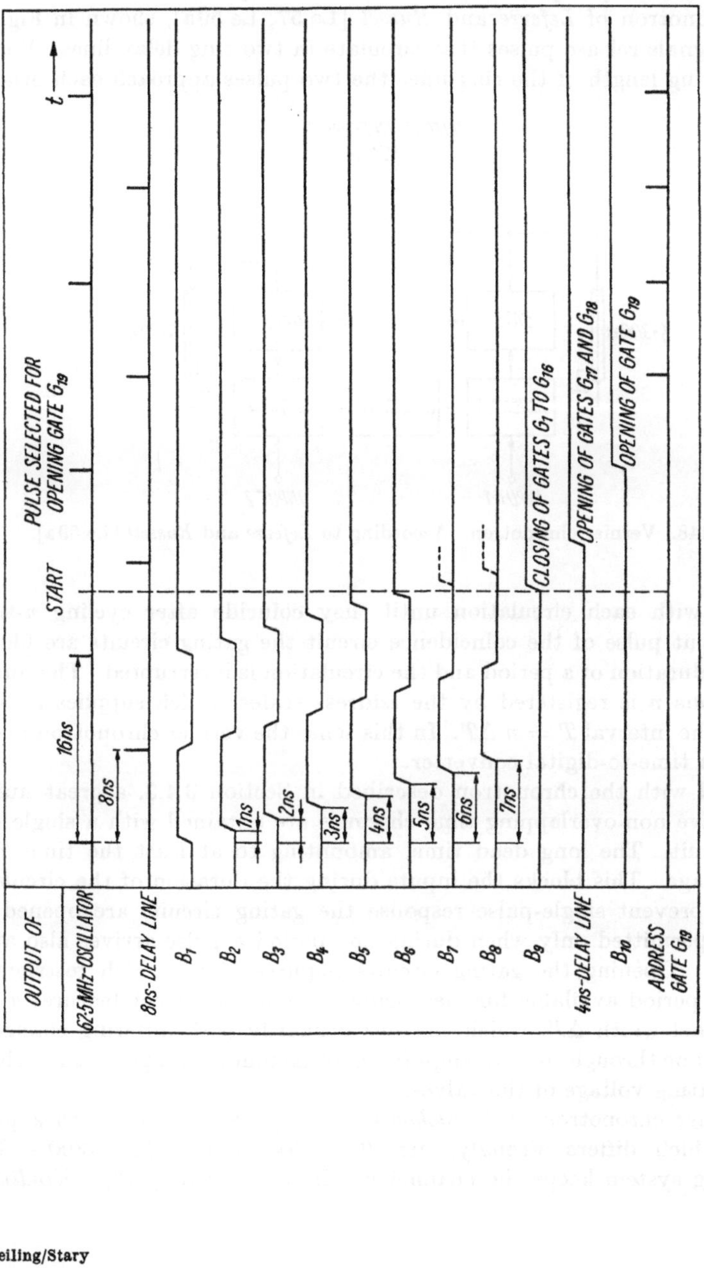

Fig. 147. Interpolation system of a time-to-digital converter. According to *Durand* and *Giraud* [Du 63c].
***a)* Block diagram. *b)* Voltage curve.**

VERNIER CHRONOTRONS

With the vernier principle (Fig. 145, p. 238) intervals that are shorter than the spacing of the clock pulses can be measured by converting the interval T between the two input pulses into the considerably longer time T'. In the vernier chronotron of *Lefevre* and *Russel* [Le 57; Le 59a], shown in Fig. 148, the input signals release pulses that circulate in two ring delay lines. Because of the differing length of the ring lines the two pulses approach each other by

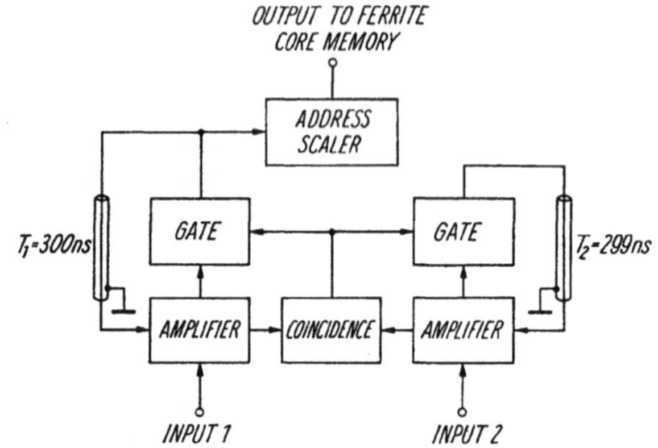

Fig. 148. Vernier chronotron. According to *Lefevre* and *Russell* [Le 59a].

$\Delta T = 1$ ns with each circulation until they coincide after cycling n-times. By the output pulse of the coincidence circuit the gating circuits are blocked beyond the duration of a period and the circulation is interrupted. The number of circulations n is registered by the address scaler, which supplies a digital readout of the interval $T = n\,\Delta T$. In this sense the vernier chronotron may be considered a time-to-digital converter.

Compared with the chronotron described in Section 3.4.3, a great number of consecutive non-overlapping time channels are obtained with a single coincidence circuit. The long dead time, amounting to at least the time T', is a disadvantage. This blocks the inputs during the duration of the circulating event. To prevent single-pulse response the gating circuits are opened and circulation permitted only when during one period a pulse arrives also at the other input. Opening the gating circuits requires 30 ns, which restricts the part of the period available for measuring. The arrangement features a very stable channel width ΔT, which changes primarily with differing changes of the transit time through the two amplifiers, for instance, by 2 ps on a 1% change of the operating voltage of the valves.

The vernier chronotron of *Kurashov* et al. [Ku 58] operates with a period duration which differs strongly, viz. $T_1 = 200$ ns and $T_2 = (800 - 1)$ ns. A stabilising system keeps the channel width constant to $\pm 1\%$. *Kindlmann* and *Sunderland* [Ki 66a] obtained a channel width instability and differential non-linearity of less than 0·1% by phase-locking each loop to a 500 MHz crystal-controlled clock. Tunnel diodes incorporated in the recirculation amplifier are biased in such a manner that the pulses returning from the delay line can only trigger the tunnel diode during a limited portion of the oscillator cycle.

Limitation to intervals of less than the duration of a period can be overcome when using the principle illustrated in Fig. 149. When both oscillators have

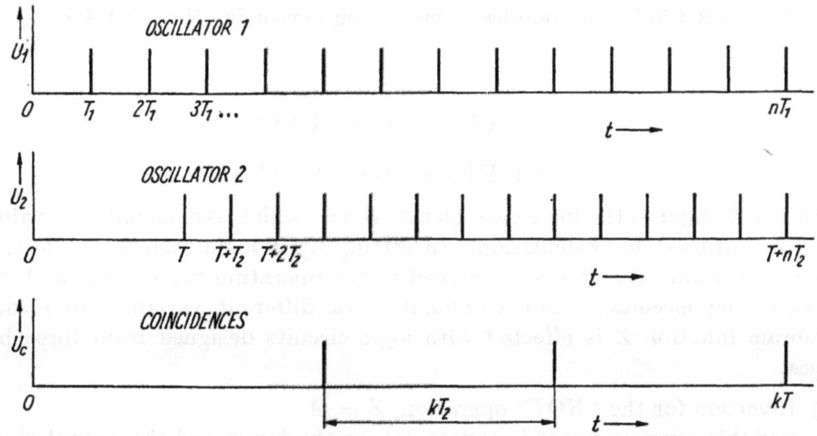

Fig. 149. Extended vernier principle for measurements exceeding the oscillator period. $k = T_1/\Delta T = 5$; $n = T/\Delta T = 13$.

run through the same number of periods n and coincidence applies additionally, with $T_2 = T_1 - \Delta T$, the following equation holds:

$$n\,T_1 = n\,T_2 + T = n\,T_1 - n\,\Delta T + T \tag{112}$$

giving

$$n = \frac{T}{\Delta T}. \tag{112a}$$

This means that the period number n supplies the interval to be measured in units of ΔT. This would require the continuous measurement of the number of periods of both oscillators, compared whenever a coincidence occurs until the same number of periods is found for both oscillators. In the vernier chronotron of *Thenard* and *Victor* [Th 66a] this comparison is made in a scaler where the pulses from the start oscillator effect count-up until the arrival of the stop pulse and where the coincidences (beat frequency nodes) cause its count-down to zero. With a period duration of $T_1 = 50$ ns and $T_2 = 49$ ns it has a measuring range of 3,000 channels of 1 ns width. Instability amounts to 3×10^{-4}/°C.

This method, however, has a considerably increased dead time, and it would be more advantageous to utilise two address scalers differing in value. In that case, the address scaler *1*, stopped by pulse *2*, counts the number of periods of the oscillator *1* up to the arrival of pulse *2*. The address scaler *2*, stopped by the coincidence circuit, counts the number of periods of oscillator *2* up to the occurrence of the first coincidence. Is the address scaler *2* advanced by one unit through every pulse and the address scaler *1* by k units, the address scaler *1* supplies the coarse address, while the address scaler *2* handles the interpolation between 0 and $(k - 1)$. Both addresses can be combined to one. The vernier chronometer of *Cresswell* and *Wilde* [Cr 64] works according to this principle. With a dynamic range of 50 ns, $T_1 = 5$ ns and a conversion factor of $k = 10$ (cp. Section 3.4.5) it only requires a maximum measuring time of 100 ns.

3.5 LOGIC CIRCUITS

3.5.1 PRELIMINARIES

In switch algebra the logic variables A, B, etc. which assume only the values 0 or L are utilised for calculation. In setting up a logic system by electronic means the values 0 and L are assigned to the operating states "on" and "off" of switching circuits. Logic combination of different variables to form an algebraic function Z is effected with logic circuits designed from three basic types:

(i) Inverters for the "NOT"-operation, $Z = \bar{A}$.
For this purpose phase inverters, where the input and the output signals are of opposed polarity, are employed.

(ii) The "OR"-gate for logic summation, $Z = A + B$.
Gating circuits, where an output signal is delivered when at least one input signal has arrived, serve for this purpose.

(iii) The "AND"-gate for logic multiplication, $Z = A \times B$.
This is handled with a gating circuit which produces an output signal when input signals arrive at every input. Its function corresponds to the coincidence circuit (cp. Section 3.4.3).

Although these circuits are also used for some applications described in Chapter 5, for instance, in counter telescopes for measurements of elementary particles (Section 5.3), their principal application is in computer technique. For digital computers identical logic circuits with diodes, tunnel diodes, transistors and rectangular ferrites are used in large numbers, with their switching speed influencing the operational speed of the whole computing machine. The complexity of these apparatuses makes demands on basic circuitry, which differ from the requirements of other applications and sometimes make a compromise necessary. Circuitry for this purpose must lend itself to easy and inexpensive manufacture, i.e. it must contain as few components as possible which must be comparably cheap. A setup must be compact, require little power and need only a few supply voltages. Wide tolerances of the components must be possible, with reliable function across a wide range of temperatures and a long service life. However, the demand for a high operational speed conflicts, for instance, with cost and power demand of a circuit, and is therefore frequently of secondary consideration. In applications where a high time resolution is required the other demands become secondary against the working speed of the circuit.

Because computer circuitry is not included in this book, only a few specialised circuits, as they apply for fast pulse-height and time analysers, shall be dealt with, namely, linear gating circuits and pile-up detection circuits.

3.5.2 LINEAR GATING CIRCUITS

A linear gate [Mi 56, chapter 14; Ne 66, chapter 11] represents an "AND"-gate with an asymmetric input (Fig. 150*a*). In ideal function, an output signal is given only if an input signal is present at both inputs, with the output pulse-height proportional to the height at the signal input (Fig. 150*b*). A certain

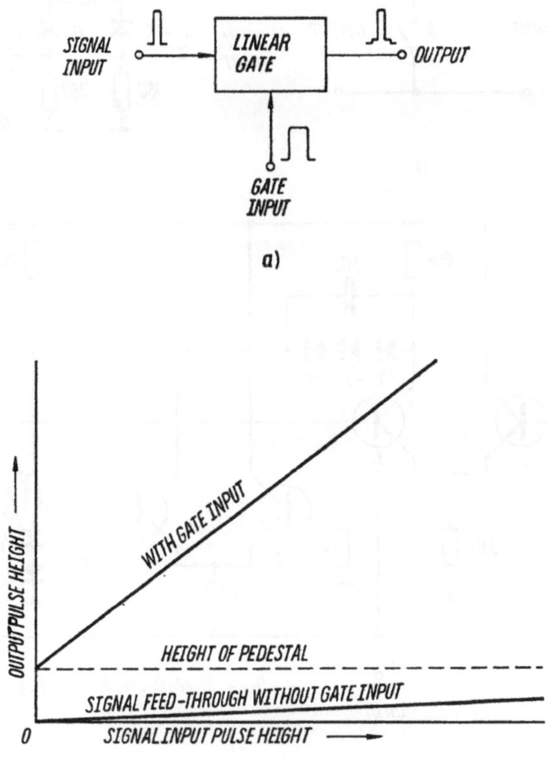

Fig. 150. Linear gate.

a) Principle. *b*) Transmission characteristic.

fraction of the gate input pulse, however, usually appears at the output in form of a pedestal, which does not interfere provided its height is constant. Parasitic capacitive coupling sometimes causes switching transients which become superposed on the pedestal [Ha 66b]. These can be eliminated by moderate integration of the output signal. In absence of a gate input a feedthrough of the input signal occurs which is also kept as low as possible.

Diode gating circuits are generally faster than transistor gates, but frequently exhibit unsatisfactory linearity and require a higher signal voltage.

PARALLEL SWITCH GATE

Fig. 151 *a* gives the principle of this gate circuit, where the switch *S* is closed in quiescent state, and opened by the gating pulse to pass the input signal.

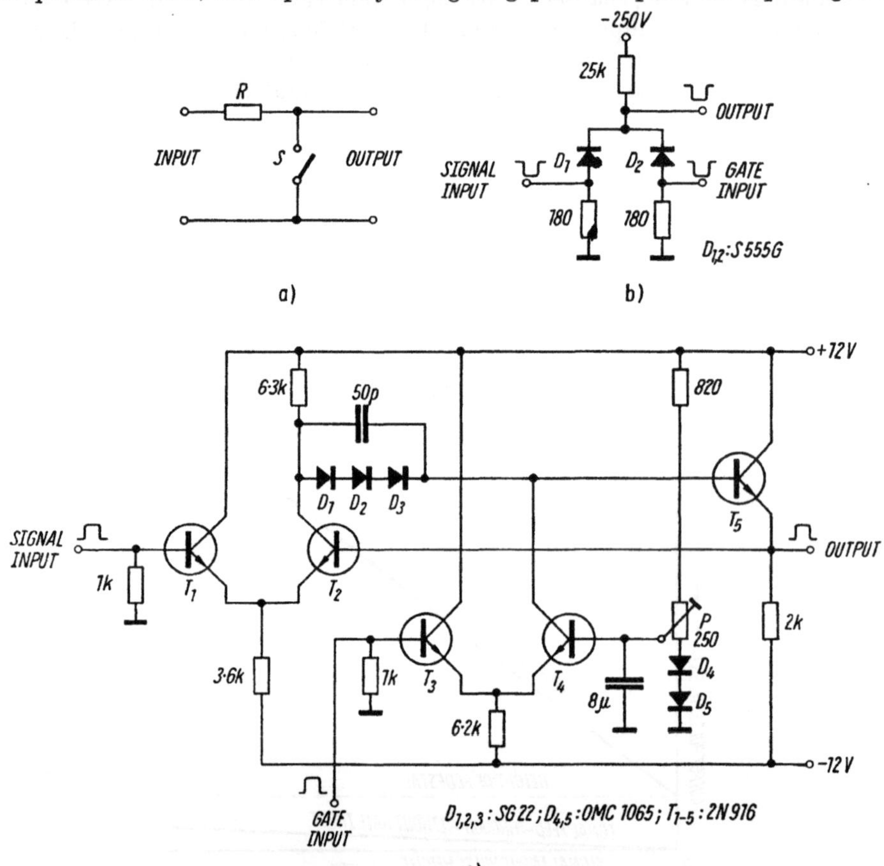

Fig. 151. Parallel switch gate.

a) Principle, circuit corresponds to open gate. *b)* Diode gate. According to *Garwin* [Ga 59a]. *c)* Transistor gate. According to *Elad* and *Rozen* [El 65].

In the very simple circuit of *Garwin* [Ga 59a] the shunt diode D_2 takes the current of D_1 when no gating pulse is given in (Fig. 151*b*). When a gating pulse of sufficient height blocks the diode D_2, the output pulse height depends linearly on input pulse height. With gating pulses of 20 ns width and input pulse heights of 0·2 to 10 V a linearity deviation occurred of only 2%. Similar behaviour obtains when a transistor serves as shunt switch [Ga 61].

The gating circuit shown in Fig. 151*c* was developed by *Elad* and *Rozen* [El 65] and contains a negative feedback amplifier with the transistors T_1, T_2 and T_5, having a highly attenuated loop gain in quiescent state through the long-tailed transistor pair current-switch T_3, T_4. The saturated transistor T_4 is shunted across the load resistor of the amplifier and is blocked by the gating pulse. With open gate the loop gain is high and total gain approaches unity, which provides for good linearity and a fast response. Due to the d.c. coupling the pulse repetition rate can amount to 1 MHz. Across a pulse height of 0·1 to 5 V non-linearity is less than 1%. On an input pulse height of 5 V, feedthrough amounts to 20 mV, while the pedestal, adjustable with the potentiometer P, has a height of 40 mV. To enable negative pulses to be handled the gate can be constructed with p-n-p transistors. The gating circuit of *Manfredi* et al. [Ma 67c] has a pedestal-compensating parallel circuit and attains a differential non-linearity of $\pm 0{\cdot}2\%$ with a dynamic range of 0 to 10 V.

SERIES SWITCH GATE

In principle, the switch S, located in the path of the signal, is open in quiescent state, but is closed by the gating pulse (Fig. 152*a*). When using a pulse transformer, the signal transmission can be prevented with blocked diodes which interrupt the current flow through the primary winding. To avoid the formation of a pedestal when the diodes open, either a compensation winding [Ch 60c] is employed or the gating pulse is symmetrically fed-in, as shown in Fig. 152*b* [Va 60]. At a minimum opening time of 40 ns this gating circuit operates linearly up to 11 V. Pedestal height was 70 mV and feedthrough amounted to 0·3%. *Coli* and *Serra* [Co 63d] use a version where a long-tailed transistor pair is utilised for switching the diode from blocked to conducting state.

A two-way switching circuit is obtained by connecting an open and a blocked transistor in parallel, which change-over for opening the gating circuit [Ha 62].

Transmission line transformers are employed by *Gorodetzky* et al. [Go 62a] for the modification given in Fig. 152*c*. Input pulses of a height under 5 V cannot pass the blocked diodes. Gating pulses exceeding 6 V height open the gate and compensate each other at the output. The residual height of the pedestal amounts to 30 mV. Positive input signals are transmitted by the diode D_2, while a negative input signal is transmitted by the diode D_1.

In Fig. 152*d* the circuit of *Papadopoulos* [Pa 65a] is shown where the transistors T_1 and T_3 are blocked in quiescent state. The gating pulse drives them into the saturation region with the transistors T_2 and T_4 acting as current generators. In the following difference amplifier T_5 and T_6 the pedestal is compensated except for the 40 mV spikes. Because of saturation, the input pulse appears at the collector of T_1 without being amplified or reversed. Between 0 and 10 V the circuit operates linearly, but is not particularly fast. Input pulses of 10 ns rise time produce a feedthrough of 5%.

A *diode bridge*, as given in Fig. 152*e*, may also serve as series switch [Ke 64]. In this circuit the diodes D_1 to D_4 are normally blocked and prevent signal transmission. At current change-over in a common emitter pair the diodes become conducting through the diodes D_5 and D_6, and positive as well as negative signals are transmitted with a rise time of approximately 2 ns. For pedestal compensation, the positive and the negative gate opening pulses should be identical. Signal pulses must remain narrower and lower than the gating pulses. A similar circuit by *Coli* and *Lupini* [Co 65a] operates linearly up to 5 V, resulting in a pedestal of less than 50 mV and a feedthrough underpassing 0·5 V. With a similar version of moderate speed *Schuster* [Sc 68] achieved a linearity deviation underpassing 0·1% for pulses up to 10 V. Pedestal was adjustable to zero, and for a 10 V pulse feedthrough amounted to 10 mV. *Smith* [Sm 67] used a completely d.c. coupled circuit with only four diodes which could handle repetition rates in excess of 50 MHz.

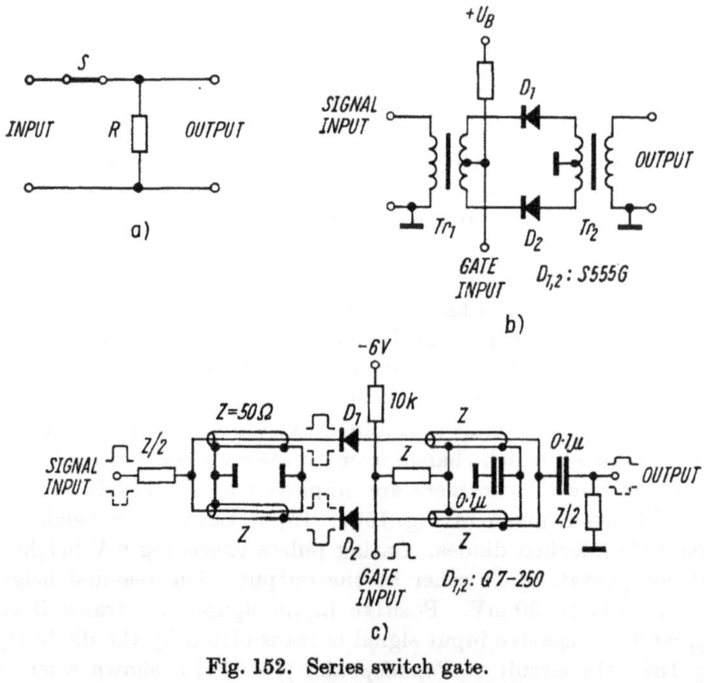

Fig. 152. Series switch gate.

a) Principle, circuit corresponds to open gate. *b*) Diode gate. According to *Valckx* and *Dymanus* [Va 60]. *c*) Diode gate with transmission line transformer. According to *Gorodetzky* et al. [Go 62a].

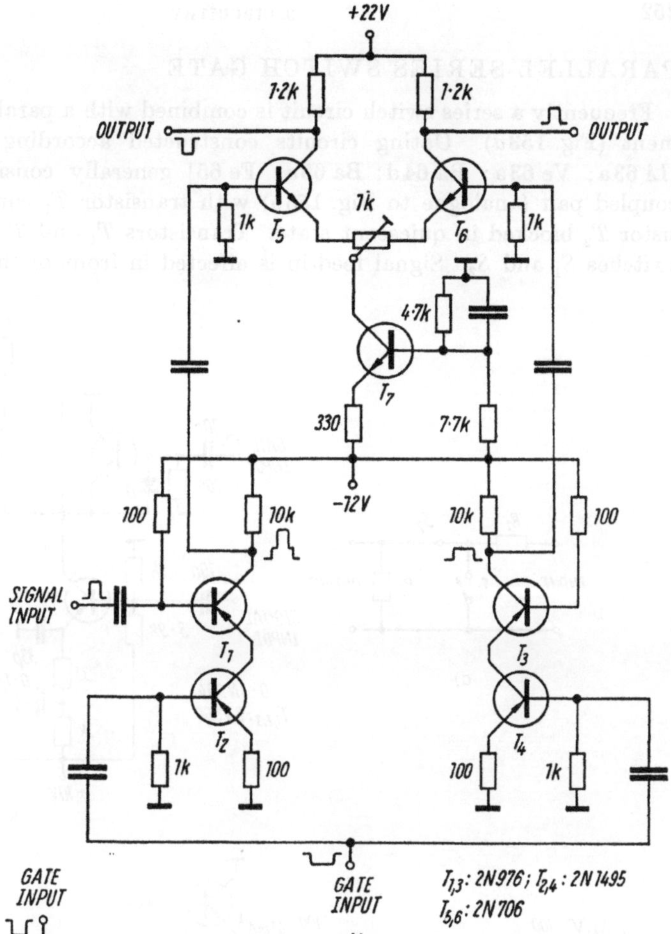

d)

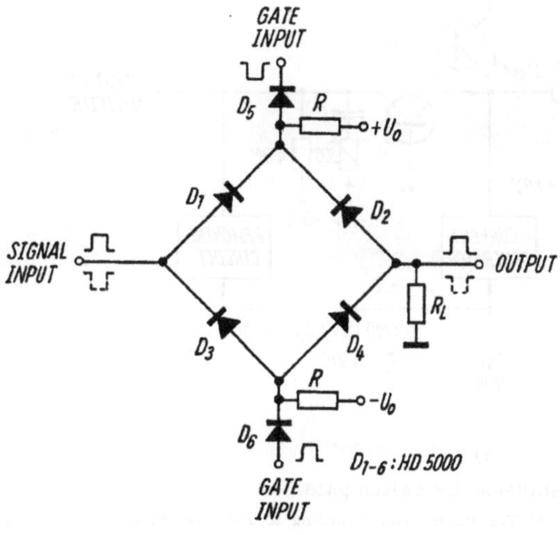

e)

Fig. 152. Series switch gate.

d) Transistor gate. According to *Papadopoulos* [Pa 65a]. e) Diode bridge gate. According to *Keller* [Ke 64].

PARALLEL-SERIES SWITCH GATE

Frequently a series switch circuit is combined with a parallel switch arrangement (Fig. 153*a*). Gating circuits constructed according to this principle [Li 63a; Ve 63a; Ba 64d; Ba 65a; Fe 65] generally consist of an emitter-coupled pair (analogue to Fig. 153*b*) with transistor T_1 conducting and transistor T_2 blocked in quiescent state. Transistors T_1 and T_2 correspond to the switches S_1 and S_2. Signal feed-in is effected in front of the common-emitter

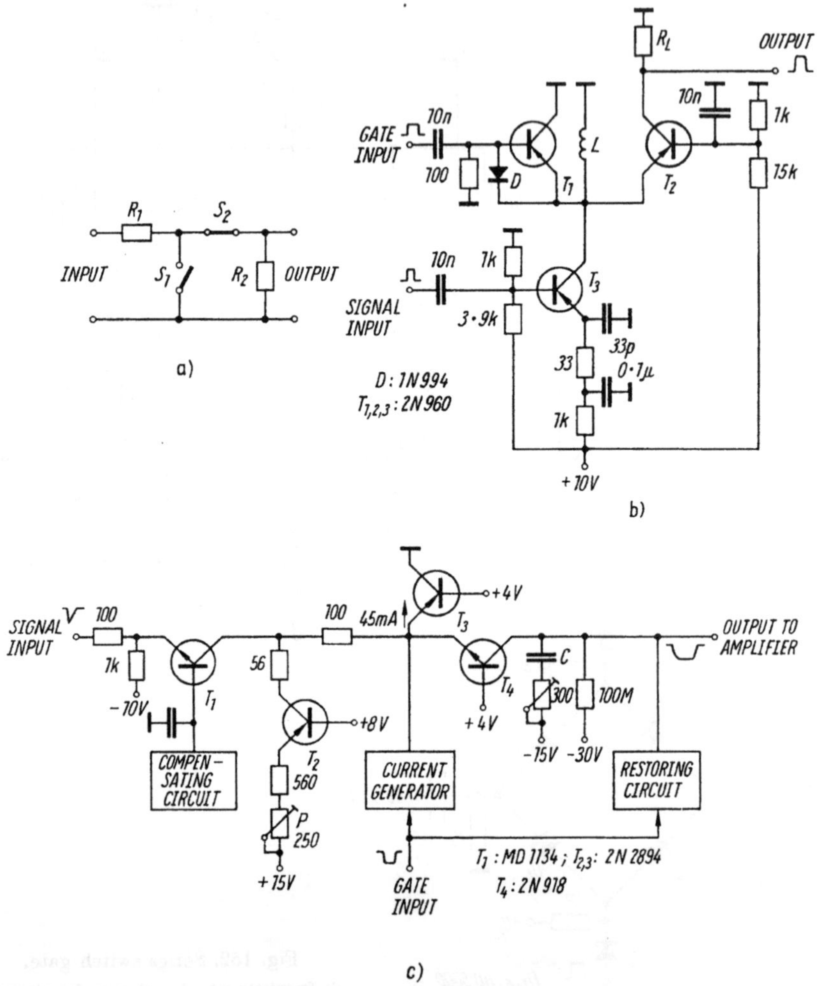

Fig. 153. Parallel-series switch gate.

a) Principle, circuit corresponds to open gate. *b*) Transistor gate, according to ***Feldman*** [Fe 65]. *c*) Current controlled transistor gate. According to ***White*** [Wh 66a].

resistor. The gating pulse is applied to the base of transistor T_1 or, with reversed polarity, also to the base of transistor T_2 [Ba 64d] and switches both transistors to the opposite state. The output signal is taken from the collector of T_2. For input impedance matching and to prevent gate pulse reaction on the signal input a transistor is employed as emitter resistor and the signal fed in at the emitter [Ve 63a; Ko 66] or at the base [Ba 64d; Ba 65a]. The resulting pedestal must be compensated at the output with a fraction of the gating pulse of opposed polarity. *Barna* and *Marshall* [Ba 64d] obtained a pedestal of less than 15 mV and a feedthrough underpassing 0·2% through a linear range of 10 mV to 2 V. Instability amounted to ± 3 mV over a period of 5 days.

In the circuit of *Feldman* [Fe 65] no pedestal compensation is necessary because the standing current through transistor T_3 is bypassed to ground through the coil L (Fig. 153*b*). Transistor T_2 is brought almost to conduction state by the 1 V high gate pulse via diode D and is then opened by the signal pulse. This gate operates linearly between 25 mV to 1 V and has a pedestal of 1 mV in height.

Fig. 153*c* gives the integrating gate circuit developed by *White* [Wh 66a], where transistor T_3 conducts and transistor T_4 is blocked in quiescent state. The gate is opened by the gating pulse which cuts off the supply of 45 mA from the current generator. The input signal is applied through transistor T_1 and can open transistor T_4 when the current generator is blocked. However, the minimum charge required for the purpose sets a lower limit of the dynamic range. On optimum adjustment of the quiescent current flowing through transistor T_1, by means of the potentiometer P, a high sensitivity obtains, but results in a higher pedestal. This gate operates in the current steering mode and is linear through approximately 1 to 25 mA. With a minimum gate opening time of 20 ns differential non-linearity amounts to about ± 1%. Opening and closing is effected at a rise time of 2 ns. Depending on sensitivity setting, pedestal height is between 1 and 20 mV, being stable to <5 mV. Output pulse height instability amounts to 0·2%/10 °C, and is maintained through a compensating circuit acting on the base of transistor T_1. Current pulses transmitted by transistor T_4 are integrated at the capacitor $C \geqq 60$ pF, which is adjustable to alter sensitivity. Width of the stretched output voltage pulses is given by the restore circuit. In the subsequent amplifier a field-effect transistor serves as input stage. Due to the direct coupling throughout a pulse repetition rate up to 10 MHz can be handled.

3.5.3 PILE-UP DETECTION CIRCUITS

When measuring statistic events, especially with nuclear radiation detectors, pulse pile-up causes spurious effects. This is the random summing up of pulses having an interval t_p of less than the recovery time of the electronic measuring system. Probability depends on pulse width and count rate and, with about 10^4 pulses per second, assumes intolerably high values. Besides the primary pulse pile-up, where the two summated pulses appear as a single pulse in analysis, the undershoot pile-up appears where the second pulse is located in the undershoot of the first pulse which is produced by differentiation and is measured with a false height [Mo 67; Bl 68].

The false result of the pulse height interferes in pulse height measurements, for example, when measuring weak-intensity lines of high energy in the presence of intense low-energy lines and vice versa. In time distribution measurements the pile-up effect causes wrong timing. With limiter timing, for instance, too little current is cut off if the preceding pulse has not decayed completely, and too much current is cut off if the following pulse arrives during undershoot. Pile-up effects can, for example, prevent the lifetime measurement, if the lifetime is less than the resolving time of the time analyser (cp. Section 5.1.3 and [Sc 63]). On zero-cross timing and short coincidence resolving time the zero-cross shift causes coincidence loss and interferes with absolute measurements [Bl 67].

In a fast-slow system (cp. Fig. 1, p. 2), the influence of the primary pulse pile-up can be reduced by making the resolving time of the slow-coincidence system sufficiently short. An undershoot pile-up is most effectively reduced through double *RC* differentiation [Sc 45] or double delay line clipping.

The formation of short pulses reduces the primary pulse pile-up. For a good energy resolution, however, it is desirable to collect the total charge released in the detector. This requires for instance, for NaI(Tl) scintillation counters, a pulse width of 1 μs, where then a pile-up occurs with a 9% probability at a rate of 10^5 pulses per second [St 63c]. If the interfering background can be separated by using a suitable selecting principle, e.g. height discrimination, $\mathrm{d}E/\mathrm{d}x$ discrimination or time correlation, the use of a fast-gated system is possible [Ka 66a]. In this arrangement, the detector current pulses, after fast current-mode amplification pass a fast linear gate which is opened by a parallel-operating electronic selector. Integration and analysis of desired rare events is effected behind the gating circuit.

More universally applicable is the later rejection of pile-up distorted pulses. For the purpose an anticoincidence gate, for instance, is used to prevent registration when a pile-up detection circuit has responded. Next to a wide *dynamic range* of the pulse height, the *residual pile-up time* (resolving time) T_r, which is the minimum spacing of two succeeding piled-up pulses that can still be determined, must be as short as compatible with the time constant of the detector pulses. The *pile-up detection time* (dead time) T_d as the maximum interval to which the pile-up detector still responds, should include the major part of the undershoot, but must not be too large to avoid excessive counting loss. Optimum selection depends on the form of the input pulse and the type of pulse shaping [Lo 66]. During this period no interfering pulse must arrive before or after the interesting event. Mere release of a dead time *after* every pulse [Sc 63], for instance, will not prevent that a fast-slow system registers fast coincidences of spurious energy, when selecting conditions are met through pile-up of a succeeding pulse.

Advantages and drawbacks of existing pile-up rejection circuits and some newly suggested versions are discussed by *Blatt* et al. [Bl 68].

COINCIDENCE SYSTEM

With the principle shown in Fig. 154*a* the integrated pulses, delayed by the time T, are brought into coincidence after limitation with the short differentiated and limited pulses. To avoid self-coincidence, the delay time T must be longer than the width of the differentiated pulses. Residual pile-up time determined in this manner amounts to $T_r = 30$ ns and $T_r < 60$ ns, respectively, in the circuits of *Rozen* [Ro 61] and *Moszyński* et al. [Mo 67]. Shaping the long pulse with a limiter causes an increase of the pile-up detection time, as also the undershoot time, with input pulse height, for instance, from $T_d = 2\,\mu$s at 140 keV to $T_d = 8\,\mu$s at 1 MeV [Mo 67]. Using a modified version of this principle *Blatt* et al. [Bl 68] obtained satisfactory operation up to 7×10^5 pps with NaI (Tl) scintillation counters.

DOUBLE PULSE DISCRIMINATOR

Weisberg [We 65c] gives a version, where a tunnel-diode univibrator is actuated by the differentiated pulses (Fig. 154*b*). The output pulses are integrated and trigger a discriminator as soon as the interval of succeeding pulses underpasses a value which depends on the integration time constant. With this circuit a residual pile-up time of $T_r = 50$ ns and a pile-up detection time of $T_d \approx 3{\cdot}5\,\mu$s was obtained. *McGervey* and *Walters* [Mc 64] used two tunnel-diode univibrators, where the output pulse producing univibrator can be triggered only if the first univibrator has been triggered shortly beforehand.

ZERO-CROSS SHIFT

Frequently the zero-cross shift of differentiated pulses caused by a pile-up is utilised. Pulse shaping is achieved either through single RC differentiation [Ma 63a] or double RC differentiation [St 63c], or with single delay line clipping [Fu 66] or, as shown in Fig. 154*c*, with double delay line clipping [Se 62a; Ba 64b; Mo 66]. The most commonly employed modification contains a fast coincidence stage, where both input signals are released by the leading edge and the zero-crossing of the shaped pulse, respectively. The leading edge signal is delayed by the width of the positive pulse part, causing the coincidence circuit to respond as long as no pile-up occurs. The circuit gives no output signal when, by a second piled-up pulse, the zero-cross is shifted by more than the resolving time of the coincidence circuit. Residual pile-up time of such a circuit amounts to $T_r \approx 30$ ns [Ma 63a; Fu 66]. To suppress the undershoot pile-up, frequently occurring with single RC shaping, *Máthé* [Ma 63a] utilised a triple coincidence which includes the output of a single-channel pulse-height analyser.

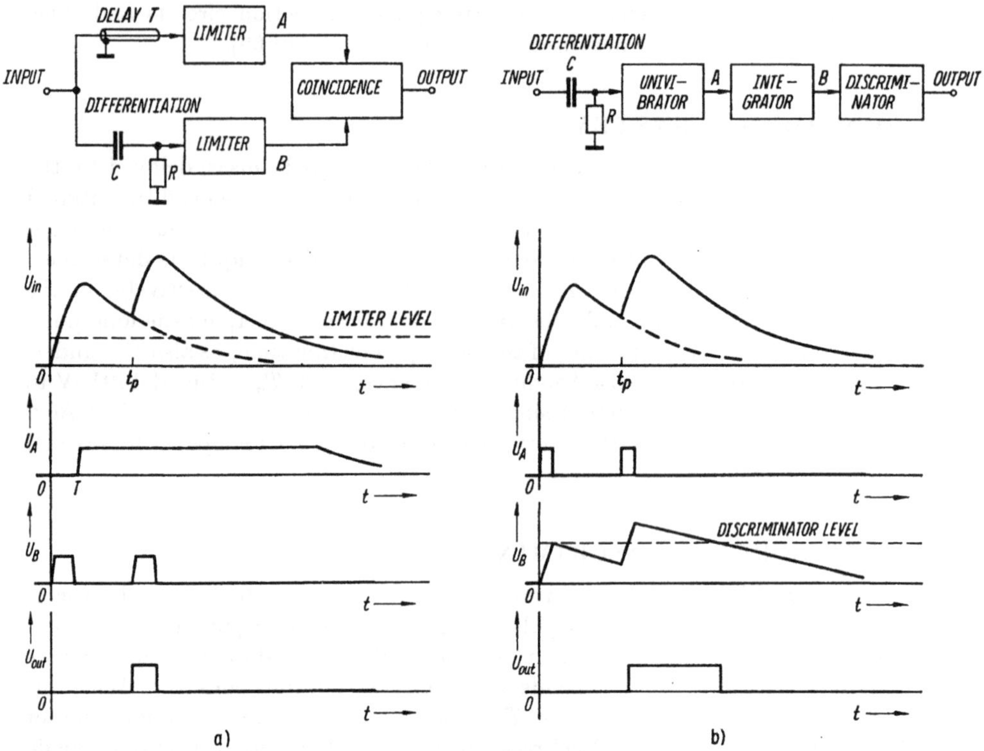

Fig. 154. Principle of pile-up detection circuits and corresponding pulse shape at various points working with:

a) Coincidence system. According to *Moszyński* et al. [Mo 67]. *b)* Double-pulse discriminator. According to *Weisberg* [We 65c].

In the circuit of *Monier* and *Tripard* [Mo 66], given in Fig. 154*c*, the positive part of the shaped pulse is limited. For its width $L \geqq L_0$ applies, with L_0 the width when no pile-up occurs. This pulse is branched and applied with different delay to two tunnel-diode discriminators. At the discriminator inputs both partial pulses add. At the discriminator *1* the mutual delay equals L_0, which locates both partial pulses side by side, with no pile-up, and no output signal is produced. A pile-up in the positive part of the pulse ($t_p < L_0$) causes an overlap of the partial pulses and will at least trigger the discriminator *1*. At the discriminator *2* the mutual delay of the two partial pulses exceeds L_0, e.g. by $\Delta t = 100$ ns. This also assures partial-pulse overlap if the pile-up occurs in the negative part of the shaped pulse ($L_0 \leqq t_p \lessapprox 2\,L_0$). The output pulses of both may, for example, block a linear gate through an "OR"-gate and prevent registration as soon as the spacing t_p of succeeding pulses is $T_r \leqq t_p \leqq T_d$. For the described circuit $T_r \approx 20$ ns, and $T_d = 2\,L_0 = 1{\cdot}5\ \mu$s applies.

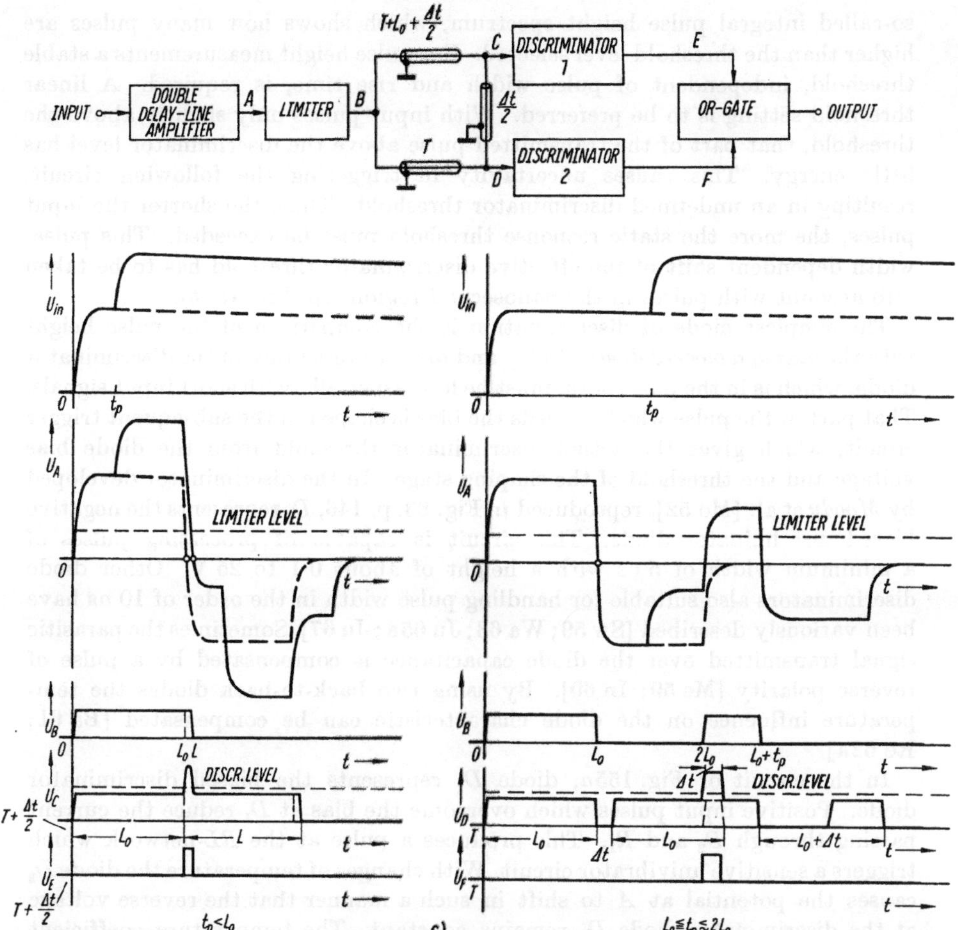

Fig. 154. Principle of pile-up detection circuits and corresponding pulse shape at various points working with:

c) Zero-cross shift. According to *Monier* and *Tripard* [Mo 66].

Using a pile-up rejection according to this principle, *Strauss* et al. [St 67a] obtained a resolution of 2 to 3·5 keV at 1·33 MeV and a spectral shift of no more than 0·1% with a Ge(Li) detector, even up to 10^5 pps.

3.6 PULSE HEIGHT ANALYSERS

3.6.1 INTEGRAL DISCRIMINATORS

The integral discriminator, also known as discriminator or amplitude discriminator, produces an output pulse as soon as the input pulse exceeds an adjustable threshold. Various threshold values enable the measurement of the

so-called integral pulse height spectrum, which shows how many pulses are higher than the threshold level selected. For pulse height measurements a stable threshold, independent of pulse width and rise time, is required. A linear threshold setting is to be preferred. With input pulses only slightly above the threshold, that part of the transmitted pulse above the discriminator level has little energy. This causes uncertainty in triggering the following circuit, resulting in an undefined discriminator threshold. Thus, the shorter the input pulses, the more the static response threshold must be exceeded. This pulse-width dependent shift of the effective discriminator threshold has to be taken into account with pulses in the nanosecond region (cp. Fig. 157*b*).

The simplest mode of discrimination is the comparison of the pulse height with the *bias of a blocked diode*. Delay and forward resistance of the discriminator diode, which is in the pulse path, must be low, especially with short input signals. That part of the pulse which exceeds the bias is shaped in the subsequent trigger circuit, which gives the actual discriminator threshold from the diode bias voltage and the threshold of the shaping stage. In the discriminator developed by *Moody* et al. [Mo 52], reproduced in Fig. 83, p. 146, D_1 represents the negative biased discriminator diode. This circuit is capable of processing pulses of a minimum width of 5 ns with a height of about 0·1 to 25 V. Other diode discriminators also suitable for handling pulse width in the order of 10 ns have been variously described [Sw 59; Wa 63; Ju 65a; Ju 67]. Sometimes the parasitic signal transmitted over the diode capacitance is compensated by a pulse of reverse polarity [Me 59; In 60]. By using two back-to-back diodes the temperature influence on the diode characteristic can be compensated [Ba 61; Ro 62a].

In the circuit of Fig. 155*a*, diode D_1 represents the biased discriminator diode. Positive input pulses which overcome the bias at D_1 reduce the current passing through D_2 and R_1. This produces a pulse at the RL-network which triggers a sensitive univibrator circuit. With changes of temperature the diode D_2 causes the potential at A to shift in such a manner that the reverse voltage at the discriminator diode D_1 remains constant. The temperature coefficient of the threshold is 0·02%/°C. Large input pulses take up the total current flowing through the resistor R_1, blocking the diode D_2. Further limiting to a maximum of 14 V is effected by the diode D_3.

The pulse under examination may also be compared with a quiescent current passing through the discriminator diode [Ba 60; We 65a]. This principle is illustrated in Fig. 155*b*. The adjustment current I passes through diode D, with transistor T_2 blocked. The input current pulse I_{in} takes up part of the current flowing through D and is removed to earth. On $I_{in} > I$ the diode D becomes blocked. The current $(I_{in} - I)$ that exceeds the threshold level I is transmitted through the transistor T_2 to the output. This circuit is employed for pulses of a height between 0·2 to 20 mA and a width of 10 ns and more.

Frequently a *valve* or a *transistor blocked in quiescent state* are used as discriminator elments. Such circuits are known as biased amplifiers or window amplifiers,

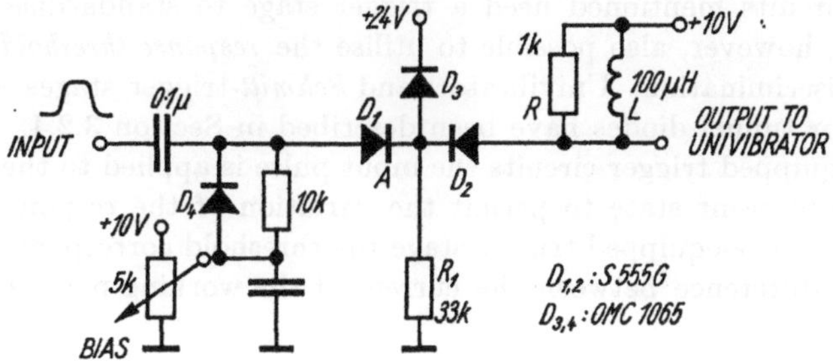

a)

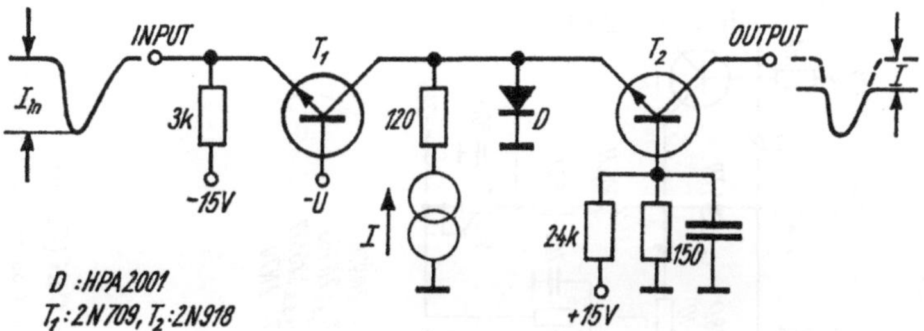

b)

Fig. 155. Diode discriminators.

a) Voltage comparison with temperature compensation. According to *Rozen* and *Altman* [Ro 62a]. *b*) Current comparison (schematic). According to *Welter* [We 65a].

depending whether the whole part of the input signal which exceeds the reverse voltage is amplified, or only the certain interval located in the window. Amplified pulse parts are applied to a trigger stage. The response threshold influence of the subsequent trigger circuit on the level of the discriminator is reduced by the pre-circuited amplifier. To enable the transmission of pulses reaching only slightly above the threshold the amplifier must have a wide band. Biased and window amplifiers have been repeatedly utilised for the discrimination of pulses with approximately 10 ns width [Fa 58; Ba 61; Sa 62; Ve 61; Co 66]. *Coli* [Co 66] developed the fast discriminator given in Fig. 156 as an example. Here, transistor T_1 serves for voltage amplification and, together with the diode D_1, limits the overdriving input pulses. The long-tailed emitter pair T_2, T_3 acts as window amplifier. In quiescent state transistor T_2 is blocked. For adjusting the bias voltage of T_2 the 2·5 kΩ helical potentiometer P_1 is used, with the voltage delivered to the window amplifier through the transistor pair T_5, T_6. Good stability is achieved by utilising long-tailed pairs. On current switching to T_2 a negative pulse is produced at the collector of transistor T_3. This is shortened by the short-circuited transmission line and shaped by the tunnel-diode univibrator *TD*.

All the circuits mentioned need a trigger stage to standardise the output pulse. It is, however, also possible to utilise the *response threshold of a trigger circuit* for discrimination. Univibrators and *Schmitt*-trigger stages with valves, transistors or tunnel diodes have been described in Section 3.2.4. In valve or transistor equipped trigger circuits the input pulse is applied to the component blocked in quiescent state to permit the variation of the response threshold.

In a *tunnel-diode*-equipped trigger stage the threshold corresponds to approximately the difference between the current at the working point and the peak

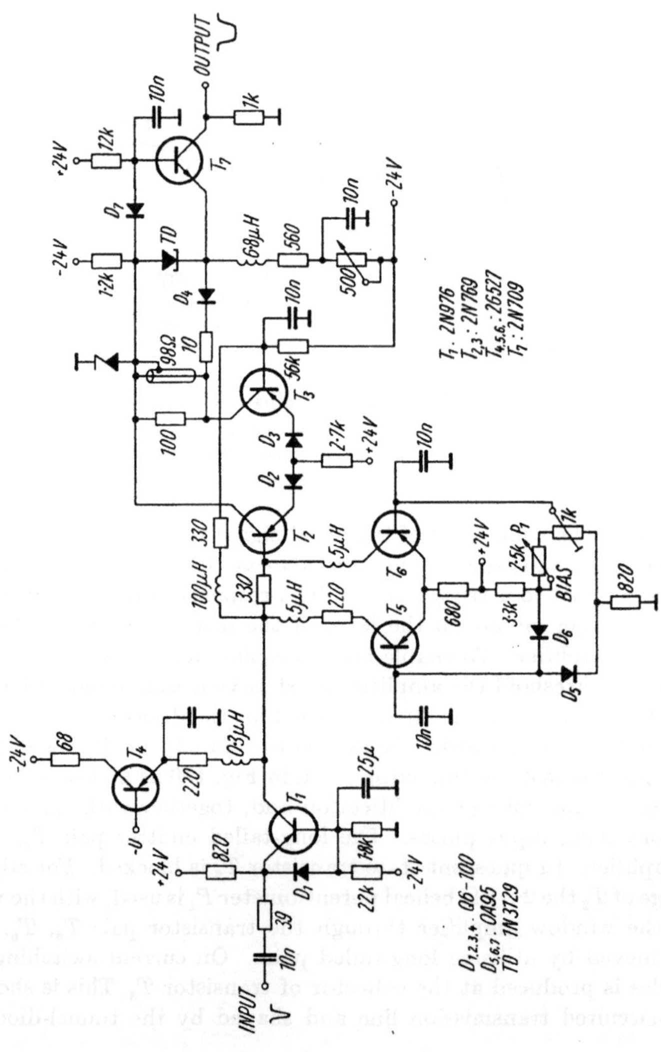

Fig. 156. Pulse height discriminator. According to *Coli* [Co 66].

Max. pulse rate 100 MHz, input pulse range 1 to 30 mA, output pulse 16 mA, width of output pulse at half-maximum 3 ns.

current. In a number of cases tunnel diodes have been employed as integral discriminators [Go 61a; Su 62a; Zu 62; Bl 63a; Ba 64c; Ri 64; Wi 64; Ve 65; Ko 66b; Iv 67b; Zu 67; Bu 68].

Maximum sensitivity is in the order of 0·1 V at a resistance of 50 Ω or about 1 mA. Because of the peak current sensitivity to temperature, less than 90% of the peak current is usually employed at the working point. High sensitivity can, therefore, be achieved by utilising tunnel diodes having a low peak current. Connecting several such stages in series yields high current output pulses of a shape and magnitude no longer dependent on the position of the trigger threshold or the height of the input pulse [Su 62b]. *Verweij* [Ve 65] described a four-stage discriminator cascade arrangement using tunnel diodes with 1 mA, 5 mA, 20 mA and 50 mA peak current, which has an input sensitivity of 50 μA. Another version, developed by *de Blust* et al. [Bl 63a], has a 5 mA tunnel diode which is biased in such a manner that pulses of 10 ns width and 20 μA have already been registered. A thermostat is used to temperature-stabilise the essential components.

A wider dynamic range is obtained if the working point is located on that part of the tunnel-diode characteristic which would correspond to reverse voltage condition [La 62a; So 64]. *Bernaola* et al. [Be 67] describe a discriminator where the tunnel diode is biased in high-voltage state. When passing the response threshold the tunnel diode is switched to low-voltage state and, independent of pulse height and threshold position, on zero-crossing of the input pulse returned to high-voltage state.

The discriminator described by *Hvam* and *Smedsdal* [Hv 63] can be considered voltage-sensitive because of its high input impedance (Fig. 157*a*). Here, the tunnel diode is included in the collector circuit of the long-tailed pair T_1, T_2. When the input signal exceeds the threshold setting the potential difference increases across the tunnel diode and switches it to high-voltage state. Voltage from the tunnel diode is then amplified by the difference amplifier T_3, T_4. Resetting of the tunnel diode is delayed by the integrating network in the collector circuit of transistor T_3 and thus determines pulse width. A d.c. coupled version is given by *Adin* and *Sabbah* [Ad 64a]. Fig. 157*b* shows the dependence of its trigger threshold on input pulse width. Non-linearity with decreasing pulse width is a characteristic of all discriminators, and stems from the limited bandwidth and from the low energy of short triggering pulses.

In the interval between input signal and output signal a variable time shift occurs, as mentioned in Sections 3.2.4 and 3.4.2. This is caused by the finite rise time of the input pulse and the internal triggering delay. The latter is especially high if the threshold is only slightly exceeded. This is particularly noticeable with single-channel analysers. Thus, when a coincidence circuit with a very low resolving time is fed with the output pulses of a discriminator, the variable time shift causes the number of coincidences to decrease [Pa 64]. Time shift can be largely eliminated if a sensitive trigger stage is arranged parallel to the discriminator (Fig. 158). This stage must have a low threshold level

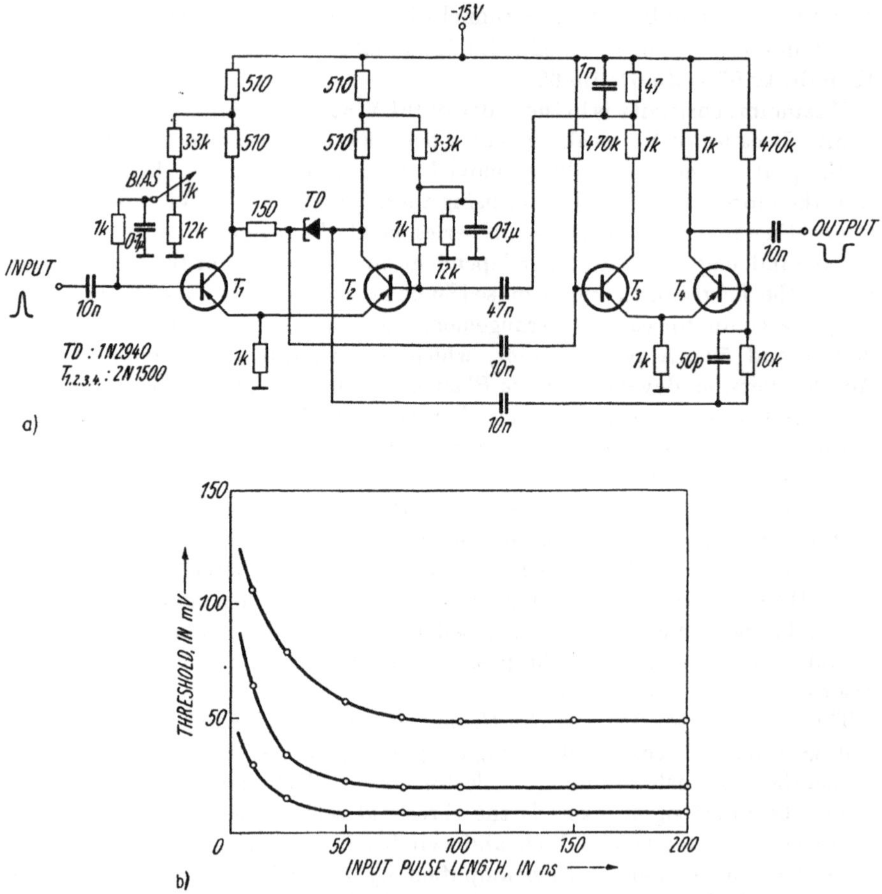

Fig. 157. Pulse height discriminator. According to *Hvam* and *Smedsdal* [Hv 63].

a) Circuit diagram: input sensitivity ≥ 3 mV, output pulse -3 V.

b) Response threshold in dependence of width of input pulse for three different threshold settings.

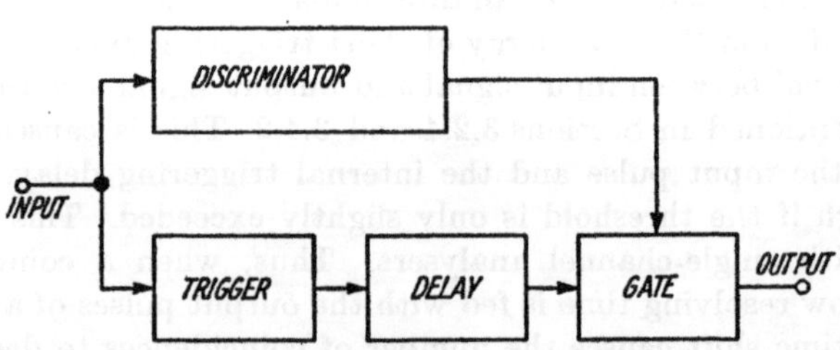

Fig. 158. Block diagram of a discriminator with nearly constant time delay between input pulse and output pulse.

or must be triggered on input pulse zero-crossing. The output pulse of this trigger stage is delayed and delivered to the output only when the gate is opened by the discriminator [Br 53; Ag 61; Or 63; Ju 65a].

3.6.2 SINGLE-CHANNEL ANALYSERS

Compared with the integral discriminator, the single-channel analyser produces an output pulse only when the height of the input pulse is between an upper and a lower response threshold. In general, the single-channel analyser consists of two differently biased discriminators (Fig. 159), where the threshold

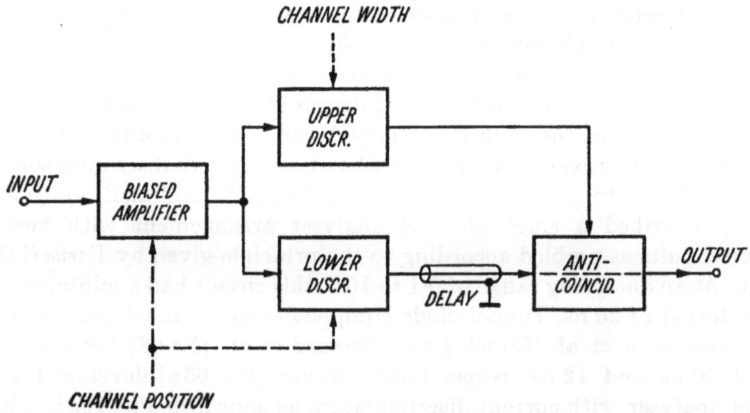

Fig. 159. Block diagram of a single-channel analyser with a window amplifier.

of the "lower" discriminator fixes the position of the channel and the channel width is determined by the "upper" discriminator. Position and width of the channel are usually independently adjustable, which allows for a constant channel width across the whole analysing range. Single-channel analysers with constant relative channel width [Al 61] are seldom used. The output of the lower discriminator is circuited in anticoincidence to the upper discriminator.

With a channel width covering only a few percent of the analysing range the uncertainty of the trigger threshold on constancy and magnitude of the channel width is particularly disturbing. In such cases a window amplifier or a biased amplifier whose variable bias also fixes the position of the channel is precircuited (Section 3.6.1). Then every pulse section which overcomes this bias is amplified and fed to both discriminators. The threshold of the lower discriminator is, in most cases, fixed. For example, with a biased amplifier having a gain of 25, and a threshold spacing of 2·5 V between both discriminators, an effective channel width of 0·1 V is obtained [He 64].

Single-channel analysers can be assembled from the stages described in the last sections. Depending on the type of discriminator the basic circuitry given in Fig. 159 requires supplementing. *Schmitt*-trigger stages and univibrators, for instance, respond to the pulse front, causing the upper threshold to be passed later than the lower. Therefore, to assure the satisfactory function of the anticoincidence stage, either the information from the lower discriminator is stored for the maximum expected delay period, which can be achieved with a delay line, or both discriminators are simultaneously sampled after a suitable time interval. With high count rates it is important that the recovery times of the upper and lower discriminator are in good agreement. In dicriminators and shaping stages using tunnel diodes, for example, it is therefore advisable to employ delay line resetting, similar to Fig. 99, p. 166 [Bu 68].

As with integral discriminators, the addition of a timing branch can make the time difference between the base of the input pulse and of the output pulse independent of the channel width [Br 53]. Zero-crossing trigger circuits are particularly suitable for such anti-walk discriminators.

In conventional circuit technique, integrated detector pulses are analysed primarily because of the high input impedance of the circuitry. Tunnel-diode discriminators, however, may be considered as current discriminators, which can be employed to construct d.c. coupled single-channel analysers. *Brafman* [Br 65c] described a single-channel analyser arrangement with two discriminator circuits assembled according to the principle given by *Verweij* (Fig. 97, p. 164). At an analysing range of 0·1 to 10 V this circuit had a minimum double-pulse interval of 20 ns. Tunnel diode equipped single-channel analysers according to *Gorodetzky* et al. [Go 66e] and *Bernard* et al. [Be 68] have a resolving time of 50 ns and 12 ns, respectively. *Welter* [We 65a] developed a single-channel analyser with current discriminators as shown in Fig. 155*b*, where the lower discriminator can process pulses of 12 ns width at a repetition rate reaching 13 MHz.

If moderate requirements of accuracy and independence of input pulse shape suffice, very simple circuits are adequate [Go 62a; Ko 63b; Ef 65a]. Two versions of this type of circuit are given in Fig. 160. In the single-channel analyser of *Gorodetzky* et al. [Go 62a] the input pulse is applied to two differently biased tunnel-diode discriminators (Fig. 160*a*). Output pulses produced by both discriminators are passed through the transmission line inverter (cp. Section 3.1.2) where they oppose each other. In this manner a negative output pulse is obtained only when the lower discriminator TD_1 responds alone. Pulse peaks left from the response of both discriminators can be eliminated by subsequent integration. This type of circuit is most suitable when handling steep slope pulses. The version of *Kononov* et al. [Ko 63b] has two series-connected tunnel diodes with the same current I_0 flowing through both (Fig. 160*b*). Channel width is set by the difference current ΔI, and tunnel diode TD_1, therefore, determines the lower threshold level. If TD_1 responds only, this produces a positive output signal through the transistor T_1. On the other hand, when both discriminators are actuated, the upper tunnel-diode discriminator TD_2 delivers a pulse which is polarity reversed in the pulse transformer Tr to give a negative compensating pulse. The delay line serves to compensate the delay developing in the tunnel-diode stage TD_2. This circuit has been tried out with slow rising pulses from a NaI(Tl) scintillation counter, but may be modified for handling nanosecond pulses.

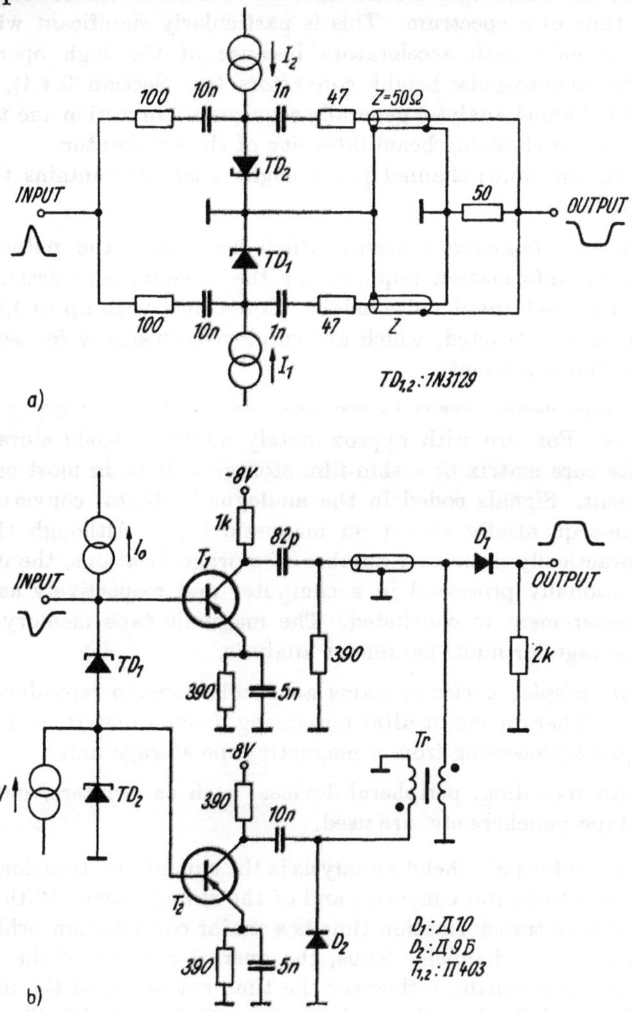

Fig. 160. Simple single-channel analyser circuits with tunnel diodes.
a) According to *Gorodetzky* et al. [Go 62a]. *b)* According to *Kononov* et al. [Ko 63b].

3.6.3 MULTI-CHANNEL ANALYSERS

With a multi-channel pulse-height analyser, tested pulses can be fed in accordance with their height into several channels simultaneously and registered. In this manner the whole pulse-height spectrum is obtainable from a single measurement, while the single-channel analyser requires a series of measurements with successively altered channel position. The advantage of the multi-channel pulse-height analyser is evident, viz. all incoming pulses are registered which, compared to the single-channel mode, leads to a considerable reduction of the measuring time of a spectrum. This is particularly significant when carrying out measurements with accelerators because of the high operating costs. Similarly to time-to-pulse-height converters (cp. Section 3.4.4), registration with a multi-channel analyser generally requires no correction due to decreasing source activity or changing beam intensity of the accelerator.

In general, the multi-channel pulse-height analyser contains the following systems [Ch 61a]:

(i) The *analogue-to-digital converter* which transforms the pulse height into the digital information required for the memory and assigns a storage location to each input pulse height. Converters with up to 4,096 channels have been constructed, which are employed especially for semiconductor detector measurements.

(ii) The *storage device* serves to register every pulse arriving at the various channels. For use with approximately 4,000 or 8,000 storage channels a ferrite core matrix or a thin-film storage unit is, in most cases, utilised at present. Signals coded in the analogue-to-digital converter may also be time-sequentially stored on magnetic tape. Although this provides for a practically unlimited number of storage locations, the data have to be additionally processed in a computer and respectively assorted after the measurement is concluded. The magnetic tape memory has proved of advantage for multi-parameter analysis.

(iii) The *data display device* contains an oscilloscope to reproduce the stored content, either during or after measuring, from a matrix memory, or after subsequent processing from a magnetic tape storage unit.

(iv) For data recording, peripheral devices, such as printers, curve plotters, paper-tape punchers etc. are used.

Time required for pulse-height analysis is the sum of the transformation time of the analogue-to-digital converter and of the storage time. With most multi-channel analysers transformation time is a major contribution, which increases with the number of channels. Thus, the operating speed of the analogue-to-digital converter decidedly influences the time resolution of the multi-channel analyser. In the following, the various conversion methods will, therefore, be viewed principally from the aspect of operating speed. With ferrite memory units the time required for the storing process is approximately 5 to 10 μs, while thin-film or tunnel-diode memories need less than 1 μs. Storage unit problems relating to the development of fast computers, as important as they may be, cannot be dealt with in the scope of this book (cp. [Ru 65a]). For a similar reason, the presentation of multi-parameter analysis has to be omitted (cp.

[Ch 61a; Sp 64; Gu 66; Za 66]), because this is essentially a question of organising the storage units, if several analogue-to-digital converters are utilised. An example of the use of multi-parameter analysis is given in Section 5.1.4.

Starting from the single-channel analyser principle, it seems convenient to construct a multi-channel analyser with n channels by connecting $n+1$ discriminators in parallel. The response threshold of the various discriminators is then adjusted in such a manner that the same width limitation applies for each of the n channels. The output pulses produced by these stacked discriminators are fed to anticoincidence and scaler stages. The advantage of a short resolving time, which has similar values as with the single-channel analysers, is offset by a number of serious disadvantages. Sets of this type are quite bulky and not very reliable, because of the many discriminators and the associated logic and storage stages. The number of channels is restricted to a maximum of 30. This would interfere little when making measurements with plastic scintillators, which have an unfavourable energy resolution. For experiments with pulsed sources the short resolving time can be an advantage. Adjustment of the various discriminator units to equidistant channel position, and thus to a constant channel width is time consuming and permits little flexibility. Another difficulty is the provision of a satisfactory stability of the channel width. For instance, with a 30-channel set and a channel width instability tolerance of 1% discriminator threshold drift must not exceed approximately 3×10^{-4}.

Gatti et al. [Ga 53; Ga 54; Co 57a] used the pulse-increment technique, which is shown in principle in Fig. 161. A constant and stable width for every analyser channel can be obtained with this arrangement where the input signal is stretched and a standard pulse of a constant time delay added. The height h of the latter is equal to or a little smaller than the distance d between the various discriminator threshold levels. Of every responding discriminator, D_1 to D_m, the leading edge of the output pulses is checked for coincidence with a strobe signal released synchronously to the superposed pulse. From the illustration, it will be seen that coincidence obtains only with the m-th channel and that the height of the standardised auxiliary pulse fixes the channel width, and that gaps may be present between the various channels.

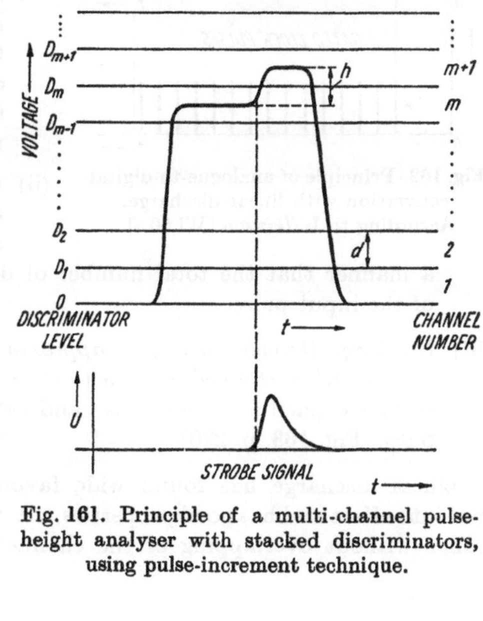

Fig. 161. Principle of a multi-channel pulse-height analyser with stacked discriminators, using pulse-increment technique.

Stacked discriminator analysers have been variously described; with 5 channels by *Swift* [Sw 63], with 10 channels by *Fiehrer* and *Pages* [Fi 59a] and with 16 channels by *Brandt* and *Cappeller* [Br 62a]. The 10-channel analyser of *Lewyn* [Le 67], using tunnel diodes, has a conversion time underpassing 10 ns.

A 20-channel analyser in pulse-increment technique is reported by *Bonitz* and *Berlowitsch* [Bo 59]. With all of these sets, resolving time is in the order of 1 μs. In general, analysers with stacked discriminators were not very successful because already a small number of channels necessitates a rather high outlay.

Designs with a large number of channels can be attained with the aid of *analogue-to-digital converters*. In these circuits the output pulses from the converters, that contain information about the input pulse height, are decoded in the storage unit. In the subsequently described principle versions, the input pulse is at first stretched in the analogue-to-digital converter by charging a storage capacitor to the peak level of the input voltage. Depending on the type of discharge of this storage capacitor the following modes are possible:

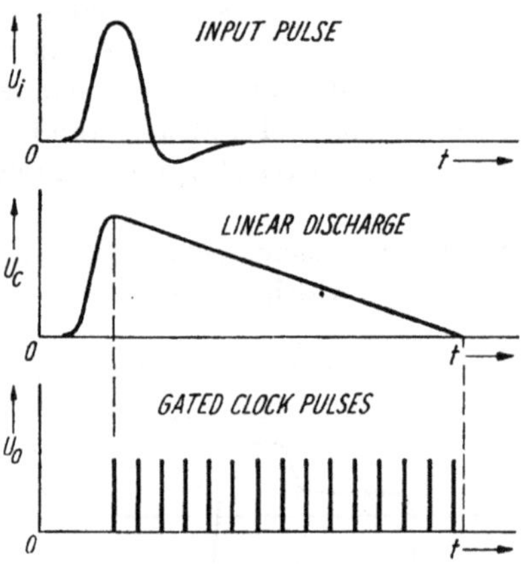

Fig. 162. Principle of analogue-to-digital conversion with linear discharge. According to *Wilkinson* [Wi 50a].

(i) *Linear discharge* according to *Wilkinson* [Wi 50a], where the storage capacitor is discharged by a constant current, which gives a discharge time proportional to the input pulse height. As shown in Fig. 162 the time difference between the beginning and the end of the discharge can be counted with pulses of constant repetition rate. Then the number of these clock pulses is a direct measure of the input pulse height and selects the position number in the storage unit for storing the input pulse.

(ii) *Staircase discharge* [Gu 53], where the storage capacitor is discharged stepwise in equal quantities in such a manner that the total number of discharge steps represents the height of the input pulse.

iii) *Discharge through successive approximation* [Ma 59], where different size pulses, either stepped at a ratio of 10:1 or in binary steps, discharge the storage capacitor, with number and value signifying the height of the input pulse (Fig. 163, p. 270).

Linear discharge has found wide favour because it is relatively simple to put into effect and has good properties, viz. total coverage of the whole analysing range without overlapping of the channel boundaries, and low integral and differential non-linearity.† The long time needed for conversion is a disadvantage, for instance, a clock pulse repetition rate of 5 MHz and n channels require a time of $n \times 0{\cdot}2\ \mu s$, which, for 256 channels, would come to a maximum of approximately 50 μs. For pulsed accelerator work analogue-to-digital converters are desirable having a short conversion time. On the other hand, an increasingly greater number of channels is aimed at in test work with semiconductor detectors to better utilise the very good energy resolution. For both, a short conversion time per channel is necessary, which can be obtained by shortening the discharge time of the storage capacitor and using a higher repetition rate of the clock pulses. At present a clock pulse rate of 100 MHz is employed, which makes fast logic and scaler circuits necessary in the analogue-to-digital converters and storage units. *Goursky* and *Guillon* [Go 63a] mention a 1,024 channel analyser operating at a clock pulse rate of 20 MHz. *Gere* and *Miller* [Ge 66] report on an analogue-to-digital converter with 4,096 channels and a clock pulse rate of 50 MHz. An analyser with 256 channels working at a clock pulse rate of 100 MHz has been developed by *Emmer* [Em 65], where a magnetic thin-film memory is utilised whose storage cycle is ended after 0·75 μs. The commercially available "Nanolyser" also handles a clock pulse rate of 100 MHz and with 256 channels has a resolving time of 4 μs. With the exception of the first channels with most of the multichannel analysers the differential non-linearity is $\leqq 1\%$.

The *successive approximation method* was introduced by *MacMahon* [Ma 59]. Here, the stepwise discharge of the storage capacitor is effected by two diode pumps, with the height of the steps corresponding to 10% and 1% of the analysing range. Initially, discharge is effected in 10% steps, however, only until the residual voltage of the capacitor reaches the height of this step. Discharge then is completed in steps of 1%. In this manner a maximum of 19 steps are needed for the discharge with a total of 100 channels, as against 100 steps with the linear method.

Binary division may also be employed for the height of the steps. Fig. 163 indicates, that at first the height of the input pulse is compared with a step pulse having the height I, which corresponds to half the analysing range. When the input pulse exceeds height I the storage capacitor is discharged by this amount and a code signal is given. The remaining voltage is then compared with the voltage step II, which corresponds to half of height I. If, on comparison, the remaining capacitor voltage is less than height II no discharge takes place and the code signal is not given. Thereafter a comparison is made with the next lower step III, with the process continued as before. Thus with s code signals $n = 2^s$ channels can be represented, i.e. $n = 128$ requires only 7 steps. If a time b is required for voltage comparison and the partial capacitor discharge, the conversion time amounts to $bs = b \log_2 n$. This is an elegant method which, however, is difficult to put into practice. Voltage comparison can only be

† Integral non-linearity denotes the deviation from the proportionality between input pulse height and number of channels, while differential non-linearity is the deviation of the actual channel width from its nominal value.

effected after the transient processes have died down completely. As with the staircase method, this mode is, consequently, much more sensitive to ringing and similar phenomena than the linear method where time intervals are counted by clock pulses. Assuming $b = 0{\cdot}25\,\mu s$, this gives a constant conversion time

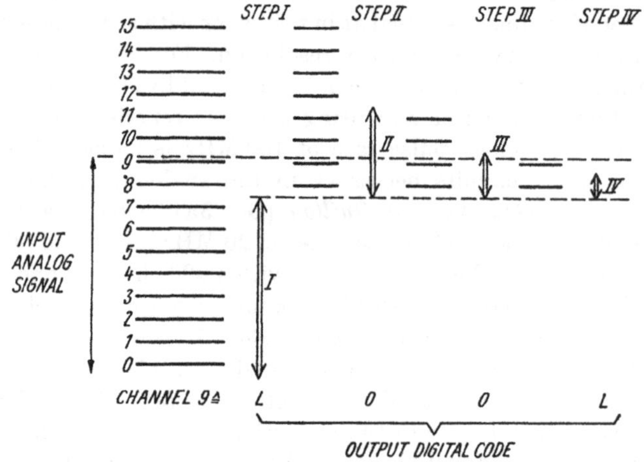

Fig. 163. Principle of analogue-to-digital conversion through successive approximation.

of $2{\cdot}5\,\mu s$ for a 1,024 channel analyser, while the linear method working with a clock pulse rate of 100 MHz requires an average of $5\,\mu s$. Thus, the rapidity of the successive approximation method becomes an advantage only in the case of a very large number of channels.

The principal disadvantage of this method is the high differential non-linearity, because it is not possible to produce the various voltage steps and the differing capacitor discharges with satisfactory precision. In final analysis, this is the reason why this principle has not found wider use.

In the 100-channel analyser of *MacMahon* [Ma 59] decade steps and single steps are used with the maximum conversion time being nearly $10\,\mu s$ at a repetition rate of 2 MHz. *Arnal* and *Borchi* [Ar 63], on the other hand, describe a 100-channel analyser where the same principle is utilised and the maximum conversion time only amounts to $0{\cdot}5\,\mu s$.

As is the case with multi-channel analysers using stacked discriminators, good differential linearity can be obtained by means of superposed pulses. In the analyser developed by *Fränz* and *Schulz* [Fr 64a] conversion takes place in binary steps and involves only the lower channel boundary to fix the channel number. Channel width is determined with a superposed pulse (cp. Fig. 161), the height of which is constant and less than the spacing between the lower channel boundaries. Although this causes gaps in the analysing range, and differing channel positions are not compensated, this does not noticeably interfere with operation. A method has been worked out by *Cottini* et al. [Co 63c;

Ga 64b], where the analysing range is covered entirely and different channel width is compensated to *an average*. In operation, an auxiliary pulse with a height A_a is added to the input signal of height A_i. The auxiliary pulse is produced with high accuracy by a digital-to-analogue converter. The sum $A_i + A_a$ is coded according to the described approximation method by the analogue-to-digital converter. From the resulting channel number, the number is then subtracted which corresponds to the height A_a of the respective auxiliary pulse. If the height A_a is now gradually increased from input pulse to input pulse, or varied statistically and independent of the height A_i, the converter will even code input pulses of the same height A_i differently, but subsequently assign them to the same storage channel. In this manner, on an average, the differences of the channel width are compensated. With 1,024 channels and a conversion time of 8 μs the differential non-linearity across the whole analysing range was less than 1%.

3.7 SCALING CIRCUITS

3.7.1 SCALING STATISTICS

The number of pulses arriving in a specific time interval can be counted with the aid of scaling circuits. Arrangements of this type consist of several series-connected scaling stages, using, in most cases decade scalers which deliver an output signal after every tenth input pulse. Highest pulse rate is handled by the first decade scaler, with pulse rate reduced by the factor of 10 or 100 etc., for each of the subsequent stages. The number of registered pulses can be read directly with a digital display giving the state of the various scaling stages at the end of a measurement.

In the following pages only such scaling circuits will be dealt with as have a resolving time of less than 100 ns. In scaling circuits, the resolving time signifies the minimum admissible time spacing between two input pulses which can still be separately registered. Scaling stages with a resolving time of 1 μs and more, which can be easily constructed, for instance, with decade counting-tubes, type E1T or S10S1, or with dekatron tubes, are not discussed here.

A pulse train can be counted without loss if the resolving time of the first scaler unit is less than the minimum occurring interval between successive pulses. This means that, for example, with a resolving time of $t_D = 10$ ns periodic pulses with a repetition frequency of nearly 100 MHz can still be registered without counting loss. Frequency measurements can be handled with high accuracy when during a specific time interval the various cycles are counted.

With *statistically time-distributed pulses*, encountered, for instance, with nuclear radiation detectors, where *Poisson* distribution applies, conditions are far more unfavourable. The probability $w(t)$ that the interval between two succeeding pulses is t, is expressed by

$$w(t) = \frac{1}{\Theta} e^{-t/\Theta} \tag{113}$$

with Θ the mean pulse interval according to

$$\Theta = \int_0^\infty t\, w(t)\, \mathrm{d}t\,. \tag{114}$$

Fig. 164 gives the curve for $w(t)$, which indicates that the probability of short intervals is highest. This means, however, that due to the influence of the finite resolving time t_D of the scaling stage a considerable number of pulses are not registered, even though the resolving time t_D is only a small fraction of the mean pulse interval Θ.

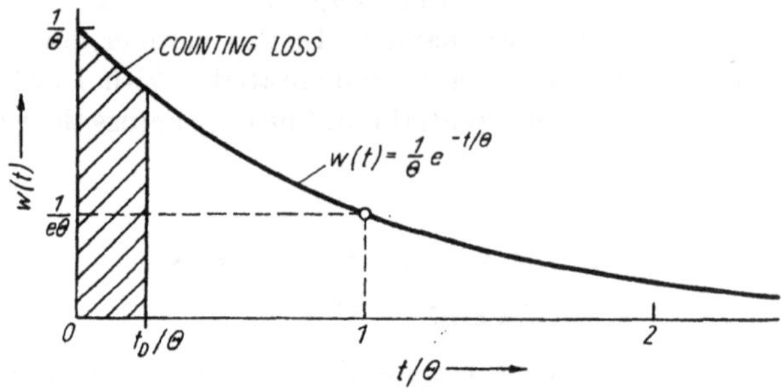

Fig. 164. Probability distribution of time intervals between two successive pulses of a statistically time-distributed pulse train (*Poisson* distribution).

The relative counting loss, which corresponds to the shaded area in Fig. 164, can be expressed with

$$\frac{\Delta N}{N} = \frac{N - n}{N} = \int_0^{t_D} w(t)\, \mathrm{d}t = 1 - \mathrm{e}^{-t_D/\Theta}\,. \tag{115}$$

In consequence, with $t_D/\Theta \ll 1$,

$$\frac{\Delta N}{N} \approx \frac{t_D}{\Theta} \tag{116}$$

holds, with N the true count rate and n the actually registered count rate. Thus with $t_D = 10$ ns a counting loss of 1% can be expected at an average pulse rate of $N = 10^6$ pps. The counting error caused by the finite resolving time of the scaler unit can be corrected by approximation (dead time correction). With $N = 1/\Theta$ for a counting loss up to about 10% equation (116) gives

$$N \approx \frac{n}{1 - t_D/\Theta} = \frac{n}{1 - N\, t_D} \approx \frac{n}{1 - n\, t_D}\,. \tag{117}$$

From this derives the significant consequence, that with statistically time-distributed pulses the resolving time of the first scaling stage has to be small against the mean pulse interval to keep the counting loss low.

For the intervals of the output pulses of a scaling unit the probability distribution shown in Fig. 165 applies. It will be seen that after two-fold scaling ($m = 2$) the probability of a very small interval tends toward zero, and that with an increasing scaling factor m interval fluctuation becomes less. The probability distribution diagrams given in Fig. 165 apply for scaler units with $t_D/\Theta = 0$. With finite resolving time, i.e. $t_D/\Theta > 0$, the probability of the occurrence of short intervals is still further reduced.

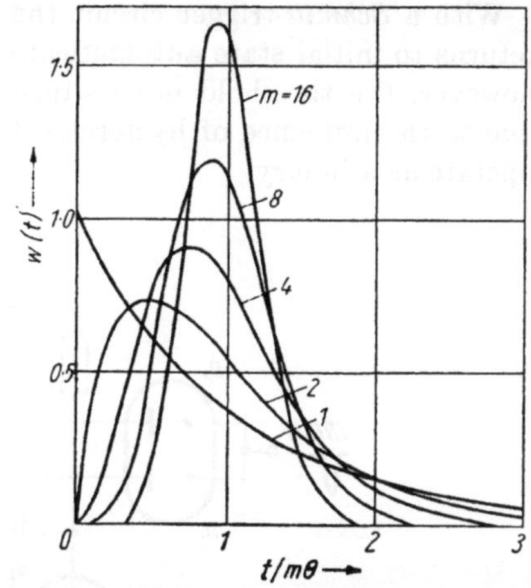

Fig. 165. Probability distribution of the time interval between two successive output pulses of a scale-of-m circuit with a random input [Se 59a; Ne 66].

Another significant conclusion derives from the described regularising action of scaling units. Let us consider several 8:1 or 10:1 scaling stages connected in series: If the resolving time of the second scaling unit approximately equals the mean interval of the pulses to be counted, the counting loss is then determined solely by the first scaling unit. With a pulse rate of 10^7 pps, for example, the resolving time of the first 10:1 scaling unit should be better than approximately 5 ns to keep the counting loss below 5%. For the next decade scaler, however, a resolving time of 100 or 200 ns is entirely sufficient for no-loss counting.

In case the resolving time of the scaler is not constant, for instance, when a pulse arriving during the dead time extends this period, or when the pulse train does not follow *Poisson's* distribution, as happens with a burst from a pulsed accelerator, mathematical treatment of counting loss becomes very complicated. Detailed treatment of the problems of scaling statistics will be found in the various papers and manuals dealing with measuring methods in the field of nuclear physics [Ev 55; Se 59a; Ne 66].

3.7.2 BISTABLE TRIGGER STAGES (BINARIES)

A bistable trigger stage, briefly called binary in the following, has two stable states, as compared to the mono-stable trigger circuit (cp. Section 3.2.4). A binary is switched to the other state by every trigger pulse and therefore the initial state of a circuit is restored after two input pulses have passed. Thus, every second trigger pulse results in an output pulse which is registered by the subsequent circuits. In the binary stage the number of pulses is scaled by the factor 2; they are therefore known as scale-of-two circuits, and represent the basic element of most of the scaling circuitry.

BINARIES USING ELECTRONIC VALVES

Essentially the binary consists of two d.c. coupled feedback valve systems similar to a *Schmitt*-trigger circuit (cp. Section 3.2.4).

With a *Schmitt*-trigger circuit the threshold is set in such a manner that it returns to initial state automatically after a trigger signal has passed. When, however, the threshold is so adjusted that the circuit does not switch back due to the influence of hysteresis, two stable states obtain — the circuit can operate as a binary.

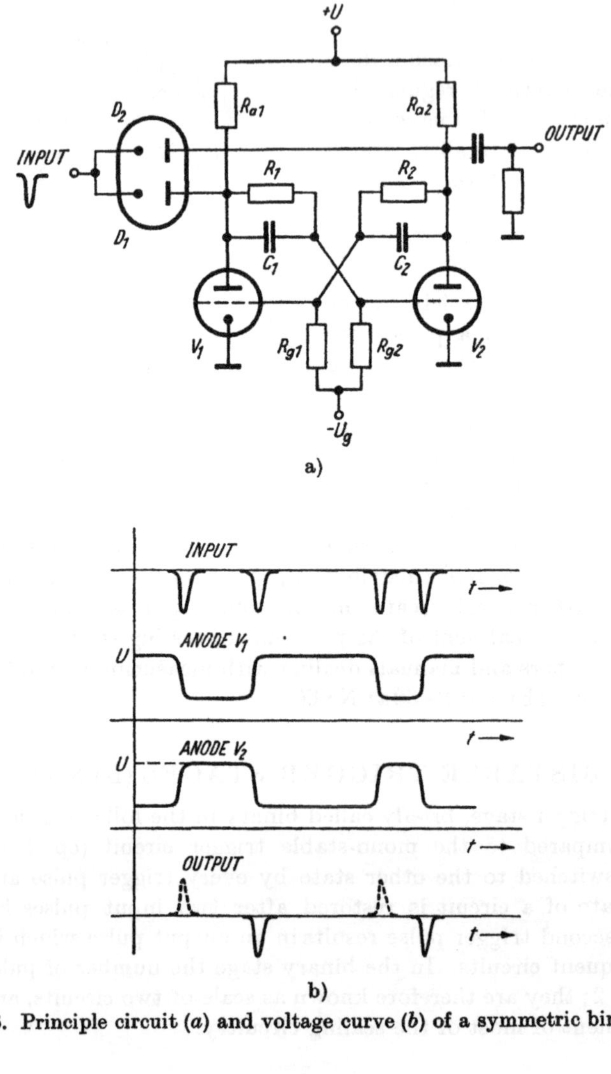

Fig. 166. Principle circuit (*a*) and voltage curve (*b*) of a symmetric binary stage.

A symmetric binary is given in Fig. 166, which also shows some pulse patterns. Apart from the transient period, one valve is always blocked and the other conducts. The circuit switches to the other stable state when a negative trigger pulse is applied to the grid of the open valve, e.g. V_1. This reduces the current flow through valve V_1; the anode potential rises and pulls the valve V_2 to conduction state. The anode potential of the latter then decreases, further reducing the current through valve V_1, until finally the initially conducting valve V_1 is blocked and the initially blocked valve V_2 becomes conducting. Details with respect to dimensioning and switching time calculations are given in a number of books [Ne 55a; Mi 56].

Low resolving time requires a fast change-over from one stable state to another. To obtain a fast transmission of the voltage change at an anode to the grid of the other valve, and to eliminate the influence of the valve input capacitance, compensated voltage dividers are utilised. In such a system, the capacitors C_1 and C_2 are connected parallel to the voltage divider resistors R_1 and R_2 (Fig. 166). The use of relatively large capacitors improves the transmission of the anode voltage changes and results in faster actual switching time. Very large capacitors, however, tend to prolong the switching time because of the heavy capacitive load exerted on the anode. After a switching event a certain recovery interval is necessary to permit the parallel capacitors to recharge to static level. This applies particularly when a valve switches from blocked state to conduction state, because in this case the capacitors are charged by the grid current.

Thus, the resolving time of a binary constitutes the sum of change-over time and recovery time. As a compromise, for short switching time and acceptable recovery time, capacitors of two or three times the value necessary for a compensated voltage divider seem to give the best results. The role of the parallel capacitors as "short-time memories" is dealt with later.

To keep the recovery time and, therefore, the resolving time, as short as possible, it is necessary:

(i) to avoid high stray capacitance;

(ii) to apply low anode time constants through low-ohmic resistors and reduce the capacitive load by means of cathode followers in the coupling branch [Fi 49; Sc 53; Na 57; Mu 59];

(iii) to use clamping diodes to limit the height of the voltage steps, which reduces the recharging time [Sc 53; Mi 56; Na 57].

Another possibility is to replace the parallel RC circuit with a glow-lamp coupling between anode and grid [Co 58; Hu 59]. This eliminates the recharging of the parallel capacitor, but requires other circuit elements to act as short-time memory.

Triggering a binary can be effected in various ways. With bistable trigger stages in gating or control circuits triggering is frequently effected asymmetric by applying the trigger pulse to one valve system only. In scaling circuits, however, the pulses must be delivered to both ends of a binary (symmetric triggering).

Triggering with a negative step at the grid of the open valve is far more sensitive than with a positive voltage step at the grid of the blocked valve. Thus, only the negative step of the anode voltage (cp. Fig. 166) triggers the subsequent second binary, which gives a scaling of the number of pulses by the factor 2. Width of the trigger pulses must approximate the switching time.

Feeding the trigger pulses directly to the grid represents an additional capacitive load and tends to slow down the transient period. Coupling the trigger pulses to the anodes is more favourable. For this purpose two diodes are used as illustrated in Fig. 166. In this case, the negative trigger pulses are transmitted only by the diode connected to the anode of the blocked valve and are passed through the voltage divider to the grid of the valve which is open.

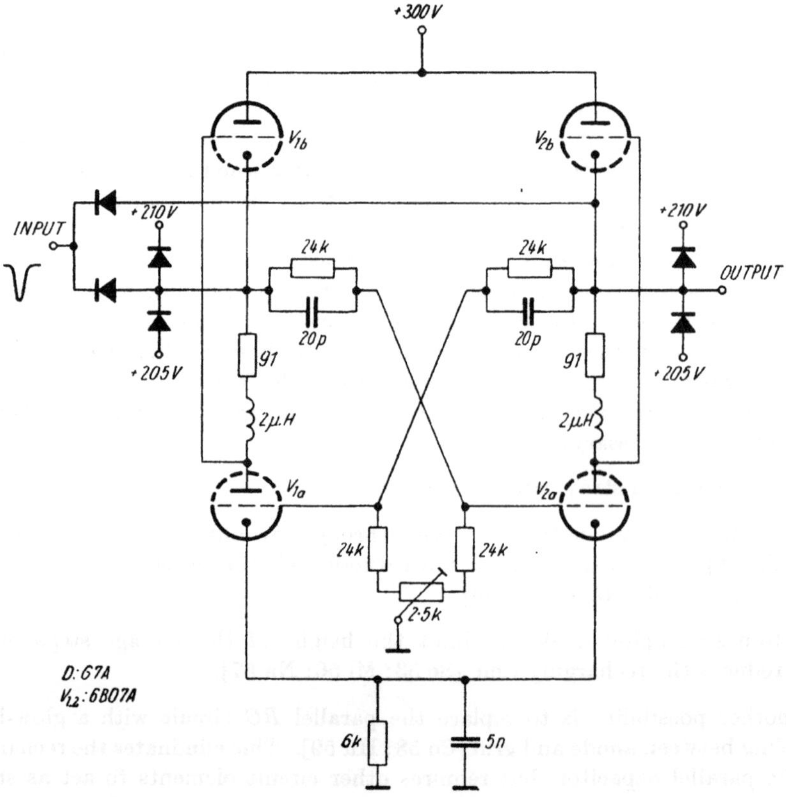

Fig. 167. Binary circuit using valves, with a resolving time of 25 ns. According to *Nakamura* [Na 57].

This coupling mode is also employed for coupling series-connected binaries, which operates very reliably and is widely utilised. Fig. 167 gives the binary developed by *Nakamura* [Na 57], which is suitable for a maximum switching rate of 40 MHz. A cathode follower is used for the coupling from the anode to the grid, with diodes inserted for limiting. *Fitch* [Fi 49] describes a similar binary which can handle a maximum of 15 MHz. *Collinge* and *Huxtable* [Co 58] give a version for 20 MHz where triggering of one and the same valve system is effected through an additional long-tailed pair, with, however, pulses of alternate polarity.

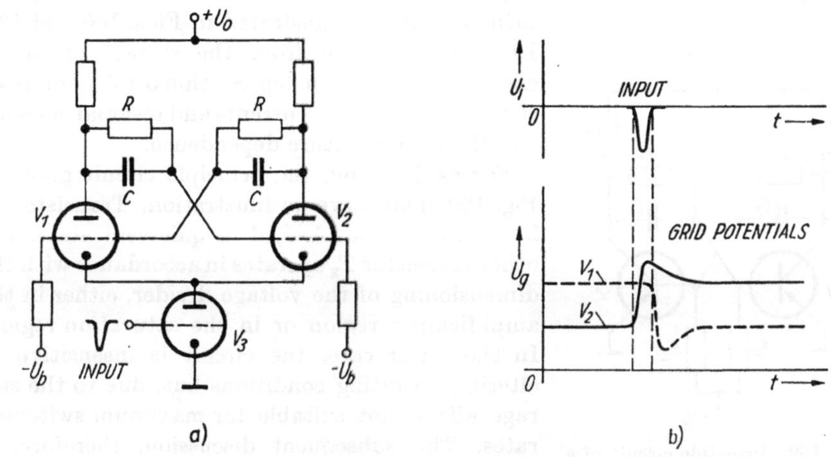

Fig. 168. Principle circuit (*a*) and voltage curve (*b*) of a binary circuit where the capacitors *C* act as short-time memory; in initial state valve V_1 conducts.

Trigger pulses may also be fed-in at the part of the anode resistor common to both valve systems or at the common cathode lead. In both cases, the parallel capacitors in the voltage divider (C_1 and C_2 in Fig. 166) have a deciding function as *short-time memory*. During the transient period they store the information regarding the initial state and ensure that the circuit switches to the other state and does not return to initial state [Mi 56]. Its principle is illustrated in Fig. 168. The negative trigger pulse blocks valve V_3, and causes the initially conducting valve, e.g. V_1, to switch to non-conduction state. The anode potential of V_1 increases. This change is transmitted almost wholly to the grid of valve V_2 by the overcompensated voltage divider *RC*, which raises the grid potential of valve V_2 above the static grid potential of valve V_1. After the trigger pulse has decayed, current again flows through the valve V_3 and is taken up by valve V_2 because of the now higher grid potential, i.e. the circuit has switched to the other state. The trigger pulse must be shorter than the action of the "memory". Instead of parallel capacitors, peaking coils or a short-circuited transmission line may serve as short-time memory in the anode circuits.

Shorter recovery time can be obtained with them. *Hutchinson* et al. [Hu 59] describe a binary circuit, similar in principle to the one given in Fig. 168, but with a short-circuited transmission line serving as memory for the initial state. This circuit has a resolving time of 11 ns.

Extremely complicated binary circuits using secondary-emission valves have been developed which permit switching rates up to 100 MHz [La 63b].

BINARY CIRCUITS USING TRANSISTORS

Transistorised binary circuits are similar in construction and function to a stage using electronic valves, as will be readily seen when comparing the principle circuits illustrated in Figs. 166 and 169. Differences evolve from the storage of charge carriers on saturation (cp. Section 3.1.5.) and from the presence of base currents and residual currents and their temperature dependence.

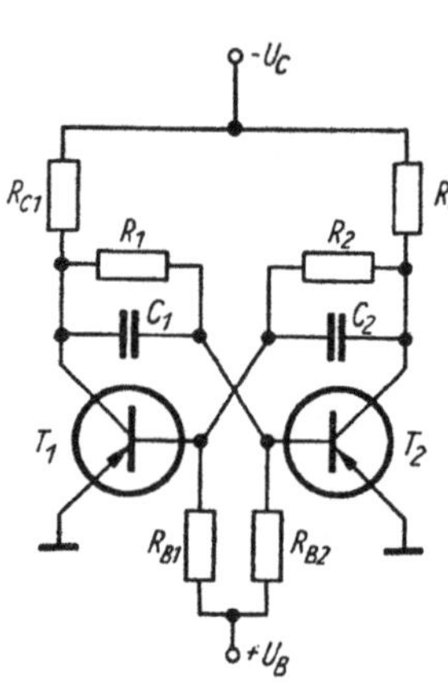

Fig. 169. Principle circuit of a transistorised symmetric binary stage.

For explanation, the principle circuit given in Fig. 169 shall serve as illustration. Transistor T_1 is assumed to be blocked in quiescent state. The other transistor T_2 operates in accordance with the dimensioning of the voltage divider, either in the amplification region or in the saturation region. In the latter case, the circuit is insensitive to altering operating conditions but, due to the storage effect, not suitable for maximum switching rates. The subsequent discussion, therefore, is limited to binary stages with unsaturated transistors. Preventing a transistor from reaching saturation can be effected by suitably dimensioned voltage dividers and the use of clamping diodes. Detailed calculations of the static and dynamic behaviour of transistorised binaries will be found, among others, with *Shea* [Sh 57], *Neeteson* [Ne 61], *Baranowski* and *Jankowski* [Ba 64], and *Rein* [Re 65a].

As with electronic valve equipped binaries, the separate voltage supply for biasing the base can be omitted if an emitter combination $R_E || C_E$ common to both transistors is utilised. On a high switching rate, however, this automatic bias voltage generation may reduce trigger sensitivity [Ba 64]. It is an advantage that, due to the d.c. feedback, the resistor R_E reduces the influence of temperature fluctuations and limits the maximum collector current.

Coupling the trigger pulses requires special care. The width of the trigger pulse must be approximately equal to the switching time. With saturated circuits duration of trigger pulse influence should not be less than the storage time of the transistors which, in turn, is reduced by large base current pulses (cp. Fig. 75, p. 132). Some circuit examples are given in Fig. 170 to illustrate the coupling of trigger pulses at the base, at the collector, and at the emitter.

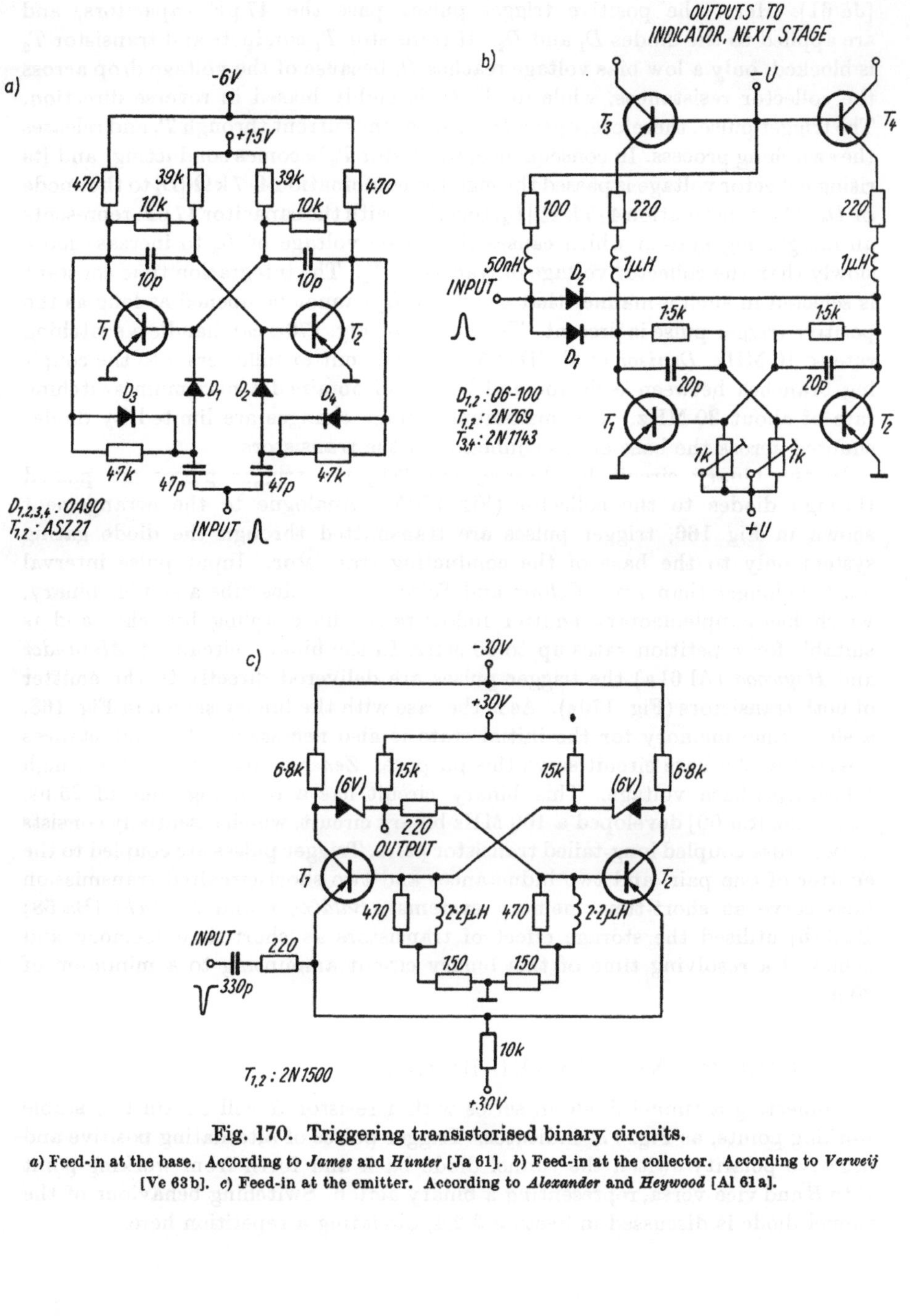

Fig. 170. Triggering transistorised binary circuits.

a) Feed-in at the base. According to *James* and *Hunter* [Ja 61]. *b)* Feed-in at the collector. According to *Verweij* [Ve 63b]. *c)* Feed-in at the emitter. According to *Alexander* and *Heywood* [Al 61a].

Fig. 170*a* presents a base-coupled binary stage as given by *James* and *Hunter* [Ja 61]. Here, the positive trigger pulses pass the 47 pF capacitors, and are applied to the diodes D_1 and D_2. If transistor T_1 conducts and transistor T_2 is blocked, only a low bias voltage reaches D_1 because of the voltage drop across the collector resistances, while diode D_2 is highly biased in reverse direction. The trigger pulse, therefore, opens D_1, reduces the current through T_1 and releases the switching process. In consequence, transistor T_2 becomes conducting, and its rising collector voltage is passed through the combination $4 \cdot 7\,\mathrm{k}\Omega || D_4$ to the anode of D_2. The combination $4 \cdot 7\,\mathrm{k}\,\Omega || D_4$, together with the capacitor 47 pF, represents an integrating system which causes the anode voltage of D_2 to increase more slowly than the collector voltage of transistor T_2. The integration time constant is selected in such a manner that the diode D_2 cannot be opened as long as the positive trigger pulse is present. The binary of Fig. 170*a* can handle a switching rate to 10 MHz. *Denisov* et al. [De 65] inserted emitter followers into the coupling branches between collector and base, and obtained a maximum switching rate of about 30 MHz. The maximum voltage changes are limited by diodes shunted across the base-emitter junction of the transistors.

In the binary circuit by *Verweij* [Ve 63b] the trigger pulses are passed through diodes to the collector (Fig. 170*b*). Analogue to the arrangement shown in Fig. 166, trigger pulses are transmitted through the diode gating system only to the base of the conducting transistor. Input pulse interval must be longer than 7 ns. *Gettner* and *Selove* [Ge 59] describe a similar binary, which has supplementary emitter followers in the coupling branches and is suitable for repetition rates up to 50 MHz. In the binary circuit of *Alexander* and *Heywood* [Al 61a] the trigger pulses are delivered directly to the emitter of both transistors (Fig. 170*c*). As is the case with the binary shown in Fig. 168, a short-time memory for the initial state is also necessary. The inductances inserted in the base circuit serve this purpose. *Zener* diodes provide for a high collector-to-base voltage. This binary circuit has a resolving time of 25 ns. *Sugarman* [Su 60] developed a 100 MHz binary circuit, which essentially consists of two cross-coupled long-tailed transistor pairs. Trigger pulses are coupled to the emitter of one pair, and two inductances and two short-circuited transmission lines serve as short-time memory systems. *Baldinger* and *Santschi* [Ba 58; Ba 59b] utilised the storage effect of transistors as short-time memory and achieved a resolving time of this binary circuit amounting to a minimum of 20 ns.

BINARIES USING TUNNEL DIODES

Connecting a tunnel diode in series with a resistor R will obtain two stable working points, as Fig. 171 indicates. Trigger pulses of alternating positive and negative polarity switch the tunnel diode back and forth from working point A to B and vice versa, representing a binary action. Switching behaviour of the tunnel diode is discussed in Section 3.2.4, obviating a repetition here.

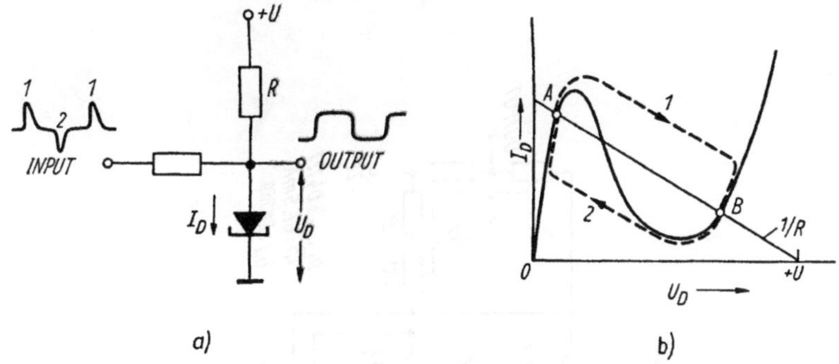

Fig. 171. Principle of a tunnel-diode binary stage.
a) Circuit diagram. *b)* Characteristic with switching trajectory.

Triggering the circuit with bipolar pulses, as suggested in Fig. 171*a*, with a storage inductance in series with the resistor *R*, causes this arrangement to operate as a binary stage [Ba 62a].

Faster and more reliable *binary circuits with one tunnel diode* [Iv 67 b] triggered by pulses of uniform polarity are shown in Fig. 172.

In the binary circuit of *Hazoni* [Ha 61] trigger pulses are alternately delivered to the two terminals of the tunnel diode through the diode gating system (Fig. 172*a*). Potentials have been selected that diode D_1 conducts and diode D_2 is blocked when the tunnel diode is in low-voltage state. The positive trigger pulse then switches the tunnel diode to high-voltage state through the diode D_1. Subsequently D_1 is blocked and D_2 conducts. The inductances *L* prevent the tunnel diode from returning to initial state after a trigger pulse has passed. With this circuit a double-pulse resolving time of 10 ns was obtained. *Harrison* and *Foote* [Ha 61 b] obtained similar results with transistor coupled pulses.

Combining tunnel diodes and transistors utilises the advantages of both circuit components, i.e. with the tunnel diode its current amplification and fast switching, and with the transistor its unilaterality and its high switching speed in common-base connection. *Amodei* and *Kosonocky* [Am 61] used this version to develop a number of fast logic circuits, of which one example is shown in Fig. 172*b*. The two stable working points of the tunnel diode are positioned near the peak and the valley of the tunnel diode characteristic. When the tunnel diode is in low-voltage state the transistor T_2 is blocked because of the low emitter-base voltage, and current flows through transistor T_1. The positive trigger pulse switches the tunnel diode to high-voltage state and only the valley current flows through the transistor T_1, while the major part of *I* flows through transistor T_2. The next trigger pulse, which must have a width that is longer than the inductive time constant, increases the current through the inductance *L*. At the end of the trigger pulse the current through *L* is reduced and the resulting induced voltage resets the tunnel diode to low-voltage state. For this version 200 MHz is mentioned as a possible switching rate. It is an ad-

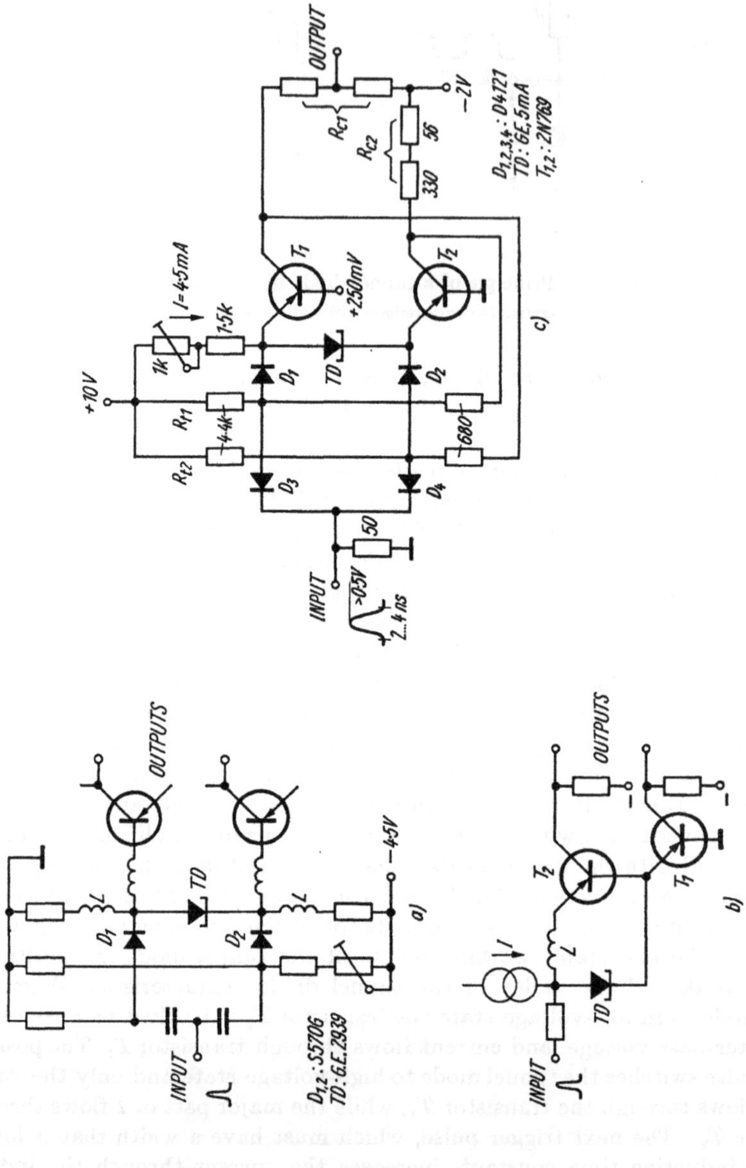

Fig. 172. Fast tunnel-diode binary circuit.

a) According to *Hazoni* [Ha 61]. *b)* According to *Amodei* and *Kosonocky* [Am 61]. *c)* According to *Radeka* [Ra 62].

vantage that the output voltage level is essentially determined by I and not by the properties of the transistors. When switching to the low-voltage state the current starts to flow through transistor T_1 only after the trigger signal has passed. Under certain conditions, this may be viewed as a disadvantage.

Radeka [Ra 62] improved this circuit and avoided the disadvantage by feeding the trigger pulses through a diode gate (Fig. 172*c*). The diodes D_1 and D_2, in this case, are slightly biased in reverse direction. When the tunnel diode is in low-voltage state the current I flows through the transistor T_2. Due to the voltage drop across R_{C2} diode D_3 conducts, while transistor T_1 and diode D_4 are blocked. The positive trigger pulse blocks the diode D_3 and the current from R_{t1} passes through diode D_1 to the tunnel diode, which then switches to high-voltage state. Transistor T_1 and diode D_4 now become conducting and only the valley current flows through the transistor T_2. It is an advantage that the trigger pulse must only block the diode D_3 or D_4, while switching is triggered by the constant current through the resistor R_{t1} or R_{t2}. With this arrangement a switching rate of approximately 250 MHz was achieved.

Binary circuits with two tunnel diodes are frequently employed. Their simple construction and reliable function are characteristic. The principle of this version is given in Fig. 173. Of the two identical tunnel diodes, one is in low-voltage state and the other in high-voltage state. The parallel resistors R have been selected in such a manner that the two stable working points of one tunnel diode are positioned near the peak and near the valley. The greater part of the peak current is taken up by R when the tunnel diode is in high-voltage state, i.e. $R \approx U_v/I_p$. The difference of the currents flowing through the two tunnel diodes passes through the storage inductance L. The operating voltage must be high enough to allow a high current ($\approx I_p$) to flow through one of the two diodes, but should not exceed a maximum value, to permit only one tunnel diode to remain in high-voltage state. For the extreme values of the operating voltage a rough approximation is given by

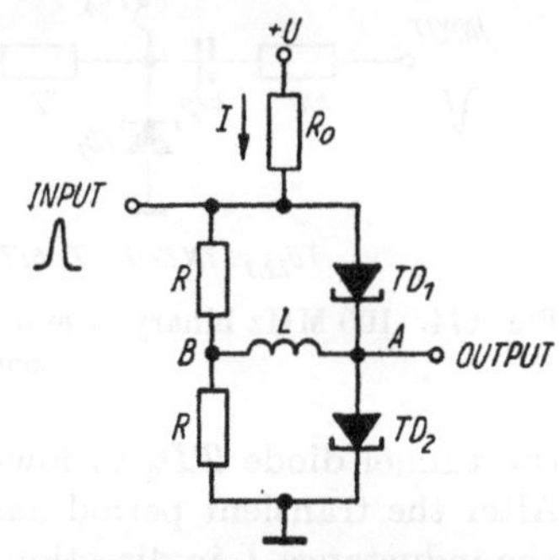

Fig. 173. Principle circuit of a two tunnel-diode binary stage.

$$\left.\begin{aligned} U_{\min} &\approx 2\,U_p + \left(I_p + \frac{U_p}{R}\right) R_0\,, \\ U_{\max} &\approx 2\,U_v + \left(I_v + \frac{U_v}{R}\right) R_0\,, \end{aligned}\right\} \quad (118)$$

where U_p and U_v are the peak and valley voltages, respectively, while I_p and I_v are the analogue current designations. More accurate equations are given by *Denisov* et al. [De 65]. In consideration of a high input impedance of the circuit the resistor R_0 should be suitably large. The inductive time constant

L/R_{eq}, with $R_{eq} = U_p/I_p + R(R_0 + R)/(R_0 + 2R)$, must be longer than the length of the trigger pulse to permit the inductance L to act as memory of the initial state. On the other hand, L/R_{eq} should be as low as possible to assure a short recovery time and consequently a high switching rate. When, for example, the tunnel diode TD_1 is in low-voltage state and the tunnel diode TD_2 in high-voltage state, the difference current through the inductance L flows in direction $A \to B$. A positive trigger pulse then switches the tunnel diode TD_1 to high-voltage state. This causes the current through the inductance L to decrease. The induced voltage is directed in such a manner that it switches

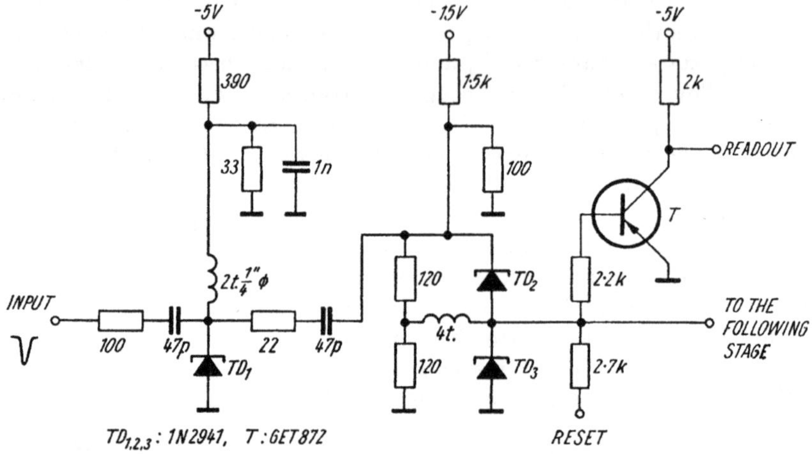

Fig. 174. 100 MHz binary circuit (TD_2, TD_3) with tunnel-diode pulse shaper (TD_1). According to *Whittaker* [Wh 64].

the tunnel diode TD_2 to low-voltage state after the trigger pulse has passed. After the transient period has died down the difference current flows through the inductance L in direction $B \to A$.

Binary circuits operating on this principle are mentioned by *Chow* [Ch 60b], with other versions given by *Baldinger* et al. [Ba 62], *Izokh* and *Efimchik* [Iz 62], *Whittaker* [Wh 64] and *Denisov* et al. [De 65]. *Sugarman* et al. [Su 62a] developed a similar arrangement. Depending on types of tunnel diodes and shape of the trigger pulse, switching rates were obtained ranging from 100 to 300 MHz. Switching rates of over 500 MHz have been reported in one instance [Ba 66b]. A practically designed binary combined with a tunnel-diode univibrator for shaping trigger pulses is shown in Fig. 174. Five of these stages were connected in series to form a scale-of-32 circuit for a pulse repetition rate of 100 MHz.

In construction of fast binaries the main difficulty consists in finding a short-time memory of suitably low recovery time. *Ortel* [Or 64] improved the principle illustrated in Fig. 173, where the memory inductance L simply becomes a loop formed by the leads of the other elements. Inductive coupling is used to the next stage by utilising this loop as primary of a pulse transformer. Three binaries connected in series achieved count rates of 1·2 GHz. In the binary arrangement of *Jungmeister* [Ju 67a] open ended homogeneous strip lines represent the short-time memory. Obtainable switching rate is largely limited by the rise time of the tunnel diode. Using sine input signals, frequencies up to 4 GHz could be handled. Similar high rates may be expected with superconductive logic elements based on the *Josephson* effect [Ma 67b].

Weber [We 64a] describes a binary circuit where the trigger pulses are fed-in through a diode gate (Fig. 175). Assuming the tunnel diode TD_1 in high-voltage

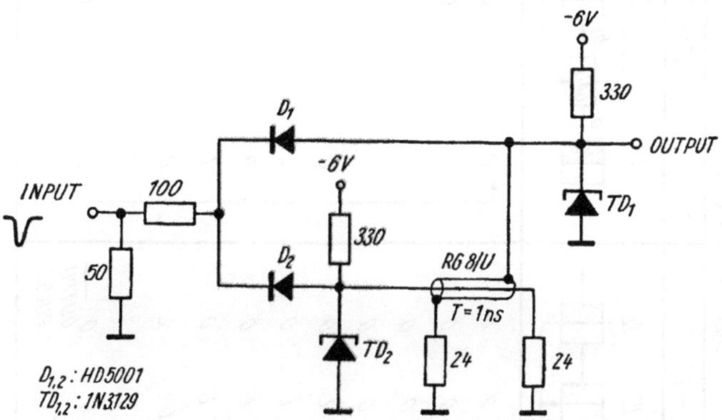

Fig. 175. Tunnel-diode binary circuit. According to *Weber* [We 64a].

state and the tunnel diode TD_2 in low-voltage state, the diode D_1 is highly biased while the diode D_2 is low-biased in reverse direction. In consequence, the negative trigger pulse is only transmitted by the diode D_2 and also switches the tunnel diode TD_2 to high-voltage state. After a delay time of 1 ns the transmission line inverter delivers a positive pulse to the tunnel diode TD_1 causing it to switch to low-voltage state. A switching rate to 300 MHz is possible with this binary circuit. *Nentwich* [Ne 65a] used backward diodes, type BD-3, as diodes D_1 and D_2, and achieved switching rates up to 420 MHz.

3.7.3 DUAL AND DECADE SCALING CIRCUITS

SERIES CONNECTED BINARIES

When connecting n binary stages in series, a dual scaling circuit is obtained. After every 2^n input pulses the circuit resets to initial state and delivers an output pulse. The pulse repetition rate is thus scaled by the factor 2^n (scale-of-2^n circuit). Because the switching rate decreases along the binary chain, the first binary must have the shortest resolving time.

On the left in Fig. 176 the example of a scale-of-8 circuit illustrates the various states assumed successively by the binary stages. The state where

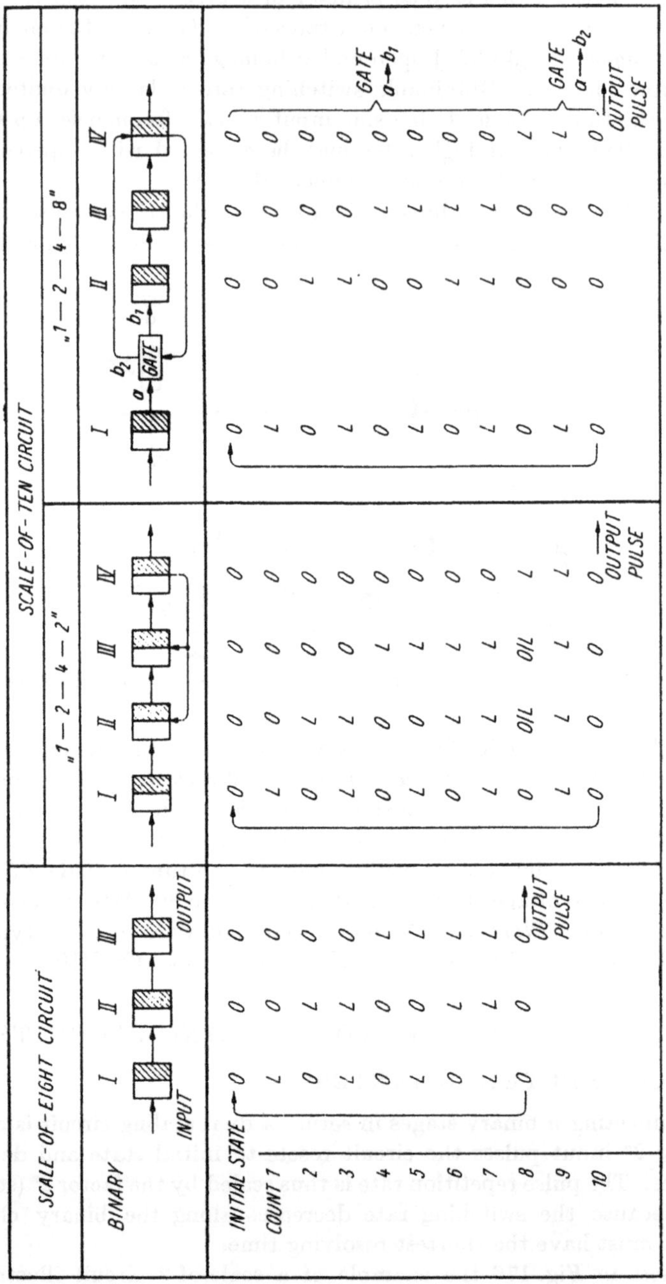

Fig. 176. Diagram of states of a scale-of-eight circuit and of two versions of a scale-of-ten circuit.

the right half of a binary stage of two active elements (cp. Fig. 166 or 169) conducts and the left half is blocked, is designated with 0 while L signifies the reverse condition. An output signal is produced after the last binary stage of the chain has been switched back from L to 0 by the 2^n-th pulse.

More convenient in handling are decade scalers which deliver an output signal after every 10th pulse. Four series-connected binary stages form a scale-of-16 circuit. By means of feedback branches the result of count is advanced by 6 which allows resetting to initial state already after 10 input pulses. With suitable selection of the feedback system any desired scaling relation between 2^{n-1} and 2^n can be obtained with n binary stages. Fig. 176 gives two of the most commonly used feedback arrangements of decade scalers.

With a coding of 1-2-4-2, the numbers giving the value of the four binaries, feedback takes place from binary IV to stages II and III. The 8th pulse switches the binaries I, II, III to the 0-state and stage IV to the L-state, then the stages II and III are returned to the L-state through the feedback system. This process corresponds to an advance of the scaler by the required $2^1 + 2^2 = 6$ steps. Decade scalers with a coding of 1-2-4-8 have a gate between the binary stages I and II. Output pulses are delivered from I to II as long as stage IV is in 0-state, and are passed from I to IV as soon as the latter is in L-state. In this manner the 10th pulse switches the stages I and IV only. Delay time between the 10th input pulse and the output pulse is only about half as with the 1-2-4-2 coding. Sometimes coding 1-2-2-4 is used which is characterised by feedback from III to II and from IV to III.

With the described decade scalers the time between the 8th and the 10th input pulse is available for feedback, while the 9th pulse switches only the binary stage I. Maximum count rate is achieved with the coding 1-2-4-8, because gate switching requires less time. In general, the maximum count rate that can be registered with the described decade scalers is about 50% lower than with a scale-of-16 circuit without feedback. For this reason dual scalers are often used for fast counting, or ring scalers, which are subsequently discussed, are employed.

With fast prescalers, reset and read-out is often omitted because the occurring absolute read-out error is less than 10 pulses with a decade scaler and can be neglected when a large number of pulses is registered.

In the construction of complete scaling circuits problems arise with respect to feedback and coupling of the various binary stages which, with fast scalers, frequently is effected with interposed amplifiers and pulse shapers. Details will be found in the publications cited in the following.

Scale-of-4 circuits with valve-equipped binary stages for count rates from 10 to 20 MHz are described by *Fitch* [Fi 49], *Schoenwetter* [Sc 53], and *Collinge* and *Huxtable* [Co 58]. The scale-of-8 circuits of *Nakamura* [Na 57] and *Hutchinson* et al. [Hu 59] have a resolving time of 25 ns and 11 ns, respectively. With the very high outlay decade scaler of *Lachinov* [La 63b] a count rate of 100 MHz was achieved. Of the transistorised binary stage scaling circuits, the scale-of-8 circuit of *Verweij* ([Ve 63b], cp. Fig. 170*b*, p. 279), should be mentioned which is capable of registering pulses with a spacing of 7 ns. *Iselin* [Is 63] describes a 25 MHz decade scaler with a coding of 1-2-4-2. Scale-of-10 circuits for a repetition rate of 20 to 30 MHz and a coding of 1-2-4-8 were reported by *Alexander* and *Heywood* ([Al 61a], cp. Fig. 170*c*), and *Denisov* et al. [De 65], and for 100 MHz by *Duchemin* [Du 63a].

Tunnel-diode binary stages are preferred for the construction of fast dual scaling circuits. For a count rate of 100 MHz *Whittaker* ([Wh 64], cp. Fig. 174) developed a scale-of-32 circuit and *Sugarman* et al. [Su 62a] a scale-of-8 circuit. Scaling circuits capable of handling a count rate of about 200 MHz, consisting of two or three series-connected binary stages, are described by *Baldinger* et al. [Ba 62] and by *Denisov* et al. [De 65]. *Lebail* [Le 63a] mentioned a circuit able to operate at a pulse repetition rate of 400 MHz. Decade scalers with the coding of 1-2-4-8 for a count rate of approximately 100 MHz have been developed by *Radeka* [Ra 64a] and by *Mantsa* et al. [Ma 66c].

RING SCALERS

A ring scaler consists of a chain of n series-connected binary stages, with the last stage again coupled to the first. In principle such a ring scaler functions as follows; through suitable resetting $(n - 1)$ stages are in state 0, and the stage m is in state L. The binary stages shall consist of two active elements, for instance, valves or transistors. The trigger pulse is simultaneously fed-in at every one of the n stages, being applied to that half of the binary stage which is blocked in 0-state. With a trigger pulse of respective polarity the stages that are in 0-state are not influenced, i.e. only the stage m is switched from L to 0. This causes the following binary stage $(m + 1)$ to transfer from 0 to L. In this manner the assigned stage advances by 1 until, after n input pulses, the whole ring has finally been run through. Thus, the scaling factor corresponds to the number of stages n. *Gruhle* [Gr 58a] has developed such a ring scaler for a count rate of 1 MHz, with the binary stages equipped with the switching tube E 91 H. Tunnel diode ring scalers for a maximum count rate of about 100 MHz were also constructed according to this principle [Am 63a; Er 68].

Contrary to the previously described dual and decade scalers the switching rate of the ring scaler is the same for all stages. Switching frequency, however, is irregularly distributed between the various stages, i.e. one binary stage remains in 0-state until it is switched to L-state by the preceding stage, but the next trigger pulse already switches it back to 0-state. For this reason all the binary stages must have the same resolving time as the first binary stage of a dual scaler for the same count rate.

With the *twisted ring scalers* proposed by *Hilberg* [Hi 60a] this disadvantage is avoided by twisting one connection between the binary stages, for instance, between stage IV and stage I, as shown in Fig. 177. The incorporated gating stages effect the change-over to another state of only one binary at a time. In this manner the initial state is obtained only after two runs, i.e. after 2 *n* pulses. *Hilberg* called this twisted ring scaler "*Moebius* counter", in analogy to the *Moebius* tape. This construction has the great advantage that during one run every binary stage is changed-over only once and the switching rate,

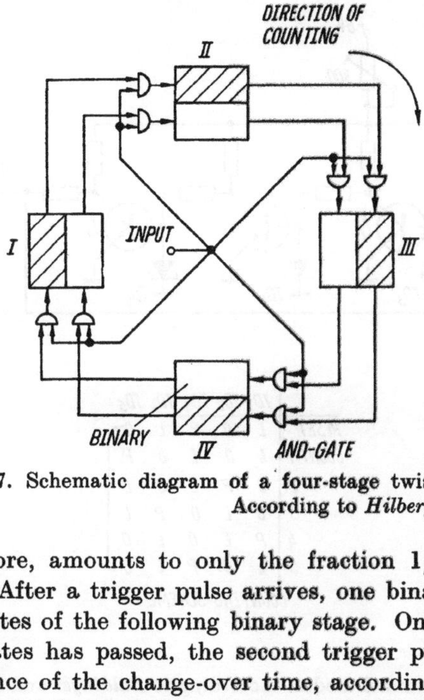

COUNTING SCHEME

BINARY	I	II	III	IV
RESET	0	0	0	0
COUNT 1	L	0	0	0
2	L	L	0	0
3	L	L	L	0
4	L	L	L	L
5	0	L	L	L
6	0	0	L	L
7	0	0	0	L
8	0	0	0	0

Fig. 177. Schematic diagram of a four-stage twisted ring scaler (scale-of-eight circuit). According to *Hilberg* [Hi 64].

therefore, amounts to only the fraction $1/n$ of the trigger pulse repetition rate. After a trigger pulse arrives, one binary stage is switched and controls the gates of the following binary stage. Only after the change-over period of the gates has passed, the second trigger pulse can be counted. Due to the influence of the change-over time, according to *Hilberg* [Hi 64], the counting speed of a three- or four-stage twisted ring scaler is not higher by the factor *n*, but only by the factor 2, than of a conventional ring scaler. The time until the whole ring has been run through is available as recovery time for the binary stages.

Hilberg continued the development of the twisted ring scaler under consideration of two aspects [Hi 64; Hi 65]. One of the gates is opened by the first arriving output pulse of the preceding binary stage, while the other of the gates of the binary stage is prepared for blocking. This is effected by the feedback from the output of one binary stage to the gating system of the preceding binary stage. It is further advantageous to use the rapidly increasing front of the trigger pulse also for the opening of the following gate, which can then be opened already before the transient period of a binary stage is concluded. In this manner *Hilberg* [Hi 64; Hi 65] achieved a resolving time of 2 ns with a three-stage twisted ring scaler, measured as minimum interval of the pulses from a double-pulse generator. With a twisted ring scaler constructed from the same components analogue to the circuit in Fig. 177, the resolving time was higher by the factor 5, and without the use of the *Moebius* principle, by the factor 10. *Tarczy-Hornoch* [Ta 61] described a five-stage twisted ring scaler which handles input repetition rates of 150 MHz, although each of the unsaturated transistor equipped binary stages can operate only at a rate of up to 35 MHz.

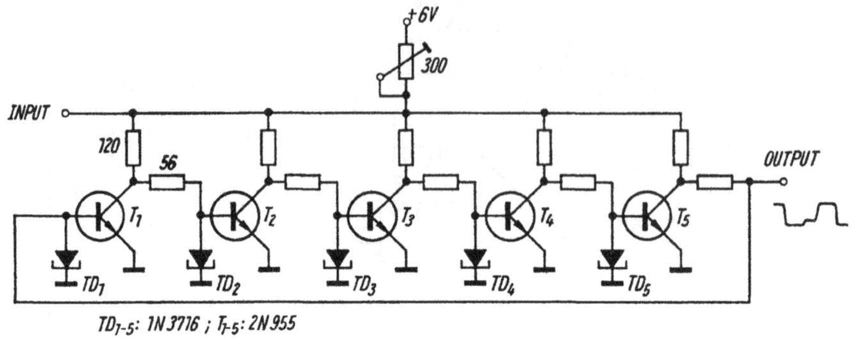

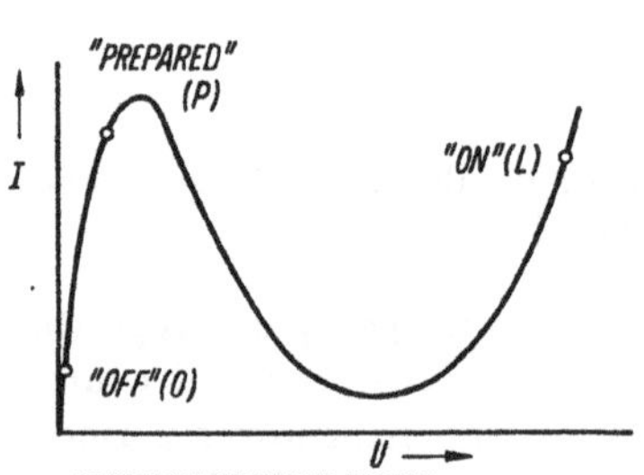

STATES OF THE TUNNEL DIODES

	TD_1	TD_2	TD_3	TD_4	TD_5
RESET	L	0	P	L	0
COUNT 1	L	0	L	0	P
2	0	P	L	0	L
3	0	L	0	P	L
4	P	L	0	L	0
5	L	0	P	L	0

COUNTING SCHEME

Fig. 178. Scale-of-five ring scaler with tunnel diodes. According to *Baldinger* and *Simmen* [Ba 65].

Frequently fast decade counters are constructed in ***bi-quinary design*** by joining a binary stage and a scale-of-5 ring scaler in tandem. In most cases, ring scalers are employed which do not consist of complete binary stages. Fig. 178 gives a very reliable five-stage ring scaler developed by *Baldinger* and *Simmen* [Ba 65]. From the potential relations a stable configuration results, where two tunnel diodes assume the state "on", two the state "off" and one tunnel diode the state "prepared". The tunnel diode in the state "on" opens the associated transistor and forces the neighbouring tunnel diode on the right into the "off"-state. The reset, not shown in Fig. 178, is selected in such a manner that the tunnel diodes TD_1 and TD_4 are initially in the "on"-state L and the tunnel diode TD_3 is in the state "prepared". The trigger pulse, which can only influence the prepared stage, switches the tunnel diode TD_3 to L; this causes the tunnel diode TD_4 to switch to 0, and the tunnel diode TD_5

shifts to the state "prepared". The tunnel diodes TD_1 and TD_2 remain unchanged. In this manner the prepared stage advances by two steps until the initial state is reached after five input pulses. This scaler circuit has the advantage that a wrong configuration caused by parasitic pulses, for example, where more than two tunnel diodes are switched to the "on"-state, is corrected by the next input signal. This avoids a permanent counting error. Other advantages are the not too close tolerance and the reliability due to the use of slightly saturated transistors, which permits production of relatively great numbers of this circuit [Ne 65a]. The fast binary stage precircuited to the ring scaler was constructed with two tunnel diodes, similar to the version given in Fig. 173, however, with a short-circuited transmission line in place of the inductance. The decade operates at a repetition rate of more than 100 MHz [Ba 65]. Silicon transistors, type 2 N 709, with shorter storage time and a better turn-on and turn-off behaviour than the germanium transistors, type 2 N 955, improve the counting capability of the ring scaler from 80 to 300 MHz [Ba 66b]. This ring scaler was used for the construction of a decade scaler capable to handle a count rate of up to 600 MHz [Ba 67b]. Here, the input pulse is formed by a combination consisting of a gallium arsenide tunnel diode, a hot carrier diode and a silicon transistor (similar as in Fig. 97, p. 164). The memory of the following binary is designed as pulse transformer. The signal taken off at the secondary end undergoes amplification before triggering the tunnel diode ring scaler. *Nentwich* [Ne 65a] combined the binary circuit given in Fig. 175 with the ring scaler according to *Baldinger* and *Simmen* [Ba 65] to obtain a decade scaler able to handle a repetition rate of 100 MHz. *Tan* [Ta 67b] describes a five-stage 500 MHz ring scaler which differs in so far as four of the five tunnel diodes are in low-voltage state. After every input pulse the stage assigned by the high-voltage state advances one step. The tunnel diodes are connected by long-tailed emitter pairs.

Weber [We 64a] developed a bi-quinary decade scaler for pulse repetition rates up to 250 MHz. In its binary stage, and in the ring scaler, tunnel diodes are used which are coupled through a transmission line inverter. The ring scaler constructed by *Engelmann* [En 63] consists of stages with two transistors each, which are switched alternately. He gives 500 MHz as maximum repetition rate. Dimensioning of the circuits, however, seems critical. The bi-quinary scaler built by *Denisov* et al. [De 65] contains a ring scaler according to *Engelmann* and is suitable for a count rate of approximately 200 MHz.

Scaling circuits can also be constructed with *series-connected tunnel diodes*. The volt-ampere characteristic of n series-connected tunnel diodes permits $(n + 1)$ stable working points (cp. Fig. 72*d*, p. 124). In this case, $(n - 1)$ trigger pulses switch $(n - 1)$ tunnel diodes successively to high-voltage state, until the n-th trigger pulse actuates a reset transistor connected parallel to the tunnel-diode chain, which returns every tunnel diode to low-voltage state. *Spiegel* [Sp 60] connected 10 tunnel diodes in series to obtain a decade scaler with a double-pulse resolution of less than 50 ns. *Foote* and *Johnson* [Fo 64] used five series-connected tunnel diodes and a following binary stage to construct a decade scaler operating at a repetition rate of more than 50 MHz.

Finally, the proposed ***sampling method*** by ***Hilberg*** [Hi 64] must be mentioned, which permits operation at count rates in excess of 1 GHz. Pulses in this circuit run through a transmission line. At a specified moment of time every pulse which happens to be in the line is sampled from measuring points evenly distributed along the line. This sampling is repeated at an interval which corresponds to the transit time through the line, allowing for a batchwise processing of the pulse train. This arrangement has a certain similarity with the chronotrons described in Section 3.4.3.

4. TESTING APPARATUS

4.1 PULSE GENERATORS FOR ELECTRICAL PULSES

4.1.1 GENERAL

For the examination of all types of pulse circuits, for the measurement of the properties of semiconductor components, for the operation of spark chambers and image intensifiers, for pulsing the particle beam of an accelerator, and similar purposes one needs current and voltage pulses with a pronounced step-type slope, especially the leading edge, e.g., rectangular pulses with adjustable width, needle pulses, etc. This section describes pulse generators producing pulses with a rise time in the order of 1 ns and a fall time which, in many instances, almost corresponds to the rise time. Generators delivering a sawtooth voltage curve, e.g. for the time base of fast oscilloscopes, are not considered.

In general, free-running pulse generators produce periodic pulses where, for instance, repetition rates of 100 MHz and more are required for scaling circuit testing. Trigger circuits, on the contrary, deliver a pulse every time they are actuated by a trigger signal. These circuits have been discussed in Section 3.2.4 and a repetition here is superfluous.

A free-running oscillator, e.g. for producing a sine or meander voltage curve, can be combined with a trigger stage to form a free-running pulse generator. The following description limits itself in the main to such free-running pulse generators where the output signals are directly utilised or fed to a subsequent shaping stage, which need not be a trigger stage, for further shaping (pulse sharpening).

In general, the problem of generating voltage steps is that of a rapid charging of capacitances which requires high currents. One of the preconditions for generating satisfactory current steps is a low-delay current switching circuit with minimum parasitic inductance.

4.1.2 UNDELAYED-FEEDBACK GENERATORS (MULTIVIBRATOR CIRCUITS, ASTABLE TRIGGER STAGES)

VALVE-EQUIPPED AND TRANSISTORISED MULTIVIBRATORS

Astable multivibrator systems consist of two feedback-coupled valves or transistors which alternately conduct or are blocked. Fig. 179 gives a schematic illustration of these multivibrators which vibrate periodically because the a.c.

coupling, unlike the bistable trigger stages shown in Figs. 166 and 169, does not permit a stable state of equilibrium. Vibrating frequency is mainly determined by the time constants $R_g C$ and $R_B C$, respectively. Detailed data on the static and dynamic behaviour of these circuits can be found in various books [Ne 55a; Mi 56; Ne 61; Ba 64].

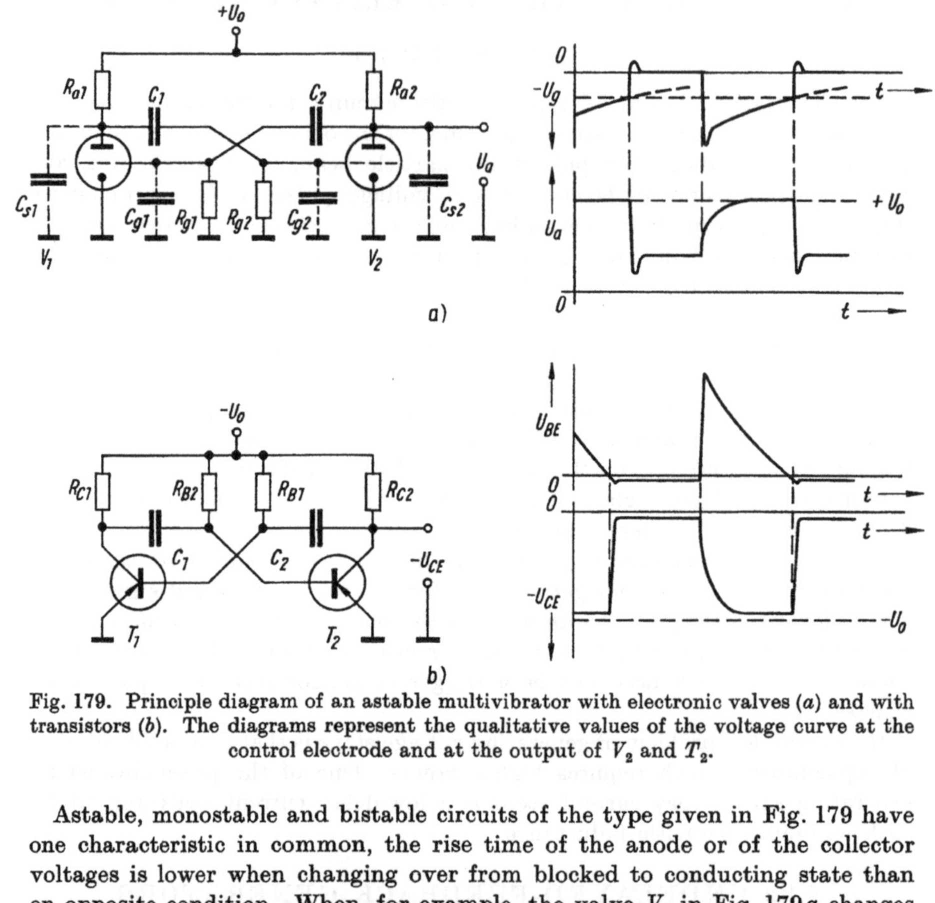

Fig. 179. Principle diagram of an astable multivibrator with electronic valves (*a*) and with transistors (*b*). The diagrams represent the qualitative values of the voltage curve at the control electrode and at the output of V_2 and T_2.

Astable, monostable and bistable circuits of the type given in Fig. 179 have one characteristic in common, the rise time of the anode or of the collector voltages is lower when changing over from blocked to conducting state than on opposite condition. When, for example, the valve V_2 in Fig. 179*a* changes from blocked to conducting state the anode capacitance of V_2,

$$C'_a = C_{s2} + \frac{C_{g1} C_2}{C_{g1} + C_2} \approx C_{s2} + C_{g1},$$

must be recharged. The maximum possible initial steepness of the negative slope results on $R_{a2} = \infty$ and amounts to I/C'_a, with I representing the switched-on anode current. Thus, a short rise time of the negative voltage step necessitates a high anode resistance R_{a2} and a high charging current I, which is usually achieved by overdriving the valve or the transistor on actuation. When changing over to blocked state the anode capacitance of V_2 is $C_a'' \approx C_{s2} + C_2 \gg C_a'$, because C_2 is charged by the grid current from valve V_1. In consequence, the positive pulse slope rises with the time constant $R_{a2} C_a''$ and the anode resistance, therefore, must be low. From this it derives then that the rise time of the positive pulse slope in valve-equipped multivibrators is longer than the negative slope by about one order of magnitude. With transistorised multivibrators conditions are analogue.

Kroebel [Kr 54] suggested connecting a secondary-emission pentode parallel to valve V_2 to shorten the rise time of the positive slope of the multivibrator in Fig. 179*a*. In this case, when V_2 is blocked, the dynode current of the secondary-emission valve charges the anode capacitance C_a'', irrespective of the value of R_{a2}. At the positive and the negative slopes anode voltage changes of 5 to 10 V/ns were obtained [Kr 54; Ru 54].

A comparison of transistorised with valve-equipped multivibrator circuits shows a better ratio of the output voltage changes to the voltage supply, as indicated in Fig. 179. The maximum obtainable switching rate of an astable multivibrator according to Fig. 179*b* is proportional to the transistor cut-off frequency, and is less strongly dependent on current gain factor and on load [Ba 64].

With high switching rates, capacitors C_1 and C_2 must be very small, and may be of the same value as the capacitances of the transistors and the wiring. In such cases it is preferable to use asymmetric multivibrators with capacitive emitter coupling, because this permits a higher value for the frequency-determining capacitance. A circuit of this type is obtained if the series network $R_d + D$ of the trigger stage given in Fig. 87, p. 151, is replaced by a second current supply [Ba 64]. *Alexander* and *Heywood* [Al 61a] designed a transistorised multivibrator for a switching rate up to 40 MHz, where inductances are utilised as frequency-determining elements. This circuit corresponds to the trigger stage shown in Fig. 170*c*, p. 279, although without the 150 Ω resistors in the base circuit which effect d.c. coupling.

USING SECONDARY-EMISSION VALVES

With the secondary-emission valves, types EFP-60 and 6B1П, mentioned in Section 3.1.3, pulse generators and trigger circuits can be easily constructed (cp. Section 3.2.4). In Fig. 180 two feedback circuits are shown, the dynode-grid and the anode-cathode type. On $U_g = 0$ both systems operate as free-running pulse generators and represent monostable trigger stages, if the valve is blocked in initial state by the grid bias $-U_g$. Function of these circuits corresponds to a blocking oscillator with pulse transformer [Mo 64].

Changes of the dynode and grid voltages are in-phase. Therefore, self-excitation can take place if the voltage amplified at the dynode is fed back to the grid through the capacitor C_1 (Fig. 180*a*). When the grid potential increases the dynode current and voltage rise avalanche-like due to the feedback action, with the grid current charging capacitor C_1. After charging, the grid current decreases, which also causes the dynode current to drop. Through the capacitor C_1 the grid is driven to negative, and the valve becomes blocked. Capacitor C_1 then discharges with the time constant $R_g C_1$ to the level of U_g. Dimensioned as given in Fig. 180*a*, pulses with a peak value of 1 A and a rise time of 10 ns were obtained at the anode [Mo 52].

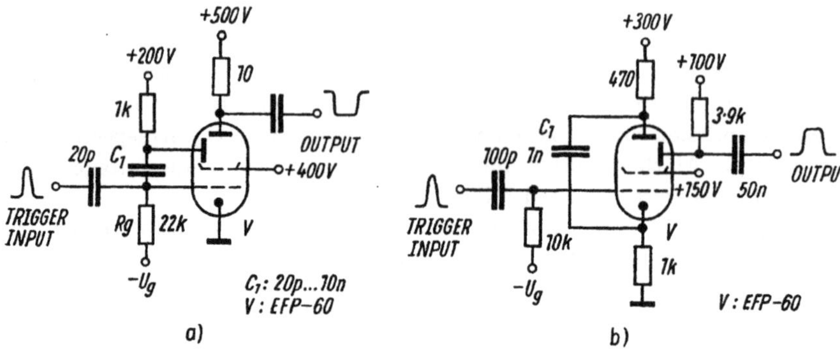

Fig. 180. Pulse generator circuits with secondary-emission pentodes.
a) Dynode-grid feedback circuit. According to *Moody* et al. [Mo 52].
b) Anode-cathode feedback circuit. According to *Wells* [We 52].

Pulse-generator function of the circuit shown in Fig. 180*b* is based on the fact that due to amplification at the dynode the anode current is higher than the cathode current. This permits a gain factor exceeding unity in the feedback loop. Circuits of this type have been used by *Wells* [We 52], *Rosenheim* and *Anderson* [Ro 57], and by *Kolotov* [Ko 62a]. *Hayes* and *Alsford* [Ha 63a] employed a pulse-generator setup with an inductance in place of the ohmic resistance in the cathode circuit. The value of the inductive time constant is selected to allow feedback action only during the rise time of the positive trigger pulse. In this manner a steep leading edge of the output pulse is obtained.

The circuits in Fig. 180 have the disadvantage of a relatively unsatisfactory stability of the pulse width and of the repetition rate (on $U_g = 0$), which depends to a great extent on the value of the feedback capacitor C_1 and the valve characteristic. Pulse width can be stabilised by means of a short-circuited transmission line in the anode circuit [We 52]. A constant pulse repetition rate is obtained by triggering or synchronising the secondary-emission valve [We 52; Ro 57; Ko 62a], or by interposing a delay line in the feedback loop, as described in Section 4.1.3.

UTILISING AVALANCHE TRANSISTORS

Analogue to secondary-emission pentodes, avalanche transistors may be used for constructing single-stage pulse generators. Avalanche transistor function is described in Section 3.1.5. The trigger stage shown in Fig. 78*a*, p. 137, is free running if the condition $U_B > I_A (R_B + r_b)$ required for monostable operation does not apply any more [see equation (100)]. In most cases, avalanche transistors are employed for monostable trigger circuits, of which some examples are given in Section 3.2.4.

Avalanche transistors are utilised to advantage when pulses between 0·1 and 1 A with a rise time in the order of 1 ns are desirable. *Tansal* and *Sobol* [Ta 63a], for instance, used an avalanche transistor as free-running pulse generator, with an open-circuited transmission line in the collector circuit for pulse width determination. An impedance transformer with a ratio of 16 : 1 is arranged in the emitter circuit. With a repetition rate of 5 MHz, limited by the maximum collector dissipation of the avalanche transistor, current pulses of 1·7 A and a halfwidth of 8 ns were obtained at a load resistance of 1 Ω. *Hamilton* et al. [Ha 62a] achieved peak currents of 7 A with type 2 N 706 transistors.

By connecting several avalanche transistors in parallel high-current nanosecond pulses can be produced. Fig. 181 gives the circuit developed by *Prince* [Pr 65] which illustrates the time-isolation technique used for the purpose. Transmission lines L_1 and L_2, each with a delay of 1 ns, are used which ensure that, either through the common trigger line or the load resistors, the reaction of the neighbouring transistor comes into effect only after 2 ns when every

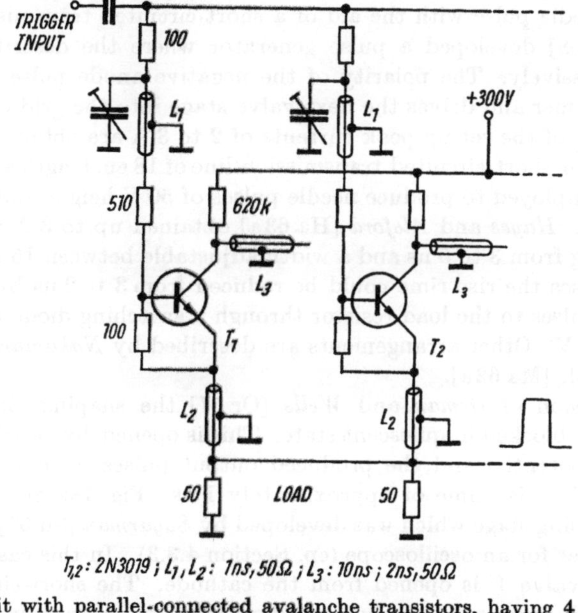

Fig. 181. Circuit with parallel-connected avalanche transistors, having 4 transistors and output pulses of 60 V height on a 10 Ω load resistor, with a rise and fall time of ≈1 ns. According to *Prince* [Pr 65].

transistor has already fired. If the transmission lines L_1 and L_2 were not employed, the first responding avalanche transistor would raise the emitter voltage of the others and probably prevent their firing until the current pulse from the first transistor had decayed. With a circuit having four parallel-connected transistors and a load resistance of 10 Ω pulses exceeding 60 V with a slope of approximately 1 ns were measured. With three avalanche transistors in parallel where the collector voltage had been individually balanced *Hansen* and *Schmidt* [Ha 67a] obtained 25 A pulses with a rise time of 5 ns and a repetition rate up to 10 kHz. Series connection of several avalanche transistors is a simple matter. In this way, *Robbins* et al. [Ro 68] produced pulses of 150 V height on a 50 Ω load with three avalanche transistors. High-voltage pulses in excess of 1 kV can also be obtained with avalanche transistors as discharge path [Ju 66; Pa 66b] in a *Marx* generator (see Fig. 191, p. 308).

STEEPENING OF PULSE SLOPES (PULSE SHARPENING)

The higher the available charging current the faster the voltage change is effected when recharging a capacitance. Driving valves with positive pulses into the grid current region obtains anode current pulses amounting to multiples of the maximum d.c. cathode current. With a suitably selected negative grid bias the steepest part of the rise of the positive control pulse can be utilised. The negative step pulse appearing at the valve output is shaped to a rectangular pulse or a needle pulse with the aid of a short-circuited transmission line.

Kohn [Ko 58] developed a pulse generator where the described overdrive is used successively. The polarity of the negative anode pulse is reversed in a 1:1 transformer and drives the next valve stage into the grid current region. At the output of the set-up peak currents of 2 to 3 A are obtained with a type PL 81 valve. A short-circuited transmission line of 18 cm length and a 60 Ω load resistor are employed to produce needle pulses of 50 V height and a base-width of under 3 ns. *Hayes* and *Alsford* [Ha 63a] obtained up to 3 A with rise time values ranging from 3 to 5 ns and a width adjustable between 15 ns and 100 μs. With 1 A pulses the rise rime could be reduced from 3 to 2 ns by transmitting the current pulses to the load resistor through a switching diode reverse biased with 30 to 50 V. Other arrangements are described by *Nakamura* [Na 59] and *Mamyrin* et al. [Ma 62a].

In the version of *Orman* and *Wells* [Or 57] the shaping circuit contains a valve that is blocked in quiescent state. This is opened by rapidly decreasing the cathode potential and the produced output pulses have a height of 7 V (on 70 Ω) and a rise time of approximately 1 ns. Fig. 182 gives an example of a pulse shaping stage which was developed by *Sugarman* [Su 57] for obtaining sampling pulses for an oscilloscope (cp. Section 4.3.3). In this case the blocked pencil triode valve V is opened from the cathode. The short-circuited transmission line L_1 serves to produce needle pulses which are further reduced through the biased discriminator diode D_1. The resulting needle pulse passes

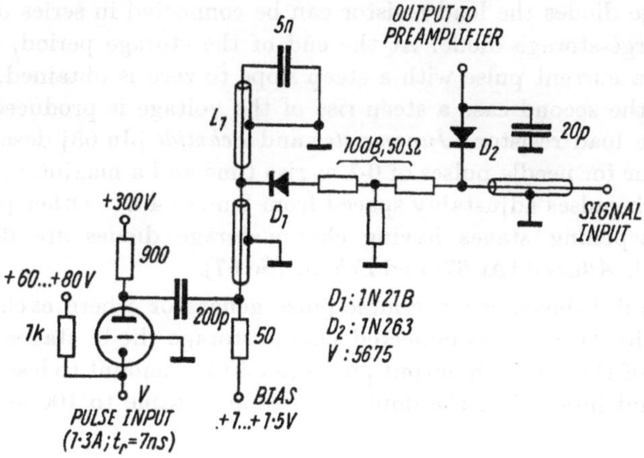

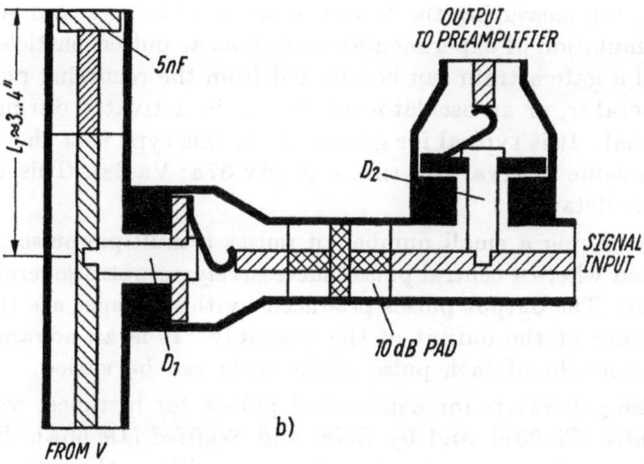

Fig. 182. Pulse shaping stage for sampling pulses of a sampling oscilloscope. According to *Sugarman* [Su 57].
a) Circuit diagram. *b)* Arrangement of output end.

through the 10 dB pad to the sampling diode D_2 and is mixed with the input signal. Pulse height is 0·1 V and width amounts to about 0·5 ns. Construction of the coaxial shaping and mixing stage is detailed in Fig. 182*b*. *Marte* [Ma 64b] analysed pulse sharpening with several series-connected valves in grounded grid circuitry.

Charge-storage diodes (Section 3.1.4) and, in case of high current intensities, the utilisation of shock waves in ferrite lines (cp. Sections 3.1.2 and 4.1.4) offer further possibilities for steepening the pulse slope. For pulse sharpening using charge-storage diodes the load resistor can be connected in series or in parallel with the charge-storage diode. At the end of the storage period, through the load resistor a current pulse with a steep slope to zero is obtained, in the first case, and in the second case a steep rise of the voltage is produced across the diode and the load resistor. *Jungmeister* and *Schmidt* [Ju 68] describe a pulse train generator for needle pulses of 0·1 ns rise time and a maximum of 60 pulses in a train with pulses adjustably spaced from 2 ns to 4 ns. Other pulse generators and sharpening stages having charge-storage diodes are described by *Fraser* [Fr 66], *Attwood* [At 67] and *Schultz* [Sc 67].

Stromer [St 65] developed a double-pulse generator where each pulse edge is shortened by two series-connected charge-storage diode stages. Rise time and fall time of the 8 V high output pulses (on 50 Ω) amount to less than 0·4 ns, with width and interval of the double pulses adjustable to 100 ns.

PULSE TRAIN GENERATORS

Compared to periodically recurrent pulses, pulse trains offer some advantages, for instance, for measuring the time resolution of trigger and scaling circuits or for the simulation of experimental conditions at pulsed particle accelerators. By means of a gate a train can be selected from the recurring pulses of a free-running generator, or an oscillator circuit can be activated during the interval of a gate signal. It is typical for generators of this type that the train contains pulses of the same interval and size, e. g. [Av 67a; Va 68]. This will, however, not be further detailed.

For trains having a small number of pulses a multiple-pulse generator can be constructed where a central pulser successively actuates several independent trigger stages. The output pulses produced by these stages are then combined to a pulse train at the output of the assembly. It is an advantage that the time and the height of each pulse of the train can be varied.

Triple-pulse generators for nanosecond pulses, for instance, were developed by *Charbonnier* [Ch 63c], and by *Deike* and *Seyfried* [De 66a]. Setups of this type initially produce a sawtooth voltage permitting the adjustment of the actuating moment with the response threshold of the three trigger stages. An avalanche transistor-equipped triple-pulse generator [De 66a], for example, delivers pulse groups up to three pulses with a delay that can be continuously adjusted from 0 to 200 ns. Repetition rate of the train amounts to 10 kHz. In the four-pulse generator of *Moody* et al. [Mo 63] *RC* networks effect the delay for each trigger stage, while delay lines are resorted to by *Rosenheim* and *Anderson* [Ro 57], and by *Iveković* [Iv 65]. As described in Section 4.1.4 (Fig. 188) a double-pulse can also be obtained when discharging a transmission line through a switch.

High-power pulse trains are produced by discharging capacitors or *LC*-lines through several spark gaps [Li 65a] or thyratrons [Ba 66c].

4.1.3 RECYCLING PULSE GENERATORS

The exterior characteristic of a recycling pulse generator set is the delay line which connects the output of an active shaping stage, such as a valve, a transistor or a trigger stage, with its input (Fig. 183). In this manner, the output pulse of the shaping stage, which is actuated by an external signal, will pass through the delay line L and again activates this stage. If the delay T is longer than the pulse width a periodic pulse repetition of a period duration $T + T'$ is produced, with T' representing the delay of the trigger circuit. The major advantage of this type of pulse generator is the high constancy of the pulse repetition rate which depends only on the transmission line, provided that $T \gg T'$ applies. Such arrangements can serve for delaying a given signal by the multiple of the recycling period $T + T'$, as, for instance, is necessary with vernier chronotrons (cp. Section 3.4.6 and [Ba 57]). When, for example, the trigger stage is repeatedly actuated during the first recycling period, the given configuration of the pulses remains unchanged during circulation. For this reason, these circuits can be used as serial storage registers in electronic computers [Le 64a].

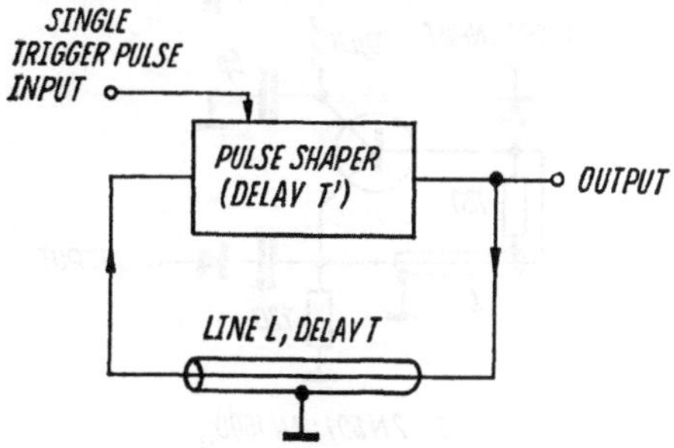

Fig. 183. Principle diagram of a recycling pulse generator.

Fig. 184 shows two examples of recycling pulse generators containing trigger stages. In the version shown in Fig. 184*a* an avalanche transistor is employed [Gy 63]. Pulse width is about 1 ns, while delay L between emitter and base amounts to 200 ns which results in a pulse repetition rate of 5 MHz. The version designed by *Efimchik* et al. [Ef 65b] is given in Fig. 184*b*. Here, the tunnel-diode univibrator TD_1 is triggered by a pulse at input A. The negative output pulse travels through the delay line L, is reflected at the diode D with the same polarity and again triggers the univibrator TD_1 after an interval of $2\,T$. When a pulse at the input B switches the tunnel diode TD_2 to high-voltage state, the pulse coming from the univibrator TD_1 is removed through the now conducting diode D which stops the pulse train. Pulses of a width of 4 ns at repetition rates up to 100 MHz are obtained. Using a modification where the tunnel diode operates without inductance a meander-shaped output voltage can be obtained [Iv 67c]. Other recycling pulse generator circuits, especially for serial storage registers, are given in various papers [Do 63a; Wi 63c; Le 64a]. *Weber* has designed a generator [We 64b] which also includes a tunnel-diode trigger stage and delivers pulses of 2 ns halfwidth at a repetition rate of 10 MHz. *Righini* [Ri 65] developed a generator with transistorised univibrators which produces pulse trains up to 130 pulses, where the pulse interval can be continuously adjusted between 60 and 300 ns.

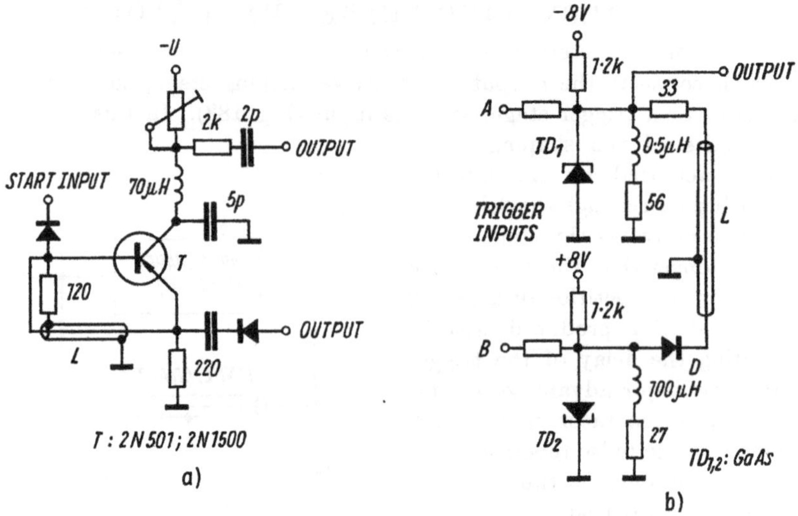

Fig. 184. Recycling pulse generators with trigger stages.
a) With an avalanche transistor. According to *Gygi* and *Schneider* [Gy 63].
b) With tunnel diodes. According to *Efimchik* et al. [Ef 65b].

The recycling pulse generator principle can also be applied if the circulating pulse is reshaped by amplifier and limiter circuits instead of trigger stages. *Yu* [Yu 53] gives the scheme shown in simplified presentation in Fig. 185*a*, where initially the valves V_1 and V_2 are blocked. At the time $t = 0$ valve V_2 is opened by the positive trigger pulse. The negative voltage step produced at the output of V_2 is fed into the delay lines L_1 and L_2. Reflexion at the short-circuited end of L_2 results in a positive voltage step which arrives at the time $2\,T_2 + T_1$ at the grid of valve V_1 and opens it. The negative step at the anode blocks valve V_1 again after a further delay T_1 until it is reopened by the signal reflected in the delay line L_2. It is a characteristic of the circuits shown in Fig. 185 that the system runs up until the pulses have reached a final shape which is determined solely by the delays and the amplification non-linearities. The number of recycling periods required until the final state is reached will be the less the better the shapes of the trigger pulse and the final pulse correspond.

If the delay line L_2 is omitted in the circuit of Fig. 185*a* one obtains a meander-shaped pulse which has a period duration of $2\,T_1$. In this manner *Borchi* [Bo 63a] obtained a repetition rate of 50 MHz with a mean rise and fall time of 0·6 and 0·85 ns with a type 2 N 1195 transistor. With 100 MHz the respective values were 0·9 and 1·25 ns. *Grigor'ev* [Gr 67a] used a secondary-emission valve, type 6B1П, whereby the delay system was interposed between anode and grid. When utilising the electron transit time in the valve (about 4 ns) a repetition rate of 230 MHz resulted.

In the circuit of Fig. 185*b* the feedback pulse is differentiated at the capacitor C_1. The differentiating system C_1R_k, with a time constant which is less than the transit time of the cable, determines the produced width. At a maximum repetition rate of 170 MHz, on 37·5 Ω load resistance, needle pulses of

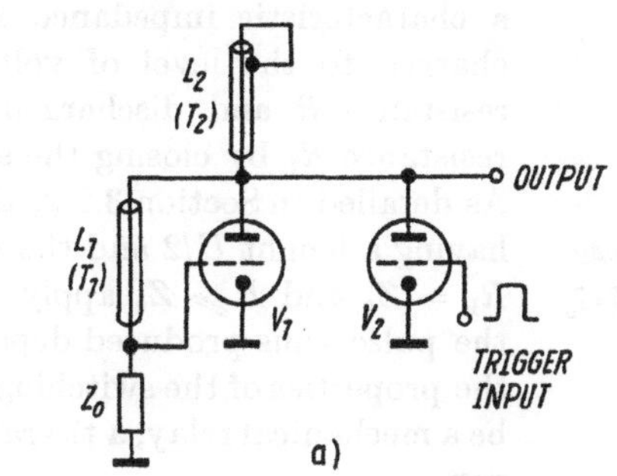

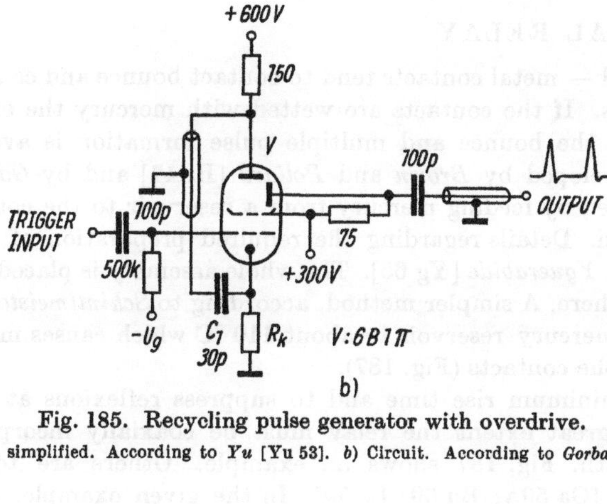

Fig. 185. Recycling pulse generator with overdrive.
a) Principle circuit, simplified. According to *Yu* [Yu 53]. *b*) Circuit. According to *Gorbachev* et al. [Go 62d].

120 V height and a width of approximately 5 ns were obtained. *Bay* and *Grisamore* [Ba 56b], *Gerharz* [Ge 56], and *Rosenheim* and *Anderson* [Ro 57] give similar circuit arrangements.

In the version designed by *Cutler* [Cu 55] the secondary-emission valves are replaced by 2 tandem-connected triodes in cathode-base circuit, and the pulses of the anode of the second valve pass through a differentiating system and a delay line to the grid of the first valve.

4.1.4 PULSE GENERATION THROUGH DISCHARGE OF A STORAGE ELEMENT (TRANSMISSION LINE, CAPACITOR)

By discharging an open-ended delay line through a switch rectangular pulses with steep edges can be obtained in a simple manner. The principle of such a discharge line generator is shown in Fig. 186. In function, the line with a characteristic impedance Z_0 and a delay T is charged to the level of voltage U through the resistance R and discharged through the load resistance R_L by closing the switching element S. As detailed in Section 3.2.2, this produces a pulse having a height $U/2$ and the width $2\,T$, provided $R_L = Z_0$ and $R \gg Z_0$ apply. Slope steepness of the pulse thus produced depends significantly on the properties of the switching element, which can be a mechanical relay, a thyratron valve or a spark gap.

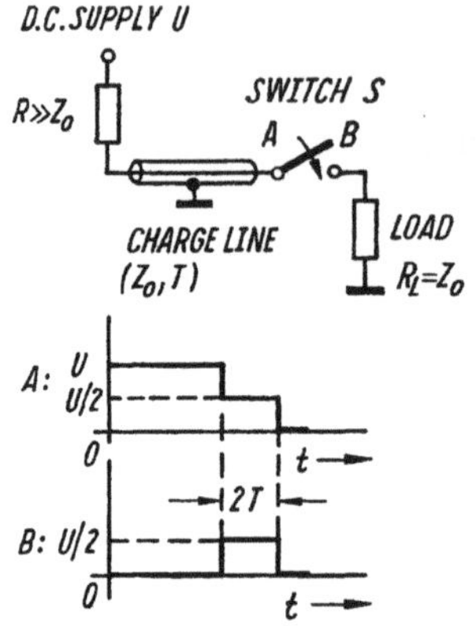

Fig. 186. Principle of a discharge line generator.

MECHANICAL RELAY

Simple metal — metal contacts tend to contact bounce and cause undesirable multiple pulses. If the contacts are wetted with mercury the electric contact is retained on the bounce and multiple pulse formation is avoïded. In the assemblies developed by *Brown* and *Pollard* [Br 47] and by *Garwin* [Ga 50a] this is performed by feeding mercury from a reservoir to the contacts through capillary action. Details regarding the required preparation of the electrodes are found with *Yguerabide* [Yg 65]. The whole assembly is placed in a 10 at hydrogen atmosphere. A simpler method, according to *Schintlmeister* et al. [Sc 61], is to heat a mercury reservoir to about 110 °C which causes mercury vapour to deposit on the contacts (Fig. 187).

To obtain minimum rise time and to suppress reflexions at the switching element to a great extent the relay must be coaxially incorporated in the conduction path. Fig. 187 shows an example. Others are to be found in various papers [Ga 50a; Bu 59; Le 59]. In the given example, the inner conductor and the vibrating reed are of steel. Opposite magnetic poles are induced across the contact gap by a magnetic alternating field produced by the driving coils. In consequence, the contact closes with the cycle of the field due to magnetic attraction. Frequency and strength of the magnetic field are selected that the relay vibrates most smoothly.

The notable attractions of mechanical switching pulse generators are that pulse height and polarity are determined by the conveniently measurable d.c.

supply U and that pulse width depends on the length of the charge line. Disadvantageous is that this type of generator cannot be triggered with constant time delay by an external pulse. The low pulse repetition rate, amounting to between 50 and 500 Hz, depending on the mechanical construction of the relay, is at times annoying.

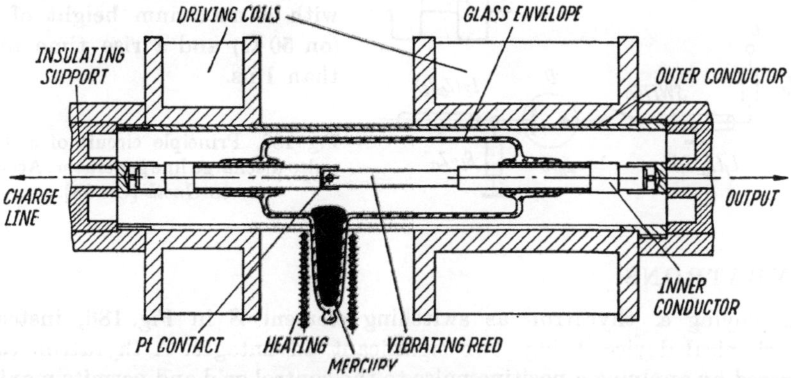

Fig. 187. Section through a coaxial vibrating switch with mercury-wetted contacts. Reed vibrates vertical to plane of presentation, vibration rate about 280 Hz. According to *Schintlmeister* et al. [Sc 61].

Mechanical relays with mercury-wetted contacts obtain pulse rise and fall times of about 0·1 to 0·5 ns. Observation has shown that with a supply of $U < 10$ to 20 V the generators give the most reliable operation, on a condition when no spark has yet developed, and with a supply of $U > 100$ V, if a defined gas discharge takes place just before mechanical contacting [Br 47; Le 59; Sc 61]. By using matching resistors the output pulse can be distributed to several outputs, with pulse height reducing to $U/4$ with two outputs and to $U/6$ with three outputs (cp. Fig. 62, p. 110). *Morugin* and *Glebovich* [Mo 64, chapter 3] give a method to produce a well defined rectangular pulse also with load resistances of $R_L > Z_0$. In this case the left end of the charge line shown in Fig. 186 is connected to ground through a correction resistance with an interposed second switch, the closing of which is delayed by the time T against that of the switching element S. This method, where the value of the correction resistance depends on the load resistance R_L, can be achieved, for instance, with two mutually delayed thyratron valves. A modification is a circuit where only one switching element is used (for details see [Mo 64]).

By discharging lumped delay lines a pulse width in the microsecond region is obtainable. When connecting such a line in series with a coaxial transmission line rise time values in the order of 1 ns can be achieved [Me 63].

Gnoth [Gn 63] developed a double-pulse generator which also operates with a mercury-wetted relay. Its principle circuit is given in Fig. 188, where a pulse having a width $2\,T_1$ is produced by the charge line L_1. The pulse is transmitted through diode D to the load resistance R_L and the open delay line L_2, where it is reflected, causing a second pulse to arrive at R_L at the time $2\,T_2$, if $T_2 > T_1$ applies. This pulse cannot be removed through the diode D. When the diode was replaced with a type 2 C 40 valve as separating stage, negative pulses were obtained with a maximum height of 10 V (on 50 Ω) and a rise time of less than 1 ns.

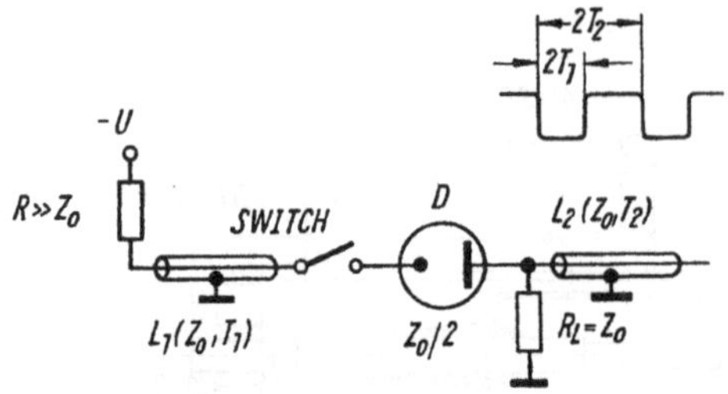

Fig. 188. Principle circuit of a double-pulse discharge line generator. According to *Gnoth* [Gn 63].

THYRATRONS

Employing a thyratron as switching element S in Fig. 186, instead of a mechanical device, brings two significant advantages. A thyratron can be triggered by applying a positive pulse to the control grid and permits maximum pulse repetition rates in the order of 10 kHz. Rise time of the output pulse amounts to some nanoseconds and, with thyratrons capable of handling a peak current of more than 100 A, about 10 ns. Occasionally the time delay between the trigger pulse and the firing of the thyratron is annoying, for instance, in spark chamber operation (cp. Section 5.3). This delay results from the built-up to breakdown and decreases with rising filament voltage and trigger pulse height. An approximation fixes this time delay at 50 to 100 ns (cp. Section 3.1.3 and [Fi 61]). Hydrogen-filled thyratrons are suitable for a maximum current of 10^3 A and an operating voltage to about 20 kV.

A typical thyratron pulse generator circuit is shown in Fig. 189. The open transmission line L is located in the anode circuit, and is charged from the supply. The line discharges through the load resistance R_L after the thyratron has fired, producing a positive rectangular pulse having a width of $2\,T$. If the line is replaced by a capacitance, a step-type exponentially decaying pulse is produced. Comprehensive representation of thyratron pulse generators, including series-connected thyratrons, are to be found in a number of books (cp. [Le 59; Mo 64]).

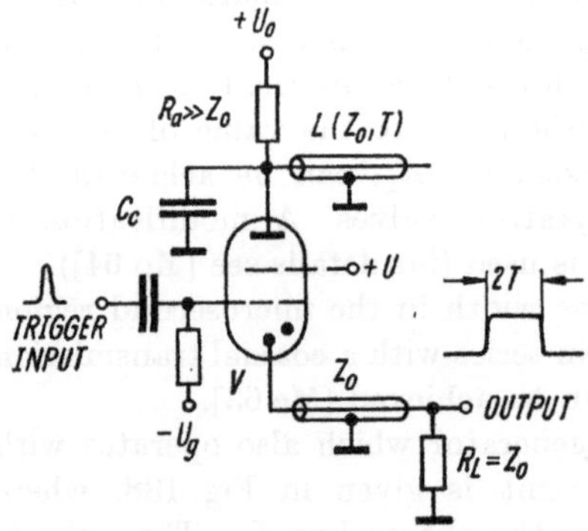

Fig. 189. Principle circuit of a thyratron-equipped discharge-type generator.

Rise and fall time of the output pulse can be reduced by interposing a correction capacitance C_c of some 10 pF between the anode of the thyratron and ground, as illustrated in Fig. 189. On firing, C_c causes a higher current and reduces the influence of the parasitic capacitance of the load resistance. Other possibilities are discussed by *Morugin* and *Glebovich* [Mo 64].

The rise time can also be reduced by using a pulse-sharpening gap (see next subsection) or by employing a non-linear ferrite line. As detailed in Section 3.1.2, a shock wave develops in the ferrite line due to attenuation, which depends on its magnetised state. This shock wave reduces the leading edge of the current pulse travelling through the line. *Meshkov* [Me 65b] feeds the output pulses of thyratrons to two ferrite lines which are circuited in opposition at the other end. This produces a rectangular pulse with a width corresponding to the transit time difference of the two transmission lines. Pulse width can be adjusted between 2 and 100 ns with the aid of the biasing current of the ferrite lines. In this manner, the ferrite line reduces the rise time from about 50 ns to approximately 1 ns. Ferrite line pulse sharpening has also been used by *Miroshnik* and *Skugarev* [Mi 62b], and by *Gel'tsel'* et al. [Ge 66a].

SPARK GAPS

Pulses of very high current or voltage, as they are required for work with spark chambers (cp. Section 5.3) or for the production of pulsed particle beams, are obtainable from the discharge of transmission lines or capacitors through spark gaps. Design of spark gap is very simple, although, very small inductances are necessary for generators producing high-current pulses in the nanosecond region. In general, delay values and rise time of the output pulses are lower than with thyratron valves. With a number of circuits the time delay between the trigger pulse and firing of the gap amounts to about 20 ns.

The principle of pulse generation with a spark gap is illustrated in Fig. 190 [Fl 49]. Here, the transmission line L_1 is charged through R and is discharged by the spark gap G_1 which is triggered by a negative pulse through the auxiliary electrode. In contrast to a thyratron circuit, this requires only low power. Ultra-violet irradiation continuously liberates electrons in the discharge path to ensure as small a time lag as possible. Pulses produced by the transmission line and gap G_1 have a rise time of about 20 ns. These are passed through a second transmission line L_2 to the pulse sharpening gap G_2 which receives an overvoltage on arrival of a pulse from G_1. This results in a very

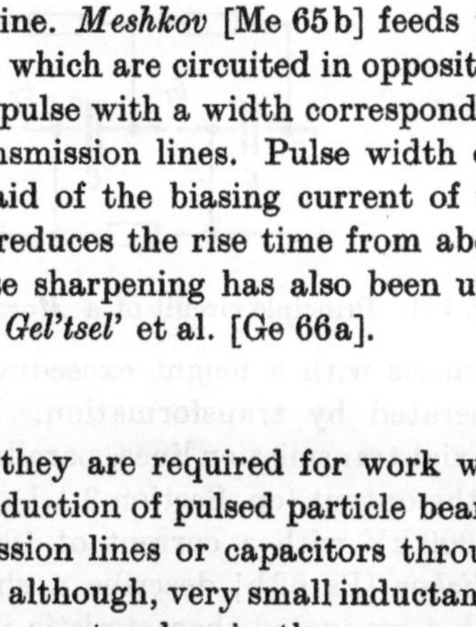

Fig. 190. Pulse generator with an isolated over-voltage spark gap for pulse sharpening. According to *Fletcher* [Fl 49].

short rise of the output pulse, amounting to less than 1 ns. *McDonald* et al. [Mc 65] developed a similar arrangement which delivered 60 kV pulses having a rise time of 0·3 ns. In several instances laser pulses have been used for triggering the spark gap [Pe 65]. Pulse-sharpening gaps may also be utilised to steepen the pulses from thyratron valves [He 54; Le 59, section 4242].

Steep slopes can also be produced by using a trigger gap which operates on gas over-pressure and has capacitance sharpening, apart from the pulse-forming line, which would correspond to the correction capacitance C_c of Fig. 189 [Fl 49; Vo 63a].

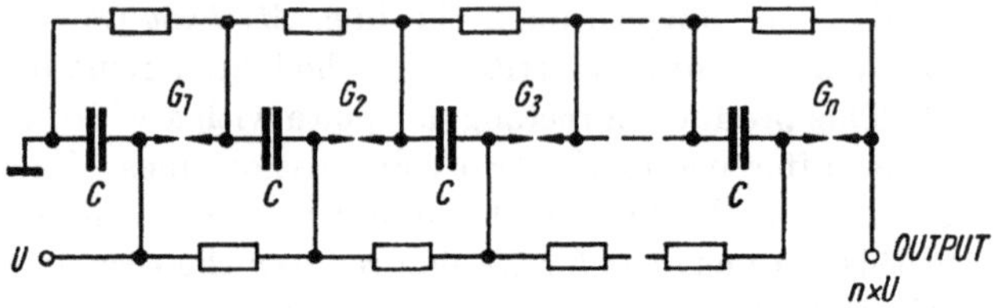

Fig. 191. Principle circuit of a *Marx* generator for high-voltage pulses. G = spark gap.

Pulses with a height exceeding the voltage of the storage elements can be generated by transformation. In this manner *Nasibov* et al. [Na 65a] use coaxial transmission lines, parallel-connected at the input and series-connected at the output (cp. Section 3.1.1), and, from 22 kV pulses, obtain a pulse height of 300 kV with a current of 100 A and a rise time of 20 ns. *Pavlovskii* and *Sklizkov* [Pa 62b] describe a similar generator. By discharging a coaxial line with a graduated characteristic impedance through a spark gap *Vorob'ev* et al. [Vo 68] produced 1 MV pulses with a rise time of 5 ns.

With the aid of a *Marx* generator very high pulse voltages are attained. In this circuit, n parallel-connected capacitors C are charged to the level of the voltage U and then series-connected through n spark gaps, which gives the voltage nU at the output (Fig. 191). *Gygi* and *Schneider* [Gy 64] constructed a *Marx*-type pulse generator with supplementary coupling capacitors. These ensure that, after the first spark gap G_1 has fired, the second spark gap receives an overvoltage of 100% irrespective of the number of stages. This results in a reduction of the firing delay and, therefore, the rise time of the output voltage. At a delay of 10 ns pulses of 200 kV were produced with 10 stages. Rise and fall time amounted to 2 ns. *Vorob'ev* and *Rudenko* [Vo 65] employed a *Marx* generator for charging a capacitor which discharges through a spark gap and obtained 500 kV pulses having a rise time of 1·5 ns.

4.2 LIGHT PULSE GENERATORS

4.2.1 GENERAL

Short light pulses are frequently required for scientific research work and for technical application [Ma 62c; Ma 62d; Ma 65a]. Nanosecond light pulses are preferably utilised for the testing of photomultipliers (cp. Section 4.4.2) and image converters, for balancing complex photon-detector assemblies and as information carriers in opto-electronic systems. With the aid of ultra-violet light pulses the fluorescence decay time of scintillators can be measured. For the purposes indicated generators with a relatively low light yield are sufficient. In laser communication systems, on the other hand, the generation of high-intensity light pulses and the light modulation with GHz-frequencies are also important [Jo 64]. This, however, is beyond the scope of this volume and will not be dealt with here.

4.2.2 GAS DISCHARGES

For the simultaneous production of fast electrical pulses and light pulses gas discharges are very suitable. The type of discharge depends on the shape of the electrode and on gas filling.

The use of *spark counters* [Sk 61; Gu 62] has the disadvantage of a statistic succession of pulses because they result from a radioactive source, e.g. ^{210}Po α-radiation. Pulse rise time decreases as the electrode gap narrows and field strength rises, and is also dependent on the gas filling. *Skachkov* [Sk 61] measured the duration of the spark in various gases of a parallel-plate spark counter with the aid of electron-optical chronography (cp. Section 2.3.3). With the plate gap of 0·1 mm, an operating voltage of 1·5 kV, a gas pressure of 15 at and on oxygen addition at a partial pressure of 0·5 at the following values were obtained with the various gases:

He: 0·4 ns, Ne: 0·66 ns, Ar: 1·6 ns, Xe: 3·3 ns.

The light yield increased by the factor 100 in the same sequence. Spark generation with a rise time of $t_r \approx 10$ ns and a high light intensity for high-speed photography has been described by *Preissinger* [Pr 67].

Náray and *Varga* [Na 61] obtained a rise time of 3 ns with *thyratrons* and *voltage stabiliser tubes*. Commercially available argon-gas lamps showed an instability of 0·3% over a period of several weeks [Wi 63 b]. Pulse rate and light intensity of thyratrons can be independently varied by external triggering. Light intensity shows a linear dependence on the charging voltage over a wide range [Ta 66 a].

From a *mercury-wetted relay switch* (cp. Section 4.1.4) *Kerns* et al. [Ke 59] developed a nanosecond light pulse generator. Pulse repetition rate is very low, amounting to 60 Hz or 160 Hz [On 61], and every contact closing could produce a group of light pulses. *Fixed mercury wetted contacts* avoid these disadvantages [Da 64 a; Da 65; Yg 65; Fr 67 a]. Similar to the example shown in Fig. 192, a free-running mode is used, with the stray capacitance C gradually charged through the large resistance R. As soon as the igniting voltage of the discharge path is attained

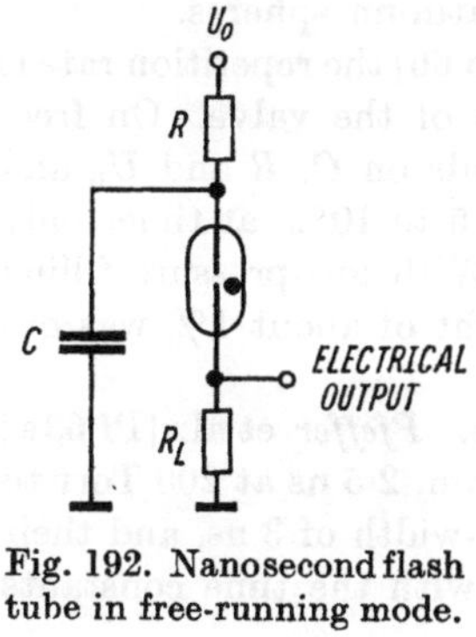

Fig. 192. Nanosecond flash tube in free-running mode.

the electric breakdown occurs. An electrical pulse is produced via the working resistance R_L by the discharge current. Low rise and fall time values require minimum stray capacitance and wiring inductance as well as coaxial construction.

As together with ionisation the gas molecules are also excited, light emission increases with rising current. After the current flow has stopped, light continues to be emitted due to further excitation from the hot electron gas and the large number of excited molecules that are present. When using biatomic gas of high mobility and high pressure frequent collision causes the rapid cooling of the electron gas and reduces the lifetime of the excited states which is primarily effected by collisions of the second kind. Therefore, to obtain minimum light pulse decay time values the bulbs are filled with hydrogen at a pressure of 10 to 20 at. This also increases the firing tension and, in consequence, the light yield. With a sampling oscilloscope having a rise time of 0·2 ns an electrical pulse of 1 kV height gave values of $t_r = 0{\cdot}25$ ns and $t_{I/2} = 0{\cdot}4$ ns by measurement [Yg 65]. Optical sampling technique (cp. Section 4.4.3), with a resolving time of 0·1 ns, obtained a light-pulse rise time of $t_r = 0{\cdot}35$ ns and a half-width of $t_{I/2} = 0{\cdot}5$ ns. Repetition rate was some 10 kHz. For mercury wetting the electrodes were electroplated with silver [Yg 65].

Corona discharge tubes have also been developed where high-voltage pulses produce a corona discharge between a tungsten wire and a surrounding aluminium housing across a solid dielectric material. At the same driving pulse height barium titanate generated a 10 times higher light yield than glass, lifetime, however, was only 150 h with 100 pps. Rise time and half-width of the light pulse amounted to about 1 ns [Ke 61]. Inconstancy of pulse height is 3% [Ga 65a].

The *flash tube* developed by *Malmberg* [Ma 57b] is frequently used [Ca 61; Bi 62b; Ny 63; Pf 63a; Ec 64; Sp 64a; Ec 65]. Its properties are similar to those of the mercury-wetted relay systems [Yg 65]. The tube consists of two rod electrodes embedded in the glass of a gas-filled bulb with a gap of ≈0·1 to 2 mm. Tungsten serves as electrode material, in most cases, but other materials are also utilised [Oz 62]. In the arrangement from *Nygaard* and *Malmskog* [Ny 63] the discharge is effected between two platinum spheres.

With thyratron triggering [Ma 57b; Bi 62b; Oz 62; On 66] the repetition rate is restricted to about 10 kHz because of the dead time of the valve. On free-running operation (Fig. 192) the repetition rate depends on C, R and U_b and can rise to 10 MHz [Pf 63a]. Pulse height instability is 5 to 10%, at times only after a burning-in period of several hours [Ny 63]. With low-pressure filling (100 Torr) an inconstancy of frequency and pulse height of about 1% was obtained [Pf 63a].

A hydrogen filling attains the shortest light pulses. *Pfeffer* et al. [Pf 63a] observed a reduction of the electrical pulse rise time from 2·5 ns at 200 Torr to 1 ns at 800 Torr. The obtained light pulses had a half-width of 3 ns, and their decay was a composite of two exponential functions with the time constants of 1·2 ns and 5·5 ns. Measurements made by *Yguerabide* [Yg 65] yielded time constants of 0·2 ns and 0·7 ns.

According to *Eckardt* et al. [Ec 64] the rise time t_r of light pulses is proportional to d^2, with d representing the gap width. A cone-shaped anode and a semi-spherical cathode yield optimum function conditions [Hu 67]. After extended operation, e.g. 200 hours, pulse height jitter increases considerably because of the whisker formation at the anode and the resulting instability of the discharge channel. A removable bulb permits contact cleaning and convenient variation of the gap width [Ec 65]. Pulse height depends on the type of gas used and increases with C, the gap width and gas pressure. An investigation of several filling gases, such as air, oxygen, nitrogen, rare gases and gas mixtures [Bi 62b; Ny 63; Sp 64a; Yg 65; Hu 67] showed that light pulses with a half-width of $t_{1/2} \approx 0{\cdot}5$ ns can also be produced with a filling of air and pure oxygen, although satisfactory stability of pulse height is attained only at high pressures. Xenon produces a high yield of ultra-violet light, while nitrogen gives an especially high light output, which is surpassed by a mixture of argon or neon with a few percent nitrogen [Sp 64a]. It is a serious drawback that neither long rectangular pulses with steep edges [Yg 65] nor slow-decay exponential pulses (>50 ns), e.g. for the simulation of NaI(Tl) scintillators, can be obtained. A favourable mixture, for instance, is given with 99% argon and 1% nitrogen which yields a rise time of 4·5 ns and a decay time of 21 ns at a pressure of 565 Torr [Sp 64a].

4.2.3 ELECTROLUMINESCENCE

Charge carrier recombinations occurring on current flow through p-n junctions may proceed in the various ways illustrated in Fig. 193*a* with the example of gallium-phosphide material. The direct band-to-band transition *1* is seldom with the GaP material. In most cases, recombination *3* is effected at an impurity *K* with the released energy of 2·25 eV transformed into heat. At other impurities *R* (zinc and oxygen) and *G* (zinc) part of the energy is emitted in form of electroluminescent light. On transition *2* red light with a wavelength of 700 nm and on transition *4* a yellow-green light with a wavelength of 565 nm is emitted. Depending on the purity of the material and the type of doping, effectivity of the electroluminescence can amount to 1% [Gr 63a; Ro 65a].

Light emission intensity depends on the magnitude of the injection current and follows, practically without delay, the rapid current changes in the case of low-lifetime carriers. For this reason, electroluminescence diodes are suitable for producing fast light pulses of any desired shape and high repetition rate. Apart from the small dimensions and the low operating voltage required the diodes have the advantage that several of them can be driven simultaneously by the same electrical pulse generator.

Whetstone [Wh 59] used a p-n silicon junction driven to avalanche region, while *Kendall* [Ke 60] and *Loude* et al. [Lo 65] did the same with silicon transistors. This produced light pulses where the single-electron measurement (cp. Section 4.4.2 and 4.4.3) obtained a half-width of $t_{1/2} \approx 1$ ns [Lo 65].

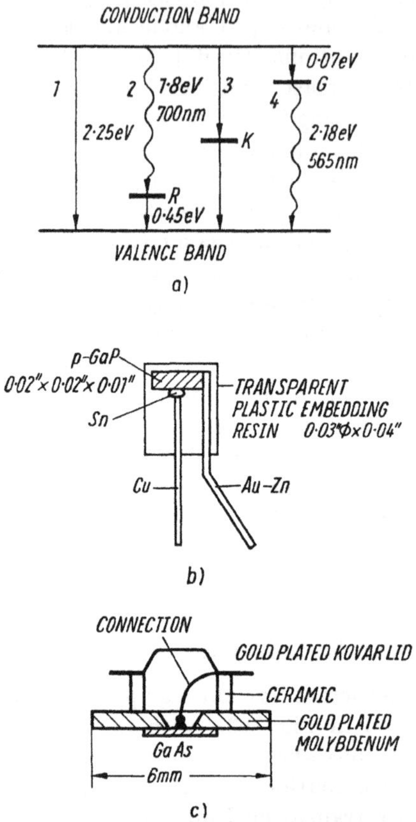

Fig. 193. Electroluminescence diode.

a) Scheme of charge-carrier recombination in gallium-phosphide. Straight arrows represent non-radiation transition. Wavy arrows represent transition associated with light emission. *b*) Construction of a gallium-phosphide diode. According to *Rowland* and *Bottomley* [Ro 65a]. *c*) Construction of a gallium-arsenide diode for coaxial installation. According to *Leclerc* and *Zajde* [Le 65].

Gallium-phosphide diode luminescence has been used among others by *Allen* and *Gibbons* [Al 61b], *Tolfree* [To 64b] and *Vorob'ev* et al. [Vo 66]. Using the single-electron method, *Rykalin* et al. [Ry 65] obtained a half-width of the light pulse $t_{1/2} = 0{\cdot}72$ ns by biasing the diode in reverse direction. For the investigation of the load characteristic of photomultipliers *Gorbenko* et al. [Go 67] developed a light-pulse generator producing pulse trains having a width between 100 ns and more than 100 ms, with facility for varying the pulse repetition rate from 2 to 200 MHz. With a rate of 30 MHz the base-width of the light pulse is ≈ 3 ns. The industrial design of a GaP diode is shown in Fig. 193*b* [Ro 65a].

Silicon-carbide diodes can produce blue light because of the wide band spacing, amounting to $\approx 2{\cdot}9$ eV [Le 51], and are, therefore, especially suitable for the investigation of photomultipliers. *Lapshin* et al. [La 65a; La 66b] measured a light pulse half-width of $t_{1/2} = 1{\cdot}3$ to $1{\cdot}8$ ns with electrical pulses of a base width of 1 ns, with reverse and forward biasing direction of the diode showing hardly any difference.

Luminescence diodes made of *gallium arsenide* yield infra-red light which admirably matches the optical spectral response of silicon and is consequently well suited for testing silicon semiconductors. These diodes may also be employed for the production of light pulses in the nanosecond region [Ke 62]. *Amsel* and *Zajde* [Am 64] obtained a rise and fall time of the light pulse below 0·5 ns. The time jitter between current and light pulse amounted to less than 30 ps, with the pulse height spread underpassing 0·3%. To attain a steep rise of the light pulse it is necessary to keep the series inductance of the diode leads small.

The diode mounting shown in Fig. 193*c* gives a series inductance of less than 1 nH when coaxially mounted in a BNC-plug [Le 65].

At the temperature of liquid nitrogen the light yield is higher by one order of magnitude than at room temperature and proportional to current pulse height up to 100 A [Gu 63; Ku 65]. Temperature sensitivity between 20° and 50 °C is about 10%/10 °C [Na 65c]. The peak wavelength shifts from 910 nm at 300 °K to 840 nm at 77 °K [Ku 65] and rise time is lower [Le 68].

For testing photodiodes in the medium infrared region *Brown* et al. [Br 67a] used an *indium-arsenide* laser diode which radiates at 3·15 μm and, when operated at 77 °K, produces an exponentially decaying pulse with a rise time of 1 ns. Since the laser diode impedance is only a fraction of one ohm, on operation with a discharge-line generator (see Fig. 186, p. 304), it represents a short-circuit. To prevent the destruction of the diode by the reverse-polarity pulse reflected by the end of the cable a long lossy cable is employed.

4.3 OSCILLOSCOPES

4.3.1 GENERAL

Cathode-ray oscilloscopes probably represent the most versatile and by far the most widely utilised measuring device in pulse testing technique. With their aid pulse curves are made visible and pulse intervals and pulse heights conveniently measured. Basic construction and function is generally known (cp. [Cz 59; Le 59; Mo 64; Mi 65]).

The signal under examination must be delayed by an interval of 50 to 100 ns against the trigger or synchronising pulses for the time-base to obtain linear horizontal display. This delay can be achieved in two ways (Fig. 194):

Mode A: A generator first releases the time-base of the oscilloscope and then the circuit under test, with respective delay. Sometimes the delayed pulse for releasing the circuit may be taken from the pulse oscilloscope.

Mode B: The signal to be measured, or a respectively synchronous pulse, triggers the time-base and is passed through a transmission line, with respective delay, to the vertical input ("internal" triggering).

Mode A avoids any deformation of the test signal that may be caused by the transmission line, but presupposes that the tested circuit can be triggered; a condition which often cannot be met, for instance, in the case of a radiation detector as pulse source. For these cases, Mode B is preferable. To restrict test pulse distortion to a minimum the attenuation of the required transmission line must be kept as low as possible. Commercially available pulse oscilloscopes have an inbuilt pulse delay line and can be used immediately for the mode B measurements.

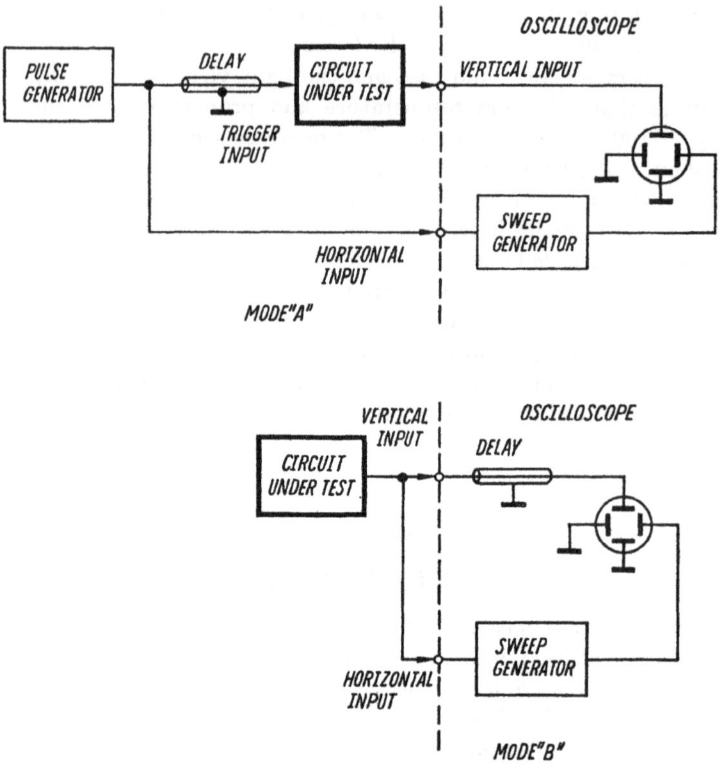

Fig. 194. Illustration of the two principle types of operation for measurements with an oscilloscope.

Conventional pulse oscilloscopes for the microsecond region have a *vertical deflexion* with an upper cut-off frequency in the order of 10 MHz. With the aid of distributed amplifiers or other wideband amplifier systems this frequency is extended to about 100 to 200 MHz, which corresponds to a rise time of a displayed step pulse of about 2 to 5 ns. Maximum deflexion sensitivity is approximately 50 mV/cm [Ke 50; Yu 51; Ke 64a].

The upper cut-off frequency of a conventional cathode-ray tube is about 300 MHz, and in consequence the vertical deflexion bandwidth is primarily determined by the amplifier with the oscilloscopes mentioned. The frequency where the vertical deflexion sensitivity has dropped by 3 dB to 70% of the low-frequency value is usually designated as the upper cut-off frequency of a cathode-ray tube. To enable the display of very short rise times the vertical amplifier must be omitted because amplifier systems able to handle a rise time of less than 2 to 3 ns at an output pulse height in the order of 100 V are not available. This is all the more serious because the increase of the upper cut-off frequency of a cathode-ray tube causes a material reduction of the vertical deflexion sensitivity. A way out of this difficulty is provided by the development of cathode-ray tubes having *travelling-wave deflexion systems* (cp. Section 4.3.2) and by the construction of so-called *sampling oscilloscopes* (cp. Section 4.3.3). With both types of oscilloscopes, however, the minimum rise time that can be displayed is at times restricted by the properties of the delay line required for the mode B testing. The reason being, that the rise time of a step pulse after having run through the required 10 to 20 m transmission line, is in the order of 0·1 ns (cp. Section 3.1.2).

Horizontal deflexion is, as a rule, less difficult. Horizontal deflexion speed can be increased as desired by using respectively high deflexion voltages. This speed may even exceed the velocity of light. During a time period of 0·1 to 1 ps the beam can be deflected by a distance corresponding to spot diameter. The resulting high time resolution can only be utilised, however, for the measurement of single pulses, because with recurrent pulses the time jitter of the time-base generator occurs additionally. When the various pulses are not traced to coincide exactly and are laterally displaced the time jitter between trigger pulse and time-base release blurs the display. Time-jitter amounts to some 10 ps and a higher deflexion speed than a few cm/ns with recurrent pulses will, therefore, offer no advantage. The same applies also for the time base of a sampling oscilloscope.

Time-base generators for a high horizontal deflexion speed necessitate a considerable outlay. Recharging the capacitance of the deflexion plates in a cathode-ray tube with several hundred volts at an interval of a few nanoseconds, requires sawtooth pulses of several ampere peak current. Thyratron equipped generators seem to be a solution because they deliver high output currents, however, their long trigger delay and their timejitter preclude their use for nanosecond pulse oscilloscopes. Time base-generators containing electron valves are detailed in various works [Kr 57; Le 59; Mo 64]. In comparison, the sampling oscilloscope has the advantage that a high horizontal deflexion speed of the electron beam is not necessary, making the use of time-base generators of low-frequency oscilloscopes possible.

4.3.2 CATHODE-RAY TUBES WITH HIGH CUT-OFF FREQUENCY

The transit time t_f of the electrons through the deflexion system in a conventional cathode-ray tube is 1 to 10 ns. In consequence, the electron beam cannot follow the very rapid voltage changes. This causes a step of the deflexion plate voltage to be displayed on the screen with the rise time t_f [Le 59]. Apart from this transit time distortion of display another effect occurs in the cathode-ray tube which also materially influences the bandwidth of the tube. Together with the lead inductance L the capacitance C of the deflexion plates represents a resonant circuit. With the aid of a damping resistance in the order of 100 Ω this resonance can be suppressed to such an extent that, when a step pulse is applied, the plate voltage increases with a minimum rise time without noticeable overshoot. With an optimum value damping resistance this minimum rise time amounts to $0{\cdot}5/f_0$, with $f_0 = 1/2\,\pi\,\sqrt{LC}$ representing the resonance frequency of the deflexion system [Le 59].

With the oscilloscopes described in the following it is recommended that the vertical deflexion sensitivity be given as deflexion voltage per spot width. Spot width can be defined as the spacing between the half peak intensity points. From this derives that with the same deflexion voltage or the same deflexion angle of the electron beam the sensitivity can actually be enhanced by reducing the spot diameter. By means of careful dimensioning, spot widths of between 0·01 and 0·04 mm are obtainable [Ar 39; Pi 49a; Ha 57a; Ed 60]. In the precision oscilloscope of *von Ardenne* [Ar 57] the spot diameter was reduced to 0·003 mm through magnetic focussing. Spot width cannot be infinitely minimised because of the resulting decrease in electron beam intensity and consequent reduction of brightness. In general a vertical deflexion through 50 to 100 spot widths will suffice.

REDUCING GEOMETRY AND DIMENSION OF THE DEFLEXION SYSTEM

The cut-off frequency of conventional cathode-ray tubes can be raised by shortening the electron transit time t_f, i.e. by using narrower deflexion plates and increasing the acceleration voltage of the electrons, and by reducing the effective inductance. *Lee* [Le 46], for instance, achieved a cut-off frequency of 8·6 GHz with very short deflexion plates and an acceleration voltage of 50 kV. Deflexion sensitivity is relatively low, amounting to 10 V/spot width. This corresponds to a deflexion of 1 kV/mm at a spot width of 0·01 mm, making microscopic viewing of the display necessary. Oscilloscope tubes having short plates are also mentioned by *Connor* [Co 56b] and by *Haine* and *Jervis* [Ha 57a]. *Akimov* and *Stepanov* [Ak 62] used a 75 Ω strip-line as deflexion plates and connected both ends to a coaxial cable in a low-reflexion manner. This obtained a cut-off frequency of the tube amounting to 1·5 GHz and gave a deflexion sensitivity of 22 V/mm. *Bartenev* et al. [Ba 61b] pass the electron beam through a longitudinal slot between the inner conductor and part of the outer conductor of a coaxial cable. This had a transit time of 40 ps. In place of the narrow deflexion plates, two parallel wires arranged vertically to the electron beam may also be employed [Ru 45]. In this manner, *Haine* and *Jervis* [Ha 57a] achieved a cut-off frequency of approximately 10 GHz. Other examples are listed by *Lewis* and *Wells* [Le 59].

All these arrangements have the serious disadvantage of low vertical deflexion sensitivity. Cathode-ray tubes of this type are therefore preferably used for the investigation of high-voltage pulses and in high-frequency engineering (radar pulses).

TRAVELLING-WAVE DEFLEXION SYSTEM

Travelling-wave deflexion systems have been developed with a cut-off frequency reaching approximately 3 GHz. Against the mentioned systems, the vertical deflexion sensitivity is considerably higher and amounts to 0·01 to 1 V/spot width. The general idea is to arrange several deflexion fields in series similar to a lumped delay line arrangement. When propagation speed of the deflexion voltage and of the electrons coincide, transit time distortion disappears. Fig. 195*a* gives the deflexion system developed by *Pierce* [Pi 49a].

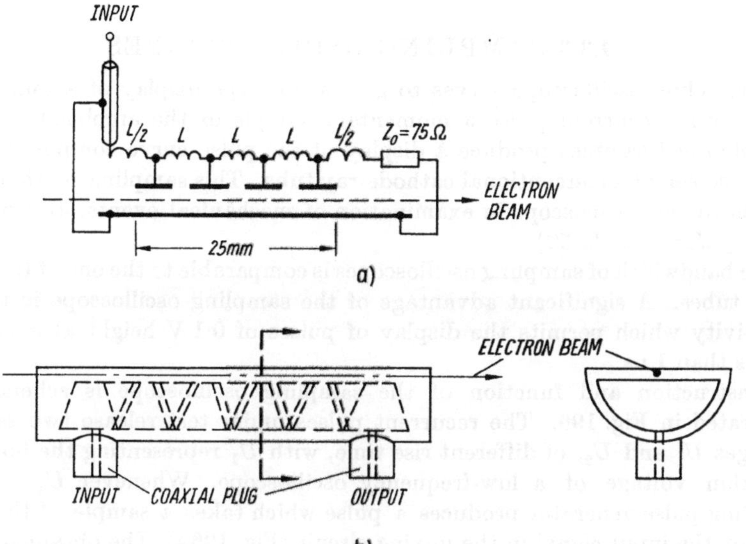

Fig. 195. Travelling-wave deflexion systems in cathode-ray tubes.
a) With four deflecting plates. According to *Pierce* [Pi 49a]. *b*) Distributed travelling-wave deflexion structure. According to *Smith* et al. [Sm 52].

It represents a lumped delay line having a characteristic impedance of $Z_0 = \sqrt{L/C} = 75\ \Omega$ and is calculated for a cut-off frequency of 1·33 GHz. With an accelerating voltage of 1 kV the obtained deflexion sensitivity is 0·037 V/spot width which corresponds to 18·5 V/cm at a spot width of 0·02 mm.

In travelling-wave systems with distributed constants the propagation speed of the electric wave is higher than that of the electron beam. Therefore, a transmission line consisting of two parallel flat strips is coiled up so that the speed component of the electrical pulse in direction of the electron beam coincides with the speed of the latter. An arrangement developed by *Smith* et al. [Sm 52] is shown in Fig. 195*b*, where the line consists of a metal strip of 1 cm width which is coiled to a five-turn helix with a radius of 1 cm. Together with the earthed shield this forms a 50 Ω line. Cut-off frequency of this structure is

2·6 GHz, with a deflexion sensitivity of 0·6 V/spot width at an accelerating voltage of 10 kV. *Edgerton* et al. [Ed 60] developed travelling-wave tubes which were used for industrial oscilloscopes. At a bandwidth of 2 GHz a rise time of 0·2 ns can be displayed with a vertical sensitivity of 0·055 to 0·0055 V/spot width. Other travelling-wave oscilloscopes are described by *Winnigstad* [Wi 58], *Gorbachev* et al. [Go 63b], *Abercrombie* et al. [Ab 61], *Bartenev* and *Piskunov* [Ba 66d], and in the survey by *Nahman* [Na 67]. *Bigelis* et al. [Bi 67] report on a travelling-wave tube having a deflexion sensitivity of $\geqq 100$ mV/mm and a cut-off frequency of 1·5 GHz.

4.3.3 SAMPLING OSCILLOSCOPES

A sampling oscilloscope serves to give a dot-type display of a pulse curve. From every recurrent pulse a momentary sample of the amplitude is taken. The obtained samples produce a display of the pulse curve composed of dots on the screen of a conventional cathode-ray tube. This sampling method corresponds to the stroboscope in examination of mechanical events, and was first used by *Janssen* [Ja 50].

The bandwidth of sampling oscilloscopes is comparable to the one of travelling-wave tubes. A significant advantage of the sampling oscilloscope is the high sensitivity which permits the display of pulses of 0·1 V height at a rise time of less than 1 ns.

Construction and function of the sampling oscilloscope is schematically illustrated in Fig. 196. The recurrent pulses under test release two sawtooth voltages U_1 and U_2, of different rise time, with U_1 representing the horizontal deflexion voltage of a low-frequency oscilloscope. Whenever $U_1 = U_2$ the sampling-pulse generator produces a pulse which takes a sample of the amplitude of the input signal in the mixing circuit (Fig. 196a). The obtained sample pulses are stretched, amplified and displayed on the screen of the cathode-ray tube in form of light spots. The different rise times and the voltage comparison $U_1 = U_2$ ensure that the sampling pulses successively cover the whole pulse under test and that the displayed light spots combine on the screen to an expanded oscillogram of proper sequence. A staircase waveform may also be utilised in place of the sawtooth voltage U_1 [Su 57; Bu 59a]. With steepening sawtooth voltage U_2, the time-base extends respectively. In this manner a time-scale of some cm/ns can be achieved. The actual horizontal sweep of the cathode-ray tube is less by some orders of magnitude and amounts to $v_x = l/t_0$, with l the diameter of the screen, as is evident from Fig. 196*a*. For a pulse display at least 100 dots are necessary, and therefore the period duration t_0 of the horizontal deflexion voltage must be greater by the factor 100 than the interval of the recurrent input pulses.

With sampling oscilloscopes, no wideband vertical amplification and no highly sensitive wideband cathode-ray tubes are necessary. The problem is essentially reduced to two comparatively easy technical problems: the production of very

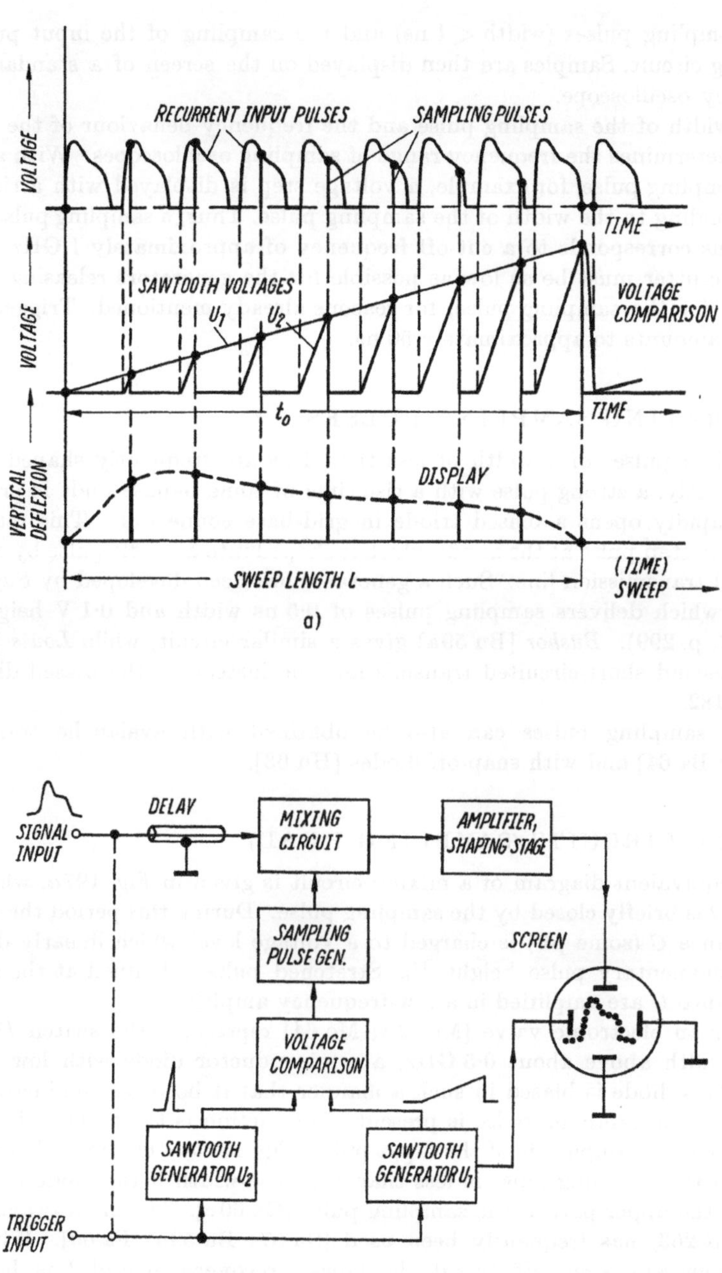

Fig. 196. Sampling oscilloscope.

***a)* Pulse curve display (schematic). *b)* Block diagram, simplified.**

short sampling pulses (width < 1 ns) and the sampling of the input pulses in a mixing circuit. Samples are then displayed on the screen of a standard low-frequency oscilloscope.

The width of the sampling pulse and the frequency behaviour of the mixing circuit determines the frequency range of sampling oscilloscopes. With a triangular sampling pulse, for example, a voltage step is displayed with a rise time corresponding to the width of the sampling pulse. Thus, a sampling pulse width of 0·35 ns corresponds to a cut-off frequency of approximately 1 GHz. Delay and time-jitter must be as low as possible for the generators releasing U_1 and U_2, and for the sampling pulse, for reasons already mentioned. Trigger delay usually amounts to approximately 50 ns.

GENERATING SAMPLING PULSES

Sampling pulses of a width of less than 1 ns are frequently shaped in two steps. Firstly, a strong pulse with a rise time of some nanoseconds is produced which rapidly opens a biased triode in grid-base connection. This results in a negative step-pulse at the anode which is shaped into a needle pulse by a short-circuited transmission line. Such a generator has been developed by *Sugarman* [Su 57] which delivers sampling pulses of 0·5 ns width and 0·1 V height (cp. Fig. 182, p. 299). *Bushor* [Bu 59a] gives a similar circuit, while *Louis* [Lo 60] uses a second short-circuited transmission line instead of the biased diode D_1 of Fig. 182.

Short sampling pulses can also be obtained with avalanche transistors [Am 60; Ba 64] and with snap-off diodes [Hu 63].

MIXING CIRCUITS (SAMPLING HEAD)

The equivalent diagram of a mixing circuit is given in Fig. 197*a*, where the switch D is briefly closed by the sampling pulse. During this period the storage capacitance C (some pF) is charged to a voltage level which linearly depends on the momentary pulse height U_i. Stretched pulses obtained at the storage capacitance C are amplified in a low-frequency amplifier.

Either an electronic valve [Mc 52a; Mo 64] represents the switch D or, at a bandwidth above about 0·5 GHz, a semiconductor diode with low reverse delay. This diode is biased in such a manner that it becomes conducting only as long as the sampling pulse is present. Conduction is the higher, the higher the momentary amplitude of the input pulse (Fig. 197*b*). The actually effective width of the sampling pulse is less than the base-width if the diode is opened only by the upper part of the sampling pulse [Go 60c]. The micro-wave diode, type 1 N 263, has frequently been used [Su 57; Bu 59a; Fa 60]. When the forward current is cut off, this diode shows a recovery time of 2 ns, however, in the present case the recovery time is considerably shorter, because during the brief event of the sampling pulse no minority charge carriers are stored.

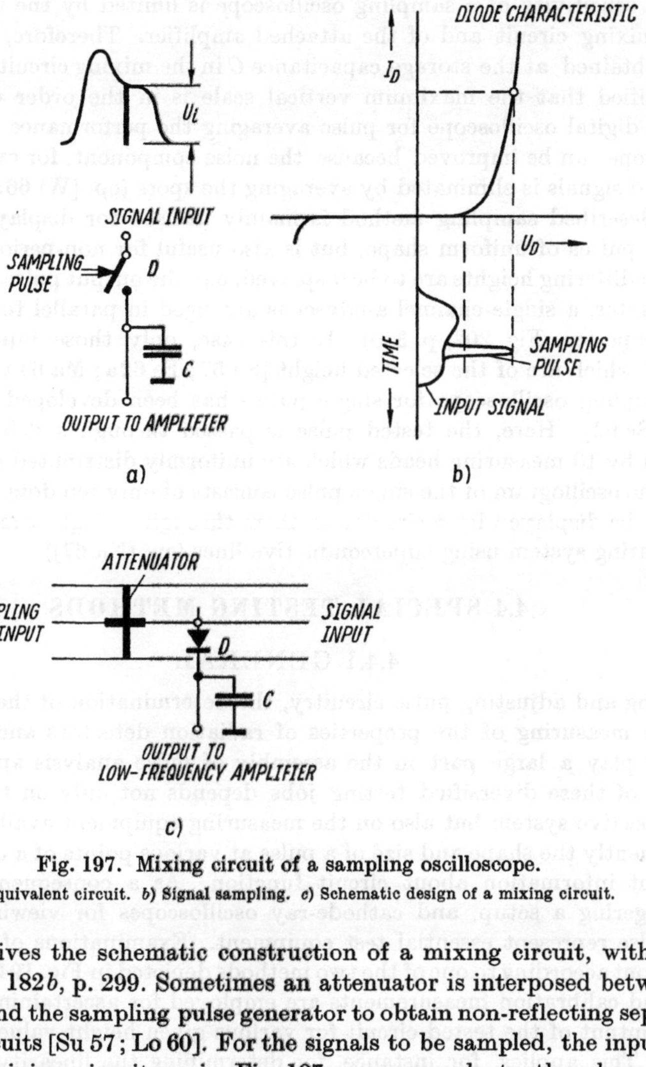

Fig. 197. Mixing circuit of a sampling oscilloscope.
a) Equivalent circuit. *b)* Signal sampling. *c)* Schematic design of a mixing circuit.

Fig. 197*c* gives the schematic construction of a mixing circuit, with details shown in Fig. 182*b*, p. 299. Sometimes an attenuator is interposed between the signal input and the sampling pulse generator to obtain non-reflecting separation of the two circuits [Su 57; Lo 60]. For the signals to be sampled, the input impedance of a mixing circuit as in Fig. 197*c* corresponds to the characteristic impedance of the coaxial system (mostly 50 Ω). The input impedance can be raised to about 100 kΩ by means of an impedance converter.

Starke [St 61] gives a detailed description of some commercially available sampling oscilloscopes having a cut-off frequency approaching 1 GHz. *Nahman* [Na 67] has compiled numerous references of sampling oscilloscopes. *Goodall* and *Dietrich* [Go 60c] received an oscilloscope rise time of 60 ps with a GaAs point-contact rectifier in the mixing circuit, while *Ragsdale* [Ra 63a] mentioned a rise time of 35 ps that was experimentally obtained.

Input sensitivity of a sampling oscilloscope is limited by the inherent noise of the mixing circuit and of the attached amplifier. Therefore, the stretched pulses obtained at the storage capacitance C in the mixing circuit need only be so amplified that the maximum vertical scale is in the order of 10 mV/cm. Using a digital oscilloscope for pulse averaging the performance of a sampling oscilloscope can be improved because the noise component, for example, of the displayed signals is eliminated by averaging the spots (cp. [Wi 66a]).

The described sampling method is mainly utilised for displaying recurring periodic pulses of uniform shape, but is also useful for non-periodic pulses. If pulses of differing heights are to be displayed, e.g. the output pulses of a scintillation counter, a single-channel analyser is arranged in parallel to the sampling oscilloscope (cp. Fig. 202, p. 330). In this case, only those input pulses are sampled which are of the selected height [Su 57; Fe 62a; Ma 64a].

A sampling oscilloscope for single pulses has been developed by *Sedlmeyer* et al. [Se 62]. Here, the tested pulse is passed through a delay line and is sampled by 10 measuring heads which are uniformly distributed along the line. Thus, the oscillogram of the single pulse consists of only ten dots. Single pulses can also be displayed by recirculating them through a single sampler, with the recirculating system using superconductive lines (cp. [Na 67]).

4.4 SPECIAL TESTING METHODS

4.4.1 GENERAL

Testing and adjusting pulse circuitry, the determination of their parameters and the measuring of the properties of radiation detectors and circuit components play a large part in the assembly of pulse analysis apparatus. The manner of these diversified testing jobs depends not only on the purpose of the respective system but also on the measuring equipment available.

Frequently the shape and size of a pulse at various points of a circuit provide sufficient information about circuit function. As a consequence generators for triggering a setup, and cathode-ray oscilloscopes for viewing the course of a pulse represent essential test equipment. Examinations of this kind are carried out according to one of the two methods depicted in Fig. 194, p. 314. Other tests and calibration measurements are employed for ascertaining the relation of the output of the tested circuit for various given height values of the input signal. This applies, for instance, for determining the linearity of amplifiers or of the response threshold of integral discriminators. As these testing methods do not differ in principle from testing "slow" pulse circuits for the microsecond region, they need no further detailing here. Hence, discussion is confined to some specific test problems characteristic for the nanosecond region. This includes the measurement of the shape of fast electrical pulses and light pulses, as delivered by radiation detectors, for example, ascertaining the delay time of transmission lines and calibrating methods for time analysers. The many problems involved in the determination of parameters of semiconductor devices, especially transistors, are dealt with only to the extent that they concern parameters which are particularly important for nanosecond pulse technique.

4.4.2 TESTING PHOTOMULTIPLIERS

Two aspects are involved in the testing of photomultipliers: Finding the most favourable operating conditions for practical work and determining the parameters relating to the time resolution theory of photomultipliers (cp. Sections 2.3.1 and 2.3.2). Examinations of this nature are preferably performed by illuminating the photocathode with fast light pulses from a pulser (Section 4.2) and measuring the pulse from the multiplier with an oscilloscope or a time analyser. Compared to the use of scintillators for measuring, the uniform height and shape of the light pulse is an advantage. One must check, however, that the differing shapes of light pulses from a generator and a scintillator do not lead to quantitatively varying results. This applies especially in case of time resolution measurements, which, as noted in Section 2.3.2, depend to a large extent on the rise and fall time of the light pulse.

Most light pulsers simultaneously produce electrical signals which can be employed for triggering the oscilloscope or as time reference signal for the time analyser. Such signals can also be taken from a second photomultiplier exposed to the same light source, e.g. a scintillator, as the multiplier under examination.

With needle-shaped light pulses, the rise time, the width and peak current of the multiplier pulses, the pulse distortion occurring at the onset of space charge limitation and the transit time of the electrons through the photomultiplier can be conveniently measured with the arrangements shown in Fig. 198 and Fig. 199. Measurements of this sort generally require little time. This permits the individual adjustment of the multiplier to optimum operating conditions with respect to the significant parameters by successively testing various potential distributions of the dynodes and focussing electrodes ("optimising"). Most practical for this purpose is a voltage divider chain consisting of potentiometers [Ec 62; Be 63; Bo 64a; Vi 64]. To ensure the independent balancing of the respective sections the use of a high-voltage source delivering a stabilised output current is recommended [Be 63].

TESTING WITH LIGHT PULSER AND OSCILLOSCOPE

By displaying multiplier pulses resulting from fast light pulses on the screen of an oscilloscope, one can directly observe the pulse shape from the multiplier, the maximum current pulse height and the influence of space charge saturation, and optimise the operating conditions of the photomultiplier with respect to these parameters (Fig. 198). The use of high-frequency modulated light allows conclusions to be made on the transit time from the phase shift at the multiplier output and on the value of the transit time spread from the decrease of the modulation depth [Ma 62b]. *Cernigoi* et al. [Ce 60a] analysed photomultipliers of the types 1 P 21, 6810 A and 56 AVP with light, the intensity of which was modulated with the aid of an ultrasonic standing-wave field in a liquid by frequencies of up to 48 MHz. The resolution of short-time photomultipliers was tested with light pulses at intervals in the order of 10 ns, either in form of double light pulses or as short pulse trains. This showed, e.g., that the types 56 AVP and XP 1020 were able to handle maximum repetition rates of 100 to 200 MHz, depending whether processing was effected analogue or digitally [Ce 60; Ra 63; Bi 64a; La 66c; Go 67].

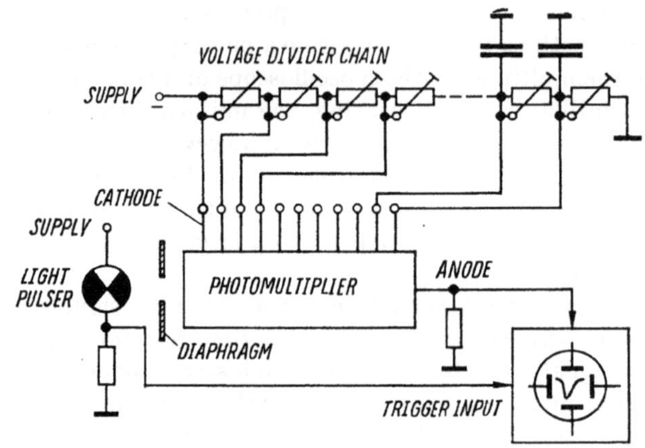

Fig. 198. Circuit for optimising the operation of a photomultiplier with the aid of a light pulser and an oscilloscope.

With point-like light illumination of the photocathode, for instance, by using a sliding diaphragm, the transit time differences of the photoelectrons between the various points of the cathode and the first dynode can be determined from the oscillogram [Ce 60; Ro 60a; Ra 63; Bi 64a]. Examples of such transit-time profiles of the cathode region are given in Fig. 12, p. 24. As the following indicates, transit-time values and transit-time profiles can also be measured with the aid of a time analyser.

TESTING WITH LIGHT PULSER AND TIME ANALYSER

A schematic test setup is given in Fig. 199. The photomultiplier pulse is delivered to a coincidence circuit or to a time-to-pulse-height converter. Electrical signals supplied by the light pulser or, in case of scintillator measurements, the output pulses of a second photomultiplier serve as time-reference signals. With the aid of the coincidence circuit the total transit-time of electrons through the photomultiplier can be ascertained from the difference of the delay values $h = T_1 - T_2$ [Gu 62]. When measuring the transit-time profile of the cathode region on condition of point-like illumination, only the relative time values are of interest which makes it more advantageous to use a time-to-pulse-height converter in place of the coincidence circuit [Ag 62; Ku 67; Pi 67a].

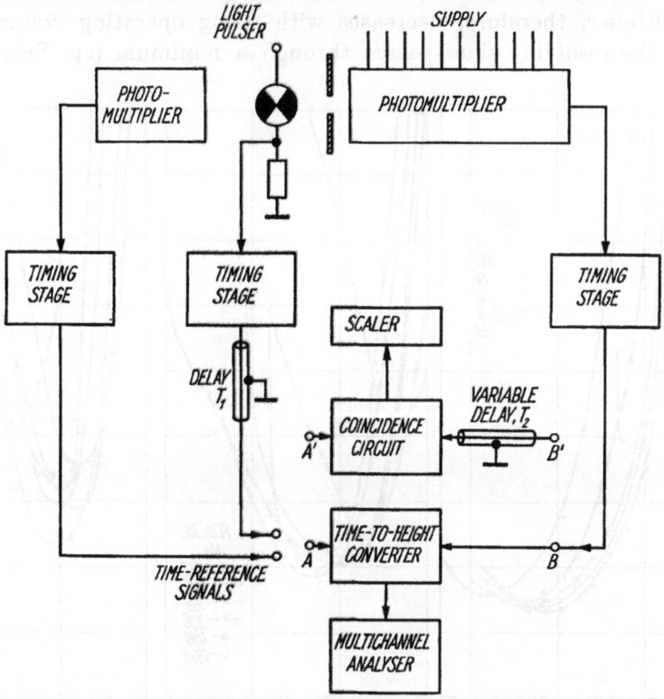

Fig. 199. Circuit for testing the behaviour of a photomultiplier with the aid of a light pulser and a time analyser.

TESTING TIME RESOLUTION

Using the setup illustrated in Fig. 199 gives coincidence curves where the halfwidth $2\tau_0$ represents a measure of the time resolution obtainable with the respective multiplier unit. An examination of this kind has some advantages against scintillator measurements, for instance, with fast-slow arrangements as shown in Fig. 1. The influence of the differing height of the scintillator pulse is practically eliminated; little time is needed for the registration of a coincidence curve because every light pulse is registered as a coincidence event when using the time-to-pulse-height converter. *Bonitz* et al. [Bo 64; Bo 64a] and *Vorob'ev* et al. [Vo 66] systematically investigated the halfwidth of coincidence curves as a function of the operating voltage of the photomultiplier, with a hydrogen-discharge tube or a gallium-phosphide diode. Similar investigations were also undertaken by *Caplan* and *Stewart* [Ca 61], *Wieber* [Wi 63a], *Miehe* et al. [Mi 66] and *Porter* et al. [Po 66b].

An example of such time resolution measurements is given in Fig. 200, where the potentials of all multiplier electrodes of each photomultiplier were adjusted to the highest and steepest anode current pulse. A time-to-pulse-height converter was employed, with a limiter as a timing unit. The utilised fraction C/R of the multiplier, therefore, decreases with rising operating voltage and, as expected, the resolving time passes through a minimum (cp. Section 2.3.2).

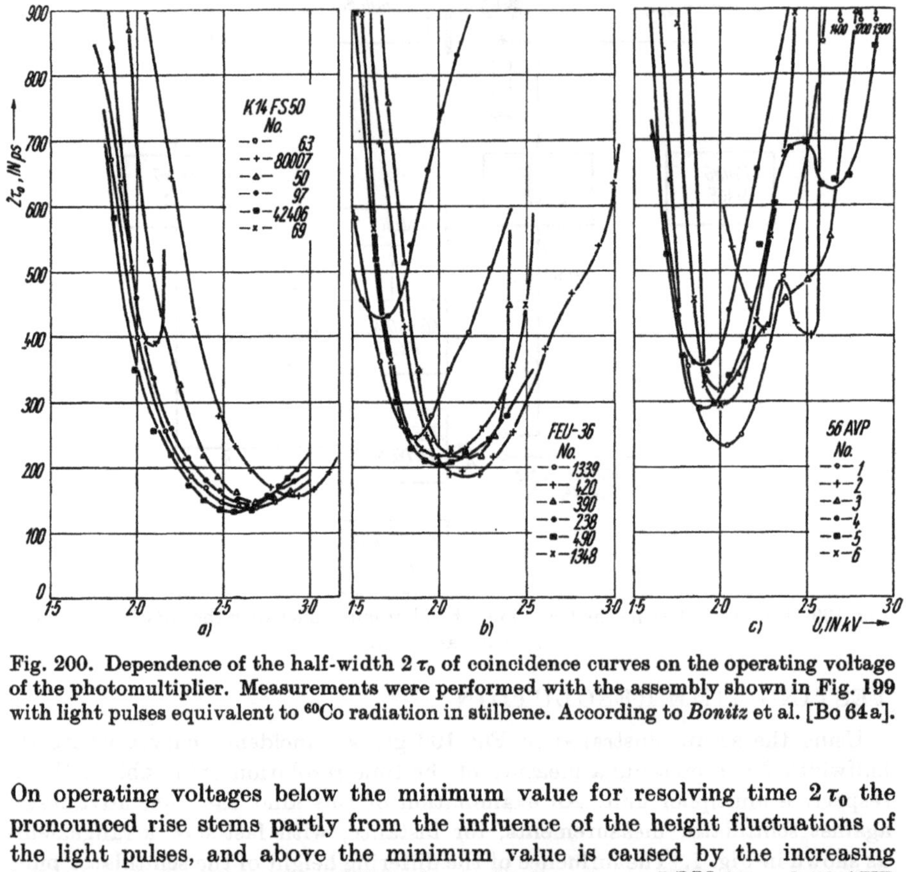

Fig. 200. Dependence of the half-width $2\tau_0$ of coincidence curves on the operating voltage of the photomultiplier. Measurements were performed with the assembly shown in Fig. 199 with light pulses equivalent to ^{60}Co radiation in stilbene. According to *Bonitz* et al. [Bo 64a].

On operating voltages below the minimum value for resolving time $2\tau_0$ the pronounced rise stems partly from the influence of the height fluctuations of the light pulses, and above the minimum value is caused by the increasing influence of the noise pulses. With the multipliers type ФЭУ-36 and 56 AVP the prepulses represent an additional influence, which increase the minimum value of $2\tau_0$, especially with type 56 AVP, and cause an intermediate maximum in some of the curves shown in Fig. 200.

These time resolution values which were measured with fast light pulses, give approximately the value obtainable with fast-decay organic scintillators, but do not apply for a slow-decay scintillator of the type NaI(Tl). Generating light pulses to simulate slow-decay scintillators is somewhat difficult. Experiments undertaken with gas discharge tubes with various gas fillings and at differing gas pressures did not bring the desired results [Sp 64a]; the use of luminescence diodes seems more promising.

Behaviour of controlled multipliers, such as type SKP-1 (Fig. 15, p. 31), can also be tested with the aid of a light pulser [Mu 65].

Using scintillators of different size *McDonald* and *Gedcke* [Mc 67] investigated the time resolution of large-cathode photomultipliers (XP 1040). To eliminate pulse-height influence on the measurement two-dimensional analysis was resorted to where the time spectrum was registered according to the height of the multiplier output signals (see Section 5.1.4).

SINGLE-ELECTRON METHOD

Bollinger and *Thomas* [Bo 61] were the first to utilise the so-called single-electron method for the measurement of scintillator decay times. An example of the setup required for this versatile testing mode is schematically shown in Fig. 201. An absorber is used to weaken the light pulse to such an extent that

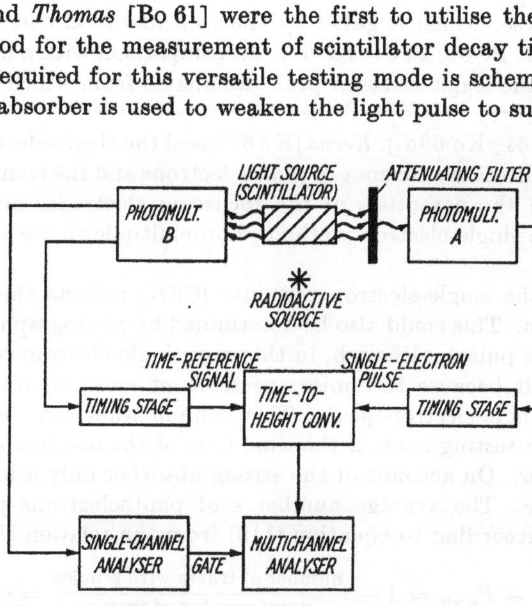

Fig. 201. Single-electron method setup for the measurement of the equivalent illumination function $I^*(t)$ of the photomultiplier A, and for measuring the shape of light pulses (decay behaviour of scintillators).

one photoelectron is released in the photomultiplier A on an average of only every 50 to 100 light pulses. This heavy weakening is necessary to keep the emission probability of two or more photoelectrons through one light pulse to a negligible level. For the emission probability of n photoelectrons the *Poisson* distribution is assumed

$$P_n = \frac{\varepsilon^n}{n!} e^{-\varepsilon} \tag{119}$$

with $\sum_{n=0}^{\infty} P_n = 1$ and ε the average number of photoelectrons per light pulse. On $\varepsilon \ll 1$ the probability for the emission of $n = 1$ electron is $P_1 = \varepsilon\, e^{-\varepsilon} \approx \varepsilon$ and for the emission of $n > 1$ electrons

$$P_{n>1} = \sum_{n=2}^{\infty} P_n = (1 - P_0 - P_1) \approx \varepsilon^2/2.$$

Because of $P_{n>1}/P_1 \approx \varepsilon/2$, e.g. $\varepsilon = 1/50$ will therefore result in only 1% of all multiplier pulses being produced by more than 1 photoelectron.

The interval between the start of a light pulse and the registration of the individual photoelectron is measured by a time analyser in the manner shown in Fig. 199. If a coincidence circuit is employed for this purpose, the single-electron pulse-height spectrum of the photomultiplier can be obtained. In this case, the output pulses of the multiplier are registered by a multi-channel pulse-height analyser which is gated by the coincidence circuit or also by the light pulser [Tu 62; Pe 64; Pi 64; Ke 67]. In comparison with a noise spectrum, in this spectrum the single-electron peak dominates from which the secondary emission factor g of the first dynode can be determined with the aid of *Poisson*'s statistics (cp. [Pe 64; Ko 66a]). *Kerns* [Ke 67] used the single-electron method to investigate the collection efficiency of photoelectrons and the transit time spread in dependence on the potentials of the focussing electrodes and of the first dynode. With the single-electron method photomultiplier noise can also be investigated [Pe 67].

The shape of the single-electron response (SER) reflects the properties of the dynode system. This could also be determined by photographing the oscillograms of the noise pulses, although, in this case single-electron response identification is difficult because the noise spectrum of most photomultipliers has no pronounced single-electron peak. This complication can be avoided with the single-electron testing mode if the time base of the oscilloscope is triggered by the light pulser. On account of the strong absorber only a few traces show a multiplier pulse. The average number ε of photoelectrons per light pulse can be obtained according to equation (119) from the relation [Bi 64a; Hy 64]

$$P_0 = 1 - P_{n>0} = 1 - \frac{\text{number of traces with a pulse}}{\text{total number of traces}} = e^{-\varepsilon}.$$

With the value ε determined in this manner the fraction of multiple-photoelectron pulses of the photographed multiplier pulses can be derived at once from $P_{n>1}/P_1 \approx \varepsilon/2$.

If the time-to-pulse-height converter is used together with a multi-channel pulse-height analyser, as in Fig. 201, the equivalent illumination function $I^*(t)$ of the photomultiplier A can be measured in a simple manner. According to equation (21) $I^*(t)$ is the convolution of the shape of the photocathode illumination function with the transit time distribution function of the single-electron pulses travelling from the photocathode to the anode. It is necessary to know

$I^*(t)$ for the calculation of time resolution (cp. Section 2.3.2). With many events the time between the start of the light pulse (time-reference signal) and the appearance of the single-electron pulse is measured and obtains the number-of-photons via time distribution in the storage unit of the multi-channel analyser which represents the course of the equivalent illumination function. Measurements of this sort were performed by *Birk* et al. [Bi 64a], by *Bertolini* et al. [Be 66; Be 66a], and by *Bertolaccini* et al. [Be 67a]. When neglecting transit time fluctuations in the photomultiplier, the measured distribution reflects the course of the light pulse, because the measured emission probability of photoelectrons is proportional to the light intensity. In this manner scintillator decay time constants were obtained, as detailed in the following.

4.4.3 TESTING THE FORM OF LIGHT PULSES

Pulses from a light pulser or from a scintillator are first converted into an electrical signal by a photon detector (cp. Section 2.3), which is then displayed on the screen of an oscilloscope or by an *X-Y* recorder. Photomultipliers are utilised as photon detector for low-intensity signals, while in case of high-intensity signals fast solid-state photon detectors can also be employed.

The measuring methods of scintillator time constants are summarised, for instance, by *Kramer* [Kr 67]. For the measurement of ultrashort light pulses generated by phase-locked lasers semiconductor photodiodes are usually employed for pulse lengths of about 0·2 ns and more (cp. Sections 2.3.4 and 5.5). For laser pulse length down to about 1 ps non-electronic methods are used.

DIRECT MEASUREMENT

Light pulses illuminate the cathode of a photomultiplier the output pulses of which are directly displayed on an oscilloscope screen. Light pulses from scintillators excited by radioactive radiation have differing heights, and to obtain pulses of the same height it is necessary to take a signal from one of the last dynodes of the photomultiplier and evaluate this with a single-channel analyser [Su 57; Fa 62; Pe 62]. Fig. 202 gives a schematic illustration of a test setup needed for this purpose.

The pulse rise time of commercial short-time photomultipliers is approximately 2 ns (cp. Table 4, p. 28), causing false presentation of light pulses with similar or shorter rise and fall time. Thus, using a sampling oscilloscope with a cut-off frequency above approximately 500 MHz for measuring multiplier pulses brings no advantage. As a point of interest, samples taken from the multiplier output pulses can also be stored in a multi-channel pulse-height analyser or in a computer and the course of the pulses is then given in digital form [St 63b; La 66c; Ku 67]. Instead of the multi-channel analyser an *X-Y* recorder may be employed [Kr 67].

The advantage of the simple test setup is counteracted by two drawbacks. On the one hand, the minimum displayable rise time of about 2 ns is too long, for instance, to measure the rise of light pulses from organic scintillators. On the other hand, the multiplier pulse is frequently distorted by ringing at the pulse top, by space charge effects in the dynode system, by after-pulses and by noise so that the decay of scintillators, for example, can only be quantitatively evaluated to approximately 5 or 10% of maximum light intensity.

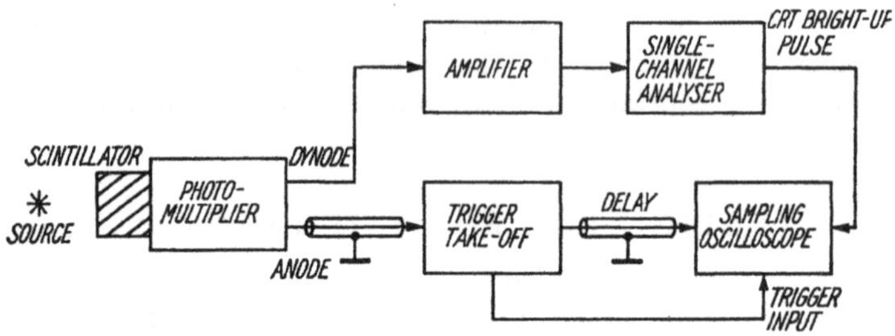

Fig. 202. Registering light pulses with an oscilloscope. With the arrangement shown only pulses of the same height are displayed.

With a fast light pulser, according to Section 4.2, the electrical pulses can be measured with a sampling oscilloscope and conclusions drawn on the rise and fall time of the light signals. No detailed information is as yet available regarding deviations of the light pulses from the shape of the electrical pulse in the case of fast light pulsers.

McGuire and *Palmer* [Mc 67 a] used electron bursts of 50 ps length and 10 MeV energy for scintillator excitation and a coaxially mounted biplanar vacuum-photodiode as detector. Together with the sampling oscilloscope this obtained overall time resolution of 150 ps which permits the direct measurement of the scintillation rise time.

SAMPLING METHOD WITH RECURRING LIGHT PULSES

This includes the *single-electron method* described in Section 4.4.2 (Fig. 201, p. 327), where the influence of the photomultiplier dynode system on the measured pulse shape is practically excluded. To enable the investigation of scintillator decay with the single-electron method only pulses having the same height are admitted through the parallel-connected single-channel analyser. The rise time of the pulses that can be displayed is in the same order as with direct measurement because the transit-time fluctuations of the single-electron pulses in the photomultiplier are not eliminated. Thus, true light pulse representation requires small transit time fluctuations in the photomultiplier and in addition a good time resolution of the time-to-pulse-height converter (Fig. 201). The single-electron method has the decided advantage that the influence of ringing, of after-pulses, of pile-up effects and part of the background noise is suppressed to such extent as permits the quantitative evaluation of the light intensity decay through 2 or 3 orders of magnitude.

The time reference pulses required for the time-to-pulse-height converter are derived from the electrical signals of the light pulser, when this is used for testing [Be 66; Be 66a]. With scintillators irradiated by a radioactive source a second photomultiplier *B*, as in Fig. 201, is utilised, which is illuminated by the same scintillator but with no interposed absorber [Bo 61; Ko 64; Mc 65a; Sj 65; Ya 66; Ki 67a]. Figs. 3 and 4, p. 8 and 9, give scintillator decay measurements which were carried out with the described method.

Sampling methods with *pulsed image intensifiers* (cp. Section 2.3.3) permit a reduction of the minimum displayable rise time. This method described by *Dreeskamp* and *Burton* [Dr 59; Dr 61; Bu 62] corresponds to the function of a sampling oscilloscope. During a time of 0·1 to 0·2 ns the image intensifier is gated and a sample taken of the light pulse. The output pulse of the image intensifier is converted into an electrical signal by the photomultiplier. With successive sampling the whole curve of the light pulse is obtained. If the light intensity is so low that on an average less than one electron is released at the intensifier cathode during a sampling event, the light intensity of every measuring point of the pulse curve must be determined by counting the samples with and without released photoelectron. The short resolving time of approximately 0·1 to 0·2 ns is an advantage. For the testing of scintillators it is a disadvantage that the time reference signal for gating the image intensifier cannot be taken from a second photomultiplier because of the long delay due to the transit time through this multiplier. For this reason scintillator investigations can be undertaken only when these are activated by a pulsed source, such as X-rays [Dr 59] or ultra-violet rays [Yg 65] because then an electrical signal can be supplied by the source.

INDIRECT MEASURING METHODS

Of the indirect measuring methods, the most familiar is the phase-fluorescence method (cp. [Li 50; Kl 61; Mü 65a]). Actuating the fluorescence of a scintillator with a periodically modulated radiation will cause the fluorescence to change with the same frequency, however, with phase-shifted maxima and minima. If the decay characteristic is known, the decay time constant of the fluorescence can be determined from the phase shift.

A scintillator can be actuated by a radioactive source and the output pulse of an attached photomultiplier can be investigated with the aid of a short-circuited transmission line [El 50; Re 60]. This method has the same disadvantage as the phase-fluorescence method, because data regarding the course of the light pulse are required for evaluation. In the last instance this is one of the causes of the often contradictory results of the scintillator decay time constants obtained with this indirect method.

4.4.4 TESTING SEMICONDUCTOR COMPONENTS

The suitability of *diodes* for use in the nanosecond pulse technique primarily depends on the time required for the transition from the conducting state to the blocked state and vice versa (cp. Section 3.1.4). The testing circuit for pulse behaviour is given in principle in Fig. 66, p. 115, the example of a test setup is shown in Fig. 203*a*. Here, the forward current I_f flows in quiescent state,

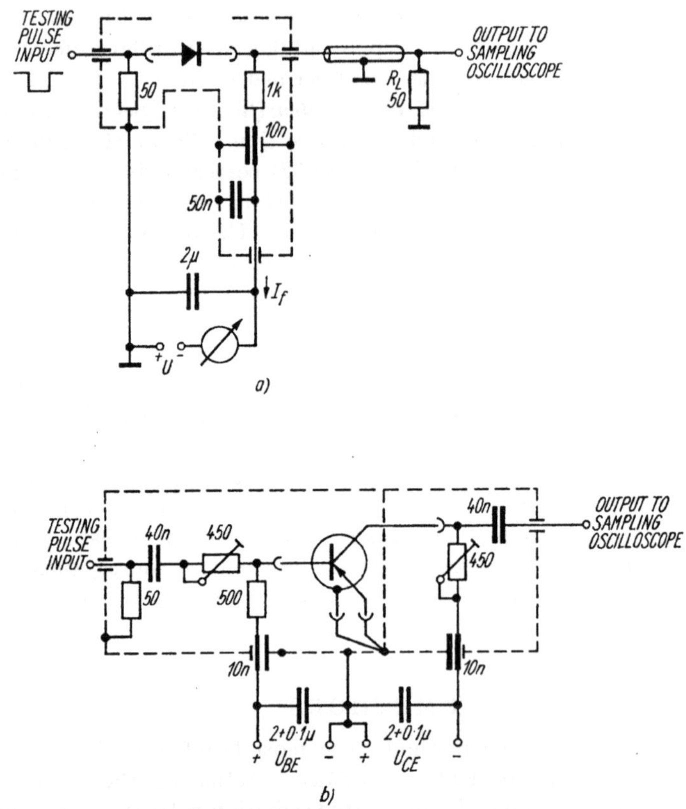

Fig. 203. Circuits for measuring the pulse behaviour of a diode (*a*) and of a transistor (*b*). According to ***Havlik*** and ***Jelinek*** [Ha 65a].

which is cut off by a negative pulse. The voltage drop across the resistance R_L represents the curve of the reverse current. The resulting oscillogram gives a reading of the reverse recovery time t_{rr} which respresents the time between the begin of the reverse pulse and the drop of the reverse current to a specific fraction of the maximum value according to its definition. A mechanical relay and a charge line serve as pulser (cp. Section 4.1.4), because this permits the generation of rectangular pulses with a short rise time ($\leq 0{\cdot}5$ ns) and easily adjustable height and width. The turn-on behaviour of the diode can, in analogue manner, be examined from the voltage drop across R_L or across the diode.

Sylvan [Sy 65] gives a simple method for measuring the stored charge of a diode, which is in linear relation to the reverse recovery time. Here, pulses with a repetition rate f block the forward current I_f of the diode and the stored charge is applied to an integrating network which causes a current I_1. Measurement is then repeated with $I_f = 0$ and the current I_2 is measured at the integrating network. This current is produced by the recharging of the junction barrier and the stray capacitances. The stored charge Q_s of the diode is then given by $Q_s = (I_1 - I_2)/f$.

In the case of *transistors* the value of the equivalent circuit components and the pulse behaviour are of interest. Fig. 203*b* gives the test setup for the transistor switching behaviour, specifically for testing p-n-p transistors in common-emitter connection (cp. Fig. 75, p. 132). Parameter measurements of the equivalent circuit are preferably made with radio-frequency techniques, which shall not be detailed here (cp. [Pa 66a]). When measuring transistors for maximum frequencies the influence of the transistor-can has to be taken into account as this affects transistor properties (cp. Fig. 73*b*, p. 126, — [Le 64b]). Pulse reflexion at the transistor may also be utilised for determining the equivalent circuit data [St 66b]. This is a method requiring little outlay and has the advantage that interference stemming from the test setup does not coincide in time with the signal reflected at the transistor. This makes interference easily discernible on the oscillogram.

When using *tunnel diodes* in pulse circuitry it is necessary to know the peak and valley currents (I_p, I_v). Both quantities can be taken from the voltage-current characteristic. If the negative part of the characteristic is not displayed, the negative impedance of germanium tunnel diodes may be assessed from

$$|R_-| \approx 0{\cdot}4 \, \frac{U_v - U_p}{I_p - I_v} \, .$$

It is possible to display the whole tunnel-diode characteristic, provided that certain stability conditions are met [Di 60; Ba 63c]. Measuring the other parameters of a tunnel-diode equivalent circuit, such as capacitance, series resistance and series inductance, is a little difficult. These testing methods have been reviewed, for instance, by *Adler* and *Selikson* [Ad 62] and by *Fistul'* and *Shvarts* [Fi 62].

4.4.5 DETERMINATION OF THE PROPAGATION SPEED OF ELECTRICAL SIGNALS IN TRANSMISSION LINES

Accurate knowledge of the propagation speed of electrical signals in transmission lines is specially significant in assemblies, incorporating a transmission line for pulse shaping, or for the calibration of a time analyser.

STANDING-WAVE TECHNIQUE

In a transmission line that is short-circuited at one end or that is open, standing-waves develop when the line is loosely coupled to an r.f. generator. Voltage nodes are $\lambda/2$ apart, while the current nodes are phase-displaced by $\lambda/4$. With the test setup shown in Fig. 204*a* those generator frequencies f_n are determined where a voltage node is produced at the input of the transmission line. From this follows for the short-circuited line of the length l

$$l = n\,\lambda/2 = n\,v/2\,f_n\,, \tag{120a}$$

where the node at the short-circuited end is not counted, and for the open-circuited line

$$l = (n - 1/2)\,\lambda/2 = (n - 1/2)\,v/2\,f_n\,. \tag{120b}$$

From these, the propagation speed v and the delay time $T = l/v$ of a transmission line are derived. When successively increasing the frequency of the generator from 0, the voltage nodes $n = 1, 2, \ldots$ are consecutively obtained with the associated frequencies f_n. The number of nodes n associated with the frequencies f_n can also be determined from $n = f_n/\Delta f$ for a short-circuited line, and from $n = (f_n/\Delta f) + 1/2$ for an open line, where $\Delta f = f_{n+1} - f_n$ is the difference to the next higher frequency with $n + 1$ nodes.

In practical work measurement of the propagation speed or line transit-time with the aid of standing-waves is usually performed with frequencies ranging from 30 to 300 MHz. According to $v \approx v_0\,(1 - \alpha_f\,v_0/2\,\pi\,f)$, with α_f denoting the attenuation constant, the line attenuation causes a slight dependence of the phase velocity v on frequency ([Fi 60] — Section 3.1.2). With the low-attenuation transmission lines used in time analysers, no frequency-dependent changes of the propagating speed within range of the measuring error ($\approx 1/2\%$) could be observed through the 30 to 300 MHz range [Cl 51; Re 62a; Be 63b; Cu 63; Me 65c; Pe 66; Jo 68a]. From this, it can be concluded that reliable values are obtainable with the standing-wave technique for testing the propagation speed of short-time photomultiplier output pulses, because the minimum rise time of about 2 ns corresponds to a cut-off frequency of approximately 200 MHz. *Cleland* and *Jastram* [Cl 51] measured the transit time in the transmission line, type RG/7U, across a range of 2 to 400 MHz and found no evidence of dispersion. The transit time obtained in this manner was used to calibrate a time analyser and to determine the propagation speed of γ-quanta, which gave $(2{\cdot}983 \pm 0{\cdot}015) \times 10^8$ m/s in good agreement with the velocity of light. This demonstrated the reliability of the standing-wave technique for testing transmission line transit time.

This method looses high accuracy if the rise and fall time of the transmitted pulses are less than 1 ns, because in this case the line attenuation flattens the pulse edges to a relatively greater extent than is the case with photomultiplier pulses. Under such conditions, and also with lumped delay lines, it is preferable to include the influence of pulse distortion when measuring the pulse transit time directly.

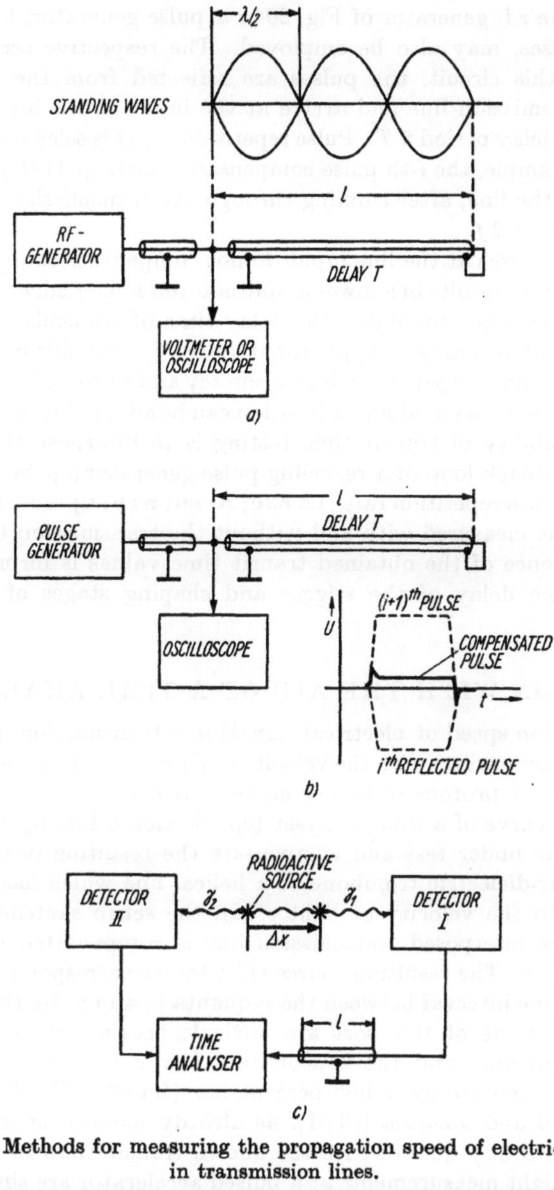

Fig. 204. Methods for measuring the propagation speed of electrical signals in transmission lines.

a) **Standing-wave method.** *b)* **Pulse superposition method.** *c)* **Comparing the propagation speed in the transmission line with the velocity of γ-quanta with the aid of a time analyser (source shifted by Δx).**

PULSE SUPERPOSITION TECHNIQUE

In place of the r.f. generator of Fig. 204*a* a pulse generator, for instance, for rectangular pulses, may also be employed. The respective setup is given in Fig. 204*b*. In this circuit, the pulses are reflected from the short-circuited end of the transmission line and arrive at the input of the line with reversed polarity after a delay period $2\,T$. Pulse repetition rate f is selected in such a manner, that, for example, the i-th pulse compensates the $(i + 1)$-th generator pulse at the input of the line, after running through the transmission line; the delay time is then $T = 1/2\,f$.

Although the pulses at the line input do not compensate completely, because the line attenuation results in a lower amplitude and longer edges of the reflected pulse than of the generator pulse, the delay time of transmission lines can be measured with an accuracy of approximately 0·1% with this method [Me 62a; Fo 63a; Mr 65; Jo 68a]. Apart from high accuracy and little outlay, it is an advantage that the pulse shape used for calibration can be adapted to later experiments.

Another possibility of transit time testing is to interpose the transmission line into the feedback loop of a recycling pulse generator (cp. Section 4.1.3) and determine the pulse repetition rate [La 63c; Ki 66; Ki 67b]. For the purpose, the repetition rate is measured with and without the transmission line under test. Then the difference of the obtained transit time values is formed and in this manner the time delay of the trigger and shaping stages of the generator eliminated.

CALIBRATION WITH THE AID OF A TIME ANALYSER

The propagation speed of electrical signals in a transmission line may also be ascertained by comparison with the velocity of light or with the velocity of particles, for instance, of protons of known energy. It is further possible to displace the coincidence curve of a time analyser (cp. Section 5.1.2) by interposing the transmission line under test and compensate the resulting displacement with the aid of an air-dielectric trombone or a helical line which has a transit time corresponding to the velocity of light c. In the setup sketched in Fig. 204*c* the delay in the interposed transmission line is compensated by shifting the radioactive source. The resulting source shift by Δx corresponds to a change of the measured time interval between the γ-quanta γ_1 and γ_2 by the value $2\,\Delta x/c$. Absolute calibrations of this sort are difficult because of the wide spacing between the detectors and the concomitant low coincidence efficiency, and frequently only accurate by a few percent (cp. [Ga 62b; Fo 63a]). In reverse manner, *Cleland* and *Jastram* [Cl 51], as already mentioned, determined the propagation speed of γ-quanta with the known transit time in a transmission line. Time-of-flight measurements at a pulsed accelerator are simpler and more accurate than the method shown in Fig. 204*c*. In this case, one input signal for the time analyser is derived from the pulse voltage of the accelerator [Ga 62b; Lo 62].

TESTING THE REFLEXION FACTOR OF COAXIAL CABLE PLUG CONNECTORS

Standing-waves may be utilised for the measurement of the reflexion factor of coaxial cable plug connectors. The factor should be in the order of 1%. For the purpose, the plug connection under test is interposed in a measuring line and, on a constant generator frequency, a shorting bar is used to shift the voltage nodes. From the difference of the displacement before and behind the plug connector, the reflexion factor can be determined [Me 62, p. 1586; Pe 66]. A comprehensive survey of measuring methods of coaxial connectors is given by *Weinschel* [We 66].

Schwartz and *Richardson* [Sc 64a] investigated the reflexion factor and the minimum transmissible pulse rise time of some commercially available plug connectors, using a pulse generator with a mercury-wetted relay (cp. Section 4.1.4). The charge line of the generator consists of two pieces of cable connected with the respective plug members. The value of the reflexion factor is derived from the pip appearing at the top of the rectangular pulses produced by the generator.

4.4.6 TESTING AND CALIBRATION OF TIME ANALYSERS

The simplest method for checking the function of a time analyser is the use of a pulse generator. With an analyser operating with scintillation counters as detectors a light pulser is recommended which illuminates both photomultipliers and thereby causes coincident multiplier pulses [Ag 62; Ec 63]. In the case of assemblies having many coincidence and anticoincidence stages, e.g. the counter telescopes utilised in high-energy physics or the large-volume scintillators with several photomultipliers, the experimental conditions can be simulated with the aid of generators delivering electrical pulses and light pulses. This sort of testing saves a part of the, at times, very expensive experimental testing of the apparatus.

For the measurement of integral and differential non-linearity of time analysers, especially of time-to-pulse-height converters, a number of methods are available (cp. [Re 67; Jo 68a]). As with multi-channel pulse height analysers (see Section 3.6.3), integral linearity mirrors the relation of the pulse height, or channel number, to time difference of both input pulses. Differential non-linearity, on the other hand, characterises the deviation of actual channel width from nominal value. Differential non-linearity measurements primarily serve for adjustment and function testing of the circuitry, while for the determination of the absolute time scale integral linearity measurements are employed.

AUTOMATIC TRACING OF CHARACTERISTIC CURVES

The characteristic of a coincidence circuit or of a time-to-pulse-height converter can be determined with the point-by-point method by successively changing the time interval of the two input pulses with an interposed delay line or a helical line. This method permits the measurement of the time resolution of coincidence circuits as well as the resolving time, linearity, measuring range, and time scale of time-to-pulse-height converters. It is a time-consuming method, especially if an investigation of the influence of various parameters on the characteristic of the circuit is desirable or if a large number of circuits have to be balanced. In such cases the automatic tracing of the characteristic curves with an oscilloscope, an *X-Y* recorder or a multichannel analyser is a decided advantage [Na 68].

Fig. 205*a* and *b* give the test setups according to *Birk* et al. [Bi 63a]. The heart of this arrangement is the delay circuit which has a delay that depends linearly on the comparison voltage U_r and which can be varied from 30 to 65 ns. During one period of the sweep generator the resolution curve of the coincidence circuit or the measuring range of the time-to-pulse-height analyser is run through and displayed as a smooth curve. Another method has been

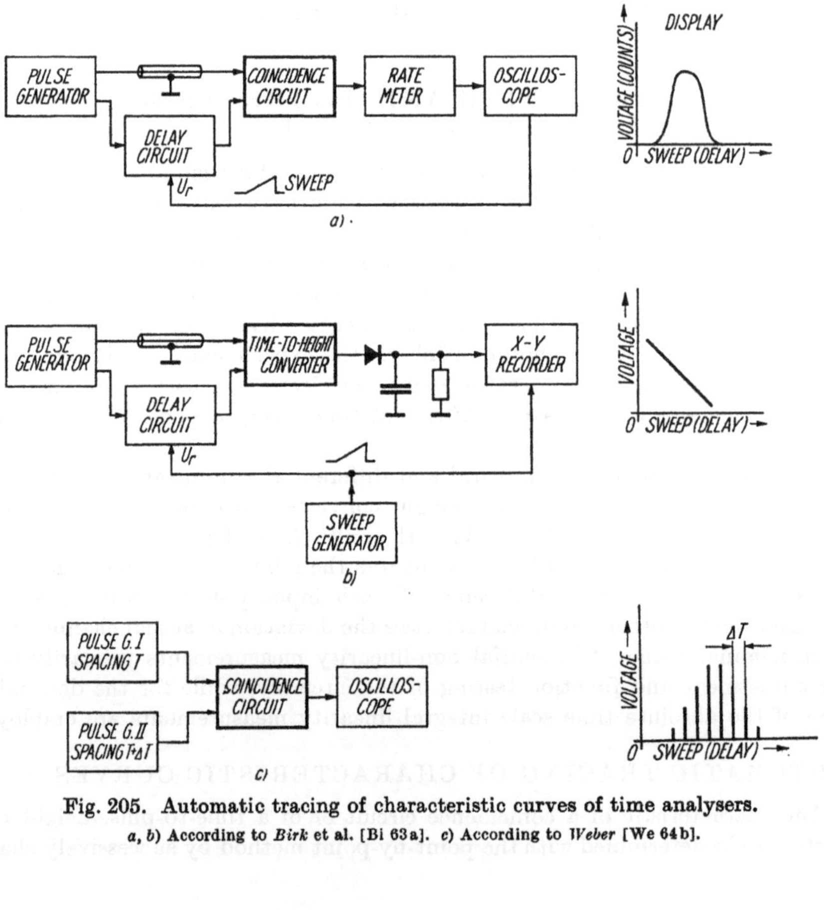

Fig. 205. Automatic tracing of characteristic curves of time analysers.
a, b) According to *Birk* et al. [Bi 63a]. *c*) According to *Weber* [We 64b].

developed by *Weber* [We 64b] where the pulses of two free-running recycling generators I and II are applied to the coincidence circuit (Fig. 205*c*). The pulse interval of the generator I, for example, amounts to 100·0 ns and that of generator II is 100·1 ns. The time difference of the input pulses of the circuit then changes successively by $\Delta T = 0{\cdot}1$ ns. On the oscilloscope screen the output pulse height of the coincidence circuit is displayed with the spacing of the measuring points corresponding to the time ΔT. *Seyfried* and *Deike* [Se 66] describe a pulse generator with height modulated output pulses. This permits the adjustment of zero-crossing timing stages to minimum time shift with pulse height variation. *Langkau* [La 66a] gives an oscilloscopic method for balancing zero-crossing trigger stages.

MEASURING DIFFERENTIAL NON-LINEARITY

If on both inputs of a time analyser the pulses are not correlated or are statistically distributed in time, according to equation (109) a „white" spectrum of chance coincidences appears, i.e. all channels of a time analyser are uniformly filled. As pulse generators two scintillation counters can be used for this which are not irradiated by the same radioactive source. More convenient is an arrangement where one time analyser input is fed from a constant pulse rate generator and the other input receives either statistically distributed pulses from a scintillation counter [Fa 65c, Ch 67] or is supplied by a second generator with a different pulse rate [Jo 68a].

DETERMINATION OF TIME SCALE

When using time-to-pulse-height converters with multi-channel pulse-height analysers the knowledge of the time scale, given in ns/channel, is important. The time scale is derived from the measurement of the integral linearity. For this purpose, delay lines are interposed to determine the relation between the interval of the signals at both inputs and the height of the converter output pulses or the channel number of the centroid of the coincidence curve. Apart from the high time consumption, this calibration method has the disadvantage that the length of the delay lines utilised for calibration is varied and does not correspond to the constant length of the cables through which the pulses pass on physical experiment. Cable attenuation influence on delays of some 100 ns and possible reflexions from plug connectors may cause systematic errors.

In the calibration arrangements shown in Fig. 206, auxiliary pulses are applied to the detector outputs. Similar to the detector signals, the auxiliary pulses pass through the *same* delay lines to reach the time-to-pulse-height converter or the coincidence circuit. *Graham* et al. [Gr 62] used a calibrating method where the time interval of the auxiliary pulses is adjusted with the aid of a helical line (Fig. 206*a*). By successive shifting of the feed-in point F the characteristic of the converter is recorded point by point. Because of the limited length of the helical line this method is restricted to a time range of 100 to 150 ns.

An elegant calibrating method has been developed by *Zinov* [Zi 63], which is given in Fig. 206*b*, where the pulse generator output pulses are fed to the time analyser through two gates. The latter are opened by statistically time-distributed pulses from two supplementary photomultipliers, e.g. noise pulses.

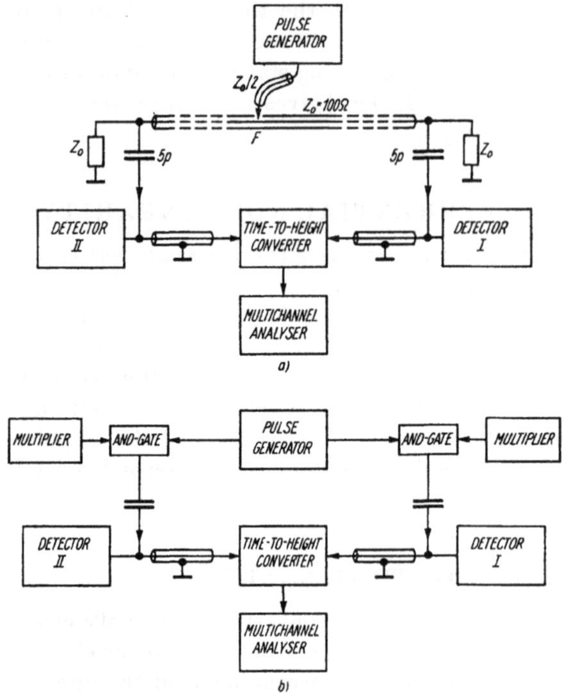

Fig. 206. Calibrating a time-to-pulse-height converter by applying auxiliary pulses. *a*) According to *Graham* et al. [Gr 62]. *b*) According to *Zinov* [Zi 63].

This delivers pulses to both inputs of the time analyser, the spacing of which differs by the multiple integer of the interval of the generator pulses, for instance, by a multiple of 10 ns. In this manner, equidistant time markers are obtained which are distributed over the whole measuring range of the time-to-pulse-height converter. It is the significant advantage of this method that it can also be employed for the calibration of time analysers in the microsecond region, where the calibration with cables of differing lengths is not possible any more. To exclude systematic error in the arrangements given in Fig. 206, shape and size of the fed-in pulses should be well matched with the detector pulses. The described method was used and improved by *Basiladze* [Ba 67c], *Cho* et al. [Ch 67], *Baker* et al. [Ba 68a], *Johnson* [Jo 68a], and *Langsford* and *Dolley* [La 68].

5. APPLICATION

5.1 LIFETIME MEASUREMENTS OF EXCITED NUCLEAR STATES

5.1.1 GENERAL

For theoretical nuclear physics the knowledge of the mean lifetime τ of radioactive nuclei and of the excited nuclear states is of great importance. Decay of nuclei as also of nuclear states is described by the radioactive decay law

$$N(t) = N_0 \, e^{-t/\tau} , \tag{121}$$

where N_0 is the number of nuclei or the number of excited nuclei at the time $t = 0$ and τ their mean lifetime. In place of τ frequent use is made of the easy to handle half-life $T_{1/2}$. The latter represents the time where half the nuclei or nuclear states have decayed. In this case

$$T_{1/2} = \tau \ln 2 = 0{\cdot}693 \, \tau \tag{122}$$

applies. For the decay probability $w(t)$ in the time interval between t and $(t + \mathrm{d}t)$ from equation (121) derives

$$w(t) \, \mathrm{d}t = -\frac{\mathrm{d}N}{N_0} = \frac{1}{\tau} e^{-t/\tau} \, \mathrm{d}t \, . \tag{123}$$

During the interval $\mathrm{d}t$ measurement obtains $\mathrm{d}N$ decay events, expressed with

$$-\frac{\mathrm{d}N}{\mathrm{d}t} = \frac{N_0}{\tau} e^{-t/\tau} = \frac{N}{\tau} \tag{124}$$

from which two methods are derived for determining the mean life τ:

(i) The decay rate $\mathrm{d}N/\mathrm{d}t$ is measured at various moments of time and the value of τ determined from the exponential decay. The method of measuring nanosecond lifetime values in the following section is based on this principle.

(ii) The decay rate of a known number N of nuclei is measured once and the equation (124) used to obtain the value for τ. This method is employed in the case of lifetime of more than several years, if the method described under (i) cannot feasibly be used because the decrease of the decay rate is too small.

The mean lifetime of excited nuclei may amount from less than 10^{-17} s to more than 10 years, spreading across 25 orders of magnitude. Mean life valuesranging from 10^{-7} to 10^{-11} s are measured with the *delayed coincidence method.* Nanosecond lifetime measurements were firstly performed by *Bell*, *Petch* and *Graham* [Be 49; Be 50], *McGowan* [Mc 49; Mc 50], and *Deutsch* and *Wright* [De 50]. These measurements required the development of fast detectors and pulse circuitry, and may be viewed as the initiation of the nanosecond pulse measuring technique. With the exception of special cases, direct time measurement cannot be employed for mean life values of less than about 10^{-11} s. These must be determined indirectly from the cross-section of *Coulomb* excitation or from the energy width of the excited level (nuclear resonance scattering, *Mössbauer* effect).

5.1.2 DELAYED COINCIDENCE METHOD

The method of delayed coincidences utilised for measuring mean life values in the nanosecond region has been described in a number of reviews [Be 55; Be 58a; Be 61a; Sc 61a; Bo 63; Be 65f].

In the decay scheme of the nucleus A shown in Fig. 207, a lifetime measurement of the excited state B^* of the daughter nucleus B is to be undertaken. This excited nucleus is produced from the parent nucleus A with emission of a β-particle (preceding event) and is de-excited to the ground state B with emission of a γ-quantum (delayed event). In principle, the measuring assembly includes two detector devices, usually scintillation counters, for the preceding and the delayed event, and the time analyser (cp. Fig. 1, p. 2). For every decaying nucleus the time analyser measures the interval t between "birth" and "death" of the excited state, i.e. its individual lifetime. Strictly speaking, the time difference T between the two input pulses of the time analyser is measured, with t and T differing only by the additive constant ΔT. The latter is of no importance for lifetime measurements, because it stems from the differing transit time of the pulses from the radiation detectors to the time analyser. Plotting the measured coincidence rate as a function of T the coincidence curve $N(T)$ is obtained which represents the decay function of the excited state and permits the determination of the mean lifetime τ.

Primarily, because of the radiation detector properties (cp. Chapter 2), in the nanosecond region the resolving time $2\,\tau_0$ of the measuring assembly is not short any more against the time difference value to be measured. The time resolution of the whole set-up is defined by the prompt curve $P(T)$. The latter has to be measured under the same conditions as the coincidence curve $N(T)$, however, with nuclei having excited states of an immeasurably small mean lifetime value; in general $\tau < 10^{-12}$ s, known as "prompt decay". Fig. 21, p. 46, gives the prompt curve measured on decay of ^{60}Co. Table 6, p. 44, contains data of the resolving time values $2\,\tau_0$ [full width at half-maximum of the curve $P(T)$] of various scintillators and photomultipliers, and of the attained slope steepness of the prompt curve designated as "instrumental half-time" $T_{1/2\,\text{instr}}$, where the coincidence rate drops by 50%.

The shape of the coincidence curve $N(T)$ is obtained by convolution of the probability density $w(t)$ with the prompt curve $P(T)$. This must be taken into account when analysing the measured data, as detailed in the following paragraph.

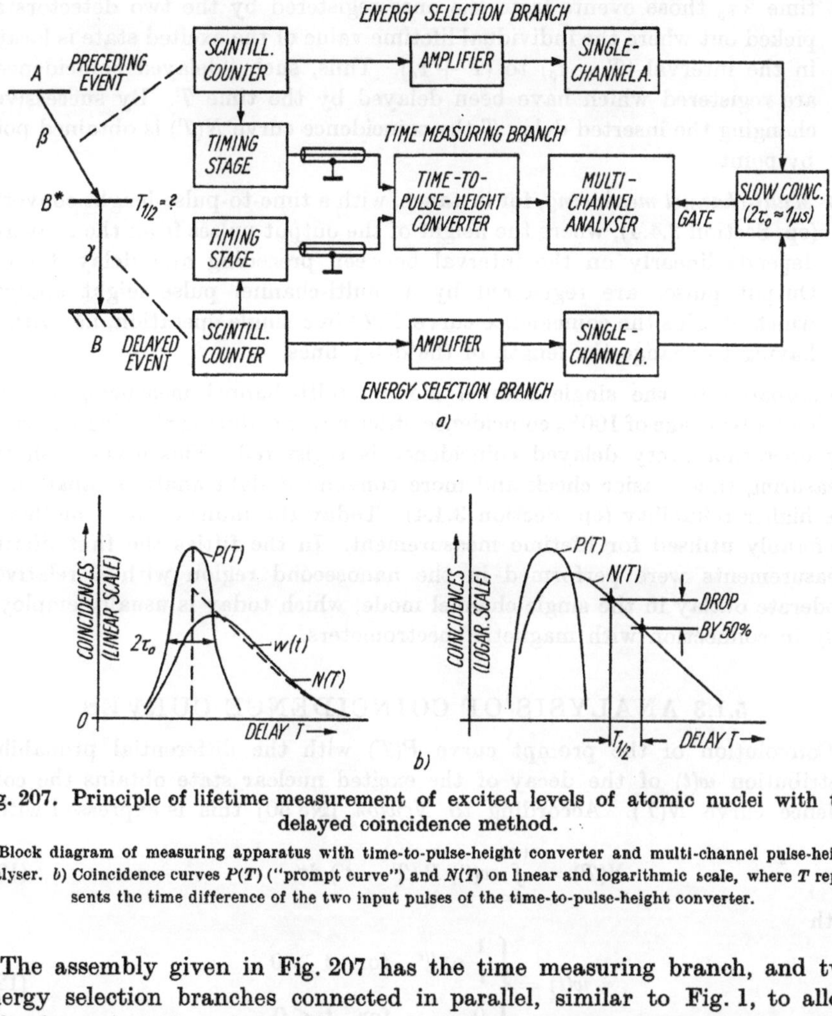

Fig. 207. Principle of lifetime measurement of excited levels of atomic nuclei with the delayed coincidence method.

a) Block diagram of measuring apparatus with time-to-pulse-height converter and multi-channel pulse-height analyser. *b)* Coincidence curves $P(T)$ ("prompt curve") and $N(T)$ on linear and logarithmic scale, where T represents the time difference of the two input pulses of the time-to-pulse-height converter.

The assembly given in Fig. 207 has the time measuring branch, and two energy selection branches connected in parallel, similar to Fig. 1, to allow only the registration of coincidences of pulses having the desired height. This energy selection permits the more or less defined assignment of the registered coincidences to the excited state that is of interest in the case of complicated decay schemes. The restriction to a specific pulse height interval also improves the resolving time because then the pulse height dependency of timing is less effective.

For the measuring procedure itself, two methods may be utilised:

(i) *Single-channel measuring* with the aid of coincidence circuits (cp. Section 3.4.3), where the preceding pulses are delayed by the time T, for example, by inserting delay lines. With the coincidence circuit having a resolving time $2\tau_0$ those events from the ones registered by the two detectors are picked out where the individual lifetime value of the excited state is located in the interval $(T - \tau_0)$ to $(T + \tau_0)$. Thus, such "delayed coincidences" are registered which have been delayed by the time T. By successively changing the inserted delay T the coincidence curve $N(T)$ is obtained point by point.

(ii) *Multi-channel measuring*, for instance, with a time-to-pulse-height converter (cp. Section 3.4.4), where the height of the output pulses from the converter depends linearly on the interval between preceding and delayed pulse. Output pulses are registered by a multi-channel pulse-height analyser which obtains the coincidence curve $N(T)$ in a single operation, i.e. without having to change the length of the delay lines.

Compared to the single-channel mode, multi-channel measuring has the decided advantage of 100% coincidence efficiency, i.e. during the single measuring operation every delayed coincidence is registered. This gives a shorter measuring time, easier check and more convenient data analysis, apart from the higher reliability (cp. Section 3.4.4). Today the multi-channel method is preferably utilised for lifetime measurement. In the fifties the first lifetime measurements were performed in the nanosecond region with a relatively moderate outlay in the single-channel mode, which today is usually employed only in connection with magnetic spectrometers.

5.1.3 ANALYSIS OF COINCIDENCE CURVES

Convolution of the prompt curve $P(T)$ with the differential probability distribution $w(t)$ of the decay of the excited nuclear state obtains the coincidence curve $N(T)$. According to *Newton* [Ne 50] this is expressed with

$$N(T) = \int_{-\infty}^{+\infty} w(t')\, P\,(T - t')\,\mathrm{d}t' \tag{125}$$

with

$$w(t) = \begin{cases} \dfrac{1}{\tau}\,\mathrm{e}^{-t/\tau} & \text{for} \quad t \geqq 0\,, \\ 0 & \text{for} \quad t < 0\,. \end{cases} \tag{126}$$

Through differentiation is obtained

$$\frac{\mathrm{d}N(T)}{\mathrm{d}T} = \frac{1}{\tau}\,[P(T) - N(T)] \tag{127}$$

and

$$\frac{\mathrm{d}\,[\ln N(T)]}{\mathrm{d}T} = -\frac{1}{\tau}\left(1 - \frac{P(T)}{N(T)}\right), \tag{128}$$

$$\frac{\mathrm{d}\,[\ln N(T)]}{\mathrm{d}T} = -\frac{1}{\tau} \quad \text{for} \quad P(T) \ll N(T)\,. \tag{129}$$

From equations (127) and (129) the following properties are evident if $N(T)$ and $P(T)$ are plotted to the same area (cp. Fig. 207 *b*):

(i) With $N(T)$ at its maximum, $N(T)$ crosses $P(T)$, but according to equation (126), under the condition that only a single lifetime τ is present and that $N(T)$ has no added prompt coincidences.

(ii) From the steepness of the tails of the ln $N(T)$ curve, where $P(T) \ll N(T)$ applies, the halfwidth $T_{1/2}$ or the mean lifetime τ can be taken directly (*tail* or *slope method*).

Bay [Ba 50] gives a general moment method for analysis of the coincidence curves $N(T)$, where the moment of the order n of a function $f(x)$ is defined with

$$M_n(f(x)) = \int_{-\infty}^{+\infty} x^n f(x) \, dx . \tag{130}$$

For example, the centroid of the function $f(x)$ is given with the first moment $M_1(f)$. The curves $P(T)$ and $N(T)$ are again normalised to the same area. After some calculations with $f(x) \equiv N(T)$, equation (130) yields [Ba 50; Ba 55; We 60; Sc 61a; Su 62c; Sp 66; Ri 67a]:

$$\text{(centroid shift method)} \quad \tau = M_1(N) - M_1(P) , \tag{131}$$

$$\text{(second moment method)} \quad \tau = [M_2'(N) - M_2'(P)]^{1/2} , \tag{132}$$

$$\text{(third moment method)} \quad \tau = \left[\frac{1}{2} M_3'(N) - \frac{1}{2} M_3'(P)\right]^{1/3} . \tag{133}$$

Thus, according to equation (131), with the *centroid shift method*, the mean lifetime τ corresponds to the distance between the centroids of the coincidence curves $N(T)$ and $P(T)$, which in this case need not be normalised to the same area. The second and third moments M' of equations (132) and (133) refer to the centroid of coincidence curves (principal moments), for instance,

$$M_2'(N) = \int_{-\infty}^{+\infty} [T - M_1(N)]^2 \, N(T) \, dT . \tag{134}$$

Another possibility is to compare the theoretical coincidence curve derived from the convolution integral of equation (125) with the measured coincidence curve and to determine the mean lifetime τ, for instance, from the least squares fit. *Gale* [Ga 62d] gives an analytical expression for $N(T)$ based on a *Gauss*ian curve $P(T)$. *Berkes* et al. [Be 64a] used the four parameters, i.e. halfwidth and location of the centroid of the prompt curve, normalising constant and mean life, for calculating $N(T)$ and performed maximum likelihood fitting. Of the prompt curve, it is merely assumed to be *Gauss*ian. The centroid shift between the curves $N(T)$ and $P(T)$ obtained by the fit, with the shift equal to the fitted value τ, can serve as test of consistency. *Boström* et al. [Bo 66] have expanded this method. The experimental prompt curve $P(T)$ is convoluted with the probability density $w(t)$, with the resulting curve $F(T)$ then fitted to the experimental delayed curve $N(T)$ through the least squares fit. Usually the curves $P(T)$ and $N(T)$ are measured separately, but under similar conditions, which may, nevertheless, show a spurious shift of both curves and lead to a faulty value of τ. The method according to *Boström* et al. has the advantage that such a spurious shift is recognised and can be corrected, by summating the least square deviation between the curves $F(T)$ and $N(T)$ in dependence of the centroid spacing of the curves $P(T)$ and $N(T)$. On "true" position of the prompt curve the resulting sum is a minimum.

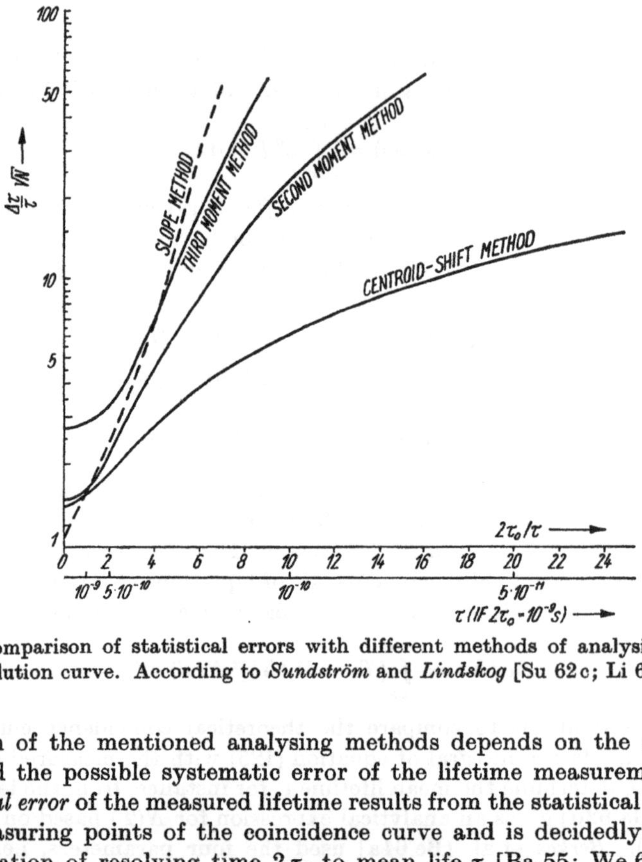

Fig. 208. Comparison of statistical errors with different methods of analysing the time resolution curve. According to *Sundström* and *Lindskog* [Su 62c; Li 63b].

Selection of the mentioned analysing methods depends on the desired accuracy and the possible systematic error of the lifetime measurement.

Statistical error of the measured lifetime results from the statistical inaccuracy of the measuring points of the coincidence curve and is decidedly influenced by the relation of resolving time $2\tau_0$ to mean life τ [Ba 55; We 60; Su 62c; Li 63b; Ol 66; Sp 66]. In Fig. 208 the fractional statistical error of the slope method [equation (129)] and of the three mentioned moment methods [cp. equations (131) to (133)] versus $2\tau_0/\tau$ is illustrated; with N the total number of all measured coincidences and $2\tau_0$ the halfwidth of a prompt curve with normal distribution. It will be seen that with mean life values $\tau > 2\tau_0$ the slope method shows the smallest statistical error. With $\tau < 2\tau_0$ the moment methods, especially the centroid shift method, give a better accuracy. With the third moment method error is highest because here the influence of the measuring points located far outside the centroid, i.e. those of smallest statistical accuracy, is most pronounced.

There are a number of *systematic errors*, which exert an influence that has to be taken into account:

(i) *Added prompt coincidences* in the coincidence curve $N(T)$. These admixtures are comparatively frequent with nuclei of a complicated decay scheme because of the unsatisfactory energy resolution of the detectors. They may also be caused by backscattering between the two detectors. All of the given equations and the diagram in Fig. 208 apply for $N(T)$ curves with no prompt component. Unfortunately the fraction of prompt coincidences in a measured $N(T)$ curve can only be assessed inaccurately or not at all, if $\tau < 2\tau_0$, i.e. if the moment methods have to be used. In most cases, an approximation based on measuring conditions and decay schemes gives only a rough value. On an intensity ratio of $(1-\alpha)/\alpha$ of the prompt and the delayed components, the centroid shift, for example, amounts only to $\alpha\tau$ instead of τ [Ga 62d]. When $\tau > 2\tau_0$ obtains, the prompt admixture can be separated from the $N(T)$ curve with comparatively high accuracy [Bi 59; Ga 62d; Kö 63c]. In this case τ is determined from the slope anyhow, which is not influenced by added prompt coincidences. When two exponential decays are mixed together, similar conditions apply [Ga 62d; Su 62c].

(ii) *Background effect.* Before an analysis can be undertaken the background must be subtracted from the measuring curve, because unrelated events, i.e. events from different nuclei, cause a constant background of *random coincidences*. These are expressed by

$$N_c = 2\tau_0 N_1 N_2, \qquad (109)$$

where N_1 and N_2 represent the count rate at the output of the energy selection branches. In a fast-slow system (cp. Fig. 1, p. 2, and Fig. 207, p. 343) random coincidences of a higher order can additionally contribute when coincidences develop between the measuring branch and the energy selection branches within the resolving time of the slow-coincidence circuit [Be 61b; Sh 61; Le 63b; Sh 63]. For example, the following case may develop: An event in which, during the resolving time of the slow-coincidence circuit, a true coincidence between two pulses, one of which is not of desired height, is followed by a single pulse satisfying height requirement. *Shera* et al. [Sh 63] showed that these contributions of a higher order will disappear against the constant background [cp. equation (109)] provided that the dead time of the branches is longer than the resolving time of the slow-coincidence circuit. *Shera* et al. further gave equations for the calculation of coincidences of a higher order with any dead time, which also apply for coincidence assemblies with three detectors. Compared with continuous sources, in experiments using pulsed sources, as with pulsed electrostatic accelerators or with cyclotrons, estimating chance coincidence rate is very complicated (cp. [Gr 66c; Hr 67]).

If a *pile-up* of detector pulses occurs during the resolving time of one of the two energy selection branches, this results in a background effect, irrespective of the dead time of these branches. This situation makes it possible that a true coincidence with too low an energy value is registered, but which has been accepted by the energy selection system due to the pile-up. The resulting coincidence curve is blurred because of the pulse height dependence of timing (cp. Section 3.4.2) and because of the poorer time resolution with low energy. The curve sits on a pedestal, so to speak, which is similar to a coincidence curve with no energy selection [St 66c].

Random coincidence rate can be approximated from equation (109). More accurate results are obtained by shifting the coincidence curve to outside the time analysis range through an additional delay system and determining the random coincidences by separate measurement. The contribution of random coincidences can also be automatically subtracted [Bl 64a; Av 67] by utilising the ambiguity prevalent in the characteristics of the majority of time-to-pulse-height converters. Background stemming from pile-up cannot be separately measured, but it can be excluded with one of the pile-up detection circuits described in Section 3.5.3.

Background correction is of some significance for such analysis methods where the fractions of a coincidence curve located far outside are quite prominent, as is the case with the slope method and the third moment method [Ri 67a]. Centroid shift is less sensitive. For some typical cases, *Olsen* and *Boström* [Ol 66] investigated the influence of background correction. Fig. 209 shows the results of a measurement undertaken by *Neal* and *Kraner* [Ne 65], where the half-time was calculated from a part of the coincidence curve slope. The half-time derived from the right hand of the curve is too low if too much background is subtracted, and too large if the background has received too little consideration; always assuming the background to be constant. Background influence makes work difficult particularly for the measurement of lifetime values which exceed about 1 μs [Ma 63b; Ra 67].

(iii) *Instrumental effect.* In most cases the time calibration error of time analysers can be neglected, in comparison to other errors (cp. Section 4.4.6). It is especially annoying when despite equal measuring conditions, a spurious shift occurs between the centroids of the $N(T)$ curve and the $P(T)$ curve, which exerts its full influence in the centroid shift method. The slope method and the two other moment methods are insensitive to this instrumental shift. The value of the shift can be determined according to the previously mentioned fitting procedure by *Boström* et al. [Bo 66]. The self-comparison method, described in the following sub-section, requires no comparison with the prompt curve. Asymmetric prompt curves may result, for instance, if one detector senses radiation of considerably higher energy than the other (cp. Section 2.3.2). This can impair the accuracy of the moment methods.

On the whole, it is evident that the statistical error is of minor importance for all methods as long as $\tau > 2\,\tau_0$ holds [Ol 66]. For this region, the slope method and the fitting of a calculated curve is superior to the moment methods.

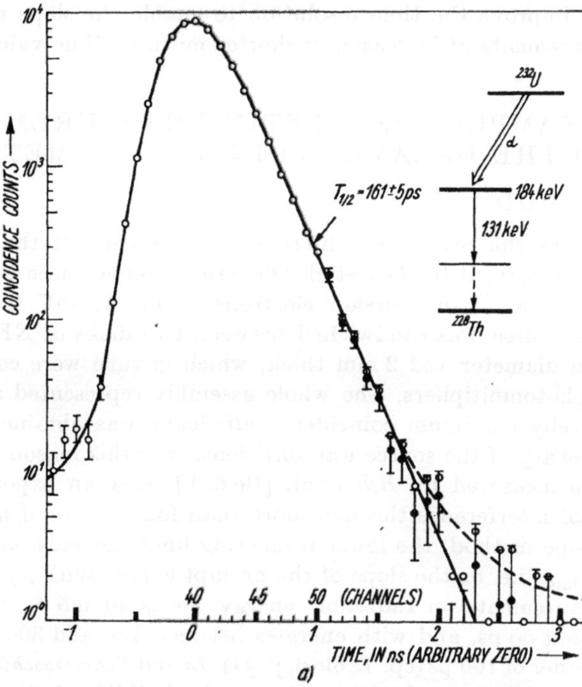

Fig. 209. Lifetime measurement of 186 keV level of ^{228}Th. According to *Neal* and *Kraner* [Ne 65].

a) αe-delayed coincidence curve, with the α-particles sensed by a semiconductor detector and the conversion electrons of the de-exciting transition in ^{228}Th by means of a Pilot *B* scintillator. *b)* Influence of background correction on measured half-life.

The slope method is very convenient to use and has the significant advantage that from the exponential slope the manner and extent of interference can be promptly recognised, which may be caused by the presence of prompt coincidences, of background or by a superposition of two exponential decays. In the case of a short mean life $\tau < 2\,\tau_0$ the moment methods may be more accurate, but the interference is primarily represented by prompt components. This applies in particular for the centroid shift method. For this reason there is a tendency to improve the time resolution to enable the slope method to be used for measurements of increasingly shorter mean lifetime values.

5.1.4 EXAMPLES OF LIFETIME MEASUREMENTS WITH THE DELAYED COINCIDENCE METHOD

SLOPE METHOD

Fig. 210*a* gives the result of a lifetime measurement of the first excited state of ^{226}Th performed by *Bell* et al. [Be 60b]. Coincidences between α-particles and the L, M, . . .-conversion electrons of the 72 keV transition were measured. The source was sandwiched between two disks of NE 102 plastics, each 20 mm in diameter and 2 mm thick, which in turn were coupled to two RCA-6342 A photomultipliers. The whole assembly represented a pair of $2\,\pi$-counters, whereby maximum coincidence efficiency was obtained and a relatively low intensity of the source was sufficient. For this reason other αe-coincidence curves measured by *Bell* et al. [Be 60b] show an exponential decay which is free of interference through more than four orders of magnitude.

With the slope method, the lower measuring limit depends on instrumental half-time $T_{1/2\,\mathrm{instr}}$, i.e. on the slope of the prompt curve. With $\beta\gamma$- and $\gamma\gamma$-coincidences, measurements on radiation energy levels of 0·5 to 1 MeV yielded $T_{1/2\,\mathrm{instr}}$ of about 50 ps, and with energies between 100 and 300 keV obtained values in the order of 100 ps (cp. Table 6, p. 44). *Li* and *Schwarzschild* [Li 63c], for instance, used the slope method to measure the half-life of the 265 keV level in ^{166}Er and obtained $T_{1/2} = 0{\cdot}12$ ns (Fig. 210*b*). The measurement was performed with γ-quanta in the energy range of 0·7 to 0·8 MeV and with conversion electrons of the 184 keV transition. Depending on radiation energy, at present the lower measuring limit of the slope method is $T_{1/2} = 0{\cdot}1$ to 0·3 ns, while the upper limit is determined by the background, the influence of which increases with $T_{1/2} > 1\,\mu$s. In a special case *Bodenstedt* et al. [Bo 66a] measured a half-life value of $T_{1/2} = 1{\cdot}1$ s with the delayed coincidence mode (1315 keV level in ^{177}Hf).

CENTROID SHIFT METHOD

A time-zero for the centroid of the coincidence curve $N(T)$ is the precondition of this method, i.e. a prompt curve $P(T)$ with no spurious centroid shift. The influence of the differing shape of the selected energy spectrum can be lessened

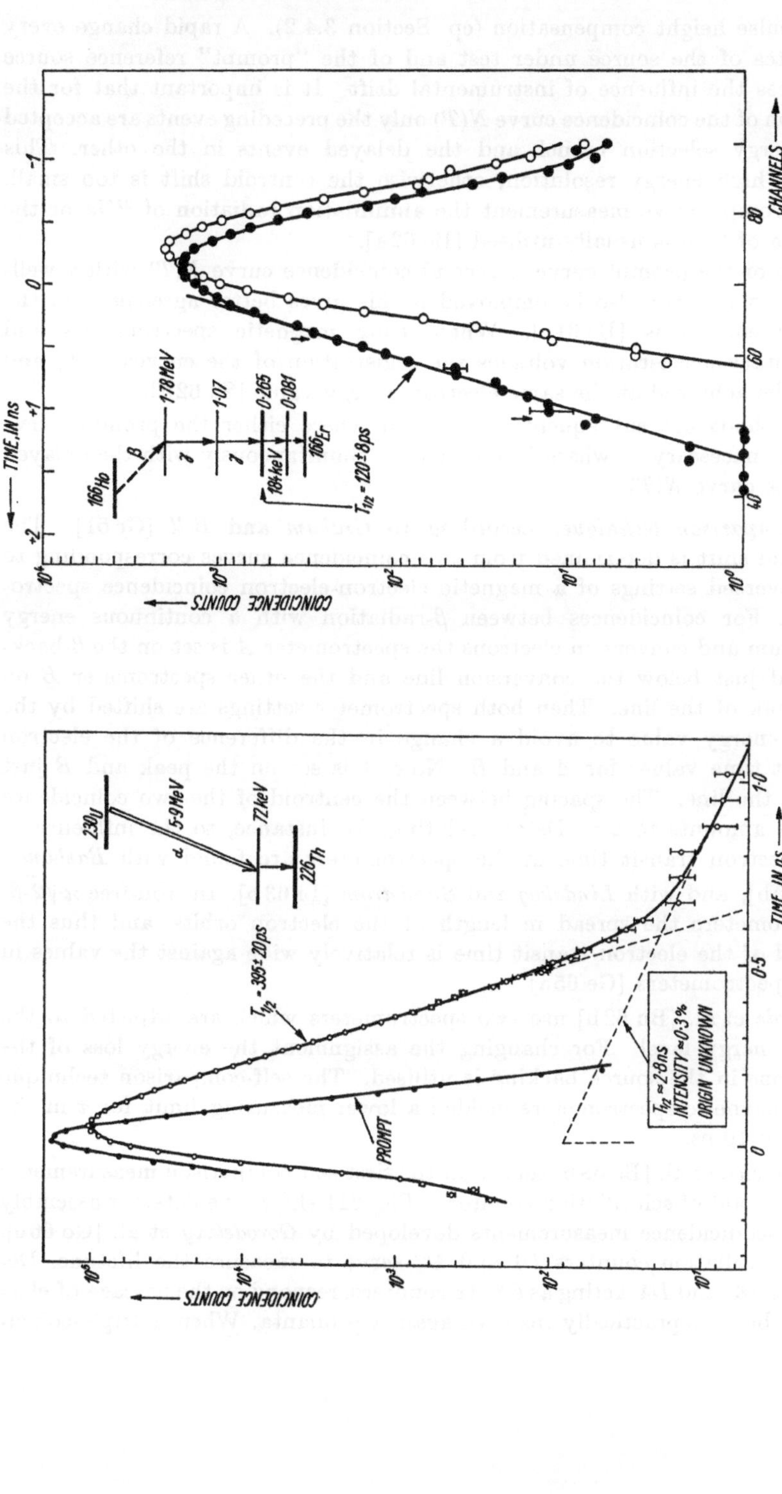

Fig. 210. Application of the slope method in measurements with scintillation counters.

a) α e-delayed coincidence curve. According to *Bell* et al. [Be 60b]. *b)* γ e-delayed coincidence curve. According to *Li* and *Schwarzschild* [Li 63c].

through pulse height compensation (cp. Section 3.4.2). A rapid change every few minutes of the source under test and of the "prompt" reference source also reduces the influence of instrumental drift. It is important that for the registration of the coincidence curve $N(T)$ only the preceding events are accepted in one energy selection branch and the delayed events in the other. This requires a high energy resolution, otherwise the centroid shift is too small. For the prompt curve measurement the annihilation radiation of ^{22}Na or the γ-radiation of ^{60}Co is usually utilised [Be 62a].

In place of the prompt curve, a second coincidence curve $N(T)$ with a well-known mean life can also be employed if this gives better agreement in the measuring conditions [Be 61c]. When using magnetic spectrometers and supplementary acceleration voltages the registration of the curves $N(T)$ and $P(T)$ can be achieved at the same electron energy value [Su 62d].

Two methods are subsequently described where either the prompt curve $P(T)$ is not necessary or where it is measured simultaneously with the delayed coincidence curve $N(T)$.

(i) *Self-comparison technique,* according to *Graham* and *Bell* [Gr 51]. The centroid shift is determined from two coincidence curves corresponding to the reversed settings of a magnetic electron-electron coincidence spectrometer. For coincidences between β-radiation with a continuous energy spectrum and conversion electrons the spectrometer A is set on the β-background just below the conversion line and the other spectrometer B on the peak of the line. Then both spectrometer settings are shifted by the same energy value to avoid a change in the difference of the electron transit time values for A and B. Now A is set on the peak and B just above the line. The spacing between the centroid of the two coincidence curves amounts to 2τ. Details relating, for instance, to the influence of the electron transit time in the spectrometers are found with *Bashandy* [Ba 60b], and with *Lindskog* and *Sundström* [Li 63b]. In iron-free $\pi\sqrt{2}$-β-spectrometers the spread in length of the electron orbits, and thus the spread of the electron transit time is relatively wide against the values in lens spectrometers [Ge 65a].

Burde et al. [Bu 62b] use two spectrometers which are adjusted to the same energy level. For changing the assignment the energy loss of the electrons in the source backing is utilised. The self-comparison technique with magnetic spectrometers yielded a lower measuring limit for τ in the order of 10 ps.

Berlovich et al. [Be 58b] report on the first self-comparison measurements with the aid of scintillation counters. Fig. 211 shows the detector assembly for $\beta\gamma$-coincidence measurements developed by *Gorodetzky* et al. [Go 66c]. The scintillation counters $D1$ and $D2$ serve to measure the lifetime. Detectors $D3$ and $D4$, acting as $\mathrm{d}E/\mathrm{d}x$-counters, respond on the passage of electrons, but are practically inactive against γ-quanta. When a triple-coincidence single-anticoincidence event occurs, for example, of the type $123\bar{4}$. the counter $D1$ must have detected an electron and the counter $D2$ a γ-quantum. If the coincidences are selectively stored in the split memory of a multichannel analyser in accordance with the two possibilities $123\bar{4}$ and $12\bar{3}4$, this results in two coincidence curves from one measurement, which are mutually displaced by 2τ. In this instance, the energy selection is adjusted to the same value. With this method the half-life of the 412 keV level in ^{198}Hg was determined, amounting to $T_{1/2} = (25 \pm 5)$ ps.

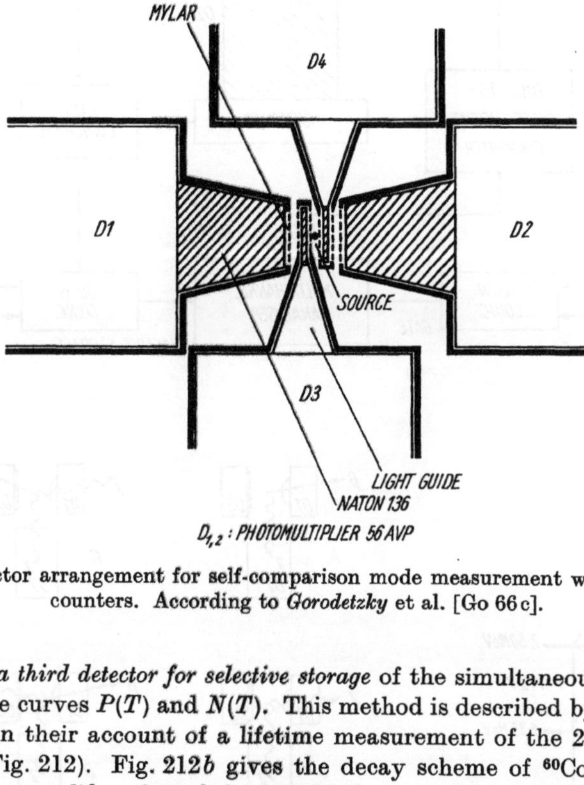

Fig. 211. Detector arrangement for self-comparison mode measurement with scintillation counters. According to *Gorodetzky* et al. [Go 66c].

(ii) *Utilising a third detector for selective storage* of the simultaneously measured concidence curves $P(T)$ and $N(T)$. This method is described by *Lee* and *Wu* [Le 63c] in their account of a lifetime measurement of the 2·50 MeV level in ^{60}Ni (Fig. 212). Fig. 212*b* gives the decay scheme of ^{60}Co. Because of the short mean life value of the 1·33 MeV level, the coincidences from the γ-quanta of the cascade represent prompt events. Therefore, the mean life value of the 2·50 MeV level can be obtained through coincidences between the β-radiation and the γ_1 or γ_2 quanta. Time measurement is made with the detectors $D1$ and $D2$ (Fig. 212*a*). Of the split memory in the multichannel analyser, one half serves for storing the double-coincidences 12, and the other half for storing the triple-coincidences 123. In Fig. 212*b* the possible double- and triple-coincidences are schematically illustrated. With plastic scintillators γ_1-quanta and γ_2-quanta are indistinguishable. Because in

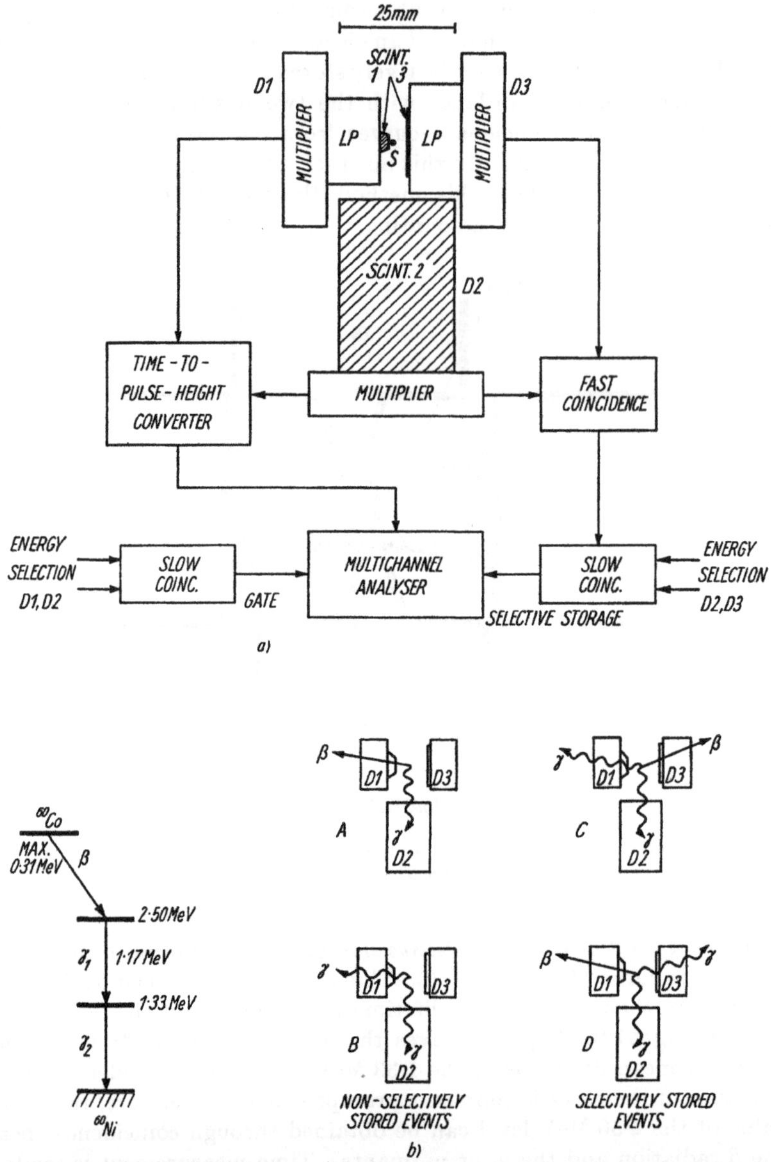

Fig. 212. Lifetime measurement of the 2·50 MeV level in ^{60}Ni. According to *Lee* and *Wu* [Le 63c].

a) Detector arrangement and simplified block diagram: *SCINT* scintillator, *LP* lucite light pipe, *S* source. *b)* Decay scheme of ^{60}Co and schematic diagram of all possible double and triple coincidence events. *A* and *C* represent the predominant main events, *B* and *D* represent the background.

plastic scintillators response probability for β-particles is much higher than for γ-quanta, the possibilities A and C dominate, while B and D can only form a small background. Therefore, the double-coincidence events A produce the $\beta\gamma$-coincidence curve $N(T)$ and the triple-coincidence events C the prompt curve $P(T)$. From the centroid shift of the two curves an upper limit of 5 ps is obtained for the mean life of the 2·50 MeV level, with respective consideration given to probable instrumental error that has gone unrecognised. This measuring assembly is so sensitive that a small displacement of the source in relation to detector $D1$ will cause a clearly discernible shift of the centroid. This shift corresponds to the time-of-flight difference between the electrons and the γ-quanta. Under particularly favourable conditions the 5 ps value represents the probable measuring limit obtainable at present for lifetime measurements with the delayed coincidence method.

A similar method was used by *Simms* et al. [Si 61b]. In cases where no suitable prompt transition is available, another source, e.g. ^{22}Na or ^{60}Co, is added to the tested source and a third detector is employed for selective storing of triple coincidences. In this manner *Li* and *Schwarzschild* [Li 63c] obtained a half life of the 309 keV level in ^{180}Hf amounting to (75 ± 10) ps with a mixture of ^{22}Na and ^{180i}Hf. A third detector can be used with great advantage in the slope method for improving the energy selection [Me 66c; Me 67].

MULTI-CHANNEL ENERGY SELECTION

With the measurements discussed so far only a certain range was selected from the pulse-height spectrum of the two detectors which serve for timing. Any of the true coincidences associated with other energy values are suppressed. *Bonitz* et al. [Bo 64c] utilised a 2048-channel two-dimensional analyser where the output pulses of the time-to-pulse-height converter were stored in 32 Y-lines, each with 64 channels, with the number of a Y-line linearly dependent on the output pulse height of one detector. From the other detector a single-channel selection of a specific pulse-height range is carried out. *Meiling* et al. [Me 67] used two four-channel analysers for pulse-height selection and obtained a quasi three-dimensional presentation of coincidence curves with $4\times4 = 16$ energy combinations in the selectively split 2048-channel memory.

For the investigation of short-lived nuclides the multi-channel energy selection gives significant advantages. Furthermore, systematic trends can be recognised with greater certainty from several simultaneously measured coincidence curves. An example of four simultaneously measured $\beta\gamma$-coincidence curves of the decay of ^{182}Ta is given in Fig. 213. Here, the four energy ranges A, B, C, D of the β-radiation between 100 and 500 keV were selected. The γ-energy exceeded 1 MeV. Prompt curves were obtained from decay of ^{160}Tb. It will be seen that with decreasing β-energy a slight increase of the decay

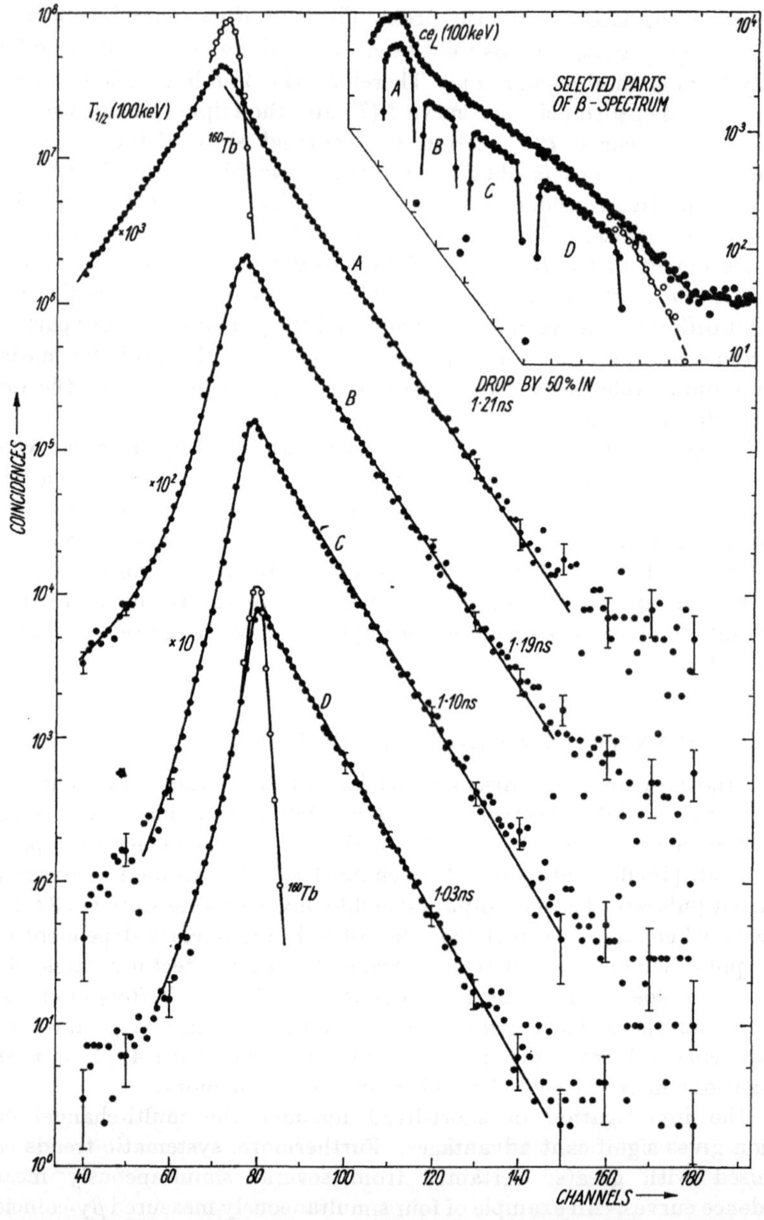

Fig. 213. Simultaneously registered $\beta\gamma$-coincidence curves of ^{182}Ta decay with four different energy ranges *A* to *D* of the β-particles. Insert shows setting of the four energy selecting channels. According to *Meiling* and *Stary* [Me 66c].

time constant of the coincidence curves occurs. Further investigation showed this to be caused by an increasing addition of a second decay with $T_{1/2} = 1{\cdot}30$ ns [Me 66c].

LIFETIME MEASUREMENTS ON NUCLEAR REACTIONS

The measurement of the mean life of energy levels excited by nuclear reaction may be roughly classified as follows:

The *pulsed beam* measurement, where the "preceding" pulse input of the time analyser is derived from the accelerating voltage, which requires only a single radiation detector for measuring. Obtainable time resolution is significantly influenced by the duration of the beam pulses. At the end of Section 5.2 examples of this type of measurement are given.

The *continuous* beam, where the outcoming particles generate the preceding input pulse for the time analyser, e.g. protons in $(d, p\gamma)$ reaction.

The γ-quanta or the corresponding conversion electrons are detected by scintillation counters, while for protons and other heavy particles silicon surface-barrier detectors are often used. *Fossan* et al. [Fo 66], for example, investigated the reaction ^{22}Ne $(d, p\gamma)$ ^{23}Ne and obtained a half-life value of $T_{1/2} = (178 \pm 10)$ ps of the 1·02 MeV level in ^{23}Ne from the slope of the coincidence curve. In the α e-coincidence curve shown in Fig. 209 the α-particles are also registered with a semiconductor detector. *Kahane* and *Vergezac* [Ka 63a] employed a fast-slow arrangement with a semiconductor detector and a scintillation counter for the time-of-flight measurement of neutrons (cp. Section 5.2).

5.1.5 OTHER LIFETIME MEASURING METHODS IN THE NANOSECOND REGION

A number of methods exist for the direct or indirect measurement of the mean lifetime (cp. [Ba 61c; Be 61a]) which are essentially limited to nuclear levels and require no fast pulse technique. These will be mentioned only in passing.

The *high-frequency method* according to *Johansson* and *Alväger* [Jo 58a; Jo 60a; Si 63c] and that of *Blaugrund* [Bl 59; Bl 60; Bl 65] correspond to the single-channel measurement of the coincidence curve $N(T)$. As with the self-comparison technique according to *Graham* and *Bell* [Gr 51] two magnetic spectrometers are used, which are placed opposite each other, with a common source used for both. The spectrometers are set to the energy level of the preceding electrons and the delayed emitted electrons, respectively. In the assembly of *Johansson*, after passing the spectrometers, the electrons have to travel through a high-frequency deflexion field to reach a slit diaphragm.

After passing the latter they are accepted by the detector, provided that the electrons have entered the deflexion field at a specific deflexion voltage phase which depends on the position of the diaphragm. The decay curve is recorded point-by-point by successively shifting the diaphragm. In the arrangement of *Blaugrund* a micro-wave resonator is placed in front of each of two spectrometers for the modulation of the energy of the electrons. In this case, only those electrons are transmitted from the spectrometer to the detectors, which pass the micro-wave field of the resonator on zero-crossing. By successively changing the phase difference between the two micro-wave fields the decay curve can be registered. Both methods have the advantage of a low resolving time (about 0·1 ns) and a good energy selection. The single-channel operation and the very low total efficiency are disadvantageous. A half-time down to 30 ps could be measured. *Tove* and *Jonasson* [To 63] developed a method where the preceding and the delayed electrons pass the same micro-wave field and are then registered with a semiconductor detector. From the measured shape of the sum line the mean life value can be calculated. The lower measuring limit of this multi-channel method is 1 to 10 ps.

The mean life values τ of states excited by nuclear reactions or by α-decay can be determined by measuring the distance travelled by *excited recoil nuclei* before they become de-excited through the emission of γ-quanta or conversion electrons (cp. [Th 53; Se 59b]). When the excited recoil nuclei move through a homogeneous electrostatic field, the emitted conversion electrons are differently accelerated, depending on the location of emission. From the distortion of the conversion lines measured with a magnetic spectrometer the value τ is obtained [No 64]. To permit the recoil nuclei to travel a mean distance in the order of 0·1 to 1 mm until they have reached de-excited state, τ must exceed about 10^{-10} s. The energy shift and the broadening of the γ-lines, which are caused by the *Doppler* effect during emission from moving nuclei, allow lifetime measurements down to 10^{-14} s. It is a disadvantage of the recoil method, that it is generally restricted to lifetime measurements of energy levels of light nuclei and of nuclei with α-decay.

The width Γ of a nuclear energy level is related to the mean lifetime τ for the emission of radiation by the *uncertainty relation* $\Gamma\tau \approx \hbar$. Thus, an experimental determination of Γ indirectly supplies the mean lifetime value τ of the respective state. Width Γ can be determined by measuring the nuclear resonance scattering and absorption, as well as the recoilless emission and absorption (*Mössbauer* effect). With these methods mean lifetime values between about 10^{-17} and 10^{-9} s could be obtained. Another indirect method is to measure the cross-section for excitation of the nuclear levels caused by the interaction between the electric field of the nucleus and that of a closely passing bombarding particle (*Coulomb excitation*). The mean lifetime τ can then be calculated from the cross-section. This measuring method can be used for the case $\tau > 10^{-12}$ s, although, *Coulomb* excitation is essentially restricted to levels which become de-excited with emission of electric quadrupole radiation.

5.2 TIME-OF-FLIGHT NEUTRON SPECTROSCOPY AND PULSED PARTICLE ACCELERATORS

TIME-OF-FLIGHT METHODS

In time-of-flight spectroscopy the particle velocity is measured and from this the energy or, if the momentum is known, the mass determined. Fig. 214*a* gives an arrangement for this purpose, where the particle under test passes

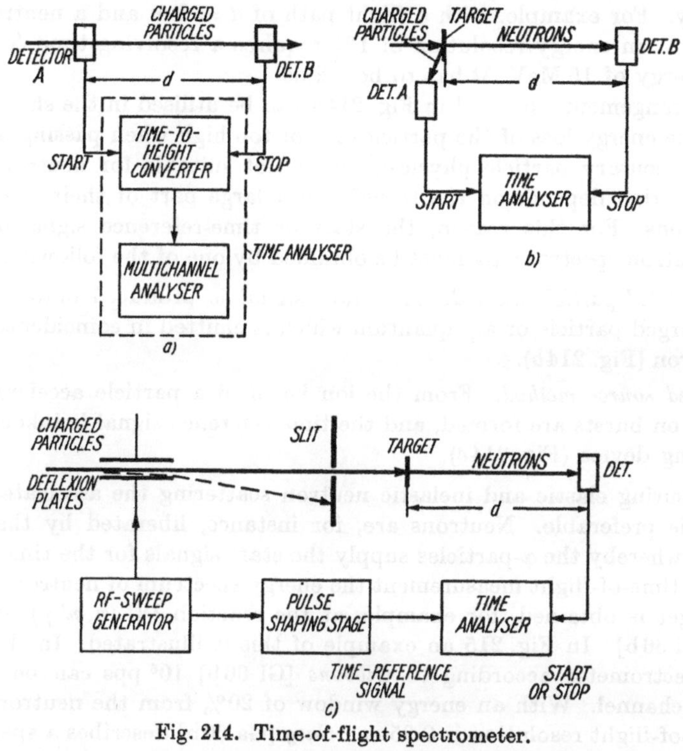

Fig. 214. Time-of-flight spectrometer.

a) Principle. *b)* Associated particle method. *c)* Pulsed source method for time-of-flight neutron spectroscopy.

two detectors A and B, spaced apart by the distance d. Time-of-flight value t is measured by a time-analyser consisting of a time-to-pulse-height converter in connection with a multi-channel analyser. For particles having a mass M, with a kinetic energy E very much lower than the rest energy Mc^2 (non-relativistic particles), the time-of-flight is

$$t = 71{\cdot}9\, d \sqrt{M/E} \tag{135}$$

and relating especially to neutrons, this becomes

$$t = 72{\cdot}3\, d/\sqrt{E} \tag{136}$$

with t in ns, E in MeV, M in atomic mass units and d in m.

For neutrons the time-of-flight spectroscopy has gained considerable importance because it permits high resolution measurements of neutron energy (cp. [Ne 60; Ha 61a]).† Energy resolution is given by

$$\left|\frac{\Delta E}{E}\right| = \frac{2\,\Delta t}{t} = \frac{2\,\Delta t\,\sqrt{E}}{72{\cdot}3\,d}, \tag{137}$$

where Δt represents the resolving time of the time-of-flight measurement. For fast-neutron investigation, time measurements with nanosecond accuracy are necessary. For example, with a flight path of $d = 2$ m and a neutron energy of 0·5 MeV an energy resolution of 1% requires a resolving time $\Delta t = 1{\cdot}0$ ns, at an energy of 15 MeV Δt has to be 0·2 ns.

The arrangement sketched in Fig. 214*a* can be utilised in the shown manner only if the energy loss of the particles is not too high when passing detector *A* (as in elementary particle physics), but is not suitable for neutrons because their detection depends on the transfer of a large part of their energy to recoil protons. For this reason, the start or time-reference signal in time-of-flight neutron spectroscopy must be obtained by one of the following methods:

(i) *Associated particle method.* The start signal is produced in detector *A* by a charged particle or a γ-quantum which is emitted in coincidence with the neutron (Fig. 214*b*).

(ii) *Pulsed source method.* From the ion beam of a particle accelerator short neutron bursts are formed, and the time-reference signal is taken from the pulsing device (Fig. 214*c*).

For measuring elastic and inelastic neutron scattering the associated particle method is preferable. Neutrons are, for instance, liberated by the reaction $t(d, n)\,\alpha$, whereby the α-particles supply the start signals for the time analyser. Through time-of-flight measurement the energy spectrum of neutrons scattered at a target is obtained, for example, at the reaction $^{12}C\ (n, n'\,\gamma)\ ^{12}C$ [De 60; Co 62; Gi 66b]. In Fig. 215 an example of this is illustrated. In the time-of-flight spectrometer according to *Gibbons* [Gi 66b] 10^6 pps can be processed in the α-channel. With an energy window of 20% from the neutron detector the time-of-flight resolution is 0·95 ns. *Garg* [Ga 60b] describes a spectrometer where on the reaction $^{27}Al\ (d, n\,\gamma)\ ^{28}Si$ the γ-quanta are utilised as reference signals. Here the advantage is that those neutrons can be selectively registered which, on the reaction, are emitted simultaneously with the γ-quanta of specific energy.

The pulsed source method can be universally employed. High resolution of time-of-flight spectroscopy of fast neutrons and high time resolution of lifetime measurements of excited nuclear states in the nanosecond region necessitate short beam pulses in the order of 1 ns. These bursts must be produced at such a repetition frequency, that in the time-of-flight spectrometer the neutrons

† Time-of-flight spectroscopy has also been successfully used in electron beam pulse experiments for energy measurements in the order of 1 eV [Na 67a].

released by the successive bursts do not mix. The advantage of the pulsed source method is that only the properties of a single scintillation counter affect the time resolution of the time-of-flight analyser.

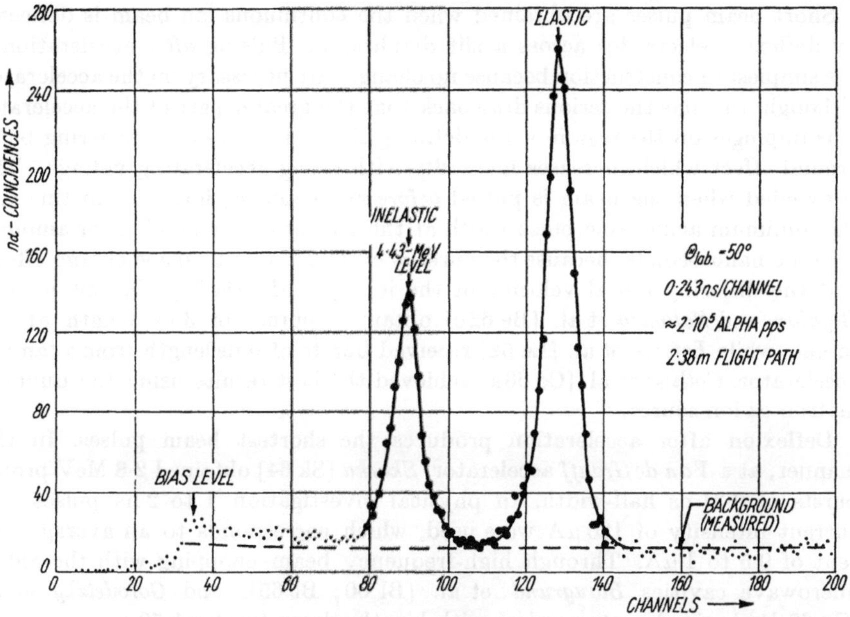

Fig. 215. Time-of-flight neutron spectrum of 14·6 MeV neutrons scattered from carbon. According to *Gibbons* [Gi 66b].

NANOSECOND-PULSED PARTICLE ACCELERATORS

A review of nanosecond-pulsed accelerators has been given by *Bloom* [Bl 62].

With a *cyclic particle accelerator*, such as a cyclotron, the natural phase grouping of the accelerated particles can be utilised. For the length of these bursts values between 1·2 and 5 ns were obtained [Ti 61; Bl 62]. The interval of the bursts corresponds to one r.f. cycle of the cyclotron and consequently amounts to about 100 ns. If external deflexion plates are used, one can arrange that only every second or fourth burst reaches the target [An 62]. In this manner the mean life τ of nuclear levels was measured which had been excited, for instance, by the nuclear reactions ^{48}Ti$(p, n\gamma)$ ^{48}V and ^{51}V$(p, n\gamma)$ ^{51}Cr. For the 305 keV level in ^{48}V the mean life is $\tau = (10{\cdot}9 \pm 0{\cdot}4)$ ns [Ba 63e]. Neutron time-of-flight measurements were also performed by utilising the natural bunching of the cyclotron beam [Kj 62]. If voltage pulses shorter than the r.f. cycle are applied to the deflexion plates individual bursts of the ion beam are obtainable [Gr 57; Be 62b].

Several possibilities exist for pulsing the ion beam of an *electrostatic particle accelerator* (single stage and tandem *Van de Graaff* accelerator):

Chopping Method

Short beam pulses are obtained when the continuous ion beam is deflected by deflexion electrodes across a slit diaphragm. Pulsing *after* acceleration is the simplest in construction because no changes are necessary on the accelerator, although, this has the serious drawback that the greater part of the accelerated ions impinges on the region of the defining slit and causes an interfering background effect which increases markedly with rising accelerating voltage. This is avoided when the beam is pulsed *before* acceleration; however, in this case the minimum achievable pulse width at the output of the accelerator amounts to some nanoseconds, because the burst is extended due to space charge effects and the differing initial velocity of the ions [Co 61; Da 64]. In this manner *Beghian* and *Salomaa* et al. [Be 62c] obtained bursts of 3 ns length at the target, while *Lefevre* et al. [Le 62] received bursts of 6 ns length from a tandem accelerator. *Collins* et al. [Co 66a] achieved the best results using the duoplasmatron as ion source.

Deflexion after acceleration produces the shortest beam pulses. In this manner, at a *Van de Graaff* accelerator, *Skorka* [Sk 64] obtained 2·8 MeV proton bursts of 0·31 ns half-width. In physical investigation 1 to 2 ns pulses at a current intensity of 120 μA were used, which corresponds to an average current of 0·5 to 1 μA. Through high-frequency beam chopping with the aid of microwave cavities *Blaugrund* et al. [Bl 60; Bl 65], and *Gorodetzky* et al. [Go 67d] obtained proton pulses with lengths down to about 50 ps.

Time Compression of Beam Pulses (Bunching)

The principal drawback of the chopping method is that only those parts of the continuous ion beam are used which correspond to the desired pulse duration. The pulse current is, therefore, the same as the beam current of the accelerator. The pulse current can be increased considerably if longer pulses are taken from the continuous beam and subsequently compressed (bunched) to short bursts.

On *klystron bunching* the ion pulse travels through an accelerating field with a field strength which increases during the passage of the pulse, so that the ions later passing receive a higher velocity. At the target the faster ions have just reached the slower ions and a short burst is produced. This bunching process is generally made before acceleration, because klystron bunching after the acceleration of ions to an energy exceeding 1 MeV requires a very high modulating voltage, which results in diffusing the particle energy by some 10 to 100 keV. The pulse to be compressed is generated either by a pulsed ion source or by means of beam deflexion through a slitted diaphragm. Frequently, post acceleration chopping of the beam pulse is employed for further shortening.

Klystron bunching has been utilised with single-stage *Van de Graaff* accelerators [Mo 64a; Vo 65a; An 66a] and with tandem machines [Le 62; An 64; Na 65b], with the pulse length shortened by a factor in the order of 10 (compression factor). In Fig. 216 a pulsed tandem generator for protons and deuterons

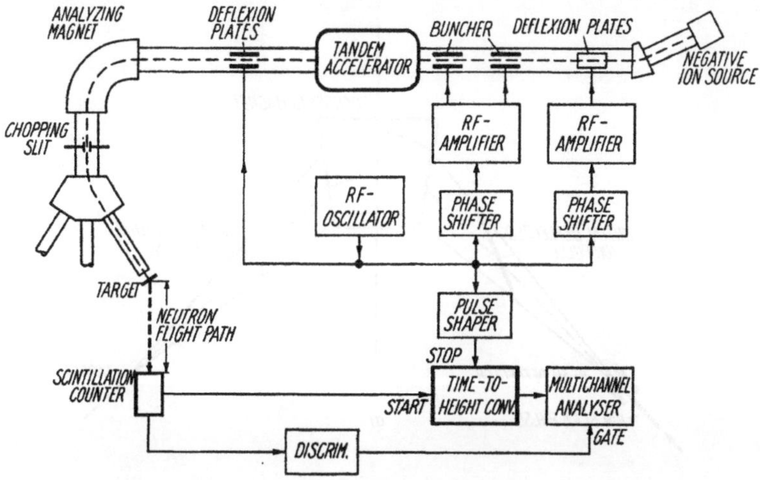

Fig. 216. Neutron time-of-flight system with pulsed tandem *Van de Graaff* accelerator. According to *Lefevre* et al. [Le 62].

with klystron bunching and beam chopping before and after acceleration is schematically illustrated [Le 62; Ba 64c; Ol 67]. Without post acceleration chopping a pulse length of 6 ns is obtained, using chopping gives a pulse length of 2 ns. Bunching increases the intensity by the factor of 8 to 10, with a peak current of 20 μA. In this arrangement the time-reference signal for the time analyser is taken from the r.f. sweep voltage. With other setups the moment of arrival of the ion burst is registered through the current at the target [Gr 57] or through a pick-up cylinder where a small charge pulse is produced when an ion burst passes [Be 62c; Ge 67 b]. Other methods have been summarised by *Rekhin* et al. [Re 67, chapter 1].

Time compression of beam pulses which are accelerated to final energy level can be achieved through *magnetic deflexion* according to *Mobley* [Mo 52b; Mo 63a]. As Fig. 217*a* indicates, the ion pulse is deflected by an electric field. Later arriving ions travel through the compression magnet on a shorter track causing the beam pulse at the target to be focussed in space and time. A disadvantage is the high outlay and the large opening angle of the beam at the target which leads to a relatively high energy spread of the produced neutrons. With *Mobley* bunching the compression factor also is about 10. With a single-stage *Van de Graaff* accelerator a pulse length of less than 1 ns was achieved in this manner [Co 61; Cr 61; Fe 62; Mo 63a; Be 65g; Ts 66].

With the *slanted-target technique* [Ka 62], a principle originating with *Devons* [De 58] and *Shapiro*, the target is positioned on a slant and the ion beam, which is deflected by a transversal electric alternating field, hits the extended target at one instant in time (Fig. 217*b*). *Kane* et al. [Ka 62] received an effective pulse length of about 0·1 ns in this manner. For the generation of intensive

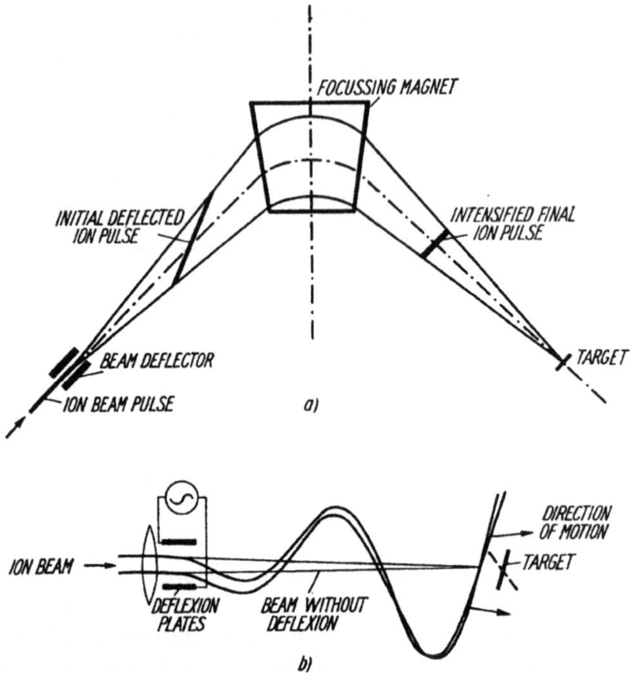

Fig. 217. Time-compression of beam pulses.

a) By magnetic deflexion. According to *Mobley* [Mo 52b]. *b*) Through slanted-target technique. According to *Kane* et al. [Ka 62].

pulsed beams of mono-energetic neutrons *Grodzins* et al. [Gr 65] proposed to slant the target in the way suggested by the broken line in Fig. 217*b*, to achieve that the ion beam crosses the target at a speed corresponding to the propagation speed of neutrons on the target plane.

With a linear electron accelerator beam pulses of less than 10 ns and of a peak current of more than 1·5 A have been obtained with pulsed electron source. Energy of the electrons could be varied from 10 to 70 MeV. With the electrons, high-energy neutron bursts were produced [Fi 66].

Electrostatic particle accelerators with beam pulsing are also utilised for lifetime measurements of excited nuclear states in the nanosecond region [Bl 60; Fe 60b; Sa 61; Ga 62b; Bl 65]. *Lowe* and *McClelland* [Lo 63] employed the slanted-target technique and obtained, for example, on reactions

$^{16}O(d,p)\,^{17*}O$, from the slope of the coincidence curve for the 811 keV level in ^{17}O, a mean life value $\tau = (263 \pm 8)$ ps.

The time-of-flight method may also be used to determine the mass number of particles produced on a reaction with the aid of pulsed accelerators. *Gemmell* [Ge 64] employed the 2 ns beam pulses of a tandem accelerator and identified particles with atomic mass numbers of 1 to 16 by simultaneously measuring the time-of-flight with a time analyser and the particle energy with a silicon semiconductor detector; from equation (135) follows $M \propto E\, t^2$. Similar measurements are mentioned by *Blignaut* et al. [Bl. 67a] and *Emerson* et al. [Em 67]. This method has the advantage that a wide range of energy is covered and no deterioration of the energy resolution occurs, as is the case in the $\mathrm{d}E/\mathrm{d}x$-method using additional detectors. The simultaneous measurement of energy and time-of-flight also allows the accurate determination of the mass distribution of fission products [Sc 65a].

5.3 ELEMENTARY PARTICLE PHYSICS

COUNTER TELESCOPES

Counter telescopes are characteristic for the work with particle accelerators which accelerate protons or electrons to energy values of more than 1 GeV. Such a telescope consists of several detectors, preferably scintillation and *Čerenkov* counters, with outputs mutually linked by coincidence and anti-coincidence circuits. Counter telescopes are employed for the determination of the trajectory, of the range and of the time-of-flight of charged particles, and in general serve for the identification of particles.

An example of a counter telescope with three scintillation counters is given in Fig. 218*a*. Similar to the setup shown in Fig. 214*a*, the time-of-flight between the counters S_1 and S_2 is measured with the aid of a time-to-pulse-height converter. The time-of-flight for the 27 m long distance between S_1 and S_2 amounts to 90 ns at the velocity of light. Particles of a specific momentum are selected by means of the bending magnet thus permitting the mass value of the particles to be determined from the time-of-flight. From the comparison of the time-of-flight of the protons p and the anti-protons $\bar{p}$ *Cocconi* et al. [Co 60] obtained the mass relation $M_{\bar{p}}/M_p = 1{\cdot}008 \pm 0{\cdot}005$.

Fig. 218*b* illustrates the extent of separating the negatively charged anti-protons from the lighter and more frequently occurring π^--mesons. Using a similar time-of-flight measurement *Chamberlain* et al. [Ch 55] proved the existence of antiprotons for the first time. *Piroue* and *Smith* [Pi 65] reported on a two-dimensional time-of-flight measurement for separating K^+-mesons from protons and π^+-mesons of a 1 GeV/c positive beam.

In Fig. 219 the counter telescope and hodoscope system of *Foley* et al. [Fo 64a; Fo 65; Li 66] is shown as another example, which operates with an on-line computer. This system was used at the Brookhaven 33 GeV alternate gradient

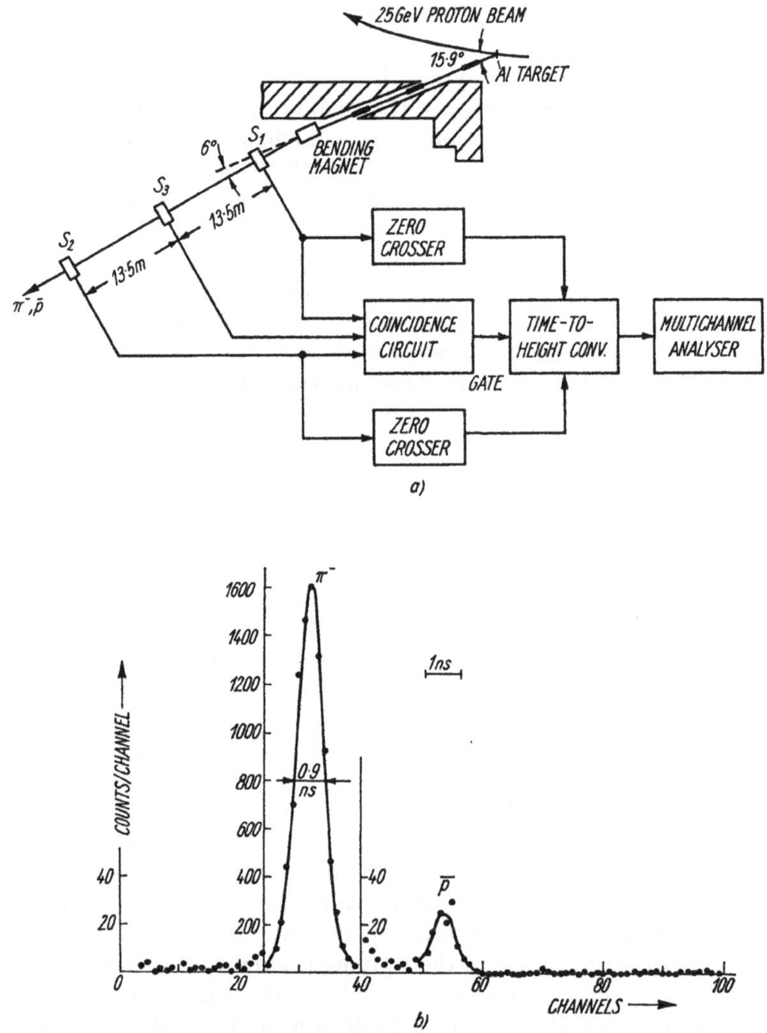

Fig. 218. Counter telescope for investigating secondary particles produced by a 25 GeV proton beam. According to *Cocconi* et al. [Co 60].

a) Schematic. *b*) Measured particle spectrum with π^--mesons and antiprotons $\bar{p}$.

synchrotron to investigate the differential cross-sections for elastic scattering of high energy protons, antiprotons, K- and π-mesons from protons. The momentum-analysed particle beam travels through a counter telescope consisting of the scintillation counters S_1, S_2, S_3 and a high resolution differential *Čerenkov* counter $\check{C}$, whereby the various types of particles of the incident beam are selected. The hodoscopes H_1 and H_2 measure the horizontal projection of the

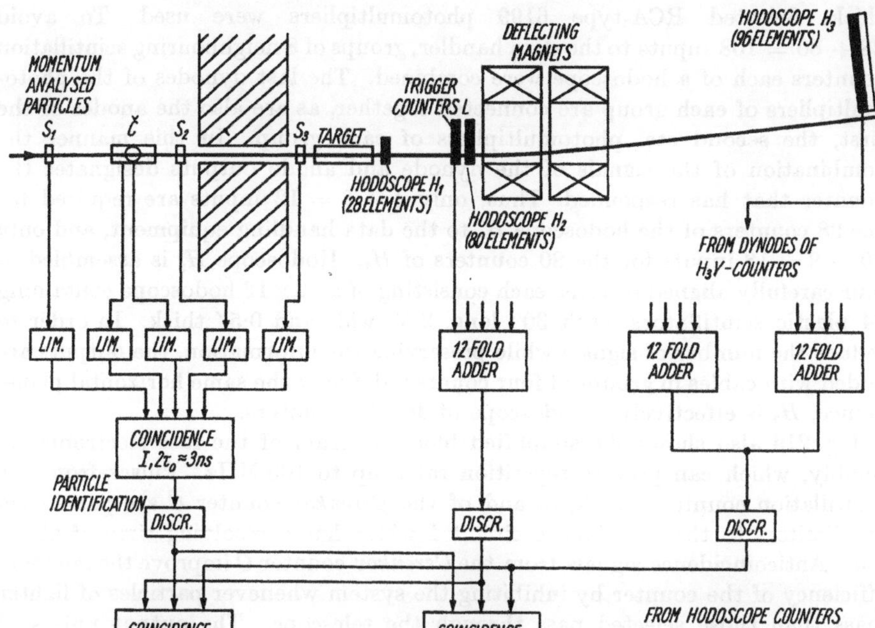

Fig. 219. Counter telescope (S_1, Č, S_2, S_3) and counter hodoscope system (H_1, L, H_2, H_3) with automatic digital data handling and on-line computer. According to *Foley* et al. [Fo 64a].

scattering angle of the particles scattered at the liquid hydrogen target. For separating the elastically scattered particles from the non-elastic background, two magnets 72″ in length are used to deflect the scattered particles. The angle of deflexion is measured with the hodoscope H_3, a two-dimensional screen which thus also measures the vertical projection of the scattering angle. The trigger counters L in front of hodoscope H_2 make it possible to limit the range of scattering angle included. Hodoscope H_1 consists of 28 vertical scintillation counters, each 0·25″ wide, 4″ high and 0·30″ thick, while H_2 has 80 vertical counters, each 0·25″ wide, 6″ high and 0·30″

thick. Selected RCA-type 6199 photomultipliers were used. To avoid $28 + 80 = 108$ inputs to the data handler, groups of 8 neighbouring scintillation counters each of a hodoscope were combined. The last dynodes of the photomultipliers of each group are connected together, as are also the anodes of the first, the second etc. photomultipliers of each group. In this manner the combination of the signals of the dynode and anode outputs designates the counter that has responded. Thus, only $4 + 8 = 12$ inputs are required for the 28 counters of the hodoscope H_1 to the data handling equipment, and only $10 + 8 = 18$ inputs for the 80 counters of H_2. Hodoscope H_3 is assembled of four carefully aligned screens, each consisting of a 12×12 hodoscope containing 24 plastic scintillators, each 30″ long, 2·5″ wide and 0·5″ thick. In order to reduce the number of signals while preserving the information, the outputs are added with cables in groups of four counters defining the same horizontal plane. Hence, H_3 is effectively a hodoscope of 48×12 counters.

Fig. 219 also shows the simplified block diagram of the fast electronic assembly, which can process repetition rates up to 100 MHz. Pulses from the scintillation counters S_1, S_2, S_3 and of the *Čerenkov* counter $\check{C}$ are passed via the limiters to the coincidence circuit I which has a resolving time of about 3 ns. Anticoincidence signals from the *Čerenkov* counter $\check{C}$ improve the rejection efficiency of the counter by inhibiting the system whenever particles of lighter mass than those selected pass through the telescope. The output pulses of the coincidence circuit I together with the pulses from the counters L and H_3 are fed to another coincidence circuit II. If the latter responds the fast gate is opened. In this manner the output pulses of the hodoscope counters reach the 96 channel buffer memory which can store a total of 32 events during the particle burst lasting 150 ms. During the pause of about 3 s until the next burst the computer processes the digital information contained in the buffer memory. Due to the technique employed, an increase in the data accumulation rate could be achieved by up to two orders of magnitude accompanied by a higher systematic accuracy than previously attained in this type of experiment.

Gettner et al. [Ge 67] used a similarly extensive arrangement as shown in Fig. 219 for the investigation of μ-meson pairs produced by the 5·2 GeV bremsstrahlung of an electron accelerator.

LIFETIME MEASUREMENTS OF ELEMENTARY PARTICLES

Mean lifetime values τ of elementary particles in the nanosecond region have been measured with similar electronic techniques as the lifetime values of excited nuclear states. π-mesons decay with $\tau_\pi = 26$ ns according to the scheme $\pi^\pm \rightarrow \mu^\pm + \nu$ in μ-mesons and neutrinos ν. The kinetic energy of the μ-mesons amounts to 4·1 MeV. The μ-mesons are also unstable, viz. $\tau_{\mu+} = (2{\cdot}197 \pm 0{\cdot}002)\ \mu s$ and $\tau_{\mu-} = (2{\cdot}198 \pm 0{\cdot}002)\ \mu s$ [Me 63c]. The mean life value of K^+-mesons, which decay according to $K^+ \rightarrow \mu^+ + \nu$ or $K^+ \rightarrow \pi^+ + \pi^0$, has been determined by *Fitch* and *Motley* [Fi 56] and amounts to $\tau_{K^+} = 12$ ns.

In Fig. 220 two arrangements are indicated which were used to determine the mean life value of π-mesons on decay in flight (*a*) and at rest (*b*). *Durbin* et al. [Du 52] employed a telescope of four scintillation counters (Fig. 220*a*), where the mesons are identified through coincidences of the counters S_1, S_2 and S_3. By utilising four-fold coincidence of all counters the proportion of the π-mesons is measured which pass through the distance L without decay. From the decrease of the π-meson intensity with increasing distance L the mean life value is given by $\tau_{\pi^+} = (24{\cdot}4 \pm 1{\cdot}8)$ ns and $\tau_{\pi^-} = (25{\cdot}5 \pm 1{\cdot}9)$ ns.

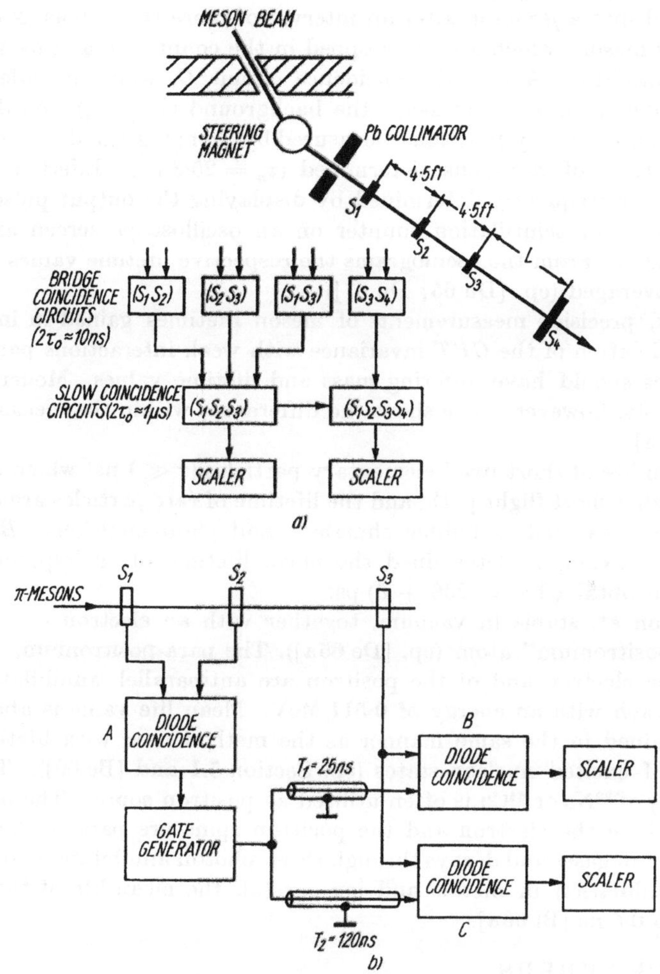

Fig. 220. Investigation of π-mesons.

a) Arrangement for measuring decay in flight. According to *Durbin* et al. [Du 52]. *b*) Registration with the aid of a π-detector. According to *Imhof* et al. [Im 58].

To reduce the random coincidence rate as far as possible coincidence circuits with a low resolving time were aimed at.

The lifetime of π-mesons was, in most cases, measured from the time difference between stopping and decay to μ-mesons. In the π-meson detector of *Imhof* et al. [Im 58] the π-mesons are sensed by the scintillation counters S_1 and S_2 and stopped in the counter S_3 (Fig. 220*b*). Because of the interposed delay line $T_1 = 25$ ns the coincidence circuit B registers only such events where the output pulse from the counter S_3 is delayed against the pulses from the counters S_1 and S_2 by at least 25 ns, that is, when the stopped π-meson has decayed into a μ-meson after an interval of more than 25 ns. Not counted are those π-mesons which are not stopped in the counter S_3 and decay outside of this scintillator. A second coincidence circuit C, which includes a delay line T_2 (120 ns), serves to measure the background ($T_2 \gg \tau_\pi$). As detailed in Section 5.1.2, the decay function is measured by changing the delay T_1 and the mean life value of π-mesons determined ($\tau_\pi = 25{\cdot}3$ ns). Lifetime values of π-mesons were frequently determined by displaying the output pulses of each event from every scintillation counter on an oscilloscope screen and photographing them. From the oscillograms the respective lifetime values were then read and averaged (cp. [Du 65; Du 67]).

Recently, precision measurements of meson lifetimes gained in interest because on violation of the *CPT*-invariance with weak interactions particles and antiparticles should have differing mass and lifetime values. Modern lifetime measurements, however, have shown no differences within the measuring error (cp. [Ro 67a]).

The mean life of short-lived elementary particles ($\tau \ll 1$ ns) which can travel only through a short flight path, and the lifetime of rare particles are calculated from ranges measured in bubble chambers and photoemulsions. *Block* et al. [Bl 63c], for example, determined the mean lifetime of Λ^0-hyperons in this manner and obtained $\tau = (236 \pm 6)$ ps.

A positron e^+, stable in vacuum, together with an electron e^-, in matter, forms a "positronium" atom (cp. [De 65a]). The para-positronium, where the spins of the electron and of the positron are antiparallel, annihilates to two γ-quanta, each with an energy of 0·511 MeV. Mean life value is about 0·2 ns and is obtained in the same manner as the method used with lifetime measurements of excited nuclear states (cp. Section 5.1 and [Be 60]). The radioactive decay of ^{22}Na or ^{64}Cu is often utilised as positron source. The ortho-positronium, where the electron and the positron spins are parallel, has a mean life of 140 ns in gases and decays through three-photon annihilation. With three-photon annihilation in metals and ion crystals the mean life of the positron amounts to 0·5 ns [Bi 66a].

SPARK CHAMBERS

At the location of the passage of a particle in a spark chamber spark discharges are caused which make the particle trajectory visible. Accuracy of trajectory measurement is at times better than 1 mm. Spark chambers have a dead time in the order of 1 to 10 ms. In a number of papers the variety of types of spark chambers have been reviewed [Ro 61a; Da 63; Ru 64a; We 64c; Ch 65].

Fig. 221 schematically shows a spark chamber with narrow gaps. In principle it consists of several parallel-plate spark counters (cp. Section 2.4.2). The telescope with counters *1* and *2* responds after an ionising particle has

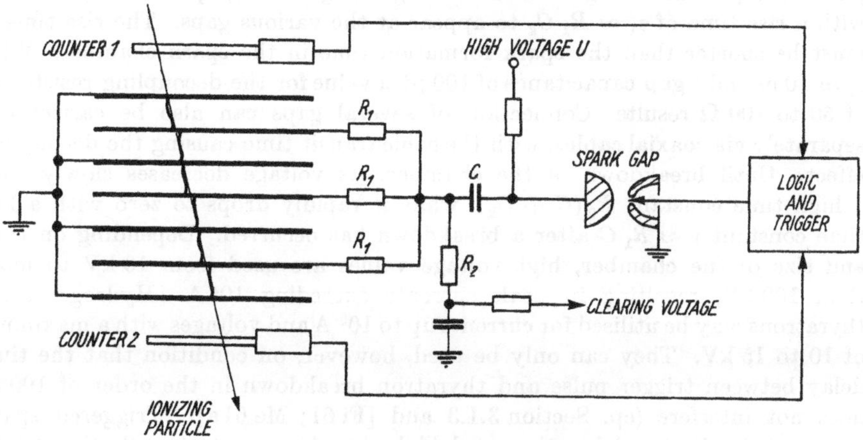

Fig. 221. Spark chamber (schematic).

passed and releases the high-voltage pulse for the spark chamber. In this controlled operation only the desired event is registered, and considerably larger electrode areas can be used than with permanently applied high-voltage. The electric "clearing" field between the electrodes extracts the free electrons during the pulse intervals and determines the storage time of the chamber. To keep the background down this storage period should be only slightly longer than the interval between particle passage and response of the spark discharge [Ch 65]. The high-voltage pulse must rise as fast as possible so that in the time interval to actual breakdown the electrons cannot have moved away too far from the place of origin. Extraction voltage and high-voltage are poled in opposition to compensate the electron drift through the clearing field.

The delay between particle passage and breakdown must be short if the spark chamber is operated in a particle beam of high intensity or in a magnetic field. In practical work the minimum delay is 0·1 to 0·2 μs. This can, however, rise to 1 μs, if a complicated processing of the output pulses from the counter telescope is necessary. Transit time in the photomultipliers of the counter telescope amounts to about 40 ns, although multipliers do exist that have a transit time of only 10 to 15 ns (cp. Table 4, p. 28 — types XP 1140, XP 1141). Transit time values of the cable connections and the trigger delay of the high-voltage switch represent further delay contributions. With a careful selection of all parameters, *Fischer* and *Zorn* [Fi 62a] and *Schrank* et al. [Sc 64c] obtained a delay of 33 ns and 57 ns, respectively, between particle passage and breakdown.

Fig. 221 gives the principle of high-voltage pulse generation [Ru 64a]. The storage capacitor C, with a value 4 to 5 times higher than the chamber capacitance C_k, is charged by the high voltage U. On triggering, the right-end plate of capacitor C is grounded causing a high-voltage pulse $U\,C/(C + C_k)$ with a rise time of $\tau_1 \approx R_1\,C_k$ to appear at the various gaps. The rise time τ_1 must be shorter than the spark formation time in the spark chamber. With $\tau_1 \approx 10$ ns and a gap capacitance of 100 pF a value for the decoupling resistor R_1 of 50 to 100 Ω results. Connection of several gaps can also be carried out separately via coaxial cables, with the cable transit time causing the decoupling effect. Until breakdown of the chamber, its voltage decreases slowly with a high time constant $R_2\,(C + C_k)$, while it rapidly drops to zero with a low time constant $\tau \approx R_1\,C$ after a breakdown has occurred. Depending on type and size of the chamber, high-voltage values are used from 10 kV to more than 100 kV, resulting in peak currents exceeding 10^4 A. Hydrogen-filled thyratrons may be utilised for currents up to 10^3 A and voltages with a maximum of 10 to 15 kV. They can only be used, however, on condition that the time delay between trigger pulse and thyratron breakdown in the order of 100 ns does not interfere (cp. Section 3.1.3 and [Fi 61; Me 61a]). Triggered spark gaps permit shorter delay time and higher peak currents (cp. Section 4.1.4). They are less expensive than thyratrons and require lower trigger power. This type of spark gap is shown in Fig. 221. With various constructions a time delay of only 20 ns could be obtained [Le 61; Sc 64c]. *Issinskii* and *Myznikov* [Is 65] describe a spark gap arrangement for a maximum current of 6×10^4 A and a maximum voltage of 150 kV; it has a delay time of 130 ns.

5.4 THEORY OF RELATIVITY

One of the two postulates of the special theory of relativity is the assumption that the velocity of electromagnetic radiation is independent of the motion of the source. This means, that a stationary observer always measures the same propagation velocity of the electromagnetic radiation, irrespective of whether the source is at rest, relative to the observer, or moves with a velocity close to the velocity of light c.

For experimental test work, this postulate can be written

$$c' = c + k\,v\,, \tag{138}$$

where v is the relative speed between source and observer, and the special theory of relativity condition is $k = 0$. Astronomical measurements, especially of binary stars, gave $k < 10^{-6}$ as best approximation [Zu 14; Th 25]. However, recent discussions, based on the extinction theorem of the dispersion theory, call in question the stringent evidence of previous experiments regarding the postulate of constant propagation speed. Isotropy of the propagation speed on condition of ultrahigh frequencies is also questioned (cp. [Al 66]). Experiments in nuclear and elementary particle physics permit the high accuracy investigation of the validity of the above postulate with sources moving with very high speed and to verify the isotropy of the propagation speed with the aid of electromagnetic radiation of very high energy, i. e. ultrahigh frequencies.

Sadeh [Sa 63c] compared the propagation speed of γ-quanta emitted on annihilation of stopped positrons with that of γ-quanta occurring on annihilation of positrons during flight. Such a measuring arrangement is shown in Fig. 222*a*, where ^{64}Cu serves as positron source. Gamma-quanta formed on annihilation of stopped positrons in the perspex layer of only 1 mm thickness and moving in opposite direction, i.e. at an angle of $\Theta = 180°$, are not registered because the two scintillation counters are positioned at an angle of $\Theta = 150°$ to 160°. On annihilation in flight, for a stationary observer, the two γ-quanta are emitted at this angle if the positron energy amounts to 0·2 to 0·6 MeV. In the arrangement given in Fig. 222*a*, on classical addition of speed, the velocity of the γ-quanta flying to the right should be higher than c and of the γ-quanta moving to the left lower than c. This would lead to a shift of the coincidence curve as compared to the curve measured under 180° geometry, i.e. with γ-quanta from the annihilation of stopped positrons. Experiments gave no evidence of such a shift, consequently $k = 0{\cdot}0 \pm 0{\cdot}1$.

Alväger et al. [Al 64a] measured the speed difference of γ-radiation emitted from a moving source ($^{12*}C$) and a source at rest ($^{16*}O$). In the cyclotron the excited ^{12}C nuclei (4·43 MeV) were generated through the reaction $^{12}C\,(\alpha, \alpha')\,^{12*}C$, and because of the recoil on emission of α-particles they received a velocity of $v = 0{\cdot}018\,c$. A similar arrangement as given in Fig. 214*c* was employed to determine the γ-velocity by a time-of-flight measurement, where, however, the natural bunching of the cyclotron beam was utilised. This resulted in a factor k of $0{\cdot}1 \pm 0{\cdot}1$.

Luckey and *Weil* [Lu 52] determined the velocity of γ-quanta of the bremsstrahlung of 310 MeV electrons, which agrees with the velocity of light c, within the estimated probable error of 1%.

Alväger et al. [Al 64; Al 66] searched for the possible breakdown of the velocity postulate with a source at very high speed. They measured the velocity of γ-radiation emitted in the decay of π^0-mesons which had an energy exceeding 6 GeV, corresponding to $v = 0{\cdot}99975\,c$. Coincidences were registered between the time-reference signals, taken from the constant r.f. frequency f during target irradiation, and the pulses from a scintillation counter, at A or B (cp. Fig. 222*b*). On condition of the velocity of light of γ-radiation and a distance between A and B being 31·450 m, a transit time is obtained which would just equal the duration of a period of the frequency f. This causes the coincidence curves associated with the detector positions A and B to become registered practically

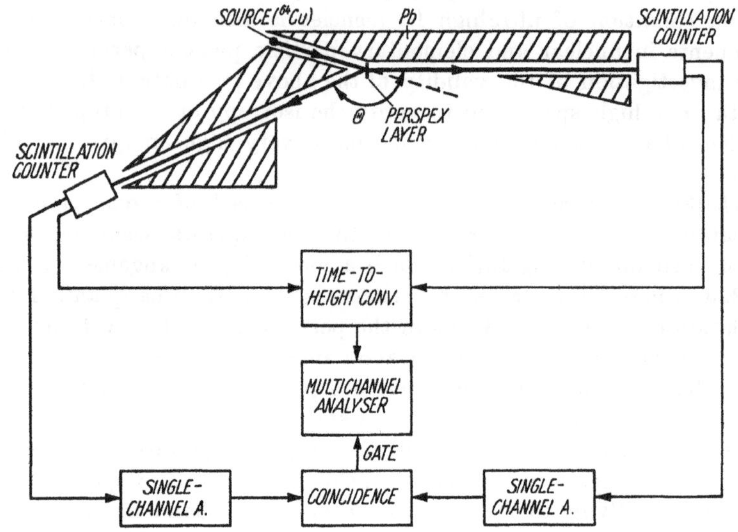

a)

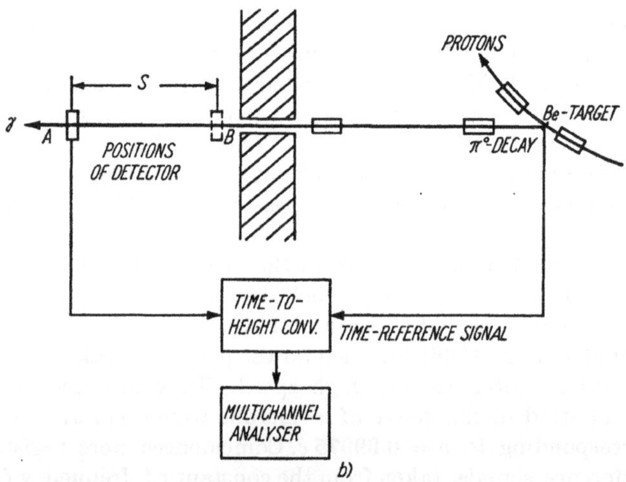

b)

Fig. 222. Arrangement for measuring the velocity of γ-radiation from a moving source.
a) Annihilation of positrons in flight. According to *Sadeh* [Sa 63c]. b) Decay of π^0-mesons. According to *Alväger* et al. [Al 64].

in the same channels of the multichannel analyser. In this manner a very accurate measurement of velocity is possible which is not very sensitive to errors of time calibration and time analyser linearity. From the coincidence curve shift $k = (0 \pm 13) \times 10^{-5}$ follows. That means the velocity postulate of special theory of relativity is valid also when the source moves practically with the velocity of light. Due to the rotation of the earth the space orientation of the flight path altered during the 12-hour measuring period. However, within the measuring error no indication of velocity anisotropy could be found for 6 GeV γ-radiation [Al 66].

5.5 MISCELLANEOUS APPLICATIONS

Although nanosecond pulse technique has found application in various fields of basic and applied research, it has not, however, attained the dominance it has in nuclear and elementary particle physics. In the subsequently cited applications nanosecond pulse technique is mainly confined to handling the production or recording of short pulses in investigation of specific phenomena.

LIFETIME OF EXCITED ATOMIC STATES

To determine the lifetime of excited atomic states, apart from a number of more or less indirect methods (cp. [St 66d]), the techniques described in Section 4.4.3 for measuring light pulse shapes can be used. Atoms to be tested are excited through electromagnetic radiation, mainly ultra-violet, by electron impact, or by laser pulses [Ro 66]. *Brewer* et al. [Br 62c] employed a phase fluorometer to measure radiative lifetimes of 10^{-5} to 10^{-9} s, where, among other things, an ultrasonic grating for frequencies to 5 MHz was utilised to modulate the excitation light. *Karstensen* [Ka 65; Ka 66b] found the decay-time constant of Na-D lines from the light-pulse oscillogram, after excitation with 15 ns pulses from an electron beam (energy 2·5 to 4·5 eV).

As with excited nuclear states, the mean lifetime can also be measured with the delayed coincidence method [He 56; Kl 66; Ba 67; Br 67; Ki 67]. Fig. 223 gives a schematic illustration of the measuring setup according to *Klose* [Kl 66], where electron-beam pulses excite the neon atoms in the excitation chamber. The resulting light pul-

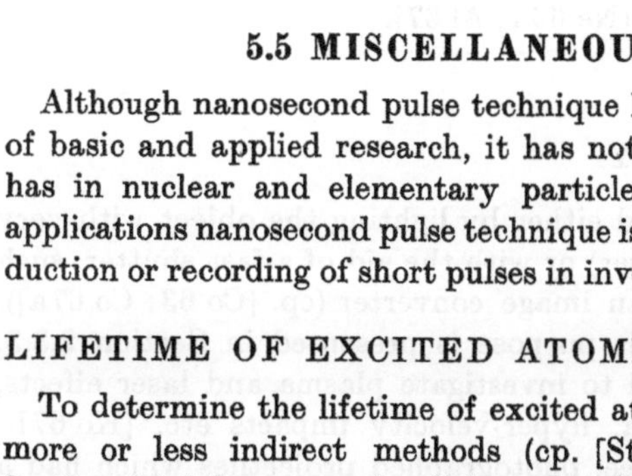

Fig. 223. Atomic lifetime measurement with the delayed coincidence method. According to *Klose* [Kl 66].

ses are fed to the photomultiplier via a monochromator. Function of this assembly corresponds to the single-electron method (Fig. 201, p. 327). In consequence it is a precondition that the de-excitation of no more than one atom is registered with each pulse of the electron beam. For various excited states of neon-I mean lifetime values ranging between 14 and 500 ns were obtained.

In conjunction with the *Mössbauer* effect, the method of delayed coincidences permits the measurement of the time evolution of the *Mössbauer* resonance following formation of the resonant nuclear level. This served for searching for metastable multiply charged ionized states, the lifetime of which is comparable to that of the resonance level (cp. [Tr 67]). The time-resolved *Mössbauer* spectroscopy was also used to investigate the time dependence of γ-radiation transmission through a resonance absorber [Ne 66a; Al 67].

HIGH-SPEED PHOTOGRAPHY

High-speed photography is achieved either by lighting the object with very short recurrent light pulses (spark, laser) or with the aid of a fast shutter, such as a rotating mirror, a *Kerr* cell or an image converter (cp. [Co 63; Co 67a]). The use of image intensifiers for this purpose is presented in Section 2.3.3. High-speed photography is employed to investigate plasma and laser effects, shock waves, detonation phenomena, hyper-velocity impacts etc. [Ko 67b; Vo 67a]. *Trammell* [Tr 65], for instance, photographed projectiles which had a speed of 6 to 7 km/s. He used a Q-switched laser as light source.

PLASMA PHYSICS

The time behaviour of discharges can be investigated by photographing the various phases. *Hopmann* [Ho 66] developed a single-frame high-speed camera having a commercially available image-orthicon television camera tube. It is suitable for pulse operation and permits shutter times down to 20 ns. This camera was used to examine spindle-cusp discharges. *Skinner* [Sk 62a] photographed the axial view of a high power theta pinch at a framing rate of 8×10^6 frames/s, showing the rotation and subsequent fission of the plasma.

Plasma diagnostics can also be effected with short electromagnetic pulses. Using light pulses of 30 ns width from a ruby laser *Kingsland* [Ki 65] utilised *Thomson* scattered laser radiation to determine the density and temperature of electrons in helium plasma. The temperature characteristic can also be found by measuring the intensity of the infra-red radiation, for instance, with the aid of a photodiode. When electrical pulses with a rise time of less than 1 ns are fed into a coaxial discharge tube, time interval and height of the reflected signals give information on the properties and shape of the ionised region [Sc 64d].

LASER

Solid state lasers (ruby laser, neodymium laser in yttrium-aluminium garnet or in glass) deliver giant pulses with Q-switching of the resonator which have a minimum length of 10 to 50 ns and a peak level of some hundred megawatts. For Q-switching either active switches are used, such as rotating mirrors, *Kerr* cells or ultrasonic cells (cp. [Bi 66b; Le 66; Su 66]) or passive switches, such as a solution of kryptocyanide which becomes transparent if the arriving laser beam intensity exceeds a specific threshold level. Further shortening of the laser pulse length is obtained by mode-locking. Using a ruby laser, *Michon* et al. [Mi 66a] achieved a train of light pulses of 4 ns length spaced 16 ns apart. *DiDomenico* et al. [Di 66] describe the generation of ultrashort optical pulses by mode-locking the YAlG:Nd laser. Light pulses are emitted at a repetition rate of 300 MHz. A photodiode gave a measured halfwidth of less than 150 ps; theoretically the lower pulse width limit is 80 ps. *DeMaria* et al. [De 67] used a neodymium laser in glass and obtained a peak level of 600 MW at pulse lengths of less than 0·5 ns. By amplification in a second laser, pulses could be produced of 20 ps length and a peak level of 40 GW.

Gas lasers have rather seldom been used for the production of pulses. With a nitrogen laser, pulses of 10 ns length and a power of 20 kW could be obtained [Ge 66b], while phase-locking of a He-Ne-laser yielded 0·6 ns pulses at a spacing of 4·46 ns [We 66a].

Semiconductor lasers primarily operate as p-n injection lasers with gallium-arsenide [Bu 64; Re 65]. This yields light pulses that are shorter than 0·2 ns [Ko 64b]. Gallium-arsenide lasers are easily handled and have the added advantage that they can be modulated either with short pulses or with very high frequencies [Go 65a]. Time and space coherence, however, are much less satisfactory than with a gas laser.

When employing an avalanche photodiode for laser beam detection, the high quantum yield gives a higher sensitivity than infrared-sensitive photomultipliers [Da 66a]. This type of diode can be utilised for investigating fast optical pulse trains and for the construction of photocoupled circuits. Pulsed lasers are used in many fields of application, which include distance measurements (see next section), pulse code modulation [We 66a; De 68], surface emission, holography, the time behaviour of chemical reactions [Wo 67], and the processing of microelements etc. [He 66a; Se 67; Vo 67a]. In many cases the high energy content of the laser pulse is a deciding factor.

RADAR

Nanosecond pulse radar is necessary to resolve discrete scatterers located on objects that may be metres long. *Bachman* [Ba 65b] reports on radar cross-section measurements and gives a block diagram for pulse radar having a pulse width of 0·6 ns at a frequency of 8·0 to 9·6 GHz and a peak power of 1·8 kW. With the aid of pulse radar it was possible to resolve the discrete scatterers of a model plane. *Young* et al. [Yo 67] produce pulses of about 1 ns length and a repetition rate of 2 MHz by keying microwave diodes at a frequency of 9·2 GHz.

Goldstein and *Dalrymple* [Go 67 a] employed a gallium-arsenide injection laser with a pulse width between 10 and 100 ns for optical radar measurements; the peak power of the light pulse amounted to 9 W with a current of 40 A and a repetition rate of 330 Hz. At night this light radar has a range of some 100 m to more than 1 km, depending on the reflector employed. Using a spark gap as a light pulse source, *Fischer* [Fi 67] achieved ranges of 10 to 75 m with an accuracy of a few centimetres.

Christie [Ch 66] describes a photographic radar with laser pulses of 20 ns width, where the reflexions at the object under test were recorded by a high-speed camera (Fig. 224). A pulsed image intensifier was used as camera. If

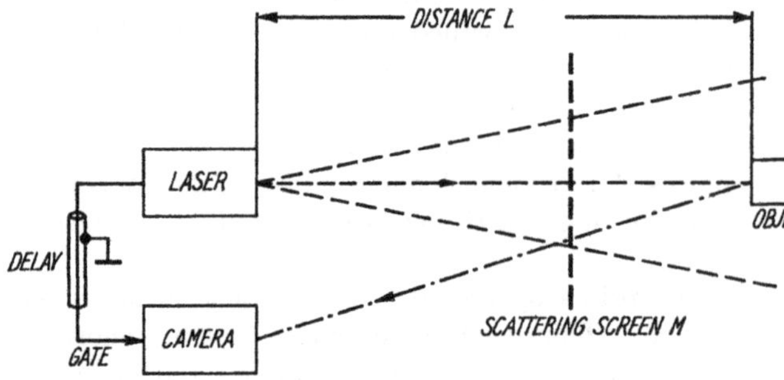

Fig. 224. Photographic radar; time-gated camera (image intensifier) records the object O in the presence of the foreground scattering screen M. Distance between O and M 6 m. According to *Christie* [Ch 66].

the delay between the emission of the light signal and the gate pulse of the image intensifier is $2\,T$, in this case, only such objects are photographed which are located at a distance of $L = T\,c$ from the light source and the camera. In this manner the scattered radiation, for example, from the diffusely reflecting muslin screen M in Fig. 224 or from fog in the foreground, is suppressed.

SATELLITE INSTRUMENTATION

The construction of fast pulse circuitry for the electronic instrumentation of satellites is difficult because this, among other things, requires a low power consumption. *Sarles* and *Roberge* [Sa 65] describe a gamma-ray telescope with 15 photomultipliers which has an analogue section using 430 transistors. This setup has a consumption of only 0·35 W and in a specific section is able to process plastic scintillation counter pulses of 3 ns width.

Schindler [Sc 63 a] developed an analogue-to-digital converter which transmits data digitally because such signals are less affected by noise over long distances.

The transmitting system has a bandwidth of 25 MHz and is suitable for the simultaneous transmission of four television channels. As Fig. 225 shows a repetition rate of 50 MHz is used for sampling, with the height of samples measured in $2^6 = 64$ stages. This 6 bit information is coded and reproduced by pulses of an interval of 3·3 ns. Charge-storage diodes are used for pulse generation and tunnel diodes as switching stages.

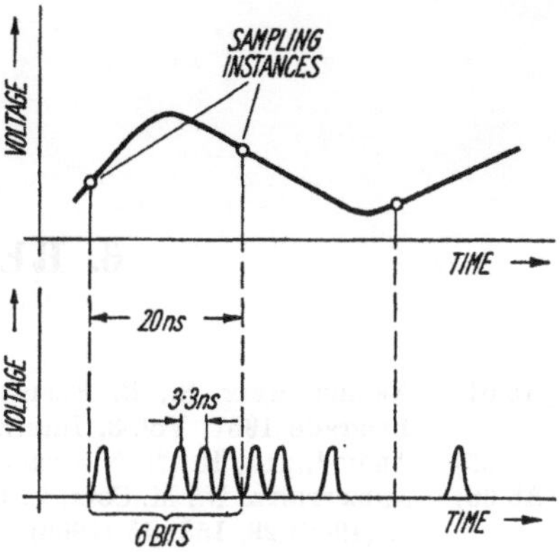

Fig. 225. V.h.f. digital converter. According to *Schindler* [Sc 63a].

COMPUTERS

Further increase in the operating speed of electronic computers necessitates components and logic circuits capable of operating with pulse rise time values and pulse intervals in the nanosecond region.

This also requires memories having a cycling time of less than 1 μs. For the construction of fast memory units the knowledge of magnetic properties is of primary importance and concerns the remagnetising speed of thin layers, ferrite cores etc. This type of measurement is frequently made with steep current pulses (rise time < 1 ns, intensity up to more than 10 A). Pulses of this type can be produced with coaxial mercury-wetted relays [Di 60a; Vo 62; Ed 65; Ni 67]. Fast integrated circuits are increasingly utilised. An integrated 16-bit storage element has, for example, been described with the short delay of 20 ns between address selection and reading or writing, which permits the utilisation in fast computers [Ng 67].

At present, large computers are being developed or have already been installed which are capable of processing far more than 10^6 operations per second, e.g. the computer IBM 360/91, that handles 16·6 million additions in one second. The logic circuits incorporated in these computers operate with a time of 4 to 5 ns. The computer Illiac-4 is in the designing stage and will carry out 10^9 operations per second [Ri 67]. At a cycle time of the processing stages of 250 ns this high speed is possible because 256 such stages operate simultaneously. Great efforts are made at present to develop faster memories with cycle times of less than 100 ns, apart from further improving the computer for faster operation [Ko 67a; Ri 68].

6. REFERENCES

[Ab 61] ABERCROMBIE, S., B. ELPHICK, H. FOSTER: in Proc. Conf. Nucl. Electronics, Belgrade 1961, Vol. 3, Internat. Atomic Energy Agency, Vienna 1962, p. 115.

[Ab 65] ABOU-LEILA, H., N. N. PERRIN, J. VALENTIN: Arkiv Fysik **29** (1965) 53.

[Ab 66] ABBATTISTA, N., M. COLI, V. L. PLANTAMURA: Nucl. Instr. Meth. **43** (1966) 383; **44** (1966) 29, 153; **45** (1966) 157.

[Ab 67] ABBATTISTA, N., V. L. PLANTAMURA, M. COLI: Nucl. Instr. Meth. **49** (1967) 155.

[Ab 68] ABBATTISTA, N., M. COLI, V. L. PLANTAMURA: Nucl. Instr. Meth. **59** (1968) 163.

[Ab 68a] ABBATTISTA, N., M. COLI, V. L. PLANTAMURA: Nucl. Instr. Meth. **60** (1968) 337.

[Ad 61] ADLER, A., M. PALMAI, V. PEREZ-MENDEZ: Nucl. Instr. Meth. **13** (1961) 197.

[Ad 62] ADLER, E., B. SELIKSON: Electronic Engng. **34** (1962) 82.

[Ad 64] ADAM, A., G. PALLA, P. KVITNER: Pribory Tekhn. Eksperim. **9** (1964) No. 4, 49 (Russ.).

[Ad 64a] ADIN, A., B. SABBAH: Nucl. Instr. Meth. **26** (1964) 355.

[Ag 61] AGENO, M.: Nucl. Instr. Meth. **12** (1961) 341.

[Ag 62] AGRINIER, B., Y. KOECHLIN, A. RAVIART: in Proc. Conf. Nucl. Electronics, Belgrade 1961, Vol. 1, Internat. Atomic Energy Agency, Vienna 1962, p. 201.

[Ag 62a] AGENO, M., C. FELICI: Nucl. Instr. Meth. **16** (1962) 59.

[Ak 57] AKIMOV, YU. K.: Pribory Tekhn. Eksperim. **2** (1957) No. 4, 98 (Russ.).

[Ak 62] AKIMOV, YU. A., B. M. STEPANOV: Pribory Tekhn. Eksperim. **7** (1962) No. 5, 128 (Russ.).

[Ak 66] AKIMOV, YU. K., S. G. BASILADZE: Dubna-Report 2803, June 1966.

[Ak 67] AKIMOV, YU. K., A. I. KALININ et al.: Poluprovodnikovye detektory yadernikh chastits i ikh primenenie (Semiconductor Nuclear Particle Detectors and Their Application), Atomizdat, Moscow 1967.

[Al 61] ALBERIGI-QUARANTA, A., B. RIGHINI: Nucl. Instr. Meth. **10** (1961) 25.

[Al 61a] ALEXANDER, T. K., D. R. HEYWOOD: Nucl. Instr. Meth. **13** (1961) 83.

[Al 61b] ALLEN, J. W., P. E. GIBBONS: Nucl. Instr. Meth. **14** (1961) 355.

[Al 63] ALBERIGI-QUARANTA, A., M. MARTINI: Nucl. Instr. Meth. **25** (1963) 125.

[Al 63a] ALON, G. I., J. R. RAUSCH: Nucl. Instr. Meth. **24** (1963) 229.

[Al 64] ALVÄGER, T., F. J. M. FARLEY et al.: Phys. Letters **12** (1964) 260.

[Al 64a] ALVÄGER, T., A. NILSSON, J. KJELLMAN: Arkiv Fysik **26** (1964) 209.

[Al 65] ALEKSANDROV, YU. A., A. V. KUTSENKO et al.: Pribory Tekhn. Eksperim. **10** (1965) No. 5, 45 (Russ.).

[Al 65a] ALBERIGI-QUARANTA, A., G. CASADEI et al.: Nucl. Instr. Meth. **35** (1965) 93.

[Al 65b] ALSTON, W. J., J. E. DRAPER: Nucl. Instr. Meth. **35** (1965) 155.

[Al 66] ALVÄGER, T., J. M. BAILEY et al.: Arkiv Fysik **31** (1966) 145.

[Al 67] ALBRECHT, K., W. NEUWIRTH: Z. Phys. **203** (1967) 420.

[Al 67a] ALBERIGI QUARANTA, A., M. MARTINI, G. OTTAVIANI: Nucl. Instr. Meth. **47** (1967) 10.

[Am 60] AMODEI, J. J.: Electronics **33** (1960) No. 26, 96.

[Am 61] AMODEI, J., W. F. KOSONOCKY: RCA-Review **22** (1961) 669.
[Am 63] AMMERLAAN, C. A. J., R. F. RUMPHORST, L. A. CH. KOERTS: Nucl. Instr. Meth. **22** (1963) 189.
[Am 63a] AMODEI, J. J., J. R. BURNS: RCA-Review **24** (1963) 355.
[Am 64] AMSEL, G., C. ZAJDE: Rev. Sci. Instr. **35** (1964) 1538.
[Am 64a] AMRAM, Y.: in Proc. Conf. Automatic Acquisition and Reduction of Nuclear Data (EANDC), Karlsruhe 1964, p. 268.
[An 61] ANDREEV, S. I.: Pribory Tekhn. Eksperim. **6** (1961) No. 4, 87 (Russ.).
[An 62] ANDERSON, J. D., C. WONG: Nucl. Instr. Meth. **15** (1962) 178.
[An 63] ANDERSON, R. L., B. D. MCDANIEL: Nucl. Instr. Meth. **21** (1963) 235.
[An 64] ANDERSON, J. H., D. SWANN: Nucl. Instr. Meth. **30** (1964) 1.
[An 66] ANDERSON, L. K., B. J. MCMURTRY: Proc. IEEE **54** (1966) 1335.
[An 66a] ANDERSON, J. H., R. BATCHELOR et al.: Nucl. Instr. Meth. **41** (1966) 30.
[An 67] ANDERSON, R. L., D. I. PORAT: Rev. Sci. Instr. **38** (1967) 1800.
[Ar 39] ARDENNE, M. VON: Hochfrequenztechn. und Elektroak. **54** (1939) 181.
[Ar 57] ARDENNE, M. VON: J. Sci. Instr. **34** (1957) 206.
[Ar 58] ARTEMENKOV, L. I., M. V. BABYKIN: Pribory Tekhn. Eksperim. **3** (1958) No. 5, 48 (Russ.).
[Ar 59] ARECCHI, F. T.: Energia Nucl. **6** (1959) 717.
[Ar 60] ARECCHI, F. T., A. DE MATTEIS, E. GATTI: Energia Nucl. **7** (1960) 697.
[Ar 60a] ARTEMENKOV, L. I., M. V. BABYKIN: Pribory Tekhn. Eksperim. **5** (1960) No. 1, 43 (Russ.).
[Ar 61] ARTIGES, J. P., J. C. BRUN: J. Phys. Radium **22** (1961) 53 A.
[Ar 63] ARNAL, R., R. BORCHI: in Proc. Internat. Symp. Nucl. Electronics, Paris 1963, p. 819.
[Ar 65] ARBEL, A. F.: Nucl. Instr. Meth. **32** (1965) 341.
[Ar 66] ARMANTROUT, G. A.: IEEE Trans. Nucl. Sci. **13** (1966) No. 3, 328.
[As 63] ASPELUND, O.: Nucl. Instr. Meth. **23** (1963) 1.
[At 67] ATTWOOD, T. V.: Proc. IEEE **55** (1967) 1215.
[Av 66] AVERCHENKOV, V. YA., A. G. BERKOVSKII et al.: Pribory Tekhn. Eksperim. **11** (1966) No. 5, 189 (Russ.).
[Av 67] AVIDA, R., S. GORNI: Nucl. Instr. Meth. **52** (1967) 125.
[Av 67a] AVEYNIER, J., C. PUTHOD: Nucl. Instr. Meth. **47** (1967) 141.
[Av 68] AVERCHENKOV, V. YA., YU. YA. NEFEDOV, V. YA. GVOZDEV: Pribory Tekhn. Eksperim. **13** (1968) No. 1, 80 (Russ.).
[Ba 48] BAY, Z., G. PAPP: Rev. Sci. Instr. **19** (1948) 565.
[Ba 48a] BALDINGER, E., P. HUBER, K. P. MEYER: Rev. Sci. Instr. **19** (1948) 473.
[Ba 50] BAY, Z.: Phys. Rev. **77** (1950) 419.
[Ba 51] BAY, Z.: Rev. Sci. Instr. **22** (1951) 397.
[Ba 55] BAY, Z., V. P. HENRI, H. KANNER: Phys. Rev. **100** (1955) 1197.
[Ba 56] BABYKIN, M. V., A. G. PLAKHOV et al.: Atomnaya Energiya **1** (1956) No. 4, 38 (Russ.).
[Ba 56a] BAY, Z.: IRE Trans. Nucl. Sci. **3** (1956) No. 4, 12; Nucleonics **14** (1956) No. 4, 56.
[Ba 56b] BAY, Z., N. T. GRISAMORE: IRE Trans. Electr. Comp. **5** (1956) 121, 213.
[Ba 57] BARON, R. G.: Proc. IRE **45** (1957) 21.
[Ba 58] BALDINGER, E., P. SANTCHI: Nucl. Instr. **2** (1958) 128.
[Ba 59] BALLENTINE, W. E., F. H. BLECHER: in Proc. Solid-State Circuits Conference, Philadelphia 1959, p. 42.
[Ba 59a] BAY, Z., F. MCLERNON: in Proc. Sec. Symp. Adv. Fast Pulse Techniques Nucl. Counting, Berkeley 1959, UCRL 8706, p. 55.
[Ba 59b] BALDINGER, E., P. SANTSHI, P. WEHRLI: Nucl. Instr. Meth. **4** (1959) 117.
[Ba 60] BAKER, S. C., H. G. JACKSON, D. A. MACK: IRE Trans. Nucl. Sci. **7** (1960) No. 2/3, 89.

[Ba 60a] Barna, A., J. H. Marshall, M. Sands: Nucl. Instr. Meth. **7** (1960) 124.
[Ba 60b] Bashandy, E.: Nucl. Instr. Meth. **6** (1960) 289.
[Ba 61] Barna, A., J. H. Marshall, M. Sands: Nucl. Instr. Meth. **12** (1961) 43.
[Ba 61a] Ballini, R., E. Pomelas: Nucl. Instr. Meth. **11** (1961) 331.
[Ba 61b] Bartenev, L. S., G. V. Glebovich, K. N. Ptitsin: Pribory Tekhn. Eksperim. **6** (1961) No. 6, 80 (Russ.).
[Ba 61c] Bashandy, E.: Nucl. Instr. Meth. **12** (1961) 227.
[Ba 62] Baldinger, E., U. Spycher, R. Bossart: Helv. Phys. Acta **35** (1962) 237.
[Ba 62a] Banzhaf, J. F., H. S. Katzenstein: Proc. IRE **50** (1962) 212.
[Ba 63] Baldin, S. A., V. V. Matveev: Pribory Tekhn. Eksperim. **8** (1963) No. 4, 5 (Russ.).
[Ba 63a] Barna, A.: Nucl. Instr. Meth. **24** (1963) 247.
[Ba 63b] Bartl, W., P. Weinzierl: Rev. Sci. Instr. **34** (1963) 252.
[Ba 63c] Baldinger, E., U. Spycher: Helv. Phys. Acta **36** (1963) 334.
[Ba 63d] Basire, A., B. de Cosnac, J. Labbe: in Proc. Internat. Symp. Nucl. Electronics, Paris 1963, p. 325.
[Ba 63e] Bauer, R. W., J. D. Anderson, L. J. Christensen: Phys. Rev. **130** (1963) 312.
[Ba 64] Baranowski, J., T. Jankowski: Tranzystorowe Układy Impulsowe, Warszawa 1964; German Translation: Transistorschaltungen in der Impulstechnik, Verlag Technik, Berlin 1966.
[Ba 64a] Baldinger, E., A. Simmen: Z. angew. Math. Phys. **15** (1964) 71.
[Ba 64b] Bass, R., W. Kessel, G. Majoni: Nucl. Instr. Meth. **30** (1964) 237.
[Ba 64c] Banner, M., J. Teiger: Nucl. Instr. Meth. **31** (1964) 205.
[Ba 64d] Barna, A., J. H. Marshall: Rev. Sci. Instr. **35** (1964) 881.
[Ba 64e] Barschall, H. H.: Nucl. Instr. Meth. **28** (1964) 44.
[Ba 65] Baldinger, E., A. Simmen: Z. angew. Math. Phys. **16** (1965) 699; Nucl. Instr. Meth. **33** (1965) 353.
[Ba 65a] Barna, A., J. H. Marshall, D. F. Torzewski: Rev. Sci. Instr. **36** (1965) 166.
[Ba 65b] Bachman, C. G.: Proc. IEEE **53** (1965) 962.
[Ba 66] Balland, J. C., M. Goyot et al.: L'Onde Électrique; Inst. Nucl. Phys. Lyon, Report Lycen 6621, April 1966.
[Ba 66a] Balland, J. C., R. Gaillard et al.: Nucl. Instr. Meth. **45** (1966) 275.
[Ba 66b] Baldinger, E., A. Simmen: Helv. Phys. Acta **39** (1966) 570.
[Ba 66c] Bachmann, K., A. Götsch, K. Schulze: Exp. Techn. Phys. **14** (1966) 303.
[Ba 66d] Bartenev, L. S., V. S. Piskunov: Pribory Tekhn. Eksperim. **11** (1966) No. 5, 161 (Russ.).
[Ba 66e] Basiladze, S. G.: Pribory Tekhn. Eksperim. **11** (1966) No. 5, 123 (Russ.).
[Ba 67] Bakos, J., J. Szigeti: Acta Phys. Hung. **23** (1967) 341.
[Ba 67a] Barna, A.: Proc. IEEE **55** (1967) 2054.
[Ba 67b] Baldinger, E., A. Simmen: Nucl. Instr. Meth. **57** (1967) 141.
[Ba 67c] Basiladze, S. G.: Pribory Tekhn. Eksperim. **12** (1967) No. 4, 85 (Russ.).
[Ba 67d] Balland, J. C., J. Pigneret, J. J. Samueli: Nucl. Instr. Meth. **52** (1967) 351.
[Ba 67e] Balland, J. C.: Inst. Nucl. Phys. Lyon, Report Lycen 6766, Sept. 1967.
[Ba 68] Baldinger, E., A. Simmen: Z. angew. Math. Phys. **19** (1968) 1.
[Ba 68a] Baker, C. A., C. J. Batty, L. E. Williams: Nucl. Instr. Meth. **59** (1968) 125.
[Be 49] Bell, R. E., H. E. Petch: Phys. Rev. **76** (1949) 1409.
[Be 50] Bell, R. E., R. L. Graham: Phys. Rev. **78** (1950) 490.
[Be 52] Bell, R. E., R. L. Graham, H. E. Petch: Can. J. Phys. **30** (1952) 35.
[Be 52a] Benedetti, S. de, H. J. Richings: Rev. Sci. Instr. **23** (1952) 37.
[Be 55] Bell, R. E., K. G. Malmfors: Measurement of Short Lifetimes of Excited States, in Beta- and Gamma-Ray Spectroscopy, ed. by K. Siegbahn, North-Holland Publ. Co., Amsterdam 1955, p. 494.
[Be 58] Beghian, L. E., G. H. R. Kegel, R. P. Scharenberg: Rev. Sci. Instr. **29** (1958) 753.

[Be 58a] Benedetti, S. de, R. W. Findley: The Coincidence Method, in Encyclopedia of Physics, Vol. 45, ed. by S. Flügge, Springer-Verlag, Berlin/Göttingen/Heidelberg 1958, p. 222.
[Be 58b] Berlovich, E. Ye., K. Grotovski et al.: Nucl. Phys. 6 (1958) 672.
[Be 60] Bell, R. E., M. H. Jørgensen: Can. J. Phys. 38 (1960) 652.
[Be 60a] Bergström, I., E. C. O. Bonacalza et al.: Nucl. Instr. Meth. 8 (1960) 151.
[Be 60b] Bell, R. E., S. Bjørnholm, J. C. Severiens: Mat. Fys. Medd. Dan. Vid. Selsk. 32 (1960) No. 12.
[Be 61] Beneteau, P. J., J. A. MacIntosh: Electronics 34 (1961) No. 41, 62.
[Be 61a] Berlovich, E. E.: Eksperimental'nye issledovaniya radiatsionnykh perekhodov v yadrakh, in Gamma luchi (Gamma Rays), ed. by L. A. Sliv, Izdatel'stvo Akademii Nauk SSSR, Moscow/Leningrad 1961, p. 85.
[Be 61b] Bell, R. E.: Nucl. Instr. Meth. 12 (1961) 199.
[Be 61c] Berlovich, E. E., M. P. Bonits, V. V. Nikitin: Izv. Akad. Nauk SSSR, Ser. Fiz. 25 (1961) 218 (Russ.).
[Be 62] Bergman, R. H., M. Cooperman, H. Ur: RCA-Review 23 (1962) 152.
[Be 62a] Berlovich, E. E., Yu. K. Gusev et al.: Zh. Eksperim. Teor. Fiz. 43 (1962) 1625 (Russ.).
[Be 62b] Beger, H., M. Giesch, W. Stockdreher: CERN-Report 62—25, August 1962.
[Be 62c] Beghian, L. E., M. K. Salomaa: Nucl. Instr. Meth. 17 (1962) 181.
[Be 63] Bellettini, G., C. Bemporad et al.: Nucl. Instr. Meth. 21 (1963) 106.
[Be 63a] Belov, N. S., A. M. Bronshtein et al.: Pribory Tekhn. Eksperim. 8 (1963) No. 4, 118 (Russ.).
[Be 63b] Beier, H.-J., K. Tradowsky: Nukleonik 5 (1963) 330.
[Be 64] Bellettini, G., C. Bemporad et al.: Nucl. Instr. Meth. 27 (1964) 38.
[Be 64a] Berkes, I., I. Dézsi et al.: Nucl. Instr. Meth. 27 (1964) 355.
[Be 64b] Beddoes, M. P.: Proc. IEE 111 (1964) 67.
[Be 65] Bertolaccini, M., S. Cova, C. Bussolati: Paper presented at Summer School, Herceg-Novi 1965.
[Be 65a] Berkovskii, A. G., S. G. Dukor: Izv. Akad. Nauk SSSR, Ser. Fiz. 29 (1965) 328 (Russ.).
[Be 65b] Beneking, H.: Nachrichtentechnik 15 (1965) 333.
[Be 65c] Bertolaccini, M., S. Cova, C. Bussolati: Nucl. Instr. Meth. 37 (1965) 297.
[Be 65d] Bell, J., S. J. Tao, J. H. Green: Nucl. Instr. Meth. 35 (1965) 213.
[Be 65e] Bell, J., J. H. Green, S. J. Tao: Nucl. Instr. Meth. 36 (1965) 320.
[Be 65f] Bell, R. E.: Coincidence Techniques and the Measurement of Short Mean Lives, in Alpha-, Beta- and Gamma-Ray Spectroscopy, Vol. 2, ed. by K. Siegbahn, North-Holland Publ. Co., Amsterdam 1965, p. 905.
[Be 65g] Beckman, L., R. B. Jansson, B. L. Nilsson: Nucl. Instr. Meth. 33 (1965) 151.
[Be 66] Bertolini, G., M. Cocchi et al.: IEEE Trans. Nucl. Sci. 13 (1966) No. 3, 119.
[Be 66a] Bertolini, G., V. Mandl et al.: Nucl. Instr. Meth. 42 (1966) 109.
[Be 66b] Bertrand, F. E., R. W. Peele et al.: IEEE Trans. Nucl. Sci. 13 (1966) No. 3, 279.
[Be 66c] Bell, R. E.: Nucl. Instr. Meth. 42 (1966) 211.
[Be 66d] Berg, A. D., R. W. Smith, R. D. Prosser: Adv. Electronics Electron Phys. 22 B (1966) 969.
[Be 67] Bernaola, O. A., A. Filevich, P. Thieberger: Nucl. Instr. Meth. 50 (1967) 299.
[Be 67a] Bertolaccini, M., C. Bussolati et al.: Nucl. Instr. Meth. 51 (1967) 325.
[Be 68] Bernard, P., J. Chambon et al.: Nucl. Instr. Meth. 60 (1968) 213.
[Bi 59] Birk, M., G. Goldring, Y. Wolfson: Phys. Rev. 116 (1959) 730.
[Bi 62] Birks, J. B.: in Proc. Conf. Nucl. Electronics, Belgrade 1961, Vol. 1, Internat. Atomic Energy Agency, Vienna 1962, p. 17.

[Bi 62a] BINNIE, D. M., M. R. JANE et al.: Adv. Electronics Electron Phys. **16** (1962) 501.
[Bi 62b] BIRKS, J. B., T. A. KING, I. H. MUNRO: Proc. Phys. Soc. **80** (1962) 355.
[Bi 63] BINNIE, D. M., M. JANE et al.: Nucl. Instr. Meth. **20** (1963) 221.
[Bi 63a] BIRK, M., Q. A. KERNS, T. A. NUNAMAKER: Rev. Sci. Instr. **34** (1963) 1026.
[Bi 64] BIRKS, J. B.: The Theory and Practice of Scintillation Counting, Pergamon Press, Oxford/London/Edinburgh/New York/Paris/Frankfurt upon Main 1964.
[Bi 64a] BIRK, M., Q. A. KERNS, R. F. TUSTING: IEEE Trans. Nucl. Sci. **11** (1964) No. 3, 129.
[Bi 66] BIRKS, J. B., J. M. DE C. CONTE, G. WALKER: IEEE Trans. Nucl. Sci. **13** (1966) No. 3, 148.
[Bi 66a] BISI, A., C. BUSSOLATI et al.: Phys. Rev. **141** (1966) 348.
[Bi 66b] BIRNBAUM, M., T. L. STOCKER: J. Appl. Phys. **37** (1966) 348.
[Bi 67] BIGELIS, E. V., O. A. GLUPOVA et al.: Materialy simpoziuma po nanosekundnoi yadernoi elektronike (Repts. Symp. Nanosecond Nucl. Electronics), Dubna 1967, Dubna-Report 13—3700, Febr. 1968, p. 662.
[Bj 61] BJERKE, A. E., Q. A. KERNS, T. A. NUNAMAKER: Nucl. Instr. Meth. **12** (1961) 25.
[Bj 62] BJERKE, A. E., Q. A. KERNS, T. A. NUNAMAKER: Nucl. Instr. Meth. **15** (1962) 249; IRE Trans. Nucl. Sci. **9** (1962) No. 3, 314.
[Bl 59] BLAUGRUND, A. E.: Phys. Rev. Letters **3** (1959) 226.
[Bl 60] BLAUGRUND, A. E., Y. DAR, G. GOLDRING: Phys. Rev. **120** (1960) 1328.
[Bl 61] BLECHSCHMIDT, H.: Diplomarbeit, Univ. Hamburg 1961.
[Bl 62] BLOOM, S. D.: Nanosecond Experimentation with Pulsed Machines, in Nuclear Reactions, Vol. 2, ed. by P. M. ENDT, P. B. SMITH, North-Holland Publ. Co., Amsterdam 1962, p. 1.
[Bl 63] BLOESS, D., F. MÜNNICH: Z. Naturforsch. **18a** (1963) 1028.
[Bl 63a] BLUST, E. DE, V. MANDL, I. DE LOTTO: in Proc. Internat. Symp. Nucl. Electronics, Paris 1963, p. 785.
[Bl 63b] BLOESS, D., A. KRUSCHE et al.: Nucl. Instr. Meth. **21** (1963) 113.
[Bl 63c] BLOCK, M. M., R. GESSAROLI et al.: Phys. Rev. **130** (1963) 766.
[Bl 64] BLOESS, D., F. MÜNNICH: Nucl. Instr. Meth. **28** (1964) 286.
[Bl 64a] BLAUGRUND, A. E., Z. VAGER: Nucl. Instr. Meth. **29** (1964) 131.
[Bl 65] BLAUGRUND, A. E., Y. DAR et al.: Nucl. Phys. **69** (1965) 423.
[Bl 66] BLANKENSHIP, J. L., C. H. NOWLIN: IEEE Trans. Nucl. Sci. **13** (1966) No. 3, 495.
[Bl 67] BLATT, S. L.: Nucl. Instr. Meth. **49** (1967) 235.
[Bl 67a] BLIGNAUT, E., W. R. MCMURRAY, A. BOTTEGA: Nucl. Instr. Meth. **51** (1967) 102.
[Bl 68] BLATT, S. L., J. MAHIEUX, D. KOHLER: Nucl. Instr. Meth. **60** (1968) 221.
[Bo 53] BORKOWSKI, C. J., R. L. CLARK: Rev. Sci. Instr. **24** (1953) 1046.
[Bo 57] BOGDANOV, G. F., N. A. VLASOV et al.: Atomnaya Energiya **3** (1957) 204 (Russ.).
[Bo 59] BONITZ, M., E. J. BERLOWITSCH: Nucl. Instr. Meth. **4** (1959) 133.
[Bo 60] BONITZ, M., E. J. BERLOVICH: Nucl. Instr. Meth. **9** (1960) 13.
[Bo 61] BOLLINGER, L. M., G. E. THOMAS: Rev. Sci. Instr. **32** (1961) 1044.
[Bo 62] BOLLINGER, L. M., G. E. THOMAS: Nucl. Instr. Meth. **17** (1962) 97.
[Bo 62a] BOŻEK, E., A. Z. HRYNKIEWICZ et al.: Nukleonika **7** (1962) 669.
[Bo 63] BONITZ, M.: Nucl. Instr. Meth. **22** (1963) 238.
[Bo 63a] BORCHI, R.: Nucl. Instr. Meth. **24** (1963) 123.
[Bo 64] BONITZ, M., W. MEILING, F. STARY: Nucl. Instr. Meth. **29** (1964) 314.
[Bo 64a] BONITZ, M., W. MEILING, F. STARY: Nucl. Instr. Meth. **29** (1964) 309.
[Bo 64b] BOWES, R. C., J. W. GRANVILLE: Proc. IEEE **52** (1964) 1598.
[Bo 64c] BONITZ, M., P. WINKLER, R. KÄSTNER: Nucl. Instr. Meth. **31** (1964) 357.
[Bo 66] BOSTRÖM, L., B. OLSEN et al.: Nucl. Instr. Meth. **44** (1966) 61.
[Bo 66a] BODENSTEDT, E., J. RADELOFF et al.: Z. Physik **190** (1966) 60.
[Bö 66b] BÖHM, E., U. J. ROOSE et al.: Nucl. Instr. Meth. **40** (1966) 73.
[Bo 67] BOSSHARD, R., C. ZAJDE, H. ZYNGIER: Rev. Sci. Instr. **38** (1967) 942.
[Br 47] BROWN, J. T. L., C. E. POLLARD: Electrical Engng. **66** (1947) 1106.

[Br 53] Breitenberger, E.: Philos. Mag. 44 (1953) 987.
[Br 61] Brown, W. L.: IRE Trans. Nucl. Sci. 8 (1961) No. 1, 2.
[Br 61a] Bret, G. C., E. F. Shrader: Nucl. Instr. Meth. 13 (1961) 177.
[Br 62] Bramson, M.: IRE Trans. Nucl. Sci. 9 (1962) No. 4, 35.
[Br 62a] Brandt, B., U. Cappeller: in Proc. Conf. Nucl. Electronics, Belgrade 1961, Vol. 2, Internat. Atomic Energy Agency, Vienna 1962, p. 167.
[Br 62b] Brun, J. C., R. Breuil, C. Victor: in Proc. Conf. Nucl. Electronics, Belgrade 1961, Vol. 3, Internat. Atomic Energy Agency, Vienna 1962, p. 99.
[Br 62c] Brewer, L., C. G. James et al.: Rev. Sci. Instr. 33 (1962) 1450.
[Br 64] Brunner, W.: Nucl. Instr. Meth. 30 (1964) 109.
[Br 65] Braunsfurth, J., H. J. Körner: Nucl. Instr. Meth. 34 (1965) 202.
[Br 65a] Brianti, G.: CERN-Report 65—10, May 1965.
[Br 65b] Brafman, H.: Nucl. Instr. Meth. 34 (1965) 239.
[Br 65c] Brafman, H.: Nucl. Instr. Meth. 32 (1965) 321.
[Br 67] Bridgett, K. A., T. A. King: Proc. Phys. Soc. 92 (1967) 75.
[Br 67a] Brown, M. A. C. S., P. Porteous, D. J. Solley: J. Sci. Instr. 44 (1967) 419.
[Bu 59] Burrus, C. A.: Rev. Sci. Instr. 30 (1959) 295.
[Bu 59a] Bushor, W. E.: Electronics 32 (1959) No. 31, 69.
[Bu 62] Burton, M., J. Yguerabide: in Proc. Conf. Nucl. Electronics, Belgrade 1961, Vol. 1, Internat. Atomic Energy Agency, Vienna 1962, p. 71.
[Bu 62a] Burrows, H. C., D. O. Caldwell et al.: in Proc. Conf. Nucl. Electronics, Belgrade 1961, Vol. 1, Internat. Atomic Energy Agency, Vienna 1962, p. 153.
[Bu 62b] Burde, J., M. Rakavy, G. Engler: Phys. Rev. 128 (1962) 325.
[Bu 64] Burns, G., M. I. Nathan: Proc. IEEE 52 (1964) 770.
[Bu 65] Butslov, M. M., V. I. Komarov et al.: Pribory Tekhn. Eksperim. 10 (1965) No. 3, 84 (Russ.).
[Bu 65a] Budyashov, Yu. G., V. G. Zinov, A. G. Morozov: Dubna-Report 2430, Nov. 1965.
[Bu 67] Bucher, W. P., C. E. Hollandsworth: Rev. Sci. Instr. 38 (1967) 1259.
[Bu 67a] Bushnin, Yu. B., A. F. Dunaitsev, V. A. Sen'ko: Pribory Tekhn. Eksperim. 12 (1967) No. 4, 74 (Russ.).
[Bu 67b] Budyashov, Yu. G., V. G. Zinov, A. G. Morozov: Pribory Tekhn. Eksperim. 12 (1967) No. 1, 99 (Russ.).
[Bu 68] Budyashov, Yu. G., V. G. Zinov: Pribory Tekhn. Eksperim. 13 (1968) No. 1, 92 (Russ.).
[Ca 61] Caplan, H. S., D. T. Stewart: J. Sci. Instr. 38 (1961) 133.
[Ca 62] Caldwell, D. O.: Adv. Electronics Electron Phys. 16 (1962) 469.
[Ca 63] Cavalleri, G., G. Fabri et al.: Nucl. Instr. Meth. 21 (1963) 177.
[Ca 64] Caini, V., P. A. Tove: Energia Nucl. 11 (1964) 186.
[Ce 60] Cernigoi, C., I. Gabrielli, G. Iernetti: Nucl. Instr. Meth. 9 (1960) 303.
[Ce 60a] Cernigoi, C., I. Gabrielli, G. Iernetti: Nucl. Instr. Meth. 6 (1960) 193.
[Ch 52] Christiansen, J.: Z. angew. Phys. 4 (1952) 326.
[Ch 55] Chamberlain, O., E. Segrè et al.: Phys. Rev. 100 (1955) 947.
[Ch 57] Chase, R. L., W. A. Higinbotham: Rev. Sci. Instr. 28 (1957) 448.
[Ch 59] Christiansen, J.: Nucl. Instr. Meth. 5 (1959) 115.
[Ch 59a] Chaplin, G. B. B., C. J. N. Candy: Nucl. Instr. Meth. 5 (1959) 242.
[Ch 60] Chase, R. L.: Rev. Sci. Instr. 31 (1960) 945.
[Ch 60a] Chery, R.: Comptes Rendus 251 (1960) 694.
[Ch 60b] Chow, W. F.: IRE Trans. Electr. Comp. 9 (1960) 295.
[Ch 60c] Chaplin, G. B. B., A. J. Cole: Nucl. Instr. Meth. 7 (1960) 45.
[Ch 61] Chase, R. L., W. A. Higinbotham, G. L. Miller: IRE Trans. Nucl. Sci. 8 (1961) No. 1, 147.

[Ch 61a] CHASE, R. L.: Nuclear Pulse Spectrometry, McGraw-Hill Book Co., New York/Toronto/London 1961.
[Ch 61b] CHASE, R. L., V. SVELTO: IRE Trans. Nucl. Sci. **8** (1961) No. 3, 45.
[Ch 62] CHRISTIANSEN, J.: in Proc. Conf. Nucl. Electronics, Belgrade 1961, Vol. 3, Internat. Atomic Energy Agency, Vienna 1962, p. 93.
[Ch 63] CHOW, P., J. CUBERT: Electronics **36** (1963) No. 42, 42.
[Ch 63a] CHRIST, K.: Archiv elektr. Übertragung **17** (1963) 42.
[Ch 63b] CHERRY, E. M., D. E. HOOPER: Proc. IEE **110** (1963) 375.
[Ch 63c] CHARBONNIER, R.: Electronics **36** (1963) No. 28, 46.
[Ch 65] CHARPAK, G., L. MASSONNET, J. FAVIER: The Development of Spark Chamber Techniques, in Progr. in Nucl. Techn. Instr., ed. by F. J. M. FARLEY, Vol. 1, North-Holland Publ. Co., Amsterdam 1965, p. 321.
[Ch 66] CHRISTIE, R. H.: J. Sci. Instr. **43** (1966) 524.
[Ch 67] CHO, Z. H., L. GIDEFELDT, L. ERIKSSON: Nucl. Instr. Meth. **52** (1967) 273.
[Ch 67a] CHEN, W. K.: Proc. IEE **114** (1967) 1065.
[Cl 51] CLELAND, M. R., P. S. JASTRAM: Phys. Rev. **84** (1951) 271.
[Cl 67] CLAUSEN, K. D., T. D. NAINAN: Nucl. Instr. Meth. **49** (1967) 22.
[Co 56] COTTINI, C., E. GATTI, G. GIANELLI: Nuovo Cimento **4** (1956) 156.
[Co 56a] COTTINI, C., E. GATTI: Nuovo Cimento **4** (1956) 1550.
[Co 56b] CONNOR, F. R.: Proc. IEE **B 103** (1956) 178.
[Co 57] COLOMBO, S., E. GATTI, M. PIGNANELLI: Nuovo Cimento **5** (1957) 1739.
[Co 57a] COLOMBO, S., C. COTTINI, E. GATTI: Nuovo Cimento **5** (1957) 748.
[Co 58] COLLINGE, B., G. B. HUXTABLE: Nucl. Instr. **3** (1958) 116.
[Co 58a] COTTINI, C., E. GATTI: Nucl. Instr. **2** (1958) 88.
[Co 59] COTTINI, C., E. GATTI, F. VAGHI: in Proc. Internat. Symp. Nucl. Electronics, Paris 1958, Vol. 1, Internat. Atomic Energy Agency, Vienna 1959, p. 177.
[Co 59a] COTTINI, C., E. GATTI et al.: in Proc. Sec. Symp. Adv. Fast Pulse Techniques Nucl. Counting, Berkeley 1959, UCRL 8706, p. 49.
[Co 60] COCCONI, V. T., T. FAZZINI et al.: Phys. Rev. Letters **5** (1960) 19.
[Co 61] CONNOR, R. J.: Nucl. Instr. Meth. **11** (1961) 122.
[Co 62] COOK, C. F.: Nucl. Instr. Meth. **15** (1962) 137.
[Co 63] COLEMAN, K. R.: Rep. Progr. Phys. **26** (1963) 269.
[Co 63a] COOPERMAN, M., R. H. BERGMAN: Electronics **36** (1963) No. 28, 42.
[Co 63b] COBB, W. C.: Nucl. Instr. Meth. **23** (1963) 353.
[Co 63c] COTTINI, C., E. GATTI, V. SVELTO: Nucl. Instr. Meth. **24** (1963) 241.
[Co 63d] COLI, M., A. SERRA: in Proc. Internat. Symp. Nucl. Electronics, Paris 1963, p. 731.
[Co 64] CONRAD, R., T. FRIESE: Nucl. Instr. Meth. **30** (1964) 203.
[Co 65] COLI, M., S. LUPINI, V. SILVESTRINI: Nucl. Instr. Meth. **33** (1965) 298.
[Co 65a] COLI, M., S. LUPINI: Nucl. Instr. Meth. **34** (1965) 235.
[Co 66] COLI, M.: Nucl. Instr. Meth. **39** (1966) 297.
[Co 66a] COLLINS, L. E., D. DANDY, P. T. STROUD: Nucl. Instr. Meth. **42** (1966) 206.
[Co 66b] CONRAD, R.: Elektronische Rdsch. **20** (1966) 298.
[Co 67] COMPTON, P. D., W. A. JOHNSON: IEEE Trans. Nucl. Sci. **14** (1967) No. 1, 116.
[Co 67a] COLEMAN, K. R.: J. Sci. Instr. **44** (1967) 321.
[Co 67b] COCCHI, M., A. ROTA: Nucl. Instr. Meth. **46** (1967) 136.
[Co 67c] COCCHI, M., A. ROTA: Nucl. Instr. Meth. **55** (1967) 365.
[Co 67d] CONRAD, R.: Nucl. Instr. Meth. **48** (1967) 229.
[Co 67e] COOPERMAN, M.: RCA-Review **28** (1967) 424.
[Co 67f] COTTINI, C., I. DE LOTTO et al.: Energia Nucl. **14** (1968) 704.
[Cr 54] CROUCH, M. F.: Rev. Sci. Instr. **25** (1954) 924.
[Cr 61] CRANBERG, L., R. A. FERNALD et al.: Nucl. Instr. Meth. **12** (1961) 335.
[Cr 64] CRESSWELL, J., P. WILDE: in Proc. Conf. Automatic Acquisition and Reduction of Nuclear Data (EANDC), Karlsruhe 1964, p. 300.

[Cu 55] CUTLER, C. C.: Proc. IRE 43 (1955) 140.
[Cu 60] CULLIGAN, G., N. H. LIPMAN: Rev. Sci. Instr. 31 (1960) 1209.
[Cu 61] CURRIE, W. M., R. E. AZUMA, G. M. LEWIS: Nucl. Instr. Meth. 13 (1961) 215.
[Cu 63] CURRIE, W. M.: Nucl. Phys. 47 (1963) 551.
[Cu 65] CUMMINGS, A. J.: Rev. Sci. Instr. 36 (1965) 1503.
[Cu 66] CUMMINGS, A. J., A. R. WILSON: IEEE Trans. Nucl. Sci. 13 (1966) No. 1, 321.
[Cz 59] CZECH, J.: Oszillographen-Meßtechnik, Verlag Radio-Foto-Kinotechnik, Berlin 1959.
[Da 63] DAION, M. I., G. A. LEKSIN: Usp. Fiz. Nauk 80 (1963) 281 (Russ.).
[Da 64] DANDY, D., D. P. HAMMOND: Nucl. Instr. Meth. 30 (1964) 23.
[Da 64a] D'ALESSIO, J. T., P. K. LUDWIG, M. BURTON: Rev. Sci. Instr. 35 (1964) 1015.
[Da 65] D'ALESSIO, J. T., P. K. LUDWIG: IEEE Trans. Nucl. Sci. 12 (1965) No. 1, 350.
[Da 66] DANGL, J. R., K. P. STEELE: Electronics 39 (1966) No. 3, 72.
[Da 66a] D'ASARO, L. A., L. K. ANDERSON: Electronics 39 (1966) No. 11, 94.
[Da 67] DAVIDENKO, V. A., B. A. DOLGOSHEIN, S. V. SOMOV: Pribory Tekhn. Eksperim. 12 (1967) No. 1, 167 (Russ.).
[Da 67a] DARDINI, C., G. IACI et al.: Nucl. Instr. Meth. 47 (1967) 233.
[De 50] DEUTSCH, M., W. E. WRIGHT: Phys. Rev. 77 (1950) 139.
[De 58] DEVONS, S.: in Proc. Rehovoth Conf. Nucl. Structure, North-Holland Publ. Co., Amsterdam 1958, p. 547.
[De 60] DECONNINCK, G., A. MARTEGANI: Nucl. Phys. 20 (1960) 33.
[De 63] DÉSI, S., A. LAJTAI, L. NAGY: Acta Phys. Hung. 15 (1963) 185.
[De 64] DELANEY, C. F. G., P. W. WALTON: Nucl. Instr. Meth. 25 (1964) 353.
[De 64a] DEARNALEY, G.: Progr. Nucl. Phys. 9 (1964) 29.
[De 64b] DEAN, K. J.: Transistors. Theory and Circuitry, McGraw-Hill Book Co., New York/London/Toronto 1964.
[De 65] DENISOV, YU. N., V. M. LACHINOV, V. J. PRILIPKO: Pribory Tekhn. Eksperim. 10 (1965) No. 4, 83; 11 (1966) No. 5, 105 (Russ.).
[De 65a] DEUTSCH, M., S. BERKO: Positron Annihilation and Positronium, in Alpha-, Beta- and Gamma-Ray Spectroscopy, Vol. 2, ed. by K. SIEGBAHN, North-Holland Publ. Co., Amsterdam 1965, p. 1583.
[De 66] DENTON, R. T., T. S. KINSEL, F. S. CHEN: Proc. IEEE 54 (1966) 1472.
[De 66a] DEIKE, S., P. SEYFRIED: Nucl. Instr. Meth. 40 (1966) 348.
[De 66b] DEARNALEY, G., D. C. NORTHROP: Semiconductor Counters for Nuclear Radiations, Spon Ltd., London 1966.
[De 67] DEMARIA, A. J., R. GAGOSZ et al.: J. Appl. Phys. 38 (1967) 2693.
[De 67a] DEMIDOV, B. A., S. D. FANCHENKO: Pribory Tekhn. Eksperim. 12 (1967) No. 6, 133 (Russ.).
[De 68] DENTON, R. T., T. S. KINSEL: Proc. IEEE 56 (1968) 140, 146.
[Di 47] DICKE, R. H.: Rev. Sci. Instr. 18 (1947) 907.
[Di 60] DILL, H. G., M. R. MACPHERSON: Electronics 33 (1960) No. 32, 62.
[Di 60a] DIETRICH, W., W. E. PROEBSTER, P. WOLF: IBM Journal 4 (1960) 189.
[Di 66] DIDOMENICO, M., J. E. GEUSIC et al.: Appl. Phys. Letters 8 (1966) 180.
[Do 59] DOSSE, J.: Der Transistor. Ein neues Verstärkerelement, R. Oldenbourg-Verlag, München 1959.
[Do 63] DORIKENS, M., J. DEMUYNCK et al.: J. Phys. Radium 24 (1963) 866.
[Do 63a] DOEMLAND, H. H., W. D. MCCAA: Proc. IEEE 51 (1963) 1767.
[Do 66] DONOVAN, P. F.: IEEE Trans. Nucl. Sci. 13 (1966) No. 3, 21.
[Do 67] DONATI, S., E. GATTI, V. SVELTO: Nucl. Instr. Meth. 46 (1967) 165.
[Dr 59] DREESKAMP, H., M. BURTON: Phys. Rev. Letters 2 (1959) 45.
[Dr 60] DRAPER, J. E., A. A. FLEISCHER: Rev. Sci. Instr. 31 (1960) 49.
[Dr 61] DREESKAMP, H., A. K. GHOSH, M. BURTON: Rev. Sci. Instr. 32 (1961) 304.
[Dr 64] DRAPER, J. E.: Nucl. Instr. Meth. 30 (1964) 148.
[Dr 65] DRAZHEV, M. N.: Dubna-Report R-1997, Febr. 1965.

[Dr 67] DRAZHEV, M. N., S. S. PARZHITSKII: Materialy simpoziuma po nanosekundnoi yadernoi elektronike (Repts. Symp. Nanosecond Nucl. Electronics), Dubna 1967, Dubna-Report 13—3700, Febr. 1968, p. 313.
[Du 52] DURBIN, R. P., H. H. LOAR, W. W. HAVENS Jr.: Phys. Rev. **88** (1952) 179.
[Du 63] DUMAS, J. C., C. AUBRET, P. BENOIT: Nucl. Instr. Meth. **21** (1963) 323.
[Du 63a] DUCHEMIN, J. P.: in Proc. Internat. Symp. Nucl. Electronics, Paris 1963, p. 755.
[Du 63b] DURAND, P., P. GIRAUD: in Proc. Internat. Symp. Nucl. Electronics, Paris 1963, p. 643.
[Du 63c] DURAND, P., P. GIRAUD et al.: in Proc. Internat. Symp. Nucl. Electronics, Paris 1963, p. 651.
[Du 64] DUNAITSEV, A. F.: Pribory Tekhn. Eksperim. **9** (1964) No. 6, 77 (Russ.); Nucl. Instr. Meth. **28** (1964) 341.
[Du 64a] DUNAITSEV, A. F., YU. D. PROKOSHKIN: Pribory Tekhn. Eksperim. **9** (1964) No. 5, 93 (Russ.).
[Du 64b] DUNAITSEV, A. F.: Pribory Tekhn. Eksperim. **9** (1964) No. 5, 119 (Russ.).
[Du 65] DUNAITSEV, A. F., R. N. PIKHTELEV et al.: Dubna-Report R-2534, Dec. 1965.
[Du 66] DUBROVSKII, I. A.: Pribory Tekhn. Eksperim. **11** (1966) No. 2, 5 (Russ.).
[Du 66a] DUCHET, M.: Adv. Electronics Electron Phys. **22 A** (1966) 499.
[Du 67] DUNAITSEV, A. F., V. M. KUT'IN et al.: Yadernaya Fizika **5** (1967) 826 (Russ.).
[Ea 66] EARNSHAW, J. B., Z. C. TAN: Electronic Engng. **38** (1966) 242.
[Eb 54] EBY, F. S., W. K. JENTSCHKE: Phys. Rev. **96** (1954) 911.
[Ec 62] ECKARDT, A., P. GÖRLICH et al.: Nucl. Instr. Meth. **16** (1962) 44.
[Ec 62a] ECKART, F.: Elektronenoptische Bildwandler und Röntgenbildverstärker, Johann Ambrosius Barth, Leipzig 1962.
[Ec 63] ECKARDT, A., R. PRAGER: Nucl. Instr. Meth. **24** (1963) 152.
[Ec 64] ECKARDT, A., W. DIETEL, R. PRAGER: Exp. Techn. Phys. **12** (1964) 63.
[Ec 65] ECKARDT, A., R. PRAGER: Exp. Techn. Phys. **13** (1965) 475.
[Ed 60] Edgerton, Germeshausen & Grier, Inc.: Rev. Sci. Instr. **31** (1960) 799.
[Ed 65] EDWARDS, J. G.: Proc. IEE **112** (1965) 1081.
[Ef 65] EFIMCHIK, M. K., V. V. IZOKH et al.: Pribory Tekhn. Eksperim. **10** (1965) No. 2, 74 (Russ.).
[Ef 65a] EFIMCHIK, M. K., V. V. IZOKH, L. D. SOSHIN: Pribory Tekhn. Eksperim. **10** (1965) No. 2, 84 (Russ.).
[Ef 65b] EFIMCHIK, M. K., V. V. IZOKH, A. F. CHERNYAVSKII: Pribory Tekhn. Eksperim. **10** (1965) No. 4, 234 (Russ.).
[Ef 68] EFIMCHIK, M. K., A. M. ZAITSEVA, A. F. CHERNYAVSKII: Pribory Tekhn. Eksperim. **13** (1968) No. 2, 102 (Russ.).
[El 50] ELLIOT, J. O., S. H. LIEBSON et al.: Rev. Sci. Instr. **21** (1950) 631.
[El 62] EL-WAHAB, M. A., J. V. KANE: Nucl. Instr. Meth. **15** (1962) 15.
[El 65] ELAD, E., S. ROZEN: Nucl. Instr. Meth. **37** (1965) 58.
[El 67] EL-WAHAB, M. A., M. GAZZALY: Nuovo Cimento **47 B** (1967) 56.
[El 68] EL-WAHAB, M. A., M. SAKKA, M. A. EL-SALAM: Nucl. Instr. Meth. **59** (1968) 344.
[Em 62] EMMER, T. L.: IRE Trans. Nucl. Sci. **9** (1962) No. 3, 305.
[Em 65] EMMER, T. L.: IEEE Trans. Nucl. Sci. **12** (1965) No. 1, 329.
[Em 67] EMERSON, S. T., W. DWAIN SIMPSON et al.: Nucl. Instr. Meth. **52** (1967) 229.
[En 62] ENGBERT, W.: Nachrichtentechn. Z. **15** (1962) 381.
[En 63] ENGELMANN, R.: Electronics **36** (1963) No. 46, 34.
[En 64] ENGELMANN, R.: Electronics **37** (1964) No. 5, 34.
[Ep 63] EPSTEIN, R. J.: Nucl. Instr. Meth. **24** (1963) 333.
[Er 66] EREMIN, A. S.: Pribory Tekhn. Eksperim. **11** (1966) No. 4, 5 (Russ.).
[Er 68] ERGLIS, K. E., V. T. SUBBOTIN: Pribory Tekhn. Eksperim. **13** (1968) No. 1, 96 (Russ.).
[Es 58] ESAKI, L.: Phys. Rev. **109** (1958) 603.

[Et 67] ETTINGER, G. M.: Electronics 40 (1967) No. 2, 108.
[Eu 64] EULING, R.: J. Appl. Phys. 35 (1964) 1391.
[Ev 55] EVANS, R. D.: The Atomic Nucleus, McGraw-Hill Book Co., New York/Toronto/London 1955, p. 785.
[Ew 64] EWAN, G. T., A. J. TAVENDALE: Can. J. Phys. 42 (1964) 2286.
[Ew 66] EWAN, G. T., R. L. GRAHAM, I. K. MACKENZIE: IEEE Trans. Nucl. Sci. 13 (1966) No. 3, 297.
[Fa 47] FANO, U.: Phys. Rev. 72 (1947) 26.
[Fa 56] FAIRSTEIN, E.: Rev. Sci. Instr. 27 (1956) 475.
[Fa 58] FARLEY, F. J. M.: Rev. Sci. Instr. 29 (1958) 595.
[Fa 60] FARBER, A. S.: Rev. Sci. Instr. 31 (1960) 15.
[Fa 61] FANCHENKO, S. D.: Pribory Tekhn. Eksperim. 6 (1961) No. 1, 5 (Russ.).
[Fa 62] FALK, W. R., L. KATZ: Can. J. Phys. 40 (1962) 978.
[Fa 63] FANG, F. F., H. N. YU: Proc. IEEE 51 (1963) 860.
[Fa 65] FALK, K., P. A. TOVE, M. MADAKBAS: Nucl. Instr. Meth. 34 (1965) 157.
[Fa 65a] FABRI, G., V. SVELTO: Nucl. Instr. Meth. 35 (1965) 83.
[Fa 65b] FAIRSTEIN, E., J. HAHN: Nucleonics 23 (1965) No. 7, 56; No. 9, 81; No. 11, 50; 24 (1966) No. 1, 54; No. 3, 68.
[Fa 65c] FALK, W., G. JONES, R. ORTH: Nucl. Instr. Meth. 33 (1965) 345.
[Fa 67] FALLSIDE, F., D. T. BICKLEY: Proc. IEE 113 (1966) 263.
[Fe 60] FEDOTOV, YU. A., YU. V. SHMARTSEV: Tranzistory, Izdatel'stvo Sovetskoe Radio, Moscow 1960; German Translation: Verlag Technik, Berlin 1963.
[Fe 60a] FELDMAN, M.: Rev. Sci. Instr. 31 (1960) 1356.
[Fe 60b] FERGUSON, A. T. G., M. A. GRACE, J. O. NEWTON: Nucl. Phys. 17 (1960) 1.
[Fe 62] FERGUSON, A. T. G., P. R. ORMAN, J. H. MONTAGUE: in Proc. Conf. Nucl. Electronics, Belgrade 1961, Vol. 3, Internat. Atomic Energy Agency, Vienna 1962, p. 335.
[Fe 62a] FELDMAN, M.: Rev. Sci. Instr. 33 (1962) 1283.
[Fe 65] FELDMAN, M.: Rev. Sci. Instr. 36 (1965) 241.
[Fi 49] FITCH, V.: Rev. Sci. Instr. 20 (1949) 942.
[Fi 52] FISCHER, J., J. MARSHALL: Rev. Sci. Instr. 23 (1952) 417.
[Fi 56] FITCH, V., R. MOTLEY: Phys. Rev. 101 (1956) 496.
[Fi 59] FITCH, V. L.: in Proc. Internat. Conf. High-Energy Accelerators and Instrumentation, CERN 1959, p. 601.
[Fi 59a] FIEHRER, M., A. PAGES: in Proc. Internat. Symp. Nucl. Electronics, Paris 1958, Vol. 1, Internat. Atomic Energy Agency, Vienna 1959, p. 249.
[Fi 60] FIDECARO, G.: Nuovo Cimento, Suppl. 15 (1960) 254.
[Fi 60a] FISCHER, J., A. LUNDBY: Rev. Sci. Instr. 31 (1960) 10.
[Fi 61] FISCHER, J., G. T. ZORN: Rev. Sci. Instr. 32 (1961) 499.
[Fi 62] FISTUL', V. I., N. Z. SHVARTS: Usp. Fiz. Nauk 77 (1962) 109 (Russ.); German Translation in: Fortschritte Phys. 10 (1962) 617.
[Fi 62a] FISCHER, J., G. T. ZORN: IRE Trans. Nucl. Sci. 9 (1962) No. 3, 261.
[Fi 63] FITCH, V. L.: Nucl. Instr. Meth. 20 (1963) 247.
[Fi 65] FIRSOV, E. I., A. F. CHERNYAVSKII, S. S. SHUSHKEVICH: Pribory Tekhn. Eksperim. 10 (1965) No. 2, 5 (Russ.).
[Fi 66] FIRK, F. W. K.: Nucl. Instr. Meth. 43 (1966) 312.
[Fi 67] FISCHER, H.: Proc. IEEE 55 (1967) 100.
[Fl 49] FLETCHER, R. C.: Rev. Sci. Instr. 20 (1949) 861.
[Fl 62] FLINN, E. A.: Adv. Electronics Electron Phys. 16 (1962) 155.
[Fl 64] FLAUGER, W., H. SCHNEIDER: Nucl. Phys. 55 (1964) 207.
[Fo 63] FORTE, M.: in Proc. Internat. Symp. Nucl. Electronics, Paris 1963, p. 105.
[Fo 63a] FOSSAN, D. B., B. HERSKIND: Nucl. Phys. 40 (1963) 24.
[Fo 64] FOOTE, R. S., D. JOHNSON: Rev. Sci. Instr. 35 (1964) 1126.
[Fo 64a] FOLEY, K. J., S. J. LINDENBAUM et al.: Nucl. Instr. Meth. 30 (1964) 45.

[Fo 65] Foley, K. J., R. S. Jones et al.: IEEE Trans. Nucl. Sci. **12** (1965) No. 4, 222.
[Fo 66] Fossan, D. B., R. E. McDonald, L. F. Chase: Phys. Rev. **141** (1966) 1018.
[Fr 58] Fraser, J. S., J. C. D. Milton: Nucl. Instr. **2** (1958) 275.
[Fr 61] Franzini, P.: Rev. Sci. Instr. **32** (1961) 1222.
[Fr 64] Fränz, K., H. Paucksch: Nucl. Instr. Meth. **27** (1964) 125.
[Fr 64a] Fränz, K., J. Schulz: in Proc. Conf. Automatic Acquisition and Reduction of Nuclear Data (EANDC), Karlsruhe 1964, p. 317.
[Fr 66] Fraser, H. J.: Nucl. Instr. Meth. **40** (1966) 106.
[Fr 67] Fränz, K.: Nucl. Instr. Meth. **47** (1967) 217.
[Fr 67a] Franke, H. G., H. Schmeing: Nucl. Instr. Meth. **52** (1967) 171.
[Fu 60] Fullwood, R.: Rev. Sci. Instr. **31** (1960) 1186.
[Fü 65] Fülle, R., G. Máthé, D. Netzband: Nucl. Instr. Meth. **35** (1965) 250.
[Fu 66] Fuschini, E., C. Maroni, P. Veronesi: Nucl. Instr. Meth. **41** (1966) 153.
[Ga 50] Garwin, R. L.: Rev. Sci. Instr. **21** (1950) 569.
[Ga 50a] Garwin, R. L.: Rev. Sci. Instr. **21** (1950) 903.
[Ga 53] Gatti, E., F. Piva: Nuovo Cimento **10** (1953) 984.
[Ga 54] Gatti, E.: Nuovo Cimento **11** (1954) 153.
[Ga 58] Gaeta, R., F. Manero: in Proc. Sec. Internat. Conf. Peaceful Uses Atomic Energy, Geneva 1958, Vol. 14, United Nations, Geneva 1958, p. 331.
[Ga 59] Gatti, E., V. Svelto: Nucl. Instr. Meth. **4** (1959) 189.
[Ga 59a] Garwin, E. L.: Rev. Sci. Instr. **30** (1959) 373.
[Ga 60] Gatti, E., V. Svelto: Energia Nucl. **7** (1960) 736.
[Ga 60a] Garg, J. B.: Nucl. Instr. Meth. **6** (1960) 187.
[Ga 60b] Garg, J. B.: Nucl. Instr. Meth. **6** (1960) 72.
[Ga 61] Gabriel, R., A. M. Segar: Nucl. Instr. Meth. **12** (1961) 307.
[Ga 62] Gaddy, O. L., D. F. Holshouser: Proc. IRE **50** (1962) 207.
[Ga 62a] Gatti, E., V. Svelto: in Proc. Symp. Nucl. Instr., Harwell 1961, ed. by J. B. Birks, Heywood, London 1962, p. 35.
[Ga 62b] Gale, N. H., J. B. Garg, J. M. Calvert: Nucl. Phys. **38** (1962) 222.
[Ga 62c] Gatti, E., F. Vaghi, E. Zaglio: in Proc. Conf. Nucl. Electronics, Belgrade 1961, Vol. 3, Internat. Atomic Energy Agency, Vienna 1962, p. 105.
[Ga 62d] Gale, N. H.: Nucl. Phys. **38** (1962) 252.
[Ga 63] Gatti, E.: in Proc. Internat. Symp. Nucl. Electronics, Paris 1963, p. 717.
[Ga 64] Gatti, E., V. Svelto: Nucl. Instr. Meth. **30** (1964) 213.
[Ga 64a] Garvey, J.: Nucl. Instr. Meth. **29** (1964) 137.
[Ga 64b] Gatti, E.: in Proc. Conf. Automatic Acquisition and Reduction of Nuclear Data (EANDC), Karlsruhe 1964, p. 259.
[Ga 65] Gatti, E., V. Svelto: Nucleonics **23** (1965) No. 7, 62.
[Ga 65a] Gazaryan, K. A., V. S. Pantuev, M. N. Khachaturyan: Pribory Tekhn. Eksperim. **10** (1965) No. 1, 161 (Russ.).
[Ga 66] Gatti, E., V. Svelto: Nucl. Instr. Meth. **43** (1966) 248.
[Ge 56] Gerharz, R.: Z. angew. Phys. **8** (1956) 531.
[Ge 59] Gettner, M., W. Selove: Rev. Sci. Instr. **30** (1959) 942.
[Ge 64] Gemmell, D. S.: IEEE Trans. Nucl. Sci. **11** (1964) No. 3, 409.
[Ge 65] Gel'tsel', M. Yu., A. D. Panfilov et al.: Pribory Tekhn. Eksperim. **10** (1965) No. 2, 121 (Russ.).
[Ge 65a] Geiger, J. S., R. L. Graham et al.: Nucl. Phys. **68** (1965) 352.
[Ge 66] Gere, E. A., G. L. Miller: IEEE Trans. Nucl. Sci. **13** (1966) No. 3, 508.
[Ge 66a] Gel'tsel', M. Yu., A. D. Panfilov et al.: Pribory Tekhn. Eksperim. **11** (1966) No. 3, 101 (Russ.).
[Ge 66b] Geller, M., D. E. Altman, T. A. DeTemple: J. Appl. Phys. **37** (1966) 3639.
[Ge 67] Gettner, M., R. Weinstein et al.: Nucl. Instr. Meth. **51** (1967) 1.
[Ge 67a] Gedcke, D. A., W. J. McDonald: Nucl. Instr. Meth. **55** (1967) 377.
[Ge 67b] Gedcke, D. A., W. J. McDonald: Nucl. Instr. Meth. **56** (1967) 148.

[Ge 68] GEDCKE, D. A., W. J. McDONALD: Nucl. Instr. Meth. 58 (1968) 253.
[Gi 54] GIACOLETTO, L. J.: RCA-Review 15 (1954) 506.
[Gi 62] GILDEMEISTER, O., R. GIESE: Adv. Electronics Electron Phys. 16 (1962) 113.
[Gi 63] GILDEMEISTER, O., R. GIESE: Nucl. Instr. Meth. 20 (1963) 233.
[Gi 66] GIBSON, W. A.: Rev. Sci. Instr. 37 (1966) 631.
[Gi 66a] GIBSON, W. M.: IEEE Trans. Nucl. Sci. 13 (1966) No. 3, 162.
[Gi 66b] GIBBONS, F. L.: IEEE Trans. Nucl. Sci. 13 (1966) No. 1, 305.
[Gl 64] GLEBOVICH, G. V., O. V. ZHUKOV: Pribory Tekhn. Eksperim. 9 (1964) No. 1, 102 (Russ.).
[Gl 66] GLEBOVICH, G. V.: Pribory Tekhn. Eksperim. 11 (1966) No. 3, 132 (Russ.).
[Gl 67] GLEBOVICH, G. V.: Pribory Tekhn. Eksperim. 12 (1967) No. 4, 103 (Russ.).
[Gn 63] GNOTH, B.: Elektronik 12 (1963) 85.
[Go 60] GORODETZKY, S., R. RICHERT et al.: Nucl. Instr. Meth. 7 (1960) 50.
[Go 60a] GOULDING, F. S., R. W. NICHOLSON, J. B. WAUGH: Nucl. Instr. Meth. 8 (1960) 272.
[Go 60b] GOULDING, F. S., R. A. McNAUGHT: Nucl. Instr. Meth. 8 (1960) 282.
[Go 60c] GOODALL, W. M., A. F. DIETRICH: Proc. IRE 48 (1960) 1591.
[Go 61] GOODRICH, G. W., W. C. WILEY: Rev. Sci. Instr. 32 (1961) 846.
[Go 61a] GORODETZKY, S., A. MUSER et al.: Nucl. Instr. Meth. 13 (1961) 282.
[Go 61b] GORODETZKY, S., A. MUSER et al.: Nucl. Instr. Meth. 14 (1961) 205.
[Gö 62] GÖRLICH, P., A. KROHS, H.-J. POHL: Exp. Techn. Phys. 10 (1962) 247.
[Go 62a] GORODETZKY, S., A. MUSER et al.: Nucl. Instr. Meth. 17 (1962) 353.
[Go 62b] GORBACHEV, V. M., N. A. UVAROV, L. D. USENKO: Pribory Tekhn. Eksperim. 7 (1962) No. 2, 92 (Russ.).
[Go 62c] GORDIENKO, A. I., A. A. IOGANSEN: Pribory Tekhn. Eksperim. 7 (1962) No. 6, 77 (Russ.).
[Go 62d] GORBACHEV, V. M., V. N. KOROLEV, N. A. UVAROV: Pribory Tekhn. Eksperim. 7 (1962) No. 6, 81 (Russ.).
[Go 63] GOTTLIEB, E., J. GIORGIS: Electronics 36 (1963) No. 27, 26.
[Go 63a] GOURSKY, V., H. GUILLON: in Proc. Internat. Symp. Nucl. Electronics, Paris 1963, p. 313.
[Go 63b] GORBACHEV, V. M., V. N. KOROLEV, N. A. UVAROV: Pribory Tekhn. Eksperim. 8 (1963) No. 1, 98 (Russ.).
[Gö 64] GÖRLICH, P., A. KROHS, H.-J. POHL: in Proc. Third Internat. Measurement Conf. (IMEKO), Stockholm 1964, Vol. 1, Ung. Wiss. Verein Meßtechnik und Automatisierung, Budapest 1964, p. 317.
[Go 64a] GOSTREM, R. V., G. S. ZINOV'EV: Tunnel'nye diody i ikh primenenie (Tunnel Diodes and Their Application), Akad. Nauk SSSR, Sibirskoe Otdelenie, Novosibirsk 1964.
[Go 64b] GOLUTVIN, I. A., YU. V. ZANEVSKII: Pribory Tekhn. Eksperim. 9 (1964) No. 5, 114 (Russ.).
[Go 64c] GOLUTVIN, I. A., YU. V. ZANEVSKII: Dubna-Report 1558, Febr. 1964.
[Go 65] GOLUTVIN, I. A., YU. V. ZANEVSKII: Pribory Tekhn. Eksperim. 10 (1965) No. 1, 86 (Russ.).
[Go 65a] GOLDSTEIN, B. S., R. M. WEIGAND: Proc. IEEE 53 (1965) 195.
[Gö 66] GÖRLICH, P., H.-J. POHL et al.: IEEE Trans. Nucl. Sci. 13 (1966) No. 3, 112.
[Go 66a] GOULDING, F. S.: Nucl. Instr. Meth. 43 (1966) 1.
[Go 66b] GOULDING, F. S., D. A. LANDIS et al.: IEEE Trans. Nucl. Sci. 13 (1966) No. 3, 514.
[Go 66c] GORODETZKY, S., N. SCHULZ et al.: Nucl. Phys. 85 (1966) 529.
[Gö 66d] GÖRLICH, P.: Photoeffekte. Vol. 3: Erzeugung photoelektromotorischer Kräfte, Geest und Portig, Leipzig 1966.
[Go 66e] GORODETZKY, S., P.-L. WENDEL et al.: Nucl. Instr. Meth. 45 (1966) 72.
[Go 67] GORBENKO, V. G., V. J. LAPSHIN et al.: Dubna-Report 13–3095, Jan. 1967; Pribory Tekhn. Eksperim. 12 (1967) No. 3, 161.

[Go 67a] GOLDSTEIN, B. S., G. F. DALRYMPLE: Proc. IEEE **55** (1967) 181.

[Go 67b] GOYOT, M., J.-J. SAMUELI, A. SARAZIN: Nucl. Instr. Meth. **46** (1967) 149.

[Go 67c] GOYOT, M., J. PIGNERET et al.: Nucl. Instr. Meth. **53** (1967) 87.

[Go 67d] GORODETZKY, S., J. GERBER et al.: Nucl. Instr. Meth. **57** (1967) 152.

[Go 67e] GORNI, S., G. HOCHNER et al.: Nucl. Instr. Meth. **53** (1967) 349.

[Gr 51] GRAHAM, R. L., R. E. BELL: Phys. Rev. **84** (1951) 380.

[Gr 57] GRISMORE, R., W. C. PARKINSON: Rev. Sci. Instr. **28** (1957) 245.

[Gr 58] GREEN, R. E., R. E. BELL: Nucl. Instr. **3** (1958) 127.

[Gr 58a] GRUHLE, W.: Nucl. Instr. **3** (1958) 204.

[Gr 59] GRUHLE, W.: Nucl. Instr. Meth. **4** (1959) 112.

[Gr 60] GRUHLE, W.: Elektronische Hilfsmittel des Physikers, Springer-Verlag, Berlin/Göttingen/Heidelberg 1960.

[Gr 61] GRIN, G. A., C. JOSEPH et al.: Helv. Phys. Acta **34** (1961) 490.

[Gr 62] GRAHAM, R. L., J. S. GEIGER et al.: Nucl. Instr. Meth. **15** (1962) 40.

[Gr 63] GRIN, G. A., C. JOSEPH: Nucl. Instr. Meth. **24** (1963) 331.

[Gr 63a] GRIMMEISS, H. G., W. KISCHIO, H. SCHOLZ: Philips' Techn. Rdsch. **25** (1963/64) 386.

[Gr 64] GRABER, H.: Exp. Techn. Phys. **12** (1964) 24.

[Gr 65] GRODZINS, L., P. H. ROSE, R. J. VAN DE GRAAFF: Nucl. Instr. Meth. **36** (1965) 202.

[Gr 66] GREENBERG, A., M. FURST, H. KALLMANN: in Internat. Lumineszenz-Symp. Phys. Chem. Szintillatoren, München 1965, ed. by N. RIEHL, H. KALLMANN, Karl Thiemig, München 1966, p. 71.

[Gr 66a] GRAHAM, R. L., I. K. MACKENZIE, G. T. EWAN: IEEE Trans. Nucl. Sci. **13** (1966) No. 1, 72.

[Gr 66b] GRUNBERG, J., L. TEPPER: IEEE Trans. Nucl. Sci. **13** (1966) No. 1, 389.

[Gr 66c] GRIFFITHS, R. J., E. A. MCCLATCHIE et al.: Nucl. Instr. Meth. **40** (1966) 181.

[Gr 67] GREBENYUK, V. M., I. F. KOLPAKOV: Dubna-Report 13–3271, April 1967.

[Gr 67a] GRIGOR'EV, A. D.: Pribory Tekhn. Eksperim. **12** (1967) No. 1, 114 (Russ.).

[Gr 67b] GRABARI, V.: Nucl. Instr. Meth. **53** (1967) 136.

[Gr 68] GREUPNER, H., H. KERKOW, S. SCHWABE: Kernenergie **11** (1968) 10.

[Gu 53] GUILLON, H.: J. Phys. Radium **14** (1953) 128.

[Gu 62] GUPTA, S. L., N. K. SAHA: Nucl. Instr. Meth. **15** (1962) 95.

[Gu 63] GUTKIN, A. A., A. A. ROGACHEV et al.: Pribory Tekhn. Eksperim. 8 (1963) No. 4, 187 (Russ.).

[Gu 66] GUILLON, H.: Nucl. Instr. Meth. **43** (1966) 240.

[Gu 66a] GUGGENBÜHL, W., W. WUNDERLIN: Archiv elektr. Übertragung **20** (1966) 299.

[Gy 63] GYGI, E., F. SCHNEIDER: Nucl. Instr. Meth. **20** (1963) 352.

[Gy 64] GYGI, E., F. SCHNEIDER: CERN-Report 64–46, Nov. 1964.

[Ha 56] HAMILTON, T. D. S., G. T. WRIGHT: J. Sci. Instr. **33** (1956) 36.

[Ha 57] HARTMANN, W., F. BERNHARD: Fotovervielfacher und ihre Anwendung in der Kernphysik, Akademie-Verlag, Berlin 1957.

[Ha 57a] HAINE, M. E., M. W. JERVIS: Proc. IEE **B 104** (1957) 379, 385.

[Ha 61] HAZONI, Y.: Nucl. Instr. Meth. **13** (1961) 95.

[Ha 61a] HAVENS, W. W.: Time-of-Flight Methods, in Methods of Experimental Physics, Vol. 5 A, ed. by L. MARTON, Academic Press, New York/London 1961, p. 495.

[Ha 61b] HARRISON, W. V., R. S. FOOTE: Electronics **34** (1961) No. 32, 154.

[Ha 61c] HARTMANN, H. J., M. MICHELITSCH, W. STEINHÄUSER: Archiv elektr. Übertragung **15** (1961) 125.

[Ha 62] HAYASHI, I., N. YOSHIMURA: in Proc. Conf. Nucl. Electronics, Belgrade 1961, Vol. 3, Internat. Atomic Energy Agency, Vienna 1962, p. 157.

[Ha 62a] HAMILTON, D. J., P. G. GRIFFITH, F. H. SHAVER: Electronic Engng. **34** (1962) 808.

[Ha 63] HAHN, J., V. GUIRAGOSSIAN: IEEE Trans. Nucl. Sci. **10** (1963) No. 3, 44.

[Ha 63a] HAYES, R. E., R. W. ALSFORD: Electronic Engng. **35** (1963) 92.
[Ha 65] HAFT, G.: Elektronik **14** (1965) 177.
[Ha 65a] HAVLIK, L., F. JELINEK: Nachrichtentechnik **15** (1965) 383.
[Ha 66] HATCH, K. F.: IEEE Trans. Nucl. Sci. **13** (1966) No. 1, 351.
[Ha 66a] HALLING, H., F. MAY: Nucl. Instr. Meth. **40** (1966) 325.
[Ha 66b] HART, B. L.: Electronic Engng. **38** (1966) 815.
[Ha 67] HALLING, H., S. MAYERHOFER et al.: Nucl. Instr. Meth. **50** (1967) 22.
[Ha 67a] HANSEN, J. P., W. A. SCHMIDT: Proc. IEEE **55** (1967) 216.
[Ha 67b] HASSELBACH, F., G. MACK: Nucl. Instr. Meth. **57** (1967) 121.
[Ha 67c] HANSEN, R. D., J. W. SEDLMEYER: IEEE Trans. Nucl. Sci. **14** (1967) No. 1, 222.
[Ha 67d] HAUSER, U., G. KNISSEL et al.: Rev. Sci. Instr. **38** (1967) 1220.
[Ha 68] HART, B. L.: Electronic Engng. **40** (1968) 40, 72.
[He 54] HEARD, H. G.: Rev. Sci. Instr. **25** (1954) 454.
[He 56] HERON, S., R. W. P. McWHIRTER, E. H. RHODERICK: Proc. Royal Soc. **A 234** (1956) 565.
[He 61] HERMSEN, H. P.: Diplomarbeit, Univ. Hamburg 1961.
[He 61a] HENEBRY, W. M.: Rev. Sci. Instr. **32** (1961) 1198.
[He 64] HEEK, K. H. v., H. D. SCHILLING, L. WALLEK: Nucl. Instr. Meth. **29** (1964) 35.
[He 66] HEATH, R. L., W. W. BLACK, J. E. CLINE: IEEE Trans. Nucl. Sci. **13** (1966) No. 3, 445; Nucleonics **24** (1966) No. 5, 52.
[He 66a] HEAVENS, O. S.: Brit. J. Appl. Phys. **17** (1966) 287.
[He 66b] HENEBRY, W. M., A. RASIEL: IEEE Trans. Nucl. Sci. **13** (1966) No. 2, 64.
[He 67] HENCK, R., A. COCHE: IEEE Trans. Nucl. Sci. **14** (1967) No. 1, 478.
[Hi 56] HIGINBOTHAM, W. A.: in Proc. First Internat. Conf. Peaceful Uses Atomic Energy, Geneva 1955, Vol. 4, United Nations, New York 1956, p. 52.
[Hi 58] HILLAS, A. M., R. M. TENNENT: Nucl. Instr. **3** (1958) 344.
[Hi 60] HINE, G. J., J. A. CARDARELLI: Nucleonics **18** (1960) No. 9, 92.
[Hi 60a] HILBERG, W.: Telefunken-Ztg. **33** (1960) No. 128, 98.
[Hi 64] HILBERG, W.: Nachrichtentechn. Z. **17** (1964) 24.
[Hi 65] HILBERG, W.: Nucl. Instr. Meth. **33** (1965) 322.
[Ho 50] HOFSTADTER, R., J. A. McINTYRE: Rev. Sci. Instr. **21** (1950) 52.
[Ho 59] HORNYAK, W. F., H. YOSHIKI: in Proc. Internat. Symp. Nucl. Electronics, Paris 1958, Vol. 1, Internat. Atomic Energy Agency, Vienna 1959, p. 195.
[Ho 63] HOFSTEIN, S. R., F. P. HEIMAN: Proc. IEEE **51** (1963) 1190.
[Ho 66] HOPMANN, W.: Kernforschungsanlage Jülich, Report 290 PP, August 1966.
[Ho 68] HOUDAYER, A., S. K. MARK, R. E. BELL: Nucl. Instr. Meth. **59** (1968) 319.
[Hr 67] HRYNKIEWICZ, A. Z., S. KOPTA et al.: Nucl. Instr. Meth. **49** (1967) 229.
[Hu 56] HUNTER, L. P.: Handbook of Semiconductor Electronics, McGraw-Hill Book Co., New York/Toronto/London 1956.
[Hu 59] HUTCHINSON, G. W., W. H. WELLS, D. H. WHITE: Nucl. Instr. Meth. **4** (1959) 228.
[Hu 60] HUTCHINSON, G. W.: Progr. Nucl. Phys. **8** (1960) 195.
[Hu 62] HUSTON, A. E., F. WALTERS: Adv. Electronics Electron Phys. **16** (1962) 249.
[Hu 63] HU, K. C.: Electronics **36** (1963) No. 7, 68.
[Hu 65] HUTH, G. C., J. B. TRICE et al.: IEEE Trans. Nucl. Sci. **12** (1965) No. 1, 275.
[Hu 66] HUTH, G. C.: IEEE Trans. Nucl. Sci. **13** (1966) No. 1, 36.
[Hu 66a] HUSTON, A. E.: Adv. Electronics Electron Phys. **22 B** (1966) 957.
[Hu 67] HUNDLEY, L., T. COBURN et al.: Rev. Sci. Instr. **38** (1967) 488.
[Hu 67a] HURLEY, J. P., V. L. DAGRAGNANO, J. M. MATHIESEN: Nucl. Instr. Meth. **51** (1967) 299.
[Hv 63] HVAM, T., M. SMEDSDAL: Nucl. Instr. Meth. **24** (1963) 55.
[Hy 64] HYMAN, L. G., R. M. SCHWARCZ, R. A. SCHLUTER: Rev. Sci. Instr. **35** (1964) 393.
[Hy 65] HYMAN, L. G.: Rev. Sci. Instr. **36** (1965) 193.

[Il 68] Il'in, Yu. I., V. A. Chirkin, M. V. Savin: Pribory Tekhn. Eksperim. **13** (1968) No. 2, 99 (Russ.).
[Im 58] Imhof, W., R. Kalibjian, V. Perez-Mendez: Rev. Sci. Instr. **29** (1958) 476.
[In 60] Infante, C.: Nucl. Instr. Meth. **9** (1960) 102.
[In 61] Infante, C., F. Pandarese: Electronics **34** (1961) No. 46, 133.
[In 62] Infante, G., F. Pandarese: in Proc. Conf. Nucl. Electronics, Belgrade 1961, Vol. 3, Internat. Atomic Energy Agency, Vienna 1962, p. 29.
[Ir 66] Iredale, P., G. W. Hinder et al.: Adv. Electronics Electron Phys. **22 B** (1966) 801.
[Is 62] Ispiryan, K. A., A. G. Oganesyan: Pribory Tekhn. Eksperim. **7** (1962) No. 2, 61 (Russ.).
[Is 63] Iselin, F.: Nucl. Instr. Meth. **20** (1963) 330.
[Is 65] Issinskii, I. B., K. P. Myznikov: Pribory Tekhn. Eksperim. **10** (1965) No. 3, 131 (Russ.).
[Iv 62] Ivanov, A. A., V. S. Nesterenko: Pribory Tekhn. Eksperim. **7** (1962) No. 6, 75 (Russ.).
[Iv 65] Iveković, D.: Nucl. Instr. Meth. **32** (1965) 339.
[Iv 66] Ivanov, A. A.: Pribory Tekhn. Eksperim. **11** (1966) No. 1, 111 (Russ.).
[Iv 67] Iveković, D.: Electronic Engng. **39** (1967) 680.
[Iv 67a] Iveković, D.: Electronic Engng. **39** (1967) 739.
[Iv 67b] Ivanov, A. A., I. F. Kolpakov, L. A. Timokhin: Primenenie tunnel'nykh diodov v priborakh yadernoi fiziki (Application of Tunnel Diodes in Devices of Nuclear Physics), Atomizdat, Moscow 1967.
[Iv 67c] Ivanov, A. A., V. S. Nesterenko, L. A. Timokhin: Pribory Tekhn. Eksperim. **12** (1967) No. 3, 77 (Russ.).
[Iz 62] Izokh, V. V., M. K. Efimchik: Pribory Tekhn. Eksperim. **7** (1962) No. 3, 86 (Russ.).
[Iz 63] Izokh, V. V., V. P. Bagrintsev: Pribory Tekhn. Eksperim. 8 (1963) No. 5, 120 (Russ.).
[Iz 65] Izokh, V. V., A. F. Chernyavskii: Pribory Tekhn. Eksperim. **10** (1965) No. 6, 97 (Russ.).
[Ja 50] Janssen, J. M. L.: Philips' Techn. Rdsch. **12** (1950) 52, 73.
[Ja 61] James, P., W. A. Hunter: Mullard Technical Commun. **6** (1961) No. 51.
[Ja 65] Jackson, H. G.: Nucl. Instr. Meth. **33** (1965) 161.
[Ja 66] Jastrzebski, J., M. Moszyński, A. Zgliński: Nukleonika **11** (1966) 471.
[Je 53] Jelley, J. V.: Progr. Nucl. Phys. **3** (1953) 84.
[Je 58] Jenny, D. A.: Proc. IRE **46** (1958) 717.
[Jo 56] Jones, G., J. B. Warren: J. Sci. Instr. **33** (1956) 429.
[Jo 57] Johansson, B.: Nucl. Instr. **1** (1957) 274.
[Jo 58] Johansson, B.: Nucl. Instr. **2** (1958) 61.
[Jo 58a] Johansson, B., T. Alväger: Nucl. Instr. **3** (1958) 49.
[Jo 60] Jones, G.: J. Sci. Instr. **37** (1960) 318.
[Jo 60a] Johansson, B., T. Alväger: Arkiv Fysik **17** (1960) 163.
[Jo 62] Jones, L. W., M. L. Perl: in Proc. Conf. Nucl. Electronics, Belgrade 1961, Vol. 1, Internat. Atomic Energy Agency, Vienna 1962, p. 165.
[Jo 63] Jones, G.: Rev. Sci. Instr. **34** (1963) 938.
[Jo 64] Jones, O. C.: J. Sci. Instr. **41** (1964) 653.
[Jo 65] Johnson, A. M.: IEEE J. Quantum Electronics **1** (1965) 99.
[Jo 65a] Jones, G., W. R. Falk: Nucl. Instr. Meth. **37** (1965) 22.
[Jo 68] Jones, G., P. H. R. Orth: Nucl. Instr. Meth. **59** (1968) 309.
[Jo 68a] Johnson, F. A.: Nucl. Instr. Meth. **59** (1968) 237.
[Ju 63] Jung, H.: Nucl. Instr. Meth. **24** (1963) 197.
[Ju 65] Jungclaussen, H.: Kernenergie **8** (1965) 514.
[Ju 65a] Jung, H., M. Brüllmann, D. Meier: Nucl. Instr. Meth. **32** (1965) 109.

[Ju 66] JUNG, E. A., R. N. LEWIS: Nucl. Instr. Meth. **44** (1966) 224.

[Ju 67] JUNG, H., M. BRÜLLMANN, D. MEIER: Nucl. Instr. Meth. **55** (1967) 301.

[Ju 67a] JUNGMEISTER, H. G.: Archiv elektr. Übertragung **21** (1967) 447.

[Ju 68] JUNGMEISTER, H. G., D. SCHMIDT: Frequenz **22** (1968) 112.

[Ka 62] KANE, J. V., M. A. EL-WAHAB et al.: in Proc. Conf. Nucl. Electronics, Belgrade 1961, Vol. 1, Internat. Atomic Energy Agency, Vienna 1962, p. 345.

[Ka 63] KATZENSTEIN, H. S., R. M. ALEXANDER: in Proc. Internat. Symp. Nucl. Electronics, Paris 1963, p. 685.

[Ka 63a] KAHANE, J., P. VERGEZAC: J. Physique **24** (1963) 962.

[Ka 65] KARSTENSEN, F.: Z. Phys. **187** (1965) 165.

[Ka 66] KALIBJIAN, R.: IEEE Trans. Nucl. Sci. **13** (1966) No. 3, 54.

[Ka 66a] KATZENSTEIN, H. J.: IEEE Trans. Nucl. Sci. **13** (1966) No. 3, 527.

[Ka 66b] KARSTENSEN, F., J. SCHRAMM: Z. Phys. **195** (1966) 370.

[Ka 66c] KALASHNIKOVA, V. I., M. S. KOZODAEV: Detektory elementarnykh chastits (Elementary Particle Detectors), Izdatel'stvo Nauka, Moscow 1966.

[Ke 48] KEUFFEL, J. W.: Phys. Rev. **73** (1948) 531.

[Ke 49] KEUFFEL, J. W.: Rev. Sci. Instr. **20** (1949) 202.

[Ke 49a] KEUFFEL, J. W.: Rev. Sci. Instr. **20** (1949) 197.

[Ke 50] KELLEY, R. G.: Rev. Sci. Instr. **21** (1950) 71.

[Ke 55] KEUFFEL, J. W., W. H. SANDMANN, R. K. STITT: Rev. Sci. Instr. **26** (1955) 1191.

[Ke 59] KERNS, Q. A., F. A. KIRSTEN, G. C. COX: Rev. Sci. Instr. **30** (1959) 31.

[Ke 60] KENDALL, H. W.: IRE Trans. Nucl. Sci. **7** (1960) No. 2/3, 202.

[Ke 61] KERNS, Q. A., G. C. COX: Nucl. Instr. Meth. **12** (1961) 32.

[Ke 62] KEYES, R. J., T. M. QUIST: Proc. IRE **50** (1962) 1822.

[Ke 64] KELLER, K. B.: Rev. Sci. Instr. **35** (1964) 1360.

[Ke 64a] KEPERNIK, KH.: in Proc. Symp. Nucl. Electronics, Budapest 1963, Dubna-Report 1677, May 1964, p. 201.

[Ke 67] KERNS, C. R.: IEEE Trans. Nucl. Sci. **14** (1967) No. 1, 449.

[Ki 62] KIBLER, L. U.: Proc. IRE **50** (1962) 1834.

[Ki 65] KINGSLAND, D. O.: Proc. IEEE **53** (1965) 196.

[Ki 66] KINDLMANN, P. J.: Electronics **39** (1966) No. 4, 87.

[Ki 66a] KINDLMANN, P. J., J. SUNDERLAND: Rev. Sci. Instr. **37** (1966) 445.

[Ki 67] KIBBLE, B. P., G. COUPLEY, L. KRAUSE: Phys. Rev. **153** (1967) 9.

[Ki 67a] KIRKBRIDE, J., E. C. YATES, D. G. CRANDALL: Nucl. Instr. Meth. **52** (1967) 293.

[Ki 67b] KIESLER, R., B. RIGHINI: Nucl. Instr. Meth. **56** (1967) 357.

[Kj 62] KJELLMANN, J., A. NILSSON: Arkiv Fysik **20** (1962) 507.

[Kl 61] KLOSS, H. G., G. WENDEL: Z. Naturforsch. **16a** (1961) 61.

[Kl 62] KLEINKNECHT, H. P.: Nachrichtentechn. Z. **15** (1962) 394.

[Kl 66] KLOSE, J. Z.: Phys. Rev. **141** (1966) 181.

[Kl 67] KLEBANOV, YU. D., L. A. LOZINA: Pribory Tekhn. Eksperim. **12** (1967) No. 2, 127 (Russ.).

[Kn 55] KNOOP, E.: Z. angew. Phys. **7** (1955) 366; 575.

[Ko 58] KOHN, G.: Archiv elektr. Übertragung **12** (1958) 109.

[Ko 61] KO, W. H.: Solid State Electronics **3** (1961) 59.

[Ko 62] KOHN, G.: Nachrichtentechn. Z. **15** (1962) 531.

[Ko 62a] KOLOTOV, O. S.: Pribory Tekhn. Eksperim. **7** (1962) No. 3, 79 (Russ.).

[Ko 63] KOLPAKOV, I. F.: Dubna-Report 1362, Sept. 1963.

[Ko 63a] KONONOV, B. N., YU. A. CHURIN: Pribory Tekhn. Eksperim. **8** (1963) No. 4, 67 (Russ.).

[Ko 63b] KONONOV, B. N., A. S. SIDOROV, V. F. LEONOV: Pribory Tekhn. Eksperim. **8** (1963) No. 5, 103 (Russ.).

[Kö 63c] KÖRNER, H. J., J. RADELOFF, E. BODENSTEDT: Z. Phys. **172** (1963) 279.

[Ko 64] KOECHLIN, Y., A. RAVIART: Nucl. Instr. Meth. **29** (1964) 45.

[Ko 64a] KOLPAKOV, I. F.: Pribory Tekhn. Eksperim. **9** (1964) No. 5, 99 (Russ.).

[Ko 64b] KONNERTH, K., C. LANZA: Appl. Phys. Letters 4 (1964) 120.
[Ko 66] KOLPAKOV, I. F.: Dubna-Report 2778, June 1966.
[Ko 66a] KOVALEVA, T. A., A. E. MELAMID et al.: Pribory Tekhn. Eksperim. 11 (1966) No. 5, 5 (Russ.).
[Ko 66b] KOLPAKOV, I. F.: Pribory Tekhn. Eksperim. 11 (1966) No. 5, 127 (Russ.).
[Ko 67] KOTT, V. M., G. K. GAVRILOV, S. F. BAVAROV: Tunnel'nye diody v vychislitel'noi tekhnike (Tunnel Diodes in Computer Techniques), Izdatel'stvo Sovetskoe Radio, Moscow 1967.
[Ko 67a] KOHN, G., W. JUTZI et al.: IBM Journal 11 (1967) 162.
[Ko 67b] KOROBKIN, V. V., A. M. LEONTOVICH et al.: Zh. Eksperim. Teor. Fiz. 53 (1967) 16 (Russ.).
[Kr 54] KROEBEL, W.: Z. angew. Phys. 6 (1954) 293.
[Kr 57] KROEBEL, W., G. OLK: Z. angew. Phys. 9 (1957) 394.
[Kr 63] KRAKAUER, S. M., S. W. SOSHEA: Electronics 36 (1963) No. 29, 53.
[Kr 65] KRUSCHE, A., D. BLOESS, F. MÜNNICH: Nucl. Instr. Meth. 33 (1965) 177.
[Kr 65a] KRUSCHE, A.: Thesis, Techn. Hochschule Braunschweig 1965.
[Kr 67] KRAMER, K.: Z. Naturforsch. 22a (1967) 1108.
[Kr 67a] KRUSCHE, A., D. BLOESS, F. MÜNNICH: Nucl. Instr. Meth. 51 (1967) 197.
[Ku 30] KUBETSKII, L. A.: Patent No. 24040 (USSR), Aug. 4, 1930.
[Ku 58] KURASHOV, A. A., A. F. LINEV et al.: Atomnaya Energiya 5 (1958) 135 (Russ.); German Translation in: Kernenergie 2 (1959) 728.
[Kü 59] KÜHN, B.: Annalen Phys. 7. Folge, 3 (1959) 241.
[Ku 60] KUTSENKO, A. V.: Pribory Tekhn. Eksperim. 5 (1960) No. 1, 3 (Russ.).
[Ku 63] KULLANDER, S.: Nucl. Instr. Meth. 24 (1963) 342.
[Ku 65] KUCKUCK, R. W., J. C. LEE: IEEE Trans. Nucl. Sci. 12 (1965) No. 1, 356.
[Ku 66] KUCHLY, J. M., P. SIFFERT, A. COCHE: Nucl. Instr. Meth. 44 (1966) 239.
[Ku 67] KULINICH, YU. A., V. G. RUKAVISHNIKOV, R. S. SHUVALOV: Pribory Tekhn. Eksperim. 12 (1967) No. 3, 141; No. 6, 140 (Russ.).
[La 62] LANDE, K., A. K. MANN: in Proc. Conf. Nucl. Electronics, Belgrade 1961, Vol. 1, Internat. Atomic Energy Agency, Vienna 1962, p. 145.
[La 62a] LACOUR, J.: in Proc. Conf. Nucl. Electronics, Belgrade 1961, Vol. 3, Internat. Atomic Energy Agency, Vienna 1962, p. 179.
[La 63] LAUTERJUNG, K. H., J. POKAR et al.: Nucl. Instr. Meth. 22 (1963) 117.
[La 63a] LAVAITTE, A.: in Proc. Internat. Symp. Nucl. Electronics, Paris 1963, p. 761.
[La 63b] LACHINOV, V. M.: Pribory Tekhn. Eksperim. 8 (1963) No. 1, 53 (Russ.).
[La 63c] LAINE, E. F.: Electronics 36 (1963) No. 5, 39.
[La 64] LANGMANN, H. J., O. MEYER: Nucl. Instr. Meth. 30 (1964) 135.
[La 65] LARSEN, R. N., E. B. SHERA: Nucl. Instr. Meth. 32 (1965) 168.
[La 65a] LAPSHIN, V. G., M. N. OMEL'YANENKO et al.: Dubna-Report 1977, Jan. 1965.
[La 66] LAPSHIN, V. G., V. I. RYKALIN, Z. TSISEK: Dubna-Report 13–2987, Nov. 1966.
[La 66a] LANGKAU, R.: Nucl. Instr. Meth. 45 (1966) 351.
[La 66b] LAPSHIN, V. G., M. N. OMEL'YANENKO et al.: Pribory Tekhn. Eksperim. 11 (1966) No. 1, 149 (Russ.).
[La 66c] LANGKAU, R., H.-H. RÜHL: Nucl. Instr. Meth. 43 (1966) 368.
[La 67] LAN'KO, E. V., G. S. DOMBROVSKAYA: Pribory Tekhn. Eksperim. 12 (1967) No. 5, 218 (Russ.).
[La 67a] LACROIX, G.: Comptes Rendus 265 (1967) B 346.
[La 68] LANGSFORD, A., P. E. DOLLEY: Nucl. Instr. Meth. 59 (1968) 120.
[Le 46] LEE, G. M.: Proc. IRE 34 (1946) 121.
[Le 51] LEHOVEC, K., C. A. ACCARDO, E. JAMGOCHIAN: Phys. Rev. 83 (1951) 603.
[Le 55] LEPRI, F., L. MEZZETTI, G. STOPPINI: Rev. Sci. Instr. 26 (1955) 936.
[Le 57] LEFEVRE, H. W., J. T. RUSSELL: Bull. Am. Phys. Soc. II, 2 (1957) 175.
[Le 59] LEWIS, I. A. D., F. H. WELLS: Millimicrosecond Pulse Techniques, Pergamon Press, London/New York/Paris/Los Angeles 1959.

[Le 59a] LEFEVRE, H. W., J. T. RUSSELL: Rev. Sci. Instr. **30** (1959) 159.
[Le 61] LEDERMAN, L. M.: Rev. Sci. Instr. **32** (1961) 523.
[Le 62] LEFEVRE, H. W., R. R. BORCHERS, C. H. POPPE: Rev. Sci. Instr. **33** (1962) 1231.
[Le 63] LENNARTZ, H., W. TAEGER: Transistor-Schaltungstechnik, Verlag Radio-Foto-Kinotechnik, Berlin 1963.
[Le 63a] LEBAIL, P.: in Proc. Internat. Symp. Nucl. Electronics, Paris 1963, p. 769.
[Le 63b] LECOMTE, J. L.: Nucl. Instr. Meth. **24** (1963) 165.
[Le 63c] LEE, Y. K., C. S. WU: Phys. Rev. **132** (1963) 1200.
[Le 64] LEBEDEV, V. N.: Pribory Tekhn. Eksperim. **9** (1964) No. 2, 133 (Russ.).
[Le 64a] LEFEVRE, H. W.: IEEE Trans. Nucl. Sci. **11** (1964) No. 3, 426.
[Le 64b] LEMKE, K. P.: Archiv elektr. Übertragung **18** (1964) 509.
[Le 65] LECLERC, P., G. ZAJDE: Rev. Sci. Instr. **36** (1965) 1056.
[Le 66] LE DUFF, Y., P. PERETTI, R. DUPEYRAT: Rev. Physique Appl. **1** (1966) 117.
[Le 67] LEWYN, L. L.: IEEE Trans. Nucl. Sci. **14** (1967) No. 1, 126.
[Le 68] LECLERC, P., M. VOOS: J. Sci. Instr., Ser. 2, **1** (1968) 478.
[Li 50] LIEBSON, S. H., M. E. BISHOP, J. O. ELLIOT: Phys. Rev. **80** (1950) 907.
[Li 61] LINDENBAUM, S. J., L. C. L. YUAN: Čerenkov Counters, in Methods of Experimental Physics, Vol. 5 A, ed. by L. MARTON, Academic Press, New York/London 1961, p. 162.
[Li 61a] LINEV, A. F.: Dubna-Report **803**, Sept. 1961.
[Li 63] LINDENBAUM, S. J., W. LOVE et al.: Nucl. Instr. Meth. **20** (1963) 256.
[Li 63a] LINDSAY, J. B.: Nucl. Instr. Meth. **20** (1963) 345.
[Li 63b] LINDSKOG, J., T. SUNDSTRÖM: Arkiv Fysik **24** (1963) 199.
[Li 63c] LI, A. C., A. SCHWARZSCHILD: Phys. Rev. **129** (1963) 2664.
[Li 65] LIKHACHEV, M. F., I. A. SAVIN, V. S. STAVINSKII: Dubna-Report R–2528, Dec. 1965.
[Li 65a] LIETTI, A.: Rev. Sci. Instr. **36** (1965) 13.
[Li 66] LINDENBAUM, S. J.: Annual Rev. Nucl. Sci. **16** (1966) 619.
[Lo 60] LOUIS, H. P.: Elektronische Rdsch. **14** (1960) 137.
[Lo 62] LOWE, J., C. L. MCCLELLAND, J. V. KANE: Phys. Rev. **126** (1962) 1811.
[Lo 63] LOWE, J., C. L. MCCLELLAND: Phys. Rev. **132** (1963) 367.
[Lo 64] LOTTO, I. DE, E. GATTI, F. VAGHI: in Proc. Conf. Automatic Acquisition and Reduction of Nuclear Data (EANDC), Karlsruhe 1964, p. 291.
[Lo 65] LOUDE, J.-F., G.-A. GRIN, J.-P. PERROUD: Helv. Phys. Acta **38** (1965) 341.
[Lo 66] LOTTO, I. DE, D. DOTTI: Nucl. Instr. Meth. **39** (1966) 281.
[Lu 51] LUNDBY, A.: Rev. Sci. Instr. **22** (1951) 324.
[Lu 52] LUCKEY, D., J. W. WEIL: Phys. Rev. **85** (1952) 1060.
[Lu 59] LUNZE, K.: Wiss. Z. Techn. Hochschule Dresden **9** (1959/60) 1239.
[Lu 60] LUCK, G.: Nucl. Instr. Meth. **8** (1960) 249.
[Lu 61] LUNZE, K.: Nachrichtentechnik **11** (1961) 146.
[Lu 63] LUNDY, R. A.: Rev. Sci. Instr. **34** (1963) 146.
[Lu 64] LUCOVSKY, G., R. F. SCHWARZ, R. B. EMMONS: J. Appl. Phys. **35** (1964) 622.
[Lu 64a] LUNZE, K.: Nachrichtentechnik **14** (1964) 82.
[Lu 64b] LUNSFORD, J. S.: Rev. Sci. Instr. **35** (1964) 1483.
[Lu 65] LUDWIG, E. J.: Rev. Sci. Instr. **36** (1965) 1175.
[Ly 66] LYNCH, F. J.: IEEE Trans. Nucl. Sci. **13** (1966) No. 3, 140.
[Ma 48] MADANSKY, L., R. W. PIDD: Phys. Rev. **73** (1948) 1215.
[Ma 50] MADANSKY, L., R. W. PIDD: Rev. Sci. Instr. **21** (1950) 407.
[Ma 54] MARSHALL, J.: Annual Rev. Nucl. Sci. **4** (1954) 141.
[Ma 54a] MANDÓ, M.: Nuovo Cimento **12** (1954) 5.
[Ma 55] MADEY, R.: Rev. Sci. Instr. **26** (1955) 971.
[Ma 57] MACDONALD SMITH, J.: Electronic Engng. **29** (1957) 184.
[Ma 57a] MALMFORS, K. G., J. KJELLMAN, A. NILSSON: Nucl. Instr. **1** (1957) 186.
[Ma 57b] MALMBERG, J. H.: Rev. Sci. Instr. **28** (1957) 1027.

[Ma 58] Maeder, D.: Nucl. Instr. **2** (1958) 299.
[Ma 59] MacMahon, J. P.: in Proc. Internat. Symp. Nucl. Electronics, Paris 1958, Vol. 1, Internat. Atomic Energy Agency, Vienna 1959, p. 291.
[Ma 60] Martin, E. J.: Electronics **33** (1960) No. 41, 60.
[Ma 62] Matveev, V. V., A. D. Sokolov: Fotoumnozhiteli v stsintillyatsionnykh schetchikakh (Photomultipliers in Scintillation Counters), Gosatomizdat, Moscow 1962.
[Ma 62a] Mamyrin, B. A., G. S. Anufriev, I. F. Kalinkevich: Pribory Tekhn. Eksperim. **7** (1962) No. 1, 99 (Russ.).
[Ma 62b] Marx, G., G. Sauerbrey: Nukleonik **4** (1962) 208.
[Ma 62c] Marshak, I. S.: Pribory Tekhn. Eksperim. **7** (1962) No. **3**, 5 (Russ.).
[Ma 62d] Marshak, I. S.: Usp. Fiz. Nauk **77** (1962) 229 (Russ.).
[Ma 63] Mayer, J. W.: in Proc. Internat. Symp. Nucl. Electronics, Paris 1963, p. 129.
[Ma 63a] Máthé, G.: Nucl. Instr. Meth. **23** (1963) 261.
[Ma 63b] Martin-Löf, J.: Nucl. Instr. Meth. **24** (1963) 31.
[Ma 64] Mantakas, Ch.: Nucl. Instr. Meth. **31** (1964) 189.
[Ma 64a] Martini, M., G. Zanarini: Nucl. Instr. Meth. **26** (1964) 343.
[Ma 64b] Marte, G.: Archiv elektr. Übertragung **18** (1964) 211.
[Ma 64c] Maeder, D.: in Proc. Conf. Automatic Acquisition and Reduction of Nuclear Data (EANDC), Karlsruhe 1964, p. 311.
[Ma 65] Maydan, D.: Nucl. Instr. Meth. **34** (1965) 229.
[Ma 65a] Mackey, R. C., S. A. Pollack, R. S. Witte: Rev. Sci. Instr. **36** (1965) 1715.
[Ma 66] Mann, H. M., H. R. Bilger, I. S. Sherman: IEEE Trans. Nucl. Sci. **13** (1966) No. 3, 252.
[Ma 66a] Mayer, J. W.: Nucl. Instr. Meth. **43** (1966) 55.
[Ma 66b] Malm, H. L.: IEEE Trans. Nucl. Sci. **13** (1966) No. 3, 285.
[Ma 66c] Mantsa, I., B. A. Zelenov et al.: Dubna-Report 13–3042, Nov. 1966.
[Ma 67] Manfredi, P. F., A. Rimini: Nucl. Instr. Meth. **49** (1967) 71.
[Ma 67a] Matalin, L. A., S. I. Chubarov, A. A. Ivanov: Mnogokanal'nye analizatory yadernoi fiziki (Multi-Channel Analysers of Nuclear Physics), Atomizdat, Moscow 1967, p. 199.
[Ma 67b] Matisoo, J.: Proc. IEEE **55** (1967) 172.
[Ma 67c] Manfredi, P. F., P. Maranesi, A. Rimini: Rev. Sci. Instr. **38** (1967) 1253.
[Ma 68] Martini, M., G. Ottaviani: Proc. IEEE **56** (1968) 224.
[Ma 68a] Matick, R. E.: Proc. IEEE **56** (1968) 47.
[Mc 49] McGowan, F. K.: Phys. Rev. **76** (1949) 1730.
[Mc 50] McGowan, F. K.: Phys. Rev. **77** (1950) 138; **79** (1950) 404; **80** (1950) 482, 923.
[Mc 52] McLusky, G. J. R., N. F. Moody: Electronic Engng. **24** (1952) 330.
[Mc 52a] McQueen, J. G.: Electronic Engng. **24** (1952) 436.
[Mc 59] McMullen, C. W.: Rev. Sci. Instr. **30** (1959) 1109.
[Mc 61] McGee, J. D.: Rept. Progr. Phys. **24** (1961) 167.
[Mc 64] McGervey, J. D., V. F. Walters: Nucl. Instr. Meth. **25** (1964) 219.
[Mc 65] McDonald, D. F., C. J. Benning, S. J. Brient: Rev. Sci. Instr. **36** (1965) 504.
[Mc 65a] McGuire, R. L., E. C. Yates et al.: IEEE Trans. Nucl. Sci. **12** (1965) No. 1, 24.
[Mc 67] McDonald, W. J., D. A. Gedcke: Nucl. Instr. Meth. **55** (1967) 1.
[Mc 67a] McGuire, R. L., R. C. Palmer: IEEE Trans. Nucl. Sci. **14** (1967) No. 1, 217.
[Me 48] Meyer, K. P., P. Huber, E. Baldinger: Helv. Phys. Acta **21** (1948) 188.
[Me 50] Meyer, K. P., E. Baldinger, P. Huber: Helv. Phys. Acta **23** (1950) 121.
[Me 57] Meunier, R., G. Davidson: Rev. Sci. Instr. **28** (1957) 1010.
[Me 58] Messenger, G. C.: Proc. IRE **46** (1958) 1116.
[Me 59] Mey, J.: Rev. Sci. Instr. **30** (1959) 282.
[Me 59a] Meunier, R., J. Teiger: Nucl. Instr. Meth. **5** (1959) 148.
[Me 60] Meunier, R., J. Teiger: Nucl. Instr. Meth. **6** (1960) 220.
[Me 61] Meiling, W., J. Schintlmeister, F. Stary: Nucl. Instr. Meth. **11** (1961) 355.

[Me 61a] MEYER, J. D., K. M. TERWILLIGER: Rev. Sci. Instr. **32** (1961) 512.
[Me 62] MEINKE, H., F. W. GUNDLACH: Taschenbuch der Hochfrequenztechnik, Springer-Verlag, Berlin/Göttingen/Heidelberg 1962.
[Me 62a] MENKE, J. L., R. R. FULLWOOD: IRE Trans. Nucl. Sci. **9** (1962) No. 4, 61.
[Me 63] MEYERS, P.: Electronics **36** (1963) No. 38, 35.
[Me 63a] MEYER, H.: in Proc. Internat. Symp. Nucl. Electronics, Paris 1963, p. 633.
[Me 63b] MEILING, W., J. SCHINTLMEISTER, F. STARY: Nucl. Instr. Meth. **21** (1963) 275.
[Me 63c] MEYER, S. L., E. W. ANDERSON et al.: Phys. Rev. **132** (1963) 2693.
[Me 64] MESYATS, G. A.: Pribory Tekhn. Eksperim. **9** (1964) No. 3, 110 (Russ.).
[Me 65] MEYER, O., H. J. LANGMANN: Nucl. Instr. Meth. **34** (1965) 77.
[Me 65a] MEINKE, H. H.: Einführung in die Elektrotechnik höherer Frequenzen, Vol. 1, Bauelemente und Stromkreise, Springer-Verlag, Berlin/Heidelberg/New York 1965.
[Me 65b] MESHKOV, A. N.: Pribory Tekhn. Eksperim. **10** (1965) No. 5, 136 (Russ.).
[Me 65c] MEILING, W., F. STARY: Nucl. Phys. **74** (1965) 113.
[Me 66] MEDVEDEV, M. N.: Usp. Fiz. Nauk **90** (1966) 143 (Russ.).
[Me 66a] MEYER, H.: IEEE Trans. Nucl. Sci. **13** (1966) No. 3, 180.
[Me 66b] MEYER, H.: Nucl. Instr. Meth. **40** (1966) 149.
[Me 66c] MEILING, W., F. STARY: Nucl. Phys. **84** (1966) 534.
[Me 67] MEILING, W., F. STARY, W. ANDREJTSCHEFF: Nucl. Phys. **A 91** (1967) 49.
[Me 67a] METZ, A. J.: Rev. Sci. Instr. **38** (1967) 1445.
[Me 67b] MELESHKO, E. A., B. A. NIKOL'SKII, B. V. SOKOLOV: Pribory Tekhn. Eksperim. **12** (1967) No. 3, 61 (Russ.).
[Me 67c] MEYERHOFER, D., A. S. KEIZER, H. NELSON: J. Appl. Phys. **38** (1967) 111.
[Mi 55] MILLER, S. L.: Phys. Rev. **99** (1955) 1234.
[Mi 56] MILLMAN, J., H. TAUB: Pulse and Digital Circuits, McGraw-Hill Book Co., New York/Toronto/London 1956; German Translation: Verlag Technik, Berlin 1963.
[Mi 56a] MINTON, G. H.: J. Res. Nat. Bureau Standards **57** (1956) 119.
[Mi 59] MILLER, R. H.: Rev. Sci. Instr. **30** (1959) 395.
[Mi 60] MILLER, G. L., W. L. BROWN et al.: IRE Trans. Nucl. Sci. **7** (1960) No. 2/3, 185.
[Mi 62] MILLER, L. G., W. M. GIBSON, P. F. DONOVAN: Annual Rev. Nucl. Sci. **12** (1962) 189.
[Mi 62a] MILLER, H. W., Q. A. KERNS: Rev. Sci. Instr. **33** (1962) 877.
[Mi 62b] MIROSHNIK, J. A., V. V. SKUGAREV: Pribory Tekhn. Eksperim. **7** (1962) No. 1, 108 (Russ.).
[Mi 64] MICHAELIS, W., H. SCHMIDT, C. WEITKAMP: Nucl. Instr. Meth. **31** (1964) 93.
[Mi 65] MILLNER, R.: Katodenstrahl-Oszillographen, Verlag Technik, Berlin 1965.
[Mi 66] MIEHE, J. A., E. OSTERTAG, A. COCHE: IEEE Trans. Nucl. Sci. **13** (1966) No. 3, 127.
[Mi 66a] MICHON, M., J. ERNEST, R. AUFFRET: Phys. Letters **21** (1966) 514.
[Mi 66b] MINEEV, YU. V.: Pribory Tekhn. Eksperim. **11** (1966) No. 5, 131 (Russ.).
[Mi 67] MILLARD, J. K.: Rev. Sci. Instr. **38** (1967) 169.
[Mi 68] MICHAELIS, W.: Nucl. Instr. Meth. **61** (1968) 109.
[Mo 49] MORTON, G. A., K. W. ROBINSON: Nucleonics **4** (1949) No. 2, 25.
[Mo 52] MOODY, N. F., G. J. R. MCLUSKY, M. O. DEIGHTON: Electronic Engng. **24** (1952) 214.
[Mo 52a] MOODY, N. F.: Electronic Engng. **24** (1952) 289.
[Mo 52b] MOBLEY, R. C.: Phys. Rev. **88** (1952) 360.
[Mo 62] MOYER, B. J.: A Survey of Čerenkov Counter Technique, in Nuclear Instruments and Their Uses, Vol. 1, ed. by A. H. SNELL, Wiley, New York/London 1962, p. 166.
[Mo 62a] MOLL, J. L., S. KRAKAUER, R. SHEN: Proc. IRE **50** (1962) 43.
[Mo 63] MOODY, N. F., B. C. DE KAT, M. N. S. SWAMY: Electronic Engng. **35** (1963) 72.

[Mo 63a] MOBLEY, R. C.: Rev. Sci. Instr. **34** (1963) 256.
[Mo 64] MORUGIN, L. A., G. V. GLEBOVICH: Nanosekundnaya impul'snaya tekhnika (Nanosecond Pulse Techniques), Izdatel'stvo Sovetskoe Radio, Moscow 1964.
[Mo 64a] MOAK, C. D., W. M. GOOD et al.: Rev. Sci. Instr. **35** (1964) 672.
[Mö 65] MÖSCHWITZER, A.: Nachrichtentechnik **15** (1965) 378; 420; Wiss. Z. Techn. Univ. Dresden **14** (1965) 707.
[Mo 66] MONIER, L. F., G. E. TRIPARD: Rev. Sci. Instr. **37** (1966) 316.
[Mö 66a] MÖSCHWITZER, A., D. LANDGRAF-DIETZ: Nachrichtentechnik **16** (1966) 249.
[Mo 67] MOSZYŃSKI, M., J. JASTRZEBSKI, B. BENGTSON: Nucl. Instr. Meth. **47** (1967) 61.
[Mo 67a] MOSER, A.: Elektronische Rdsch. **21** (1967) 109.
[Mr 65] MRUK, W. F.: Nucl. Instr. Meth. **35** (1965) 61.
[Mu 59] MULLER, F. A.: Nucl. Instr. Meth. **4** (1959) 115.
[Mu 62] MURRAY, R. B.: Scintillation Counters, in Nuclear Instruments and Their Uses, Vol. 1, ed. by A. H. SNELL, Wiley, New York/London 1962, p. 82.
[Mu 65] MUCHIN, S. V., H.-J. POHL et al.: Nucl. Instr. Meth. **33** (1965) 339.
[Mü 65a] MÜLLER, A., R. LUMRY, H. KOKUBUN: Rev. Sci. Instr. **36** (1965) 1214.
[Mu 67] MULLICK, S. K.: IBM Journal **11** (1967) 558.
[Mu 67a] MURATA, Y.: Electronic Engng. **39** (1967) 356.
[Na 57] NAKAMURA, M.: Rev. Sci. Instr. **28** (1957) 1015.
[Na 59] NAKAMURA, M.: Rev. Sci. Instr. **30** (1959) 778.
[Na 61] NÁRAY, Zs., P. VARGA: J. Sci. Instr. **38** (1961) 352.
[Na 65] NADAV, E., B. KAUFMAN: Nucl. Instr. Meth. **33** (1965) 289.
[Na 65a] NASIBOV, A. S., V. L. LOMAKIN, V. G. BAGRAMOV: Pribory Tekhn. Eksperim. **10** (1965) No. 5, 133 (Russ.).
[Na 65b] NAYLOR, H., K. H. PURSER, P. H. ROSE: IEEE Trans. Nucl. Sci. **12** (1965) No. 3, 305.
[Na 65c] NAUMOV, V. V., M. N. OMEL'YANENKO et al.: Dubna-Report R–2262, July 1965.
[Na 67] NAHMAN, N. S.: Proc. IEEE **55** (1967) 855.
[Na 67a] NAKAI, M. Y., D. A. LABAR et al.: Rev. Sci. Instr. **38** (1967) 820.
[Na 68] NADAV, E., M. PALMAI, D. SALZMANN: Nucl. Instr. Meth. **59** (1968) 173.
[Ne 47] NEDDERMEYER, S. H., E. J. ALTHAUS et al.: Rev. Sci. Instr. **18** (1947) 488.
[Ne 50] NEWTON, T. D.: Phys. Rev. **78** (1950) 490.
[Ne 55] NEILSON, G. C., D. B. JAMES: Rev. Sci. Instr. **26** (1955) 1018.
[Ne 55a] NEETESON, P. A.: Elektronenröhren in der Impulstechnik, Philips' Technische Bibliothek IX, Eindhoven 1955.
[Ne 59] NEILSON, G. C., W. K. DAWSON, F. A. JOHNSON: Rev. Sci. Instr. **30** (1959) 963.
[Ne 60] NEILER, J. H., W. M. GOOD: Time-of-Flight Techniques, in Fast Neutron Physics, Part 1, ed. by J. B. MARION, J. L. FOWLER, Interscience Publ., New York/London 1960; Russ. Translation: Gosatomizdat, Moscow 1963.
[Ne 61] NEETESON, P. A.: Flächentransistoren in der Impulstechnik, Philips' Technische Bibliothek IX A, Eindhoven 1961.
[Ne 65] NEAL, W. R., H. W. KRANER: Phys. Rev. **137** (1965) B 1164.
[Ne 65a] NENTWICH, H.: DESY Hamburg, Report 65/17, Dec. 1965.
[Ne 66] NEUERT, H.: Kernphysikalische Meßverfahren zum Nachweis für Teilchen und Quanten, Braun, Karlsruhe 1966.
[Ne 66a] NEUWIRTH, W.: Z. Phys. **197** (1966) 473.
[Ng 67] NGUYEN-HUU, A., R. H. MURPHY: Electronic Engng. **39** (1967) 604.
[Ni 63] NICOLL, D. R., R. E. U. ROUTLEY: in Proc. Internat. Symp. Nucl. Electronics, Paris 1963, p. 79.
[Ni 67] NITZAN, D., V. W. HESTERMANN: J. Appl. Phys. **38** (1967) 1199.
[Ni 67a] NISHIZAWA, J.: Electronics **40** (1967) No. 25, 117.
[No 64] NOVAKOV, T., J. M. HOLLANDER, R. L. GRAHAM: Nucl. Instr. Meth. **26** (1964) 189.
[Nu 67] NUTT, R.: IEEE Trans. Nucl. Sci. **14** (1967) No. 1, 110.
[Ny 63] NYGAARD, K., S. MALMSKOG: Nucl. Instr. Meth. **22** (1963) 363.

[Ny 67] NYBAKKEN, T. W.: Nucl. Instr. Meth. **53** (1967) 331.

[Og 68] OGATA, A., S. J. TAO, J. H. GREEN: Nucl. Instr. Meth. **60** (1968) 141.

[Ol 66] OLSEN, B., L. BOSTRÖM: Nucl. Instr. Meth. **44** (1966) 65.

[Ol 67] OLIVER, C. J., B. COLLINGE, G. KAYE: Nucl. Instr. Meth. **50** (1967) 109.

[On 61] ONUCHIN, A. P.: Pribory Tekhn. Eksperim. **6** (1961) No. 6, 100 (Russ.).

[On 66] ONUCHIN, A. P.: Pribory Tekhn. Eksperim. **11** (1966) No. 4, 97 (Russ.).

[Op 64] OPHIR, D.: Nucl. Instr. Meth. **28** (1964) 237.

[Or 57] ORMAN, P. R., F. H. WELLS: Nucl. Instr. **1** (1957) 183.

[Or 63] ORMAN, P. R.: Nucl. Instr. Meth. **21** (1963) 121.

[Or 64] ORTEL, W. C. G.: Proc. IEEE **52** (1964) 1746.

[Or 66] ORTEL, W. C. G.: Proc. IEEE **54** (1966) 936.

[Ow 67] OWENS, A. R., G. WHITE: Nucl. Instr. Meth. **49** (1967) 291; Proc. IEE **113** (1966) 1763.

[Oz 62] OZHGYANI, L., V. S. PANTUEV, M. N. KHACHATURYAN: Pribory Tekhn. Eksperim. **7** (1962) No. 5, 80 (Russ.).

[Pa 60] PATRY, J.-P., J. SEGUINOT, M. SCHERER: Comptes Rendus **251** (1960) 2012.

[Pa 62] PAUL, H.: Rev. Sci. Instr. **33** (1962) 881.

[Pa 62a] PAGANO, R., C. J. S. DAMERELL, R. D. CHERRY: Rev. Sci. Instr. **33** (1962) 955.

[Pa 62b] PAVLOVSKII, A. J., G. V. SKLIZKOV: Pribory Tekhn. Eksperim. **7** (1962) No. 2, 98 (Russ.).

[Pa 64] PANDARESE, F.: IEEE Trans. Nucl. Sci. **11** (1964) No. 2, 16.

[Pa 65] PARDIES, J., J. PERRIN, J. L. SOULÉ: J. Physique **26** (1965) 236 A.

[Pa 65a] PAPADOPOULOS, L.: Nucl. Instr. Meth. **36** (1965) 122.

[Pa 66] PAPADOPOULOS, L.: Nucl. Instr. Meth. **41** (1966) 241.

[Pa 66a] PAUL, R.: Transistormeßtechnik, Verlag Technik, Berlin 1966.

[Pa 66b] PARKER, S. I., C. A. REY: Nucl. Instr. Meth. **43** (1966) 361.

[Pe 62] PEUCKERT, K.: Nucl. Instr. Meth. **17** (1962) 257.

[Pe 64] PERTSEV, A. N., A. N. PISAREVSKII, L. D. SOSHIN: Pribory Tekhn. Eksperim. **9** (1964) No. 3, 132 (Russ.).

[Pe 65] PENDLETON, W. K., A. H. GUENTHER: Rev. Sci. Instr. **36** (1965) 1546; Proc. IEEE **55** (1967) 1504.

[Pe 66] PETERS, H., TH. FISCHER: Proc. IEE **113** (1966) 963.

[Pe 67] PERTSEV, A. N., A. N. PISAREVSKII: Pribory Tekhn. Eksperim. **12** (1967) No. 4, 5 (Russ.).

[Pe 68] PEHL, R. H., F. S. GOULDING et al.: Nucl. Instr. Meth. **59** (1968) 45.

[Pf 63] PFEFFER, G., H. LAMI et al.: in Proc. Internat. Symp. Nucl. Electronics, Paris 1963, p. 93.

[Pf 63a] PFEFFER, G., H. LAMI et al.: Nucl. Instr. Meth. **23** (1963) 74.

[Pi 49] PIDD, R. W., L. MADANSKY: Phys. Rev. **75** (1949) 1175.

[Pi 49a] PIERCE, J. R.: Electronics **22** (1949) No. 11, 97.

[Pi 60] PISAREVSKII, A. N., L. D. SOSHIN: Pribory Tekhn. Eksperim. **5** (1960) No. 6, 3 (Russ.).

[Pi 62] PIETRI, G.: IRE Trans. Nucl. Sci. **9** (1962) No. 3, 62.

[Pi 64] PIETRI, G.: IEEE Trans. Nucl. Sci. **11** (1964) No. 3, 76.

[Pi 65] PIROUE, P. A., A. J. S. SMITH: IEEE Trans. Nucl. Sci. **12** (1965) No. 4, 249.

[Pi 66] PIGNERET, J., J. J. SAMUELI, A. SARAZIN: IEEE Trans. Nucl. Sci. **13** (1966) No. 3, 306.

[Pi 67] PIETRI, G.: Bull. d'informations scientifiques et techniques, suppl. (1967) No. 115, 36.

[Pi 67a] PINTSCHOVIUS, L., R. LANGKAU: Nukleonik **10** (1967) 154.

[Po 50] POST, R. F., L. I. SCHIFF: Phys. Rev. **80** (1950) 1113.

[Po 65] POHL, H.-J.: Jenaer Rdsch. **10** (1965) 222.

[Po 66] Porter, F. T., M. S. Freedman et al.: Nucl. Instr. Meth. **39** (1966) 35.
[Po 66a] Pohl, H.-J.: Thesis, Friedrich-Schiller-Univ. Jena 1966.
[Po 66b] Porter, D., J. McDonald, D. T. Stewart: Nucl. Instr. Meth. **44** (1966) 309.
[Po 67] Poenaru, D. N.: IEEE Trans. Nucl. Sci. **14** (1967) No. 5, 1.
[Po 67a] Poenaru, D. N.: Nucl. Instr. Meth. **54** (1967) 229, 242.
[Pr 54] Pritchard, R. L.: Proc. IRE **42** (1954) 786.
[Pr 64] Present, G., A. Schwarzschild et al.: Nucl. Instr. Meth. **31** (1964) 71.
[Pr 65] Prince, P. R.: Proc. IEEE **53** (1965) 304.
[Pr 67] Preissinger, H.: Z. angew. Phys. **22** (1967) 225.
[Ra 61] Rabson, T. A.: Nucl. Instr. Meth. **12** (1961) 127.
[Ra 62] Radeka, V.: Nucl. Instr. Meth. **22** (1962) 153.
[Ra 63] Rathje, J.: DESY Hamburg, Report A 2.99, June 1963.
[Ra 63a] Ragsdale, R. H.: Rev. Sci. Instr. **34** (1963) 450.
[Ra 64] Ramachandran, T. B., W. J. Moroney: Proc. IEEE **52** (1964) 1358.
[Ra 64a] Radeka, V.: IEEE Trans. Nucl. Sci. **11** (1964) No. 1, 296.
[Ra 67] Radeloff, J., N. Buttler et al.: Nucl. Instr. Meth. **47** (1967) 109.
[Re 60] Reichel, R.: Kernenergie **3** (1960) 1172.
[Re 61] Reynolds, G. T.: Scintillation Counters and Luminescent Chambers, in Methods of Experimental Physics, Vol. 5 A, ed. by L. Marton, Academic Press, New York/London 1961, p. 120.
[Re 61a] Reichel, R.: Kernenergie **4** (1961) 285.
[Re 62] Rethmeier, J., C. C. Jonker et al.: Nucl. Instr. Meth. **17** (1962) 273.
[Re 62a] Reyes-Suter, P., T. Suter: Arkiv Fysik **20** (1962) 415.
[Re 63] Rediker, R. H., T. M. Quist, B. Lax: Proc. IEEE **51** (1963) 218.
[Re 65] Rediker, R. H.: Physics Today **18** (1965) No. 2, 42.
[Re 65a] Rein, H.-M.: Archiv elektr. Übertragung **19** (1965) 411, 525.
[Re 67] Rekhin, E. I., A. A. Kurashov, P. S. Chernov: Izmerenie intervalov vremeni v eksperimental'noi fizike (Measurement of Time Intervals in Experimental Physics), Atomizdat, Moscow 1967.
[Re 67a] Remshardt, R.: Archiv elektr. Übertragung **21** (1967) 661.
[Rh 63] Rhoades, W. C.: Electronics **36** (1963) No. 6, 56.
[Ri 62] Riesz, R. P.: Rev. Sci. Instr. **33** (1962) 994.
[Ri 64] Righini, B.: Nucl. Instr. Meth. **29** (1964) 89.
[Ri 65] Righini, B.: Nucl. Instr. Meth. **33** (1965) 114.
[Ri 67] Riley, W. B.: Electronics **40** (1967) No. 10, 141.
[Ri 67a] Richter, F. W., J. Schütt: Z. Phys. **199** (1967) 422.
[Ri 67b] Richman, D.: Electronics **40** (1967) No. 23, 105.
[Ri 67c] Riedinger, M., F. Schmitt et al.: Nucl. Instr. Meth. 47 (1967) 100.
[Ri 68] Riley, W. B.: Electronics **41** (1968) No. 5, 165.
[Ro 53] Rodda, S.: Photo-electric Multipliers, MacDonald, London 1953.
[Ro 53a] Robinson, E.: Proc. Phys. Soc., Sect. A, **66** (1953) 73.
[Ro 57] Rosenheim, D. E., A. G. Anderson: Proc. IRE **45** (1957) 212.
[Ro 60] Roberts, A.: Nucl. Instr. Meth. **9** (1960) 55.
[Ro 60a] Roth, S. J.: IRE Trans. Nucl. Sci. **7** (1960) No. 2/3, 57.
[Ro 61] Rozen, S.: Nucl. Instr. Meth. **11** (1961) 316.
[Ro 61a] Roberts, A.: Rev. Sci. Instr. **32** (1961) 482.
[Ro 62] Rozen, S., M. Etzion: Nucl. Instr. Meth. **17** (1962) 332.
[Ro 62a] Rozen, S., E. Altman: Nucl. Instr. Meth. **17** (1962) 337.
[Ro 63] Rodda, J. L., J. E. Griffin, M. G. Stewart: Nucl. Instr. Meth. **23** (1963) 137.
[Ro 64] Roddick, R. G., F. J. Lynch: IEEE Trans. Nucl. Sci. **11** (1964) No. 3, 399.
[Ro 65] Rohde, U. L.: Transistoren bei höchsten Frequenzen, Verlag Radio-Foto-Kinotechnik, Berlin 1965.
[Ro 65a] Rowland, M. C., R. C. Bottomley: Brit. Comm. Electronics **12** (1965) 90.
[Ro 66] Rosenberger, D., I. Thumb: Z. Naturforsch. **21a** (1966) 175.

[Ro 67] ROTA, A., G. BERTOLINI, M. COCCHI: IEEE Trans. Nucl. Sci. **14** (1967) No. 1, 152.
[Ro 67a] ROSENFELD, A. H., N. BARASH-SCHMIDT et al.: UCRL-Report 3030, Sept. 1967.
[Ro 68] ROBBINS, W. P., R. M. LANTZ, B. ANCKER-JOHNSON: Rev. Sci. Instr. **39** (1968) 69.
[Ro 68a] ROSSET, G.: Nucl. Instr. Meth. **60** (1968) 93.
[Ru 45] RUDENBERG, H. G.: J. Appl. Phys. **16** (1945) 279.
[Ru 54] RUMSWINKEL, K.-E.: Z. angew. Phys. **6** (1954) 551.
[Ru 58] RUDENKO, A. A.: Pribory Tekhn. Eksperim. **3** (1958) No. 6, 60 (Russ.).
[Ru 59] RUTHROFF, C. L.: Proc. IRE **47** (1959) 1337.
[Ru 62] RUDENKO, A. A.: Pribory Tekhn. Eksperim. **7** (1962) No. 2, 59 (Russ.).
[Ru 63] RUBY, S. L., Y. HAZONI, M. PASTERNAK: Phys. Rev. **129** (1963) 826.
[Ru 64] RUSH, C. J.: Rev. Sci. Instr. **35** (1964) 149.
[Ru 64a] RUTHERGLEN, J. C.: Progr. Nucl. Phys. **9** (1964) 3.
[Ru 65] RUTZ, R. F., M. I. NATHAN et al.: Proc. IEEE **53** (1965) 1664.
[Ru 65a] RUSSEL, L. A.: Adv. Electronics Electron Phys. **21** (1965) 249.
[Ry 53] RYDER, E. J.: Phys. Rev. **90** (1953) 766.
[Ry 65] RYKALIN, V. I., T. G. KMITA et al.: Dubna-Report 2466, Nov. 1965.
[Sa 58] SAMUELI, J. J., A. SARAZIN: J. Phys. Radium **19** (1958) 109 A.
[Sa 60] SARAZIN, A., J. SAMUELI et al.: Nucl. Instr. Meth. **8** (1960) 70.
[Sa 60a] SAMUELI, J., A. SARAZIN: J. Phys. Radium **21** (1960) 390.
[Sa 61] SAMUELI, J. J., A. SARAZIN: J. Phys. Radium **22** (1961) 692.
[Sa 62] SAMUELI, J. J., A. SARAZIN: in Proc. Conf. Nucl. Electronics, Belgrade 1961, Vol. 3, Internat. Atomic Energy Agency, Vienna 1962, p. 41.
[Sa 63] SAKAI, E.: in Proc. Internat. Symp. Nucl. Electronics, Paris 1963, p. 205.
[Sa 63a] SALOW, H., H. BENEKING et al.: Der Transistor. Physikalische und technische Grundlagen, Springer-Verlag, Berlin/Göttingen/Heidelberg; J. F. Bergmann, München 1963.
[Sa 63b] SAMUELI, J. J., G. ESPI, A. SARAZIN: L'Onde Électrique **43** (1963) 1256.
[Sa 63c] SADEH, D.: Phys. Rev. Letters **10** (1963) 271.
[Sa 64] SAPP, W. W., E. J. STERNGLASS: IEEE Trans. Nucl. Sci. **11** (1964) No. 3, 108.
[Sa 64a] SAMUELI, J. J., A. SARAZIN, J. GUYON: Nucl. Instr. Meth. **26** (1964) 71.
[Sa 64b] SAMUELI, J. J., J. PIGNERET, A. SARAZIN: Nucl. Instr. Meth. **28** (1964) 330.
[Sa 65] SARLES Jr., F. W., J. K. ROBERGE: IEEE Trans. Nucl. Sci. **12** (1965) No. 1, 46.
[Sa 67] SANIN, A. A., P. N. SHAREIKO: Pribory Tekhn. Eksperim. **12** (1967) No. 6, 70 (Russ.).
[Sc 45] SCHINTLMEISTER, J.: Die Elektronenröhre als physikalisches Meßgerät. 4th Edition. Springer-Verlag, Wien 1945, p. 146; Russ. Translation: Gosudarstvennoe Izdatel'stvo Tekhniko-Teoreticheskoi Literatury, Moscow/Leningrad 1949.
[Sc 53] SCHOENWETTER, H. K.: Rev. Sci. Instr. **24** (1953) 515.
[Sc 56] SCHUMANN, R. W.: Rev. Sci. Instr. **27** (1956) 686.
[Sc 58] SCIVER, W. J. VAN, L. BOGART: IRE Trans. Nucl. Sci. **3** (1958) No. 3, 90.
[Sc 61] SCHINTLMEISTER, J., F. STARY, E. STOSCHEK: Kernenergie **4** (1961) 531.
[Sc 61a] SCHINTLMEISTER, J., F. STARY: Acta Phys. Austr. **14** (1961) 358.
[Sc 63] SCHWARZSCHILD, A.: Nucl. Instr. Meth. **21** (1963) 1.
[Sc 63a] SCHINDLER, H. R.: Electronics **36** (1963) No. 35, 37.
[Sc 64] SCIVER, W. J. VAN: Phys. Letters **9** (1964) 97.
[Sc 64a] SCHWARTZ, R. B., A. C. B. RICHARDSON: Nucl. Instr. Meth. **29** (1964) 83.
[Sc 64b] SCHAPPER, M. A.: Nucl. Instr. Meth. **27** (1964) 172.
[Sc 64c] SCHRANK, G., G. HENRY et al.: Rev. Sci. Instr. **35** (1964) 1326.
[Sc 64d] SCHMITT, H. J.: IEEE Trans. Nucl. Sci. **11** (1964) No. 1, 125.
[Sc 65] SCHWEIMER, W.: Nucl. Instr. Meth. **32** (1965) 190.
[Sc 65a] SCHMITT, H. W., W. E. KIKER, C. W. WILLIAMS: Phys. Rev. **137** (1965) B 837.
[Sc 66] SCIVER, W. J. VAN: IEEE Trans. Nucl. Sci. **13** (1966) No. 3, 138.
[Sc 67] SCHULTZ, W.: Elektronik **16** (1967) 7.

[Sc 68] Schuster, H. J.: Nucl. Instr. Meth. **58** (1968) 179.

[Se 59] Séguinot, J., M. Schérer, J. Teiger: Nucl. Instr. Meth. **5** (1959) 170.

[Se 59a] Segré, E.: Radioactive Decay, in Experimental Nuclear Physics, Vol. 3, ed. by E. Segré, Wiley, New York; Chapman & Hall, London 1959, p. 1.

[Se 59b] Severiens, J. C., F. W. Richter: Nucl. Phys. **13** (1959) 239.

[Se 62] Sedlmeyer, J. W., R. B. Patten, L. Fussell: in Proc. Conf. Nucl. Electronics, Belgrade 1961, Vol. 3, Internat. Atomic Energy Agency, Vienna 1962, p. 145.

[Se 62a] Segel, R. E.: Bull. Am. Phys. Soc. II, **7** (1962) 542.

[Se 62b] Seitzer, D.: Archiv elektr. Übertragung **16** (1962) 263.

[Se 63] Sear, B. E.: Proc. IEEE **51** (1963) 1215.

[Se 66] Seyfried, P., S. Deike: Nucl. Instr. Meth. **39** (1966) 313.

[Se 67] Seifert, G.: Atompraxis **13** (1967) 403.

[Se 68] Sen, P., A. P. Patro: Nucl. Instr. Meth. **59** (1968) 289.

[Se 68a] Sen, P., A. P. Patro: Nucl. Instr. Meth. **60** (1968) 335.

[Sh 57] Shea, R. F.: Transistor Circuit Engineering, Wiley, New York 1957; German Translation: Verlag Technik, Berlin; Berliner Union, Stuttgart 1961.

[Sh 58] Sharp, J., E. E. Thomson: in Proc. Sec. Internat. Conf. Peaceful Uses Atomic Energy, Geneva 1958, Vol. 14, United Nations, Geneva 1958, p. 311.

[Sh 59] Sharpless, W. M.: Bell System Techn. J. **38** (1959) 259.

[Sh 61] Shera, E. B.: Nucl. Instr. Meth. **12** (1961) 198.

[Sh 63] Shera, E. B., K. J. Casper, B. L. Robinson: Nucl. Instr. Meth. **24** (1963) 482.

[Sh 66] Shoffner, B. M., E. F. Shrader: IEEE Trans. Nucl. Sci. **13** (1966) No. 1, 394.

[Si 61] Sims, R. C., E. R. Beck, V. C. Kamm: Proc. IRE **49** (1961) 136.

[Si 61a] Simms, P. C.: Rev. Sci. Instr. **32** (1961) 894.

[Si 61b] Simms, P. C., N. Benczer-Koller, C. S. Wu: Phys. Rev. **121** (1961) 1169.

[Si 62] Simmons, J. E., R. B. Perkins: in Proc. Symp. Nucl. Instr., Harwell 1961, ed. by J. B. Birks, Heywood, London 1962, p. 62.

[Si 63] Sirotinin, E. I.: Pribory Tekhn. Eksperim. **8** (1963) No. 4, 122 (Russ.).

[Si 63a] Siffert, P., H. Rougeot, A. Coche: in Proc. Internat. Symp. Nucl. Electronics, Paris 1963, p. 155.

[Si 63b] Sidi, M., U. Sold: Nucl. Instr. Meth. **21** (1963) 89.

[Si 63c] Sigfridsson, B.: Arkiv Fysik **24** (1963) 311.

[Si 67] Sigfridsson, B.: Nucl. Instr. Meth. **54** (1967) 13.

[Sj 65] Sjölin, P. G.: Nucl. Instr. Meth. **37** (1965) 45.

[Sk 61] Skachkov, Yu. F.: Pribory Tekhn. Eksperim. **6** (1961) No. 6, 41 (Russ.).

[Sk 62] Skilling, J. K.: Electronics **35** (1962) No. 50, 49.

[Sk 62a] Skinner, A.: J. Sci. Instr. **39** (1962) 336.

[Sk 64] Skorka, S. J.: Nucl. Instr. Meth. **28** (1964) 192.

[Sm 51] Smith, L. G.: Rev. Sci. Instr. **22** (1951) 166.

[Sm 52] Smith, S. T., R. V. Talbot, C. H. Smith: Proc. IRE **40** (1952) 297.

[Sm 67] Smith, B.: Nucl. Instr. Meth. **55** (1967) 138.

[So 61] Souchere, G.: Centre d'Études Nucléaires Saclay, Report CEA-1844, Jan. 1961.

[So 64] Sold, U., S. Brojdo: Nucl. Instr. Meth. **26** (1964) 147.

[So 65] Soboleva, N. A., A. G. Berkovskii et al.: Fotoelektronnye pribory (Photo-electric Instruments), Izdatel'stvo Nauka, Moscow 1965.

[Sp 56] Spenke, E.: Elektronische Halbleiter, Springer-Verlag, Berlin/Göttingen/Heidelberg 1956.

[Sp 57] Sparkes, J. J., R. Beaufoy: Proc. IRE **45** (1957) 1740.

[Sp 58] Spenke, E.: Z. angew. Phys. **10** (1958) 65.

[Sp 60] Spiegel, P.: Rev. Sci. Instr. **31** (1960) 754.

[Sp 62] Sprinz, H., W. Brunner: Nachrichtentechnik **12** (1962) 294.

[Sp 63] Spicer, W. E., F. Wooten: Proc. IEEE **51** (1963) 1119.

[Sp 64] Spinrad, R. J.: Annual Rev. Nucl. Sci. **14** (1964) 239.

[Sp 64a] Spahn, W.: Diplomarbeit, Techn. Univ. Dresden 1964.

[Sp 66] SPARRMAN, P., F. FALK: Arkiv Fysik **32** (1966) 447.
[St 53] STRAUCH, K.: Rev. Sci. Instr. **24** (1953) 283.
[St 61] STARKE, L.: Elektronik **10** (1961) 117, 353.
[St 62] STADLER, H. L.: Rev. Sci. Instr. **33** (1962) 761.
[St 62a] STUCKENBERG, H. J.: DESY Hamburg, Report A 2.92, July 1962.
[St 62b] STUCKENBERG, H. J.: DESY Hamburg, Report A 2.93, July 1962.
[St 63] STOLZ, W.: Wiss. Z. Techn. Univ. Dresden **12** (1963) 1167.
[St 63a] STRAUSS, M. G.: Rev. Sci. Instr. **34** (1963) 1248.
[St 63b] STEINGRABER, O. J., I. B. BERLMAN: Rev. Sci. Instr. **34** (1963) 524.
[St 63c] STRAUSS, M. G.: Rev. Sci. Instr. **34** (1963) 335.
[St 65] STROMER, E. G.: Z. angew. Math. Phys. **16** (1965) 298.
[St 65a] STEGMEIER, H.: Archiv elektr. Übertragung **19** (1965) 251.
[St 66] STANLEY, V. A.: IEEE Trans. Nucl. Sci. **13** (1966) No. 3, 63.
[St 66a] STRAUSS, M. G., R. N. LARSEN, L. L. SIFTER: IEEE Trans. Nucl. Sci. **13** (1966) No. 3, 265; Nucl. Instr. Meth. **46** (1967) 45.
[St 66b] STROMER, E.: Z. angew. Math. Phys. **17** (1966) 313.
[St 66c] STARY, F., W. MEILING: Zentralinst. Kernforsch. Rossendorf, Report PhA 19, Febr. 1966.
[St 66d] STROKE, H. H.: Physics Today **19** (1966) No. 10, 55.
[St 67] STROMER, E.: Z. angew. Math. Phys. **18** (1967) 529.
[St 67a] STRAUSS, M. G., I. S. SHERMAN et al.: Rev. Sci. Instr. **38** (1967) 725.
[St 67b] STEFANSKI, R. J., R. K. ADAIR et al.: Rev. Sci. Instr. **38** (1967) 1792.
[Su 57] SUGARMAN, R.: Rev. Sci. Instr. **28** (1957) 933.
[Su 58] SURAN, J. J.: Proc. IRE **46** (1958) 1260.
[Su 60] SUGARMAN, R. M.: IRE Trans. Nucl. Sci. **7** (1960) No. 1, 23.
[Su 62] SUGARMAN, R.: in Proc. Conf. Nucl. Electronics, Belgrade 1961, Vol. 3, Internat. Atomic Energy Agency, Vienna 1962, p. 57.
[Su 62a] SUGARMAN, R., W. A. HIGINBOTHAM, A. H. YONDA: in Proc. Conf. Nucl. Electronics, Belgrade 1961, Vol. 3, Internat. Atomic Energy Agency, Vienna 1962, p. 3.
[Su 62b] SUGARMAN, R., F. MERRITT, W. A. HIGINBOTHAM: Brookhaven National Lab., Report BNL-711, Febr. 1962.
[Su 62c] SUNDSTRÖM, T.: Nucl. Instr. Meth. **16** (1962) 153.
[Su 62d] SUTER, T., P. REYES-SUTER et al.: Arkiv Fysik **20** (1962) 431.
[Su 66] SUPRYNOWICZ, A. A.: J. Appl. Phys. **37** (1966) 778.
[Sw 59] SWIFT, D. F., V. PEREZ-MENDEZ: Rev. Sci. Instr. **30** (1959) 1004.
[Sw 63] SWIFT, C. D.: Rev. Sci. Instr. **34** (1963) 306.
[Sw 67] SWIFT, J.: Electronic Engng. **39** (1967) 490.
[Sy 65] SYLVAN, T. P.: Electronics **38** (1965) No. 13, 84.
[Ta 61] TARCZY-HORNOCH, Z.: Electronic Design **9** (1961) No. 2, 34.
[Ta 63] TAYLOR, J. M.: Semiconductor Particle Detectors, Butterworths, London 1963.
[Ta 63a] TANSAL, S., H. SOBOL: Rev. Sci. Instr. **34** (1963) 1075.
[Ta 65] TAVENDALE, A. J.: IEEE Trans. Nucl. Sci. **12** (1965) No. 1, 255.
[Ta 66] TAVENDALE, A. J.: IEEE Trans. Nucl. Sci. **13** (1966) No. 3, 315.
[Ta 66a] TAWARA, H.: Nucl. Instr. Meth. **42** (1966) 318.
[Ta 67] TAMM, U., W. MICHAELIS, P. COUSSIEU: Nucl. Instr. Meth. **48** (1967) 301.
[Ta 67a] TADA, K., J. L. R. LARAYA: Proc. IEEE **55** (1967) 2064.
[Ta 67b] TAN, Z. C.: Rev. Sci. Instr. **38** (1967) 1415.
[Ta 67c] TAN, Z. C., J. B. EARNSHAW: Electronic Engng. **39** (1967) 624.
[Th 25] THIRRING, H.: Z. Phys. **31** (1925) 133.
[Th 53] THIRION, J., V. L. TELEGDI: Phys. Rev. **92** (1953) 1253.
[Th 62] THUY, H.-J.: Nachrichtentechn. Z. **15** (1962) 425.
[Th 62a] THIEBERGER, P.: Arkiv Fysik **22** (1962) 127.
[Th 63] THORN, R.: Nucl. Instr. Meth. **23** (1963) 237.
[Th 65] THOMMEN, W., M. J. O. STRUTT: Archiv elektr. Übertragung **19** (1965) 169.

[Th 66] THIEBERGER, P.: Nucl. Instr. Meth. **44** (1966) 349.
[Th 66a] THENARD, J., G. VICTOR: Nucl. Instr. Meth. **40** (1966) 318.
[Th 67] THENARD, J., G. VICTOR: Nucl. Instr. Meth. **53** (1967) 285.
[Ti 61] TINTĂ, F., N. MARTALOGU et al.: Nucl. Instr. Meth. **12** (1961) 138.
[Ti 62] TINTĂ, F.: in Proc. Conf. Nucl. Electronics, Belgrade 1961, Vol. 3, Internat. Atomic Energy Agency, Vienna 1962, p. 71.
[To 60] TODD, C. D.: Electronics **33** (1960) No. 34, 59.
[To 61] TOVE, P. A., K. FALK: Nucl. Instr. Meth. **12** (1961) 278.
[To 63] TOVE, P. A., L. G. JONASSON: in Proc. Internat. Symp. Nucl. Electronics, Paris 1963, p. 699; Nucl. Instr. Meth. **27** (1964) 45.
[To 64] TOVE, P. A., K. FALK: Nucl. Instr. Meth. **29** (1964) 66.
[To 64a] TOMLINSON, R. B., R. L. BROWN: IEEE Trans. Nucl. Sci. **11** (1964) No. 2, 28.
[To 64b] TOLFREE, D. W. L.: J. Sci. Instr. **41** (1964) 788.
[To 67] TOJO, A.: Nucl. Instr. Meth. **50** (1967) 45.
[To 67a] TOVE, P. A., W. SEIBT, W. LEITZ: Nucl. Instr. Meth. **51** (1967) 304.
[To 67b] TOVE, P. A., W. SEIBT: Nucl. Instr. Meth. **51** (1967) 261.
[Tr 65] TRAMMELL, W. V.: Rev. Sci. Instr. **36** (1965) 1551.
[Tr 67] TRIFTSHÄUSER, W., P. P. CRAIG: Phys. Rev. **162** (1967) 274.
[Ts 65] TSOU, K. C.: IEEE Trans. Nucl. Sci. **12** (1965) No. 1, 28.
[Ts 66] TSUKADA, K., S. TANAKA et al.: Nucl. Instr. Meth. **39** (1966) 249.
[Tu 62] TUSTING, R. F., Q. A. KERNS, H. K. KNUDSEN: IRE Trans. Nucl. Sci. **9** (1962) No. 3, 118.
[Uz 64] UZUNOGLU, V.: Semiconductor Network Analysis and Design, McGraw-Hill Book Co., New York 1964.
[Va 60] VALCKX, F. P. G., A. DYMANUS: Nucl. Instr. Meth. **7** (1960) 197.
[Va 63] VARLAMOV, V. G., YU. M. GRASHIN et al.: Pribory Tekhn. Eksperim. **8** (1963) No. 4, 63 (Russ.).
[Va 68] VASIL'EV, E. A., A. S. DENISOV et al.: Pribory Tekhn. Eksperim. **13** (1968) No. 1, 123 (Russ.).
[Ve 61] VERWEIJ, H.: Nucl. Instr. Meth. **10** (1961) 308.
[Ve 63] VERWEIJ, H.: Nucl. Instr. Meth. **24** (1963) 39.
[Ve 63a] VERWEIJ, H.: in Proc. Internat. Symp. Nucl. Electronics, Paris 1963, p. 723.
[Ve 63b] VERWEIJ, H.: Nucl. Instr. Meth. **20** (1963) 323.
[Ve 64] VERGEZAC, P., J. KAHANE: Nucl. Instr. Meth. **26** (1964) 317.
[Ve 65] VERWEIJ, H.: Nucl. Instr. Meth. **37** (1965) 309.
[Ve 66] VERWEIJ, H.: Nucl. Instr. Meth. **41** (1966) 181.
[Ve 66a] VENABLE, W. H.: Rev. Sci. Instr. **37** (1966) 1443.
[Vi 60] VILBIG, F.: Lehrbuch der Hochfrequenztechnik, Vol. 1, Geest und Portig, Leipzig 1960.
[Vi 64] VIL'DGRUBE, G. S., ZH. M. RONKIN, YU. A. KOLOSOV: Izv. Akad. Nauk SSSR, Ser. Fiz. **28** (1964) 377 (Russ.).
[Vo 62] VOGLER, G.: Phys. Status Solidi **2** (1962) 1241.
[Vo 63] VOLTZ, R., J. KLEIN et al.: in Proc. Internat. Symp. Nucl. Electronics, Paris 1963, p. 71.
[Vo 63a] VOROB'EV, G. A., G. A. MESYATS, G. S. KORSHUNOV: Pribory Tekhn. Eksperim. **8** (1963) No. 2, 98 (Russ.).
[Vo 65] VOROB'EV, G. A., N. S. RUDENKO: Pribory Tekhn. Eksperim. **10** (1965) No. 1, 109 (Russ.).
[Vo 65a] VOROTNIKOV, P. E.: Pribory Tekhn. Eksperim. **10** (1965) No. 3, 27 (Russ.).
[Vo 66] VOROB'EV, A. A., YU. V. DOTSENKO et al.: Izv. Akad. Nauk SSSR, Ser. Fiz. **30** (1966) 135 (Russ.).
[Vo 67] VOROB'EV, V. V., L. V. SEMKOV, R. S. SHUVALOV: Pribory Tekhn. Eksperim. **12** (1967) No. 6, 186 (Russ.).

[Vo 67a] VOLLRATH, K., G. THOMER (Editors): Kurzzeitphysik, Springer-Verlag, Wien/New York 1967.
[Vo 68] VOROB'EV, G. A., N. S. RUDENKO et al.: Pribory Tekhn. Eksperim. **13** (1968) No. 1, 126 (Russ.).
[Vr 62] VRIES, H. DE: in Proc. Conf. Nucl. Electronics, Belgrade 1961, Vol. 3, Internat. Atomic Energy Agency, Vienna 1962, p. 59.
[Vy 61] VYSOCHANSKII, M., S. V. MUKHIN et al.: Pribory Tekhn. Eksperim. **6** (1961) No. 5, 67 (Russ.).
[Vy 62] VYSOCHANSKII, M., S. V. MUKHIN et al.: in Proc. Conf. Nucl. Electronics, Belgrade 1961, Vol. 3, Internat. Atomic Energy Agency, Vienna 1962, p. 49.
[Vy 63] VYSOCHANSKY, M., S. V. MUKHIN et al.: in Proc. Internat. Symp. Nucl. Electronics. Paris 1963, p. 581.
[Vy 65] VYSOCHANSKII, M., S. V. MUKHIN et al.: Pribory Tekhn. Eksperim. **10** (1965) No. 2, 88 (Russ.).
[Vy 66] VYSOČANSKÝ, M., Ľ. ONDRIŠ et al.: Nucl. Instr. Meth. **41** (1966) 157.
[Wa 61] WALTON, A. J.: J. Electr. Control **11** (1961) 341.
[Wa 61a] WAGNER, R., P. HUBER, Z. LEWANDOWSKI: Helv. Phys. Acta **34** (1961) 440.
[Wa 62] WALTER, F. J., J. W. T. DABBS, L. D. ROBERTS: in Proc. Conf. Nucl. Electronics, Belgrade 1961, Vol. 1, Internat. Atomic Energy Agency, Vienna 1962, p. 391.
[Wa 63] WARD, C. B., C. M. YORK: Nucl. Instr. Meth. **23** (1963) 213.
[Wa 63a] WALKER, R. J.: Nucl. Instr. Meth. **24** (1963) 337.
[Wa 64] WANG, L., M. R. ZATZICK, F. P. ZIEMBA: IEEE Trans. Nucl. Sci. **11** (1964) No. 1, 314.
[Wa 64a] WAHL, H.: Nucl. Instr. Meth. **25** (1964) 247.
[Wa 64b] WARD, C., A. BERICK et al.: Nucl. Instr. Meth. **30** (1964) 61.
[Wa 65] WALTHARD, B., L. SCHELLENBERG, O. HUBER: Helv. Phys. Acta **38** (1965) 299.
[We 52] WELLS, F. H.: Nucleonics **10** (1952) No. 4, 28.
[We 56] WEINZIERL, P.: Rev. Sci. Instr. **27** (1956) 226.
[We 56a] WEBER, W., C. W. JOHNSTONE, L. CRANBERG: Rev. Sci. Instr. **27** (1956) 166.
[We 60] WEAVER, R. S., R. E. BELL: Nucl. Instr. Meth. **9** (1960) 149.
[We 64] WEISBERG, H. L., S. BERKO: IEEE Trans. Nucl. Sci. **11** (1964) No. 3, 406.
[We 64a] WEBER, J.: Nucl. Instr. Meth. **26** (1964) 325.
[We 64b] WEBER, J.: Nucl. Instr. Meth. **25** (1964) 285.
[We 64c] WENZEL, W. A.: Annual Rev. Nucl. Sci. **14** (1964) 205.
[We 65] WEDDIGEN, C., E. L. HAASE: Nucl. Instr. Meth. **33** (1965) 157.
[We 65a] WELTER, L. M.: Rev. Sci. Instr. **36** (1965) 487.
[We 65b] WEISBERG, H.: Nucl. Instr. Meth. **32** (1965) 133.
[We 65c] WEISBERG, H.: Nucl. Instr. Meth. **32** (1965) 138.
[We 66] WEINZIERL, P.: Acta Phys. Austr. **23** (1966) 312.
[We 66a] WEBER, S.: Electronics **39** (1966) No. 22, 109.
[We 67] WEINSCHEL, B. O.: Proc. IEEE **55** (1967) 923.
[Wh 59] WHETSTONE, A.: Rev. Sci. Instr. **30** (1959) 447.
[Wh 62] WHETSTONE, A., S. KOUNOSU: Rev. Sci. Instr. **33** (1962) 423.
[Wh 63] WHETSTONE, A. L.: Rev. Sci. Instr. **34** (1963) 412.
[Wh 64] WHITTAKER, J. K.: Nucl. Instr. Meth. **28** (1964) 293.
[Wh 66] WHITTAKER, J. K.: IEEE Trans. Nucl. Sci. **13** (1966) No. 1, 399.
[Wh 66a] WHITE, G.: Nucl. Instr. Meth. **45** (1966) 270.
[Wh 66b] WHITTAKER, J. K.: Nucl. Instr. Meth. **45** (1966) 138.
[Wh 67] WHITE, G.: Nucl. Instr. Meth. **55** (1967) 157.
[Wi 50] WIEGAND, C.: Rev. Sci. Instr. **21** (1950) 975.
[Wi 50a] WILKINSON, D.: Proc. Cambridge Philos. Soc. **46** (1950) 508.
[Wi 57] WIGINGTON, R. L., N. S. NAHMAN: Proc. IRE **45** (1957) 166.
[Wi 58] WINNIGSTAD, C. N.: Rev. Sci. Instr. **29** (1958) 578.

[Wi 62] WILLIAMS, C. W., J. H. NEILER: IRE Trans. Nucl. Sci. **9** (1962) No. 5, 1.
[Wi 63] WILLIAMS, L. L., D. J. HAMILTON: Electronics **36** (1963) No. 23, 52.
[Wi 63a] WIEBER, D. L.: Nucl. Instr. Meth. **24** (1963) 269.
[Wi 63b] WIEGAND, C.: Nucl. Instr. Meth. **20** (1963) 313.
[Wi 63c] WINE, C. M., L. S. COSENTINO: Proc. IEEE **51** (1963) 625.
[Wi 64] WINTER, J.: Nucl. Instr. Meth. **28** (1964) 229.
[Wi 64a] WILLIAMS, C. W., J. A. BIGGERSTAFF: Nucl. Instr. Meth. **25** (1964) 370.
[Wi 66] WIEBER, D. L., H. W. LEFEVRE: IEEE Trans. Nucl. Sci. **13** (1966) No. 1, 406.
[Wi 66a] WILSON, W. E., W. C. ROESCH: Rev. Sci. Instr. **37** (1966) 1623.
[Wo 66] WOLFGANG, L. G., J. M. ABRAHAM, C. N. INSKEEP: IEEE Trans. Nucl. Sci. **13** (1966) No. 3, 46.
[Wo 67] WOLFF, CH., H. BUCHWALD et al.: Naturwiss. **54** (1967) 489.
[Ya 66] YATES, E. C., D. G. CRANDALL: IEEE Trans. Nucl. Sci. **13** (1966) No. 3, 153.
[Ya 67] YAKUSHIN, V. V.: Pribory Tekhn. Eksperim. **12** (1967) No. 2, 84 (Russ.).
[Yg 65] YGUERABIDE, J.: Rev. Sci. Instr. **36** (1965) 1734.
[Yo 67] YOUNG, M. C., H. W. PRINSEN et al.: Proc. IEEE **55** (1967) 2185.
[Yu 51] YU, Y. P., H. E. KALLMAN, P. S. CHRISTALDI: Electronics **24** (1951) No. 7, 106.
[Yu 53] YU, Y. P.: Proc. IRE **41** (1953) 228.
[Yu 61] YUAN, L. C. L., S. J. LINDENBAUM: Measurement of Velocity Using Čerenkov Counters, in Methods of Experimental Physics, Vol. 5 A, ed. by L. MARTON, Academic Press, New York/London 1961, p. 454.
[Za 55] ZAVOISKII, E. K., G. E. SMOLKIN et al.: Dokl. Akad. Nauk SSSR **100** (1955) 241 (Russ.).
[Za 56] ZAVOISKII, E. K., M. M. BUTSLOV et al.: Atomnaya Energiya **1** (1956) No. 4, 34 (Russ.).
[Za 56a] ZAVOISKII, E. K., G. E. SMOLKIN: Atomnaya Energiya **1** (1956) No. 4, 46 (Russ.).
[Za 62] ZAPEVALOV, V. A., E. M. LEIKIN: Pribory Tekhn. Eksperim. **7** (1962) No. 2, 64 (Russ.).
[Za 65] ZAVOISKY, E. K., S. D. FANCHENKO: Applied Optics **4** (1965) 1155; Kurchatov-Institute Moscow, Report IAE-1061, 1966.
[Za 66] ZABIYAKIN, G. I.: Pribory Tekhn. Eksperim. **11** (1966) No. 1, 5 (Russ.).
[Zi 63] ZINOV, V. G.: Pribory Tekhn. Eksperim. **8** (1963) No. 1, 165 (Russ.).
[Zu 14] ZURHELLEN, W.: Astr. Nachr. **198** (1914) 1.
[Zu 62] ZURK, R. VAN: Nucl. Instr. Meth. **16** (1962) 157.
[Zu 63] ZURK, R. VAN, C. E. A. FRANCE: in Proc. Internat. Symp. Nucl. Electronics, Paris 1963, p. 619.
[Zu 67] ZURK, R. VAN: Nucl. Instr. Meth. **46** (1967) 125; **53** (1967) 45.

7. INDEX (IN ENGLISH)

8. INDEX (IN GERMAN)

Bei Fragen zur Produktsicherheit wenden Sie sich bitte an:
If you have any questions regarding product safety,
please contact:

Walter de Gruyter GmbH
Genthiner Straße 13
10785 Berlin
productsafety@degruyterbrill.com